D#260212

1 June dP

not on list

DATE DUE

DEC 25 1989			
JUN 0 9 1992			
NOV 2 1995			
NOV 1 3 1995			
NOV 1 6 1995			
GAYLORD			PRINTED IN U.S.A.

Nearby Galaxies Catalog

Designed as a companion to the *Nearby Galaxies Atlas* this catalog consists of a principal table that provides information on the 2,367 galaxies that are mapped in the atlas.

The information includes positions, morphological descriptions, sizes, luminosities, red shifts, neutrohydrogen properties, and characteristics of the environment of each galaxy. Distance estimates are given for each system, and this information allows intrinsic properties to be derived.

Entries in the catalog are reordered according to group-association-cloud assignments in a separate table. In a third table, rich clusters of galaxies at large distances are assigned to features of large-scale structure called supercluster complexes.

The Atlas was designed to appeal to a wide audience. The Catalog contains technical information that will be of particular interest to professional astronomers and serious amateurs.

Nearby Galaxies
Catalog

R. Brent Tully

Institute for Astronomy
University of Hawaii

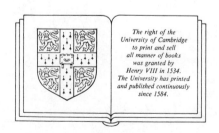

The right of the
University of Cambridge
to print and sell
all manner of books
was granted by
Henry VIII in 1534.
The University has printed
and published continuously
since 1584.

Cambridge University Press
Cambridge
New York   New Rochelle   Melbourne   Sydney

Published by the Press Syndicate of the University of Cambridge
The Pitt Building, Trumpington Street, Cambridge CB2 1RP
32 East 57th Street, New York, New York 10022 U.S.A.
10 Stamford Road, Oakleigh, Melbourne 3166, Australia

First published in 1988

Library of Congress Cataloging-in-Publication Data
Tully, R. Brent
Nearby galaxies catalog.
1. Galaxies—Catalogs.   I. Title.
QB857.T853   1987   523.1'12   87-21044

British Library Cataloguing in Publication Data
Tully, R. Brent
Nearby galaxies catalog.
1. Galaxies
I. Title
523.1'12   QB857
ISBN 0-521-35299-1

TABLE OF CONTENTS

ACKNOWLEDGMENTS

This catalog was compiled in connection with the production of the *Nearby Galaxies Atlas*, a project undertaken with Richard Fisher of the National Radio Astronomy Observatory. Although Fisher is not a coauthor of this work, he might have been. He provided essentially all the neutral hydrogen information that appears here, either because it was he that made the H I observations that we have jointly discussed in other publications or because it was he that searched the literature and made the effort to assure the compatibility of alternative sources of H I information with our own data.

Several hundred velocities included in this catalog have not been published elsewhere. These new velocities have been obtained with the Parkes Telescope in a collaborative effort with Pierre Chamaraux, Hugo van Woerden, Miller Goss, and Ulrich Mebold. This material will presumably be discussed in greater detail someday, but I thank my collaborators for allowing me to include the velocity information here.

The layout of the catalog has been in the capable hands of Louise Good and Jane Eckelman. The cartoons of the author were drawn by Jean-Pierre Petit. Thank you all.

INTRODUCTION

This compendium is a companion to the *Nearby Galaxies Atlas* (ref. 29; hereafter *NBG Atlas*). Data has been accumulated on 2367 galaxies with systemic velocities less than 3000 kilometers/second. Any galaxy was admitted to the catalog if it had a known velocity in 1978 that satisfied the specified limit of 3000 kilometers/second, or if it was subsequently observed to have a suitable velocity in surveys of the entire sky by the author and collaborators (ref. 6, 11, 22).

There are many sources of velocities, so this catalog could potentially be quite heterogeneous. However, two sources dominate. One of these is the magnitude-limited Shapley-Ames sample, which assures the inclusion of all galaxies brighter than 12th magnitude in the blue passband (ref. 25). There are 1053 Shapley-Ames galaxies within the velocity limit. The other source was already mentioned: the all-sky survey in the neutral hydrogen line by the author and collaborators. This survey was undertaken after a complete reinspection of photographic atlases of the sky. This survey was insensitive to gas-deficient systems and has severe incompletion problems at velocities beyond 2000 kilometers/second (discussed in ref. 11). However, the virtue of our survey is homogeneous coverage across the unobscured part of the sky. The neutral hydrogen survey provides 1515 velocities to the catalog. There is some overlap between the two principal sources. It is this cumulative sample that is mapped in the *NBG Atlas*.

There are three parts to the catalog. The first and by far the largest section provides information about each of the 2367 galaxies. One element of that information is a group affiliation, and in a second section there is a reordered listing that clarifies the composition of each group. The final, very short section identifies the rich clusters of galaxies that delineate the supercluster complexes mapped in the last two plates of the *NBG Atlas*.

The text of the atlas is written at a level that can be appreciated by a wide audience. The material in this catalog and the following description are of a more technical nature. This catalog is intended for an audience of professional and motivated amateur astronomers.

TABLE I
A Catalog of Galaxies within 40 Megaparsecs

If a galaxy was known to have a velocity (corrected for local motion) of less than 3000 kilometers/second when the catalog was compiled, then there is an entry for this system in Table I. A value for the Hubble Constant of 75 kilometers/second/megaparsec is assumed, so this velocity limit corresponds to a distance of 40 megaparsecs.

There are 18 columns and three rows of information associated with each entry. The following is a description of each item of information, ordered by column number (1–18) and row letter (a–c).

Column 1

(a) Galaxy Name

Entries are identified, in order of priority, by a *New General Catalogue* number preceded by an "N" (ref. 10; hereafter NGC), or by an *Uppsala General Catalogue* number preceded by a "U" (ref. 20; hereafter UGC), or by a number constructed from the equatorial coordinates of the system. Alternative names are recorded in column 18.

Columns 2 and 3

(a) Equatorial Coordinates

α, δ: right ascension in hours (2 digits) and minutes (last 3 digits in column 2) and declination in degrees (2 digits) and arcminutes (last 2 digits in column 3) at the epoch 1950.0. Positions have an accuracy of about 1 arcminute.

(b) Galactic Coordinates

ℓ, b: longitude (column 2) and latitude (column 3), in degrees. Coordinates follow the conventions of the International Astronomical Union. The Galactic poles are taken to lie at $12^h49^m0 + 27°24'$ (north) and $0^h49^m0 - 27°24'$ (south). The equator ($b = 0°$) follows the plane of the Milky Way, and the origin in longitude ($\ell = 0°$) is coincident with the Galactic Center.

(c) Supergalactic Coordinates

SGL,SGB: longitude (column 2) and latitude (column 3), in degrees. Coordinate system defined in the *Second Reference Catalogue of Bright Galaxies* (ref. 9). The Supergalactic north pole lies at $18^h52^m8 + 15°37'$. The equator (SGB = 0°) lies along the band of concentration in the distribution of nearby galaxies. The origin in longitude (SGL = 0°) lies at one of the two points of intersection of the Galactic and Supergalactic equatorial great circles, at $\ell = 137°37$. The Galactic and Supergalactic

equatorial planes are almost perpendicular to each other. The deviation from a right angle between these two planes is 6°32.

Column 4

(a) Morphological Type

The morphological type of each galaxy is given by a numeric code that is only slightly modified from the system used in the *Second Reference Catalogue* (ref. 9). Table A provides a translation from the numeric code to the revised Hubble sequence type. Morphological types are provided for all entries in the catalog. Sources for this information are ref. (7, 9, 16, 20, 25, 33, 34).

TABLE A
Morphological Type Codes

CODE NO.	REVISED HUBBLE TYPE	
−5	E	elliptical
−3	E/SO	elliptical/lenticular*
−2	SO	lenticular
0	SO/a	lenticular/spiral
1	Sa	spiral
2	Sab	spiral
3	Sb	spiral
4	Sbc	spiral
5	Sc	spiral
6	Scd	spiral
7	Sd	spiral
8	Sdm	spiral
9	Sm	spiral/irregular
10	Im	irregular
12	S	spiral/irregular*
13	P	peculiar

*Classification uncertain.

Following the morphology number code there may be a "B" indicating the existence of a bar, or an "A" indicating the absence of a bar, or an "X" indicating the intermediate case, or a blank indicating no information. In the following position there may be a "P" indicating the existence of a peculiarity.

1

(b) Local Density

ϱ_{xyz}: density of galaxies brighter than –16 mag in the vicinity of the entry, in galaxies/megaparsec³. The local density was determined on a three-dimensional grid at 0.5 megaparsec spacings. A Gaussian smoothing function of the following form was used to calculate the contribution of each member of the sample brighter than –16 mag to the density at a specific location:

$$\varrho_i (SGX, SGY, SGZ) = C \cdot \exp [-r_i^2 / 2(F^{1/3}\sigma)^2],$$

where $C = 0.52$ is a normalization constant, r_i is the distance of sample member i from the location (SGX, SGY, SGZ), $\sigma = 1.0$ megaparsec is the smoothing constant, and the factor F describes the growth of incompletion as a function of distance:

$$F = \exp [0.041 \cdot (\mu - 28.5)^{2.78}].$$

Here, μ is the distance modulus and $F = 1$ if $\mu < 28.5$.

The local density is then

$$\varrho_{xyz} = \sum_{i=1}^{n = 2189} \varrho_i,$$

where there are 2189 galaxies within 40 megaparsecs with $B_T^{b,i} \leqslant -16$. The local density recorded with each entry in Table I is the local density at the nearest position to the entry on the 0.5 megaparsec interval grid. Two-dimensional projections of the local density grid give surface densities. Contours of surface density based on a $\sigma = 0.5$ megaparsecs smoothing constant are plotted on the maps of Plates 14–21 in the atlas. The function that describes incompletion and the derivation of local density will be discussed in detail elsewhere (ref. 28). If an entry is at a distance greater than 40 megaparsecs, then the local density is not recorded.

(c) Group Affiliation

Galaxies may be affiliated with other galaxies in groups, associations, or clouds. These affiliations are described by a code in the form $AB \pm CD + EF$. A galaxy is located in cloud AB, group $-CD$ or first level association $+CD$, and second level association $+EF$. There is information in the cloud identification on the general location of the cloud, as described in Table B.

A value of B greater than zero specifies a cloud within region A. If B is zero, the system is apparently isolated within region A. Within each cloud there are groups ($-CD$) and associations ($+CD$), where $CD = 1, \ldots, N$ (N = number of groups and associations in cloud). Groups and associations in turn can be linked into loose associations ($+EF$). If there is no entry for EF, then $EF = |CD|$ is assumed. If CD is zero, the system is at-large in cloud AB. There is a description of the definition of the groups in ref. (26). Table II of this catalog provides the membership of each of the 336 groups identified in the present sample.

Column 5

(a) Observed Diameter

D_{25}: diameter at the 25 magnitude/arcsecond² isophote in blue light, in arcminutes. Conversions from diameters in the major catalogs are in accordance with the following equations:

$$\log D_{25} = 0.983 \log (D_{UGC} + 0.3) - 0.051$$
$$\text{(UGC: ref. 20),}$$

$$\log D_{25} = 0.998 \log (D_{ESO} + 0.3) - 0.132$$
$$\text{(ESO: ref. 16),}$$

$$\log D_{25} = 1.020 \log (D_{MCG} + 0.3) - 0.007$$
$$\text{(MCG: ref. 35),}$$

where all diameters are in arcminutes. These transformation equations are discussed in ref. (13, 30).

(b) Corrected Diameter

$D_{25}^{b,i}$: diameter adjusted for the effects of projection and obscuration, in arcminutes. Adjustments are made according to the equation

$$\log D_{25}^{b,i} = \log D_{25} - c \log [D/d] + A_B^b \cdot K_D^{25},$$

where D/d is the ratio of the major to minor axis diameter, $c = 0.22$, A_B^b is the Galactic absorption in the blue passband, and $K_D^{25} = 0.09$. The last term on the right accounts for the truncation of a galaxy due to Galactic obscuration. The second last term on the right accounts for an enhancement of the diameter of edge-on systems due to a projection effect. The adjustment procedures are described in ref. (30).

(c) Source

Source of diameter information. The numeric code is given in Table C, in decreasing order of priority.

TABLE B
Cloud Location Codes

FIRST DIGIT IN CLOUD ID	PLATE IN NBG ATLAS
A = 1	15
2	16
3	17
4	18
5	19
6	20
7	no detailed map

TABLE C
Diameter Source Codes

CODE NO.	REFERENCE
9	12 (standards)
2	20 (UGC)
6	16 (ESO)
4	35 (MCG)
5	7 (BGC)

Column 6

(a) Axial Ratio

d/D: ratio of the minor axis diameter to the major axis diameter.

(b) Inclination

i: inclination from face-on, in degrees. The inclination is almost always given by

$$i = \cos^{-1}\{[(d/D)^2 - 0.2^2]/(1 - 0.2^2)\}^{1/2} + 3°.$$

It is assumed that an edge-on system would be measured to have the axial ratio $d/D = 0.2$. The constant of 3° that is added is in accordance with an empirical recipe (ref. 2). In a statistically insignificant number of cases, inclinations come from other information, such as the form of rings or spiral structure.

Column 7

(a) Blue Magnitude

$B_T^{b,i}$: blue apparent magnitude adjusted for reddening:

$$B_T^{b,i} = B_T - A_B^b - A_B^{i-0},$$

where B_T is the total observed blue magnitude, A_B^b is the absorption within our Galaxy, and A_B^{i-0} is the absorption within the external galaxy beyond that expected if the galaxy were face-on.

(b) Infrared Magnitude

$H_{-0.5}^{b,i}$: apparent magnitude at 1.6 μm adjusted for reddening:

$$H_{-0.5}^{b,i} = H_{-0.5} - A_H^b - A_H^{i-0},$$

where $H_{-0.5}$ is the observed 1.6 μm magnitude within roughly $D_{25}/3$, $A_H^b = 0.1\ A_B^b$, and $A_H^{i-0} = 0.1\ A_B^{i-0}$.

(c) Color

$B-H$: blue-to-infrared color. Not a true color because the blue luminosity, $B_T^{b,i}$, and the infrared luminosity, $H_{-0.5}^{b,i}$, refer to different apertures.

Column 8

(a) Source

Sources for blue magnitudes. Numeric code is given in Table D, in order of priority.

TABLE D
Blue Magnitude Source Codes

CODE NO.	REFERENCE
1	14 (Holmberg)
2	9 (RCII)
7	8 (de Vaucouleurs et al.)
6	miscellaneous
3	36 (CGCG)
5	7 (Harvard)

For sources 3 and 5, blue magnitudes are only given to one decimal. The 1.6 μm (H band) magnitudes all come from ref. (1).

(b) Internal Absorption

A_B^{i-0}: obscuration within candidate galaxy. Reddening is given by the relationship

$$A_B^i = -2.5 \log\left[f(1 + e^{-\tau \sec i}) + (1 - 2f)(\frac{1 - e^{-\tau \sec i}}{\tau \sec i}) \right]$$

where $\tau = 0.55$ and $f = 0.25$. If $i > 80°$, a constant maximum value is assumed: $A_B^i = A_B^{80}$. The correction is to face-on orientation: $A_B^{i-0} = A_B^i - A_B^0$. The maximum value is $A_B^{80-0} = 0.67$. For elliptical and S0 galaxies, $A_B^{i-0} = 0$ is assumed. These correction procedures are described in ref. (30).

(c) Galactic Absorption

A_B^b: obscuration within Milky Way Galaxy. Corrections are those prescribed in ref. (4) and are based on measurements of galactic HI column densities along the line of sight. An HI contour corresponding to $A_B^b = 0.5$ mag is plotted across the first ten maps in the *NBG Atlas*.

Column 9

(a) Heliocentric Velocity

V_h: heliocentric velocity, in kilometers/second. Preferred sources are ref. (6, 11, 22); otherwise, the weighted mean of literature values are taken.

(b) Systemic Velocity

V_o: velocity adjusted for motion of 300 kilometers/second toward $\ell = 90°$, $b = 0°$.

(c) Source

Reference to HI observation by author and collaborators. If the source of the velocity is one of ref. (6, 11, 22), then the code is nonzero and specifies the telescope/receiver configuration involved in the detection (see Table E). The telescopes used are the National Radio Astronomy Observatory 91-m and 43-m telescopes, the Max-Planck-Institut 100-m telescope, and the Parkes 64-m telescope.

TABLE E
Telescope/Resolution Codes

NUMERIC CODE	TELESCOPE (m)	RESOLUTION (km/s)
1	91	22
2	43	22
3	91	5.5
4	43	5.5
5	100	5.5
6	100	22
7	64	4.9

Column 10

(a) Observed Line Width

W_{20}: neutral hydrogen line width at 20% of maximum intensity, in kilometers/second.

(b) Rotational Velocity

W_R^i: rotational velocity profile width parameter, in

kilometers/second. This parameter is determined from the equation

$$W_R^2 = W_{20}^2 + W_t^2 - 2W_{20}W_t(1 - e^{-(W_{20}/W_c)^2}) - 2W_t^2 e^{-(W_{20}/W_c)^2}$$

and

$$W_R^i = W_R/\sin i.$$

Rotation and random components of motion add linearly for giant galaxies ($W_{20} \gg W_c$) and in quadrature for dwarf galaxies ($W_{20} \ll W_c$). There is a transition region around $W_c = 120$ kilometers/second. The amplitude of random motions is described by $W_t = 38$ kilometers/second, the full width at 20% of maximum of a Gaussian with a dispersion of 10 kilometers/second. The parameter W_R^i is found statistically to equal *twice the maximum rotation velocity*. No value is determined for W_R^i if $i < 30°$ or if the HI line width is suspect; for example, because of poor signal/noise, or confusion, or an anomalous line shape, or extreme morphological peculiarity of the galaxy. There is a description of the derivation of the parameter W_R^i in ref. (30).

(c) **Global Velocity**

W_D^i: dynamical profile width parameter, in kilometers/second. The parameter is defined such that

$$(W_D^i)^2 = (W_R^i)^2 + 4\sigma^2,$$

where the three-dimensional random motion is assumed to be characterized by $\sigma = 10(3)^{1/2} = 17$ kilometers/second if $W_{20} > 38$ kilometers/second, and $\sigma = 10(3)^{1/2} (W_{20}/38)$ if $W_{20} < 38$ kilometers/second. The dynamical profile width parameter might be expected to correlate better with such properties as luminosity in samples that include dwarf galaxies, where random motions might be comparable to rotational motions. This parameter is also discussed in ref. (30). If W_R^i is not given, then W_D^i is not given.

Column 11

(a) **Velocity Uncertainty**

e_v: uncertainty in velocity measurement, in kilometers/second. If the galaxy was observed by the author and collaborators (ref. 6, 11, 22: nonzero entry in column 9c), then this uncertainty also applies to the value of W_{20}.

(b) **Line Width Uncertainty**

e_w: uncertainty in W_{20}, in kilometers/second. If the sole source for W_{20} is an observation by the author and collaborators, then this position is empty, and $e_w = e_v$ is assumed (see column 11a). If a literature source (with uncertainty e_ℓ) is the sole source or a contributing source for W_{20}, then the overall weighted error in W_{20} is given here. If e_v is nonzero,

$$1/e_w^2 = 1/e_v^2 + 1/e_\ell^2.$$

Otherwise, $e_w = e_\ell$.

(c) **Source**

Literature source for W_{20} and e_ℓ. The code for this entry is deciphered in Table F (page 5).

Column 12

(a) **Neutral Hydrogen Flux**

Log F_c: log of the HI flux adjusted for resolution effects, in units of million solar masses/megaparsec^2.

(b) **Source**

Literature source of the HI flux (left) and e_F, percent uncertainty in HI flux (right). The HI flux recorded in (a) is the average of the flux in ref. (11) and the flux from the literature source. Literature source codes are given in Table F.

(c) **Flux Correction Factor**

f_H: resolution correction factor. All HI observations by the author and collaborators were single-beam measurements with a beam that is frequently smaller than the size of the source. The HI flux in column 12a is the product of the observed flux and the correction factor. The statistical correction factors are discussed in ref. (11).

Column 13

(a) **Distance**

R: distance, in megaparsecs. Distances are based on velocities, an assumed value of the Hubble Constant of 75 kilometers/second/megaparsec, and the model that describes velocity perturbations in the vicinity of the Virgo Cluster discussed in ref. (31). This model assumes the Galaxy is retarded by 300 kilometers/second from universal expansion by the mass of the Virgo Cluster. Galaxies in 13 clusters with high-velocity dispersion (identified in ref. 26) are given a distance consistent with the mean velocity of the cluster. For galaxies within 28° of the Virgo Cluster there may be ambiguity in the transform between velocity and the distance prescribed by the velocity perturbation model, with three possible distances associated with one velocity. In each of such cases, the most plausible distance among the three alternatives (two of which are usually almost the same) is chosen based on independent distance estimators or association with galaxies with independently determined distances.

(b) **Distance Modulus**

μ: distance modulus. This alternative description of the distance is given by the standard formula,

$$\mu = 5 \log R + 25.$$

Column 14

(a) **Supergalactic X**

SGX: distance of galaxy parallel to one of the Cartesian axes in Supergalactic coordinates, in megaparsecs. The SGX axis is directed along the intersection

TABLE F
Hydrogen Line Width and Flux Literature References

2   R. J. Allen, B. F. Darchy, and R. Lauqué, *Astronomy and Astrophysics* **10**, 198, 1971.

3   R. J. Allen, W. M. Goss, R. Sancisi, W. T. Sullivan, III, and H. van Woerden. In *The Formation and Dynamics of Galaxies*. IAU Symposium, no. 58, ed. J. R. Shakeshaft, p. 425. Dordrecht: Reidel, 1974.

5   C. Balkowski, L. Bottinelli, P. Chamaraux, L. Gouguenheim, and J. Heidmann, *Astronomy and Astrophysics*, **34**, 43, 1974.

6   C. Balkowski, L. Bottinelli, L. Gouguenheim, and J. Heidmann, *Astronomy and Astrophysics*, **21**, 303, 1972.

7   C. Balkowski, L. Bottinelli, L. Gouguenheim, and J. Heidmann, *Astronomy and Astrophysics*, **23**, 139, 1973.

10 L. Bottinelli, P. Chamaraux, L. Gouguenheim, and J. Heidmann, *Astronomy and Astrophysics*, **29**, 217, 1973.

11 L. Bottinelli, P. Chamaraux, L. Gouguenheim, and R. Lauqué, *Astronomy and Astrophysics*, **6**, 453, 1970.

12 L. Bottinelli, R. Duflot, L. Gouguenheim, and J. Heidmann, *Astronomy and Astrophysics*, **41**, 61, 1975.

17 L. Bottinelli, L. Gouguenheim, and J. Heidmann, *Astronomy and Astrophysics*, **22**, 281, 1973.

20 N. Carozzi, P. Chamaraux, and R. Duflot-Augarde, *Astronomy and Astrophysics*, **30**, 21, 1974.

21 P. Chamaraux, J. Heidmann, and R. LauqueLA, *Astronomy and Astrophysics*, **8**, 424, 1970.

24 R. D. Davies, and B. M. Lewis, *Monthly Notices Royal Astronomical Society*, **165**, 231, 1973.

25 J. F. Dean, and R. D. Davies, *Monthly Notices Royal Astronomical Society*, **170**, 503, 1975.

36 L. Gouguenheim, *Astronomy and Astrophysics*, **3**, 281, 1969.

41 W. Huchtmeier, *Astronomy and Astrophysics*, **17**, 207, 1972.

51 B. M. Lewis, and R. D. Davies, *Monthly Notices Royal Astronomical Society*, **165**, 213, 1973.

54 W. H. McCutcheon, and R. D. Davies, *Monthly Notices Royal Astronomical Society*, **150**, 337, 1970.

56 S. D. Peterson, and G. S. Shostak, *Astronomical Journal*, **79**, 767, 1974.

60 M. S. Roberts, *Astronomical Journal*, **73**, 945, 1968.

66 B. J. Robinson, and K. J. van Damme, *Australian Journal of Physics*, **19**, 111, 1966.

67 B. J. Robinson, and J. A. Koehler, *Nature*, **208**, 993, 1965.

69 D. H. Rogstad, I. A. Lockhard, and M. C. H. Wright, *Astrophysical Journal*, **193**, 309, 1974.

72 D. H. Rogstad, and G. S. Shostak, *Astronomy and Astrophysics*, **22**, 111, 1973.

76 R. R. Sholbrook, and B. J. Robinson, *Australian Journal of Physics*, **20**, 131, 1967.

77 G. S. Shostak, *Astrophysical Journal*, **187**, 19, 1974.

79 G. S. Shostak, *Astrophysical Journal*, **198**, 527, 1975.

80 G. S. Shostak, and D. H. Rogstad, *Astronomy and Astrophysics*, **24**, 405, 411, 1973.

96 E. E. Epstein, *Astronomical Journal*, **69**, 490, 1964.

102 J. V. Hindman, F. J., Kerr, and R. X. McGee, *Australian Journal of Physics*, **16**, 570, 1963.

105 L. Volders, and J. A. Hogbom, *Bulletin of the Astronomical Institutes of the Netherlands*, **15**, 307, 1961.

106 J. Heidmann, *Bulletin of the Astronomical Institutes of the Netherlands*, **15**, 314, 1961.

201 W. K. Huchtmeier, G. A. Tammann, and H. J. Wendker, *Astronomy and Astrophysics*, **46**, 381, 1976.

202 W. K. Huchtmeier, G. A. Tammann, and H. J. Wendker, *Astronomy and Astrophysics*, **57**, 313, 1977.

203 B. Balick, S. M. Faber, and J. S. Gallagher, *Astrophysical Journal*, **209**, 710, 1976.

204 J. S. Gallagher, S. M. Faber, and B. Balick, *Astrophysical Journal*, **202**, 7, 1976.

205 G. R. Knapp, J. S. Gallagher, S. M. Faber, and B. Balick, *Astronomical Journal*, **82**, 106, 1977.

206 A. Bosma, R. D. Ekers, J. Lequeux, *Astronomy and Astrophysics*, **57**, 97, 1977.

208 D. A. Cesarsky, E. G. Falgarone, and J. Lequeux, *Astronomy and Astrophysics*, **59**, L5, 1977.

209 P. Chamaraux, *Astronomy and Astrophysics*, **60**, 67, 1977.

210 J. H. Bieging, and P. Biermann, *Astronomy and Astrophysics*, **60**, 361, 1977.

211 J. R. Dickel, and H. J. Rood, *Astrophysical Journal*, **223**, 391, 1978.

212 R. J. Allen, J. M. van der Hulst, W. M. Goss, and W. Huchtmeier, *Astronomy and Astrophysics*, **64**, 359, 1978.

214 C. Balkowski, P. Chamaraux, and L. Weliachew, *Astronomy and Astrophysics*, **69**, 263, 1978.

216 J. H. Bieging, *Astronomy and Astrophysics*, **64**, 23, 1978.

217 G. S. Shostak, *Astronomy and Astrophysics*, **68**, 321, 1978.

218 T. D. Kinman, V. C. Rubin, N. Thonnard, W. K. Ford, Jr., and C. J. Peterson, *Astronomical Journal*, **82**, 871, 1977.

220 G. D. van Albada, *Astronomy and Astrophysics*, **61**, 297, 1977.

222 A. J. Longmore, T. G. Hawarden, B. L. Webster, W. M. Goss, and U. Mebold, *Monthly Notices Royal Astronomical Society*, **184**, 97P, 1978.

223 N. Krumm, and E. E. Salpeter, *Astrophysical Journal*, **227**, 776, 1979.

225 K. Y. Lo, and W. L. W. Sargent, *Astrophysical Journal*, **227**, 756, 1979.

226 T. X. Thuan, and P. D. Seitzer, *Astrophysical Journal*, **231**, 327, 1979.

300 G. Helou, C. Giovanardi, E. E. Salpeter, and N. Krumm, *Astrophysical Journal Supplement*, **46**, 267, 1981.

301 K. Reif, U. Mebold, W. M. Goss, H. van Woerden, and B. Siegman, *Astronomy and Astrophysics Supplement*, **50**, 451, 1982.

5

of the Galactic and Supergalactic planes, where $\ell = 137°37$, $b = 0°$, and SGL = 0°, SGB = 0°.

(b) Supergalactic Y

SGY: distance of galaxy parallel to a second axis in Supergalactic coordinates. The SGY axis is in the Supergalactic plane, toward SGL = 90°, SGB = 0°, and points 6°32 from the north pole of the Galactic plane.

(c) Supergalactic Z

SGZ: distance of galaxy parallel to the third axis in Supergalactic coordinates. The SGZ axis is perpendicular to the Supergalactic plane, toward SGB = 90° and diverges from the Galactic plane at an angle of 6°32.

Column 15

(a) Absolute Magnitude

$M_B^{b,i}$: absolute blue magnitude of galaxy. This parameter follows directly from the apparent magnitude and distance modulus

$$M_B^{b,i} = B_T^{b,i} - \mu.$$

(b) Linear Diameter

$\Delta_{25}^{b,i}$: linear diameter of galaxy, in kiloparsecs. This dimension follows from the isophotal diameter and distance

$$\Delta_{25}^{b,i} = 0.292 \, D_{25}^{b,i} R.$$

Column 16

(a) Luminosity

Log L_B: the intrinsic blue luminosity of the galaxy, in solar luminosity units, $L_\odot$. This parameter is a reexpression of the absolute magnitude

$$L_B = \text{dex} \, [0.4 \, (5.48 - M_B^{b,i})].$$

(b) H I Mass

Log M_H: the mass of neutral hydrogen within the galaxy, in solar mass units, $M_\odot$. The H I mass follows from the H I flux and distance:

$$M_H = F_c \, 10^6 \, R^2.$$

(c) Total Mass

Log M_T: the total mass of the galaxy, in solar mass units, $M_\odot$. The mass is calculated from the Keplerian formula, given the rotation velocity and dimensions of the system:

$$M_T = (W_R^i)^2 \Delta_{25}^{b,i} / 8G.$$

where G is the Universal Gravitation Constant. M_T is not calculated if $i < 30°$ or the H I line width is of poor quality.

Column 17

(a) H I Mass to Light

M_H/L_B: ratio of neutral hydrogen mass to blue luminosity, in solar units.

(b) H I Mass Fraction

M_H/M_T: ratio of neutral hydrogen mass to total mass.

(c) Mass to Light

M_T/L_B: ratio of total mass to blue luminosity, in solar units.

Column 18

(a,b) Alternative Names

References to names that date back beyond 12 years can be found in ref. (7, 9). For more recent sources see Table G. At the very end of column 18b there may be a notation that indicates if the galaxy has a Seyfert (S) or LINER (L) active nucleus. The number that follows specifies whether the type is 1, or 2, or an intermediate case.

TABLE G
References for Alternative Names

IDENTIFICATION	REFERENCE
Arak	3 (Arakelian)
Kar	15 (Karachentseva)
M81 dw, CVn dw	18 (Lo and Sargent)
RMB	24 (Rubin et al.)
SAGDIG	5 (Cesarsky et al.)
SCLDIG	17 (Laustsen et al.)
ScI	23 (Rubin et al.)
Turn	32 (Turner)
UGCA	21 (Nilson)
UKS	19 (Longmore et al.)

(c) Major Catalogs

UGC, ESO, and MCG names. At the extreme left, there might be either the UGC name (a number $\leqslant$ 12921: ref. 20) or the ESO name (a hyphenated number: ref. 16). The UGC and ESO catalogs cover mutually exclusive parts of the sky, so a galaxy can never be in both of these catalogs. To the right of the position where these names might be found, there may be a double-hyphenated number. This number is the identification given the galaxy in the MCG catalog (ref. 35). To the extreme right there may be the number "2". If so, the galaxy appears in the *Second Reference Catalogue* (ref. 9).

(1) Name	(2) α / ℓ / SGL	(3) δ / b / SGB	(4) Type / Q_{xyz} / Group	(5) $D_{25}^{b,i}$ / $D_{25}^{b,i}$ / source	(6) d/D / i	(7) $B_T^{b,i}$ / $H_{-0.5}$ / $B-H$	(8) source / A_B^{i-o} / A_B^b	(9) V_h / V_o / Tel	(10) W_{20} / W_R / W_D	(11) e_v / e_w / source	(12) F_c / source/e_F / f_h	(13) R / μ	(14) SGX / SGY / SGZ	(15) $M_B^{b,i}$ / $\Delta_{25}^{b,i}$	(16) $\log L_B$ / $\log M_H$ / $\log M_T$	(17) M_H/L_B / M_H/M_T / M_T/L_B	(18) Alternate Names UGC/ESO MCG AGN? RCII
0000-80	0.2 / 305.46 / 216.15	-8037 / -36.55 / -13.29	10 / 0.15 / 55 -5	2.1 / 2.0 / 6	0.54 / 62.		/ 0.25 / 0.32	1953. / 1757. / 7		20		23.4 / 31.84	-18.4 / -13.4 / -5.4	13.7			012-14
N 7814	0.7 / 106.42 / 309.10	1552 / -45.18 / 16.39	2A / 0.12 / 65 -6	5.9 / 5.1 / 2	0.46 / 68.	10.92	2 / 0.34 / 0.09	1047. / 1250.		50		15.1 / 30.90	9.2 / -11.3 / 4.3	-19.98 / 22.5	10.18		8 03-01-020 2
U 17	1.1 / 106.21 / 308.19	1457 / -46.09 / 16.10	9 / 0.15 / 65 -6	2.9 / 2.7 / 2	0.70 / 50.	15.04	7 / 0.13 / 0.13	881. / 1081. / 1	107	20	0.23 / 31 / 1.07	13.0 / 30.57	7.7 / -9.8 / 3.6	-15.53 / 10.2	8.40 / 8.46	1.13	DDO 222 / 17 02-01-017 2
N 7817	1.4 / 108.23 / 313.86	2029 / -40.75 / 17.13	4X / 0.12 / 64 -8	3.8 / 3.0 / 2	0.31 / 79.	11.7 / 9.16 / 2.54	3 / 0.64 / 0.12	2316. / 2532. / 1	403 / 372. / 374.	15	0.58 / 37 / 1.08	31.5 / 32.49	20.9 / -21.7 / 9.3	-20.79 / 27.6	10.51 / 9.58 / 11.04	0.12 / 0.034 / 3.44	19 03-01-021 2
0003-41	3.8 / 332.83 / 253.58	-4146 / -72.92 / -1.55	7B / 0.12 / 61 -0+16	4.3 / 3.0 / 6	0.18 / 90.		/ 0.67 / 0.05	1542. / 1502. / 3	314	20	0.86 / 19 / 1.02	18.3 / 31.31	-5.2 / -17.5 / -0.5	16.0	9.38		293-34 -7-01-009 2
0005-34	5.6 / 351.54 / 260.22	-3451 / -78.11 / 0.43	10 / 0.25 / 14 -13	2.1 / 2.1 / 6	1.00		/ 0.00 / 0.05	207. / 198.	33	10 / 6 / 208	0.12 / 208 25	2.4 / 26.87	-0.4 / -2.3 / 0.0	1.5	6.88		SCLDIG / 349-31 -6-01-000
N 14	6.2 / 108.13 / 309.05	1532 / -45.83 / 15.02	10 / 0.18 / 65 -6	2.9 / 2.8 / 2	0.85 / 35.	12.1	3 / 0.05 / 0.12	860. / 1059. / 1	143 / 196. / 199.	10	0.71 / 13 / 1.08	12.8 / 30.54	7.8 / -9.6 / 3.3	-18.44 / 10.5	9.57 / 8.92 / 10.07	0.23 / 0.072 / 3.15	ARP 235, VV 80 / 75 03-01-026 2
N 24	7.4 / 43.65 / 269.42	-2515 / -80.44 / 3.23	5A / 0.12 / 19 -8 +7	6.8 / 4.8 / 6	0.19 / 90.	11.40 / 9.68 / 1.72	2 / 0.67 / 0.04	561. / 595. / 2	222 / 185. / 188.	10	1.10 / 7 10 / 1.04	6.8 / 29.17	-0.1 / -6.8 / 0.4	-17.77 / 9.5	9.30 / 8.77 / 9.98	0.29 / 0.062 / 4.74	UGCA 2 / 472-16 -4-01-018 2
0007-18	7.7 / 73.84 / 275.80	-1833 / -77.03 / 5.32	10 / 0.08 / 61 -0	2.1 / 2.1 / 6	1.00		/ 0.00 / 0.05	1546. / 1611. / 3	35	5	0.29 / 19 / 1.03	19.3 / 31.43	1.9 / -19.1 / 1.8	11.8	8.86		UGCA 3 / 538-24 -3-01-021
U 99	8.1 / 108.05 / 307.03	1325 / -48.00 / 14.11	9 / 0.20 / 64 +7	2.9 / 3.0 / 2	1.00		/ 0.00 / 0.16	1744. / 1935. / 1	100	10 / 8 / 226	0.36 / 226 12 / 1.09	23.7 / 31.87	13.8 / -18.3 / 5.8	20.8	9.11		99 02-01-023
U 122	10.7 / 110.01 / 310.52	1645 / -44.88 / 14.23	10 / 0.19 / 65 -6	2.5 / 1.8 / 2	0.19 / 90.	14.1	3 / 0.67 / 0.11	848. / 1048. / 1	113 / 89. / 96.	10	0.56 / 16 / 1.04	12.7 / 30.51	8.0 / -9.3 / 3.1	-16.41 / 6.7	8.76 / 8.77 / 9.19	1.03 / 0.382 / 2.69	122 03-01-028
N 45	11.4 / 55.89 / 271.41	-2327 / -80.64 / 2.94	8A / 0.12 / 19 -8 +7	8.3 / 7.8 / 6	0.73 / 47.	11.03 / 9.63 / 1.40	7 / 0.11 / 0.06	471. / 511. / 2	185 / 204. / 207.	7 / 6 / 25	1.78 / 25 4 / 1.09	5.9 / 28.85	0.1 / -5.9 / 0.3	-17.82 / 13.4	9.32 / 9.32 / 10.21	1.00 / 0.129 / 7.76	UGCA 4, DDO 223 / 473-01 -4-01-021 2
N 48	11.4 / 116.52 / 342.78	4757 / -14.18 / 18.40	5XP / 0.06 / 60 -0	1.7 / 1.8 / 2	0.63 / 56.	14.0	2 / 0.17 / 0.53	1776. / 2036.		10	0.63 / 217 15	26.4 / 32.11	23.9 / -7.4 / 8.3	-18.11 / 13.9	9.44 / 9.47	1.09	133 08-01-031 2

7

N 55	12.4	-3928	9B	34.1	0.19	7.50	2	140.	202	5	2.82	1.3	-0.3	-18.13	9.44	0.40	
	332.88	-75.74	0.33	24.2	90.		0.67	106.	166.	15	66	25.63	-1.3	9.2	9.05	0.152	
	256.29	-2.35	14-13	9			0.05		170.	96			-0.1		9.87	2.64	293-50 -7-01-013 2
U 156	14.2	1204	10	2.9	0.70	13.7	3	1135.	143	10	0.42	15.9	9.1	-17.30	9.11	0.51	
	109.80	-49.64	0.16	2.8	50.		0.13	1318.	149.	8	226 13	31.00	-12.5	13.0	8.82	0.079	
	306.03	12.35	65 +6	2			0.20	1	153.	226	1.07		3.4		9.92	6.47	156 02-01-029
0016-19	16.3	-1917	9	1.5	0.94			2060.		64	0.41	25.6	2.6				UGCA 5, DDO 1
	78.56	-78.95	0.06	1.5	23.		0.02	2116.			5 77	32.04	-25.4	11.2	9.23		
	275.74	3.15	60 -0	6			0.06						1.4				539-07 -3-01-027 2
U 191	17.6	1036	9	2.5	0.82	13.54	7	1144.	142	10	0.64	15.9	8.9	-17.46	9.18	0.74	DDO 2
	110.62	-51.25	0.14	2.5	39.		0.07	1320.	181.		17	31.00	-12.8	11.6	9.04	0.100	
	304.78	11.19	65 +6	2			0.24	1	184.		1.06		3.1		10.04	7.35	191 02-01-033 2
U 192	17.7	5901	10B	7.2	0.89	7.97	2	-345.	79	4	2.19	0.7	0.7	-16.26	8.70	0.15	IC 10
	118.98	-3.34	0.55	15.2	31.		0.04	-83.	120.	5	25 5	24.23	-0.1	3.1	7.88	0.058	
	354.45	17.83	14-12	2			3.69		124.	217			0.2		9.11	2.61	192 10-01-000 2
0017-77	17.8	-7722	9B	1.7	0.80			1810.		20		21.4	-16.1				
	305.22	-39.90	0.11	1.7	41.		0.08	1622.				31.66	-13.3	10.6			
	219.60	-13.34	55 +5	6			0.21	7					-4.9				028-14
0018-64	18.6	-6408	10	2.1	0.19			1764.		20		20.7	-12.3				
	308.49	-52.94	0.13	1.5	90.		0.67	1622.				31.58	-16.2	9.1			
	232.78	-10.36	55 -0	6			0.00	7					-3.7				078-22
0020-53	20.0	-5355	5	3.4	0.65			1430.		20		16.5	-7.5				
	312.40	-62.95	0.07	3.1	54.		0.16	1329.				31.09	-14.6	14.9			
	242.81	-7.86	61 -0	6			0.05	7					-2.3				150-05
N 100	21.5	1612	5	5.3	0.15	12.8	3	844.	236	10	1.08	12.5	8.0	-17.69	9.27	1.01	
	113.50	-45.89	0.17	3.8	90.	10.95	0.67	1035.	199.	6	217 7	30.49	-9.3	13.9	9.27	0.118	
	310.56	11.58	65 +6	2		1.85	0.11	1	202.	217	1.14		2.5		10.20	8.59	231 03-02-009 2
U 260	24.5	1118	5	2.9	0.27	12.7	3	2131.	291	20	0.97	28.3	16.4	-19.56	10.02	0.72	
	113.47	-50.85	0.07	2.3	82.		0.67	2305.			14	32.26	-22.6	19.0	9.87		
	305.89	9.72	64 -0	2			0.16	1			1.05		4.8				260 02-02-011
U 300	27.5	314	10	1.5	1.00			1345.	36	7	0.02	17.8	8.4				
	112.57	-58.93	0.09	1.5			0.00	1488.			226 16	31.26	-15.6	7.8	8.52		
	298.19	6.94	65 -0	2			0.03	3			1.03		2.2				300 00-02-000
N 134	27.9	-3332	4X	7.6	0.20	10.30	2	1587.	504	10	1.56	19.0	-2.3	-21.09	10.63	0.31	
	338.22	-82.37	0.22	5.4	90.	7.51	0.67	1572.	466.		10	31.39	-18.8	30.0	10.12	0.069	
	262.89	-3.55	61-18	6		2.79	0.05	4	467.		1.07		-1.2		11.28	4.44	350-23 -6-02-012 2
N 147	30.4	4814	-5	17.8	0.62	9.74	1	-163.		19		0.7	0.6	-14.49	7.99		DDO 3
	119.81	-14.25	0.55	18.3	57.		0.00	89.				24.23	-0.2	3.7			
	343.36	15.26	14-12	2			0.66						0.2				326 08-02-005 2
U 328	30.8	-124	9B	1.7	1.00			1984.	149	10	0.61	25.5	10.3				
	112.72	-63.64	0.07	1.7			0.00	2107.			18	32.03	-23.2	12.7	9.42		
	293.93	4.90	61 -0	2			0.08	1			1.03		2.2				328 00-02-070

(1) Name	(2) α / ℓ / SGL	(3) δ / b / SGB	(4) Type / ϱxyz / Group	(5) D_{25} / $D_{25}^{b,i}$ / source	(6) d/D / i	(7) $B_T^{b,i}$ / $H_{-0.5}$ / B-H	(8) source / A_B^{i-o} / A_B^b	(9) V_h / V_o / Tel	(10) W_{20} / W_R / W_D	(11) e_v / e_w / source	(12) F_v / source/e_f / f_h	(13) R / μ	(14) SGX / SGY / SGZ	(15) $M_B^{b,i}$ / $\Delta_{25}^{b,i}$	(16) $\log L_B$ / $\log M_H$ / $\log M_T$	(17) M_H/L_B / M_H/M_T / M_T/L_B	(18) Alternate Names — UGC/ESO MCG AGN? RCII
0031-31	31.7	-3103	9	1.9	1.00	13.65	7	1585.	79	8	0.63	19.0	-1.5	-17.74	9.29	0.79	UGCA 6, DDO 224
	347.90	-84.75	0.29	1.9			0.00	1579.			12	31.39	-18.9	10.5	9.19		
	265.50	-3.62	61-18	6			0.05	2			1.01		-1.2				
N 150	31.8	-2804	4BP	3.9	0.44	11.30	2	1593.	329	15	1.02	19.2	-0.5	-20.12	10.24	0.22	
	22.06	-86.14	0.27	3.3	69.	9.08	0.37	1601.	311.		14	31.42	-19.2	18.5	9.59	0.074	410-18 -5-02-016 2
	268.38	-2.82	61-20+18	6		2.22	0.07	2	313.		1.03		-0.9		10.72	2.99	
N 148	31.8	-3204	-2	2.1	0.38	13.03	2	1544.		40		18.4	-1.8	-18.29	9.51		UGCA 7
	340.59	-84.03	0.28	1.8	74.		0.00	1534.				31.32	-18.3	9.7			410-19 -5-02-018 2
	264.53	-3.93	61-18	6			0.05						-1.3				
N 157	32.2	-840	4X	3.0	0.69	10.75	1	1659.		12		20.9	6.1	-20.85	10.53		
	110.26	-70.85	0.21	2.8	51.		0.14	1751.				31.60	-20.0	17.1			410-20 -5-02-017 2
	287.03	2.56	61 -0+21	9			0.13						0.9				
0033-25	33.4	-2539	8X	3.5	0.78	12.50	7	1557.	145	20	0.88	18.8	0.3	-18.87	9.74	0.49	IC 1558, UGCA 8
	58.78	-86.10	0.21	3.4	43.		0.09	1574.	170.		11	31.37	-18.8	18.7	9.43	0.171	DDO 225
	270.80	-2.49	61-20+18	6			0.05	2	173.		1.02		-0.8		10.19	2.85	474-02 -4-02-024 2
N 185	36.2	4804	-5 P	17.0	0.84	9.36	1	-211.		10		0.7	0.6	-14.87	8.14		
	120.79	-14.48	0.55	19.0	37.		0.00	39.				24.23	-0.2	3.9			396 08-02-010 2
	343.31	14.28	14-12	2			0.71						0.2				
N 178	36.6	-1427	9B	2.1	0.50	12.77	2	1479.		90		18.4	3.8	-18.56	9.62		
	109.83	-76.73	0.16	1.8	65.		0.29	1544.				31.33	-18.0	9.7			IC 39
	281.77	-0.09	61-22+21	4			0.04						0.0				-2-02-078 2
N 205	37.6	4125	-5 P	18.7	0.60	8.63	1	-242.		12		0.7	0.6	-15.60	8.43		
	120.71	-21.13	0.52	17.5	58.		0.00	-1.				24.23	-0.3	3.6			426 07-02-014 2
	336.58	13.05	14-12	9			0.22						0.2				
N 210	38.1	-1409	3X	4.9	0.67	11.48	2	1635.		50		20.3	4.3	-20.06	10.22		
	111.62	-76.51	0.24	4.5	53.		0.15	1700.				31.54	-19.9	26.7			
	282.16	-0.36	61-22+21	4			0.02						-0.1				-2-02-081 2
0038-63	38.6	-6343	7BP	2.2	0.52			1721.		20		20.1	-11.6				
	304.94	-53.65	0.15	1.9	64.		0.27	1575.				31.51	-15.8	11.2			079-05
	233.67	-12.42	55 -0	6			0.00	7					-4.3				
N 216	39.0	-2119	13	2.0	0.33	13.74	2	1564.		28	0.70	19.1	1.8	-17.67	9.26	1.00	HARO 13
	101.98	-83.50	0.14	1.6	77.		0.00	1597.			10	31.41	-19.0	8.9	9.26		540-15 -4-02-035 2
	275.33	-2.53	61 -0	6			0.06	2			32		-0.8				
N 221	40.0	4036	-5	12.6	0.65	8.89	1	-203.		7		0.7	0.6	-15.34	8.33		M 32, ARP 168
	121.16	-21.97	0.52	12.1	54.		0.00	35.				24.23	-0.3	2.5			452 07-02-015 2
	335.83	12.46	14-12	2			0.26						0.2				
N 224	40.0	4100	3A	193.2	0.32	3.56	1	-298.	535	3	3.93	0.7	0.6	-20.67	10.46	0.14	M 31
	121.18	-21.57	0.52	155.5	78.	0.86	0.63	-59.	507.	30	25	24.23	-0.3	31.8	9.62	0.018	454 07-02-016 2
	336.24	12.53	14-12	9		2.70	0.17		508.	25	6		0.2		11.37	8.22	

0040-22	40.7	-2231	10	2.3	0.43	13.96	7	361.	81	20	0.19	4.5	0.3	-14.28	7.90	0.39	IC 1574, UGCA 9
	101.53	-84.76	0.06	1.9	70.		0.39	388.	67.		25	28.24	-4.4	2.5	7.50	0.097	DDO 226
	274.28	-3.23	14-13	6			0.05	2	75.		1.01		-0.3		8.51	4.05	474-18 -4-02-043 2
N 244	43.3	-1552	13 P	1.1	0.91	13.68	2	941.		16	0.43	11.6	2.2	-16.65	8.85	0.51	UGCA 10, HARO 14
	116.19	-78.39	0.06	1.1	28.		0.00	995.			10 34	30.33	-11.4	3.7	8.56		
	280.85	-2.03	19 -0	2			0.02						-0.4				-3-03-003 2
U 477	43.6	1913	9	3.3	0.27	13.6	3	2657.	256	15	0.91	35.5	24.8	-19.15	9.85	1.44	
	121.25	-43.36	0.06	2.5	82.		0.67	2843.	220.		14	32.75	-25.0	25.9	10.01	0.282	
	314.78	7.17	60 -0	2			0.08	1	223.		1.07		4.4		10.56	5.11	477 03-03-002
0043-11	43.6	-1148	9X	2.3	0.75	14.22	7	1617.	98	10	0.59	20.2	5.2	-17.31	9.12	1.22	DDO 5
	118.11	-74.35	0.28	2.2	46.		0.10	1688.	108.		16	31.53	-19.5	13.0	9.20	0.362	
	284.79	-1.00	61-21	4			0.10	1	113.		1.05		-0.4		9.64	3.36	-2-03-009 2
N 247	44.6	-2101	7X	22.2	0.30	8.66	1	159.	218	5	2.26	2.1	0.2	-17.98	9.38	0.33	UGCA 11
	113.87	-83.54	0.32	17.3	80.	7.62	0.67	190.	184.		25 2	26.64	-2.1	10.6	8.90	0.077	
	275.97	-3.71	14-13	6		1.04	0.06	4	187.		1.46		-0.1		10.02	4.30	540-22 -4-03-005 2
N 254	45.0	-3142	-2B	2.5	0.60	12.65	2	1441.		90		17.1	-1.3	-18.51	9.60		
	314.15	-85.61	0.15	2.3	58.		0.00	1425.				31.16	-16.9	11.5			
	265.62	-6.53	61-19+18	6			0.05						-1.9				411-15 -5-03-005 2
N 253	45.1	-2534	5X	25.8	0.23	7.34	2	249.	431	7	2.83	3.0	0.1	-20.02	10.20	0.38	UGCA 13
	97.43	-87.96	0.21	18.8	86.	4.67	0.67	260.	394.	15	41	27.36	-3.0	16.5	9.78	0.082	
	271.59	-5.00	14-13	9		2.67	0.04		395.	41			-0.3		10.87	4.68	474-29 -4-03-009 2
0045-10	45.2	-1011	5B	3.1	0.19			1346.	153	7	0.57	16.8	4.8				UGCA 14
	119.86	-72.76	0.09	2.2	90.		0.67	1423.	122.		21	31.13	-16.1	10.8	9.02	0.225	
	286.45	-0.94	61-21	2			0.15	1	127.		1.06		-0.3		9.67		-2-03-016 2
N 255	45.3	-1145	4X	3.3	1.00	12.25	2	1604.	155	7	1.03	20.0	5.2	-19.26	9.90	0.54	
	119.66	-74.32	0.28	3.4			0.00	1674.		10	217 15	31.51	-19.3	19.9	9.63		
	284.95	-1.39	61-21	4			0.10			211			-0.5				-2-03-017 2
0047-21	47.4	-2116	10	2.1	0.52			303.	47	8	0.03	3.7	0.4				UGCA 15, DDO 6
	119.55	-83.85	0.13	1.8	64.		0.27	331.	29.		5 22	27.87	-3.7	1.9	7.17	0.309	
	275.90	-4.40	14-13	6			0.04	2	45.		1.01		-0.3		7.68		540-31 -4-03-019 2
0048-66	48.1	-6649	7B	1.4	0.81			1673.		20		19.5	-12.0				
	303.14	-50.58	0.14	1.3	40.		0.07	1513.				31.45	-14.6	7.4			
	230.68	-13.87	55 -0	6			0.00	7					-4.7				079-07
N 274	48.5	-720	-2XP	1.6	0.94	12.7	5	1733.		35		21.9	7.3	-19.00	9.79		VV 81
	122.65	-69.93	0.25	1.7	23.		0.00	1820.				31.70	-20.6	10.9			
	289.42	-0.96	61-23+21	5			0.21						-0.4				-1-03-021 2
N 275	48.6	-720	6BP	1.4	0.82	12.73	2	1735.	286	25	0.86	21.9	7.3	-18.97	9.78	0.58	VV 81
	122.72	-69.93	0.25	1.4	39.		0.07	1822.		13	17	31.70	-20.6	9.0	9.54		
	289.42	-0.98	61-23+21	4			0.20	1		211	1.03		-0.4				-1-03-022 2
0049-01	49.0	-130	13 P			14.8	3	1390.		220		17.9	7.6	-16.46	8.78		ARAK 16
	123.01	-64.10	0.08				0.07	1500.				31.26	-16.2				
	295.07	0.50	61-24+21				0.12						0.2				00-03-015

(1) Name	(2) α / l / SGL	(3) δ / b / SGB	(4) Type / Q_{xyz} / Group	(5) D_{25} / $D_{25}^{b,i}$ / source	(6) d/D / i	(7) $B_T^{b,i}$ / $H_{-0.5}$ / $B-H$	(8) source / $A_B^{i,o}$ / A_B^b	(9) V_h / V_o / Tel	(10) W_{20} / W_R / W_D	(11) e_v / e_w / source	(12) F_c / source·e_f / f_h	(13) R / μ	(14) SGX / SGY / SGZ	(15) $M_B^{b,i}$ / Δ_{25}	(16) $\log L_B$ / $\log M_H$ / $\log M_T$	(17) M_H/L_B / M_H/M_T / M_T/L_B	(18) Alternate Names UGC/ESO MCG; AGN? RCII
N 278	49.3	4717	3XP	2.7	1.00	10.75	2	641	140	9	0.88	11.8	11.1	−19.62	10.04	0.10	528 08-02-016 2
	123.06	−15.32	0.11	3.2			0.00	883.		10	211 9	30.37	−3.4	11.0	9.02		
	342.86	11.98	17 −2	2			0.76			211			2.5				
0049-00	49.5	−44	13 P			14.7	3	1687		220		21.6	9.4	−16.98	8.98		ARAK 18
	123.29	−63.33	0.13				0.00	1800.				31.68	−19.5				00-03-018
	295.84	0.58	61 −24+21				0.08						0.2				
N 289	50.4	−3129	4B	8.3	0.82	11.5	5	1633	305	10	1.58	19.4	−1.3	−19.94	10.17	0.97	411-25 −5-03-010 2
	298.77	−85.90	0.22	8.0	39.		0.07	1614	425.		6	31.44	−19.2	45.3	10.16	0.060	
	266.11	−7.60	61 −19+18	6			0.05	2	426.		1.03		−2.6		11.38	16.12	
0051-73	51.0	−7306	10B	281.8	0.56	2.39	2	151		5	5.34	0.1	0.0	−16.10	8.63	5.11	SMC, A 0051
	302.79	−44.30	0.50	257.2	61.		0.23	−30.			102 27	18.49	0.0	7.5	9.34		029-21 2
	224.26	−14.80	14 −12	9			0.17						0.0				
N 300	52.5	−3757	7A	20.2	0.75	8.54	2	145	177	6	2.74	1.2	−0.2	−16.88	8.94	0.90	295-20 −6-03-005 2
	299.22	−79.42	0.45	19.1	46.		0.10	97.	200.	15	76	25.42	−1.2	6.7	8.90	0.102	
	259.84	−9.48	14 −13	9			0.05		203.	96			−0.2		9.89	8.83	
U 608	56.1	4745	8X	2.0	0.49	14.1	3	2760	215	15	0.54	38.9	36.6	−18.85	9.73	0.97	608 08-03-001
	124.24	−14.83	0.11	2.0	66.		0.30	3000.	196.		16	32.95	−10.8	22.7	9.72	0.207	
	343.55	10.94	60 −1	2			0.75	1	199.		1.04		7.4		10.40	4.69	
N 337	57.3	−751	7B	2.9	0.69	11.65	2	1651	264	15	1.12	20.7	6.9	−19.93	10.16	0.39	−1-03-053 2
	129.14	−70.35	0.24	2.9	50.		0.13	1729.		8	217 7	31.58	−19.5	17.5	9.75		
	289.50	−3.20	61 −23+21	4			0.30	1		217	1.07		−1.2				
0057-33	57.6	−3358	−5	37.0	0.85			196.		150		0.1	0.0				SCL
	287.78	−83.18	0.50	36.1	36.		0.00	162.				20.00	−0.1	1.1			351-30 −6-03-015 2
	264.01	−9.63	14 −12	6			0.05						0.0				
U 625	58.0	4725	4X	4.1	0.31	11.7	3	2614	346	10	0.97	36.9	34.8	−21.14	10.65	0.29	IC 65
	124.58	−15.16	0.11	3.7	79.	9.87	0.64	2852.	314.		15	32.84	−10.4	39.9	10.10	0.111	625 08-03-005 2
	343.28	10.56	60 −1	2		1.83	0.73	1	316.		1.14		6.8		11.06	2.56	
U 634	58.8	721	9X	1.9	0.62	14.2	3	2211	148	15	0.53	28.7	16.2	−18.09	9.43	1.04	DDO 7
	127.26	−55.17	0.12	1.8	56.		0.18	2347.	142.		20	32.29	−23.8	15.1	9.45	0.316	634 01-03-011 2
	304.25	0.53	52 −0+12	2			0.14	1	146.		1.03		0.3		9.95	3.29	
0059-07	59.0	−752	8X	6.4	0.75			1084	100	5	1.38	13.7	4.6				NGC 337A
	130.38	−70.32	0.13	6.4	46.		0.10	1161.	110.		6	30.68	−12.9	25.6	9.65	0.501	−1-03-065 2
	289.60	−3.61	17 −7	4			0.28	1	115.		1.31		−0.9		9.95		
N 357	100.8	−636		2.3	0.65	12.69	2	2541		50		32.1	11.4	−19.84	10.13		
	131.22	−69.01	0.08	2.2	54.		0.00	2622.				32.53	−29.9	20.6			−1-03-081 2
	290.94	−3.70	52 −0	4			0.21						−2.1				
U 668	102.2	151	10B	18.6	0.92	9.93	2	−238	40	7	2.09	0.7	0.3	−14.30	7.91	0.74	IC 1613, DDO 8
	129.73	−60.58	0.52	18.6	27.		0.03	−125.		5	105 10	24.23	−0.6	3.8	7.78		668 00-03-070 2
	299.18	−1.78	14 −12	2			0.03			96			0.0				

0102-06	102.6	-629	7X	4.4	0.75	11.7	5	1092.	189	8	1.20	13.8	5.0	-19.00	9.79	0.49	A 0103
	132.40	-68.83	0.14	4.3	46.		0.10	1172.	216.	5	217 5	30.70	-12.9	17.3	9.48	0.129	
	291.17	-4.10	17 -7	4			0.20	1	219.	217	1.16		-1.0		10.37	3.78	-1-03-085 2
N 406	105.7	-7009	3A	2.7	0.41	11.9	5	1509.		11		17.3	-11.3	-19.30	9.91		
	300.91	-47.18	0.08	2.3	71.		0.41	1334.				31.20	-12.3	11.6			
	227.39	-15.76	55 -0	6			0.04						-4.7				051-18 2
U 711	106.1	123	5	3.7	0.15	13.3	3	1979.	226	15	0.60	25.3	12.3	-18.72	9.68	0.53	
	131.83	-60.92	0.13	2.6	90.	12.27	0.67	2088.	189.		15	32.02	-22.1	19.2	9.41	0.128	
	299.00	-2.84	52 -0+12	2		1.03	0.04	1	192.		1.08		-1.3		10.30	4.16	711 00-04-008 2
N 404	106.7	3527	-2A	6.6	1.00	10.84	1	-35.		18		2.4	2.1	-16.11	8.64		
	127.05	-27.01	0.20	6.9			0.00	178.				26.95	-1.1	4.8			
	331.89	6.22	14+12	2			0.21						0.3				718 06-03-018 2
U 731	107.7	4920	10	2.4	0.92			639.	146	10	0.98	11.8	11.3				DDO 9
	126.13	-13.15	0.13	2.8	26.		0.03	875.			9	30.37	-2.9	9.6	9.12		
	345.53	9.39	17 -2	2			0.78	5			1.17		1.9				731 08-03-000 2
U 750	109.2	-55	13 P	1.1	0.90	13.9	3	2550.		200		32.5	14.7	-18.66	9.66		IC 1639, MARK 562
	134.22	-63.08	0.14	1.1	29.		0.00	2647.				32.56	-28.9	10.4			
	296.99	-4.21	52 -0+12	2			0.13						-2.4				750 00-04-031 2
N 428	110.4	43	9X	4.6	0.76	11.73	1	1157.	184	8	1.26	14.9	7.1	-19.14	9.85	0.57	
	134.24	-61.42	0.09	4.3	44.		0.09	1260.	215.	10	217 7	30.87	-13.1	18.7	9.61	0.161	
	298.65	-4.06	17 -0	2			0.03		217.	217			-1.1		10.40	3.56	763 00-04-036 2
N 450	113.0	-108	6X	3.3	0.82	12.29	2	1759.	201	15	0.88	22.4	10.1	-19.46	9.98	0.40	
	136.37	-63.11	0.14	3.3	39.		0.07	1853.	265.	8	217 8	31.75	-19.8	21.6	9.58	0.087	
	297.03	-5.18	52 -0	2			0.14	1	267.	79	1.10		-2.0		10.64	4.65	806 00-04-062 2
N 470	117.2	309	4A	3.2	0.61	12.37	2	2374.	373	20	0.52	30.5	15.8	-20.05	10.21	0.19	
	136.65	-58.70	0.26	2.9	57.		0.19	2481.			23	32.42	-25.9	25.8	9.49		
	301.46	-5.03	52-12	2			0.05	1			1.08		-2.7				858 00-04-084 2
N 473	117.3	1617		2.5	0.65	12.6	3	2252.		90		29.8	20.7	-19.77	10.10		
	132.77	-45.78	0.08	2.3	54.		0.00	2406.				32.37	-21.4	20.0			
	314.09	-1.34	52 -0	2			0.11						-0.7				859 03-04-022 2
N 474	117.5	309	-2A	11.6	0.91	11.94	2	2529.		37		32.5	16.9	-20.62	10.44		ARP 227
	136.79	-58.69	0.19	11.5	29.		0.00	2636.				32.56	-27.6	109.1			
	301.48	-5.10	52-12	2			0.06						-2.9				864 00-04-085 2
U 871	117.8	535	10	2.0	1.00			2178.	102	15	0.47	28.1	15.6				
	136.00	-56.29	0.24	2.1			0.00	2294.		11	226 12	32.24	-23.2	17.2	9.37		
	303.85	-4.50	52-12	2			0.12	1		226	1.05		-2.2				871 01-04-022
U 891	118.6	1209	9X	3.1	0.51	14.32	7	646.	142	7	0.64	9.4	6.1	-15.55	8.41	1.49	DDO 10
	134.26	-49.80	0.12	2.7	64.		0.28	785.	125.		12	29.87	-7.2	7.4	8.59	0.115	
	310.21	-2.83	17 -4	2			0.10	1	130.		1.07		-0.5		9.53	13.00	891 02-04-030 2
N 488	119.2	500	3A	5.4	0.74	10.97	1	2268.	451	10	0.60	29.3	16.0	-21.36	10.74	0.06	
	136.83	-56.79	0.28	5.2	46.		0.10	2380.		15	211 50	32.33	-24.3	44.5	9.53		
	303.38	-5.00	52-12	2			0.08			211			-2.6				907 01-04-033 2

(1) Name	(2) α / l / SGL	(3) δ / b / SGB	(4) Type / ρ_{xyz} / Group	(5) D_{25} / $D_{25}^{b,i}$ / source	(6) d/D / i	(7) $B_T^{b,i}$ / $H_{-0.5}$ / B-H	(8) source / A_B^{i-o} / A_B^b	(9) V_h / V_o / Tel	(10) W_{20} / W_R / W_D	(11) e_v / e_w / source	(12) F_c source/e_F / f_h	(13) R / μ	(14) SGX / SGY / SGZ	(15) $M_B^{b,i}$ / $\Delta_{25}^{b,i}$	(16) log L_B / log M_H / log M_T	(17) M_H/L_B / M_H/M_T / M_T/L_B	(18) Alternate Names UGC/ESO, MCG / AGN? RCII
N 493	119.6	41	5	4.0	0.41	12.47	2	2340.	292.	10	0.98	29.9	14.5	-19.91	10.16	0.60	914 00-04-099 2
	138.93	-60.97	0.21	3.4	71.	10.83	0.41	2436.	268.		14	32.38	-25.9	29.7	9.93	0.138	
	299.24	-6.28	52-12	2		1.64	0.06	1	270.		1.10		-3.3		10.79	4.32	
N 514	121.4	1240	5X	4.0	0.75	12.31	2	2477.		12		32.4	21.2	-20.24	10.29		947 02-04-035 2
	135.14	-49.17	0.13	3.8	45.		0.10	2615.				32.55	-24.4	36.0			
	310.91	-3.34	52-0+12	2			0.09						-1.9				
N 520	122.0	332	13 P	4.6	0.44	11.99	1	2166.	412	7	0.74	27.8	14.7	-20.23	10.28	0.22	ARP 157, VV 231
	138.71	-58.06	0.25	3.9	69.		0.00	2271.		30	217	32.22	-23.4	31.7	9.63		966 01-04-052 2
	302.16	-6.07	52-12	2			0.06			56	30		-2.9				
N 524	122.2	917	-2A	3.2	0.98	11.37	2	2470.		65		32.1	19.6	-21.16	10.66		968 01-04-053 2
	136.52	-52.44	0.15	3.2	15.		0.00	2596.				32.53	-25.3	30.0			
	307.72	-4.50	52-0+12	9			0.13						-2.5				
0123-06	123.7	-620	5	2.9	0.16			1968.	262.	10	0.66	24.6	9.4	15.1	9.44	0.126	UGCA 17
	145.95	-67.38	0.28	2.1	90.		0.67	2033.	224.		17	31.96	-22.4		10.34		-1-04-044 2
	292.69	-9.13	52-7	2			0.19	1	227.		1.05		-3.9				
U 1072	127.2	-130	-2	1.8	0.66	13.9	3	819.		220		10.6	4.9	-16.23	8.68		ARAK 50
	144.08	-62.53	0.07	1.7	53.		0.00	900.				30.13	-9.3	5.3			1072 00-04-164 2
	297.64	-8.70	17-0	2			0.05						-1.6				
N 578	128.0	-2256	5X	5.4	0.71	11.36	1	1629.	308.	15	1.26	19.5	2.1	-20.09	10.23	0.41	UGCA 18
	188.30	-80.11	0.07	5.0	49.	9.58	0.12	1622.	360.		9	31.45	-18.8	28.5	9.84	0.065	476-15 -4-04-020 2
	276.43	-13.94	52-0	6		1.78	0.02	2	362.		1.04		-4.7		11.03	6.33	
N 584	128.8	-708	-5	3.8	0.63	11.16	2	1874.		11		23.4	8.7	-20.68	10.46		IC 1712
	149.80	-67.64	0.42	3.5	55.		0.00	1931.				31.84	-21.3	23.9			-1-04-060 2
	292.23	-10.56	52-7	5			0.14						-4.3				
U 1102	130.0	420	13 P	1.7	0.89	13.9	3	1985.	181.	10	0.51	25.5	13.9	-18.13	9.44	0.76	ARP 306, VV 173
	141.89	-56.76	0.17	1.7	31.		0.00	2086.	289.		24	32.03	-21.1	12.7	9.32	0.069	1102 01-05-002 2
	303.50	-7.77	52-13+12	2			0.04	1	291.		1.03		-3.4		10.49	11.03	
N 596	130.4	-717	-5	3.4	0.63	11.66	2	1906.		10		23.8	8.8	-20.22	10.28		-1-05-005 2
	150.91	-67.62	0.40	3.2	56.		0.00	1962.				31.88	-21.6	22.2			
	292.18	-10.98	52-7	5			0.12						-4.5				
N 600	130.6	-734	7B	4.4	1.00	12.90	2	1841.	89	10	0.84	22.9	8.4	-18.90	9.75	0.64	-1-05-007 2
	151.35	-67.85	0.41	4.5			0.00	1895.			9	31.80	-20.8	30.1	9.56		
	291.92	-11.10	52-7	4			0.10	1			1.20		-4.4				
N 598	131.1	3024	6A	56.5	0.62	5.92	1	-182.	204.	4	3.41	0.7	0.6	-18.31	9.52	0.38	M 33
	133.63	-31.33	0.52	52.6	56.	4.36	0.18	3.	202.	10	2	24.23	-0.4	10.8	9.10	0.099	1117 05-04-069 2
	328.50	-0.13	14-12	9		1.56	0.16		205.	25	8		0.0		10.10	3.88	
N 613	132.0	-2940	4B	5.0	0.84	10.68	2	1487.	383.	10	1.00	17.5	-0.1	-20.53	10.40	0.12	413-11 -5-04-044 2
	229.02	-80.29	0.07	4.9	37.	7.89	0.06	1449.	577.		17	31.21	-16.8	25.0	9.49	0.013	
	269.73	-16.02	52-0	9		2.79	0.05	2	578.		1.06		-4.8		11.38	9.54	

Name			Type	Dim	Ratio/PA	mag		HI	N						mag		Alt. name
U 1133	132.6 / 143.14 / 303.48	407 / -56.77 / -8.45	10 / 0.17 / 52-13+12	3.7 / 3.3 / 2	0.53 / 63.		0.25 / 0.07	1968. / 2067. / 1	119 / 106. / 111.	10	0.50 / 18 / 1.10	25.2 / 32.01	13.8 / -20.8 / -3.7	24.3	9.30 / 9.90	0.254	DDO 12 ; 1133 01-05-005 2
N 615	132.6 / 152.57 / 292.01	-736 / -67.67 / -11.58	3A / 0.41 / 52 -7	4.1 / 3.4 / 4	0.34 / 77.	11.63	2 / 0.56 / 0.12	1857. / 1910.	443	10 / 30 / 211	0.65 / 20 / 1.19	23.1 / 31.82	8.5 / -21.0 / -4.6	-20.19 / 22.9	10.27 / 9.38	0.13	-1-05-008 2
N 625	132.9 / 273.62 / 257.32	-4140 / -73.14 / -17.72	9B / 0.16 / 14+13	5.0 / 3.9 / 6	0.31 / 79.	11.6	5 / 0.66 / 0.05	404. / 317. / 2	98 / 78. / 86.	8	0.88 / 9 / 1.06	3.9 / 27.95	-0.8 / -3.6 / -1.2	-16.35 / 4.4	8.73 / 8.06 / 8.89	0.21 / 0.147 / 1.45	297-05 -7-04-017 2
N 628	134.0 / 138.62 / 314.56	1532 / -45.70 / -5.40	5A / 0.18 / 17 -4	7.0 / 7.3 / 9	1.00	9.61	1 / 0.00 / 0.14	659. / 798. / 4	74	5	2.01 / 25 3 / 1.17	9.7 / 29.93	6.8 / -6.9 / -0.9	-20.32 / 20.7	10.32 / 9.98	0.46	M 74 ; 1149 03-05-011 2
N 636	136.6 / 155.07 / 292.10	-746 / -67.36 / -12.59	-5 / 0.36 / 52 -7	2.3 / 2.2 / 5	0.83 / 38.	12.21	2 / 0.00 / 0.09	1941. / 1990.		50		24.2 / 31.92	8.9 / -21.8 / -5.3	-19.71 / 15.5	10.08		-1-05-013 2
U 1176	137.5 / 139.74 / 314.94	1539 / -45.37 / -6.16	10 / 0.17 / 17 -4	4.6 / 4.6 / 2	0.82 / 39.		0.07 / 0.14	634. / 770. / 1	54 / 57. / 66.	5	0.92 / 7 / 1.19	9.3 / 29.84	6.5 / -6.5 / -1.0	12.5	8.86 / 9.07	0.611	DDO 13 ; 1176 03-05-000 2
U 1195	139.7 / 141.25 / 313.25	1343 / -47.08 / -7.27	10 P / 0.16 / 17 -4	3.3 / 2.7 / 2	0.36 / 75.	12.5	3 / 0.52 / 0.12	767. / 895. / 6	145	20	0.69 / 16 / 1.11	10.8 / 30.17	7.3 / -7.8 / -1.4	-17.67 / 8.5	9.26 / 8.76	0.31	1195 02-05-011
U 1197	139.8 / 139.59 / 317.40	1803 / -42.91 / -5.93	10 / 0.16 / 52 -0+14	2.0 / 1.6 / 2	0.34 / 77.	14.2	3 / 0.57 / 0.14	2800. / 2942. / 1	206 / 174. / 178.	20	0.47 / 18 / 1.03	36.9 / 32.83	27.0 / -24.8 / -3.8	-18.63 / 17.2	9.64 / 9.60 / 10.18	0.91 / 0.265 / 3.44	1197 03-05-000 2
N 660	140.3 / 141.60 / 312.98	1323 / -47.36 / -7.51	1BP / 0.12 / 17 -4	7.2 / 5.8 / 9	0.33 / 77.	10.79	1 / 0.58 / 0.13	856. / 982. / 1	327 / 296. / 298.	8	1.62 / 6 / 1.27	11.8 / 30.37	8.0 / -8.6 / -1.5	-19.58 / 20.0	10.02 / 9.76 / 10.71	0.55 / 0.114 / 4.80	1201 02-05-013 2
0140+19	140.6 / 139.23 / 319.07	1944 / -41.24 / -5.57	10 / 0.15 / 17 -4	3.0 / 2.4	0.29 / 81.		0.67 / 0.17	501. / 648. / 1	103 / 82. / 89.	15	0.33 / 20 / 1.05	8.0 / 29.52	6.0 / -5.2 / -0.8	5.6	8.14 / 9.04	0.125	UGCA 20 ; 03-05-000 2
0142-43	142.9 / 273.07 / 255.18	-4350 / -70.31 / -19.71	10 / 0.16 / 14+13	4.0 / 4.1 / 6	0.96 / 19.		0.01 / 0.05	394. / 293. / 2	86	7	1.29 / 6 / 1.02	3.6 / 27.81	-0.9 / -3.3 / -1.2	4.3	8.40		A 0143
U 1249	144.7 / 137.96 / 326.34	2705 / -33.90 / -4.03	9B / 0.15 / 17 -5	7.1 / 6.2 / 2	0.42 / 71.	11.43	1 / 0.41 / 0.27	339. / 506. / 1		20	1.27	6.4 / 29.04	5.3 / -3.6 / -0.5	-17.61 / 11.6	9.24		245-05 -7-04-032 2 ; IC 1727, VV 338
0144-58	144.8 / 289.77 / 239.14	-5855 / -57.03 / -20.03	10 / 0.06 / 50 -0	2.0 / 2.0 / 6	0.92 / 27.		0.03 / 0.00	2209. / 2055. / 7		20		26.1 / 32.08	-12.6 / -21.0 / -8.9	15.2			1249 04-05-009 2
N 672	145.0 / 138.01 / 326.46	2711 / -33.78 / -4.05	6B / 0.16 / 17 -5	6.4 / 5.6 / 2	0.43 / 70.	10.73	1 / 0.38 / 0.25	425. / 592. / 1	255	20 / 15 / 211	1.57 / 217 10	7.5 / 29.38	6.2 / -4.1 / -0.5	-18.65 / 12.3	9.65 / 9.32	0.47	114-07 ; VV 338 ; 1256 04-05-011 2

14

(1) Name	(2) α / ℓ / SGL	(3) δ / b / SGB	(4) Type / ρxyz / Group	(5) D25 / D$^{b,i}_{25}$ / source	(6) d/D / i	(7) B$^{b,i}_T$ / H$_{-0.5}$ / B-H	(8) source / A$_B^{i-0}$ / A$_B^b$	(9) V$_h$ / V$_o$ / Tel	(10) W$_{20}$ / W$_R$ / W$_D$	(11) e$_v$ / e$_w$ / source	(12) F$_c$ / source/e$_f$ / f$_h$	(13) R / μ	(14) SGX / SGY / SGZ	(15) M$_B^{b,i}$ / Δ$_{25}$	(16) log L$_B$ / log M$_H$ / log M$_T$	(17) M$_H$/L$_B$ / M$_H$/M$_T$ / M$_T$/L$_B$	(18) Alternate Names UGC/ESO MCG RCII AGN?
N 685	145.9 / 284.47 / 245.40	-5302 / -62.28 / -20.38	5X / 0.06 / 53 -0	3.9 / 3.8 / 6	0.90 / 29.	11.9	5 / 0.04 / 0.05	1358. / 1223.		10		15.2 / 30.92	-5.9 / -13.0 / -5.3	-19.02 / 16.9	9.80		152-24 2
0146-12	146.0 / 167.90 / 287.78	-1238 / -70.10 / -15.98	9 / 0.24 / 52 -9 +7	3.1 / 3.0 / 4	0.82 / 39.	13.63	7 / 0.07 / 0.00	1623. / 1644. / 1	110. / 138. / 142.	15	0.70 / 14 / 1.10	19.8 / 31.48	5.8 / -18.1 / -5.4	-17.85 / 17.3	9.33 / 9.29 / 9.98	0.91 / 0.205 / 4.46	ARP 4, DDO 14 -2-05-050 2
N 676	146.3 / 147.88 / 305.97	539 / -54.16 / -11.27	0.12 / 52-11	4.5 / 3.8 / 2	0.40 / 72.	10.8	3 / 0.00 / 0.14	1508. / 1601. / 1		20	0.56 / 42 / 1.13	19.5 / 31.45	11.2 / -15.5 / -3.8	-20.65 / 21.6	10.45 / 9.14	0.05	1270 01-05-034 2
U 1276	146.5 / 140.73 / 320.20	2027 / -40.17 / -6.64	8B / 0.25 / 52-15+14	2.0 / 1.8 / 2	0.49 / 66.	13.2 / 11.41 / 1.79	3 / 0.30 / 0.25	2750. / 2895. / 1	235. / 217. / 219.	10	0.56 / 26 / 1.03	36.3 / 32.80	27.7 / -23.1 / -4.2	-19.60 / 19.1	10.03 / 9.68 / 10.42	0.44 / 0.184 / 2.42	1276 03-05-018
0146-10	146.7 / 164.26 / 290.16	-1019 / -68.17 / -15.62	6X / 0.28 / 52 -8 +7	2.5 / 2.5 / 2	0.86 / 34.		0.05 / 0.06	1991. / 2021. / 1	220. / 326. / 327.	15	0.78 / 17 / 1.06	24.6 / 31.95	8.2 / -22.2 / -6.6	18.0	9.56 / 10.74	0.066	UGCA 21 -2-05-053
N 681	146.7 / 164.85 / 289.79	-1041 / -68.47 / -15.70	2X / 0.34 / 52 -8 +7	3.0 / 2.8 / 9	0.66 / 53.	12.50	2 / 0.15 / 0.04	1750. / 1779.		9	0.66 / 211 72	21.5 / 31.67	7.0 / -19.5 / -5.8	-19.17 / 17.6	9.86 / 9.32	0.29	-2-05-052 2
U 1281	146.7 / 136.89 / 331.43	3220 / -28.71 / -2.63	8 / 0.13 / 17 +6	4.4 / 3.3 / 2	0.25 / 85.	11.8	7 / 0.67 / 0.15	163. / 343. / 1	139. / 111. / 116.	7	1.00 / 9 / 1.11	4.6 / 28.29	4.0 / -2.2 / -0.2	-16.49 / 4.4	8.79 / 8.33 / 9.20	0.34 / 0.134 / 2.58	1281 05-05-014
0147-13	147.2 / 169.39 / 287.41	-1304 / -70.24 / -16.36	10 / 0.30 / 52 -9 +7	1.8 / 1.8 / 4	0.86 / 34.	15.27	7 / 0.05 / 0.00	1720. / 1739. / 1	148.	20	-0.02 / 47 / 1.04	21.0 / 31.61	6.0 / -19.2 / -5.9	-16.34 / 11.0	8.73 / 8.62	0.79	-2-05-057 2
N 691	147.9 / 140.74 / 321.31	2130 / -39.07 / -6.60	5A / 0.26 / 52-15+14	3.6 / 3.5 / 2	0.74 / 46.	12.0	3 / 0.10 / 0.26	2660. / 2807. / 6	343. / 422. / 423.	20	0.55 / 25 / 1.18	35.2 / 32.74	27.3 / -21.9 / -4.1	-20.74 / 36.0	10.49 / 9.64 / 11.27	0.14 / 0.023 / 6.04	1305 04-05-019 2
N 701	148.6 / 164.67 / 290.65	-957 / -67.60 / -15.99	5B / 0.37 / 52 -8 +7	2.8 / 2.4 / 4	0.48 / 67.	12.53 / 10.02 / 2.51	2 / 0.32 / 0.05	1839. / 1869. / 1	283. / 267. / 269.	30 / 17 / 217	0.70 / 217 16 / 1.06	22.7 / 31.78	7.7 / -20.4 / -6.2	-19.25 / 15.9	9.89 / 9.41 / 10.52	0.33 / 0.079 / 4.22	-2-05-060 2
N 720	150.6 / 173.04 / 286.66	-1359 / -70.35 / -17.37	-5 / 0.25 / 52 -9 +7	4.4 / 4.0 / 5	0.64 / 55.	11.15	2 / 0.00 / 0.00	1664. / 1676.		14		20.3 / 31.53	5.5 / -18.5 / -6.0	-20.38 / 23.7	10.34		-2-05-068 2
N 718	150.6 / 150.74 / 304.63	357 / -55.30 / -12.79	1X / 0.13 / 52-11	2.5 / 2.5 / 2	0.86 / 34.	12.37	2 / 0.05 / 0.08	1672. / 1755.		55		21.4 / 31.65	11.9 / -17.2 / -4.7	-19.28 / 15.6	9.90		1356 01-05-041 2
0153+05	153.6 / 150.91 / 306.23	521 / -53.72 / -13.09	13 P / 0.05 / 17 -0	0.4 / 0.4 / 2	1.00	14.6	3 / 0.00 / 0.13	814. / 900.		220		10.8 / 30.16	6.2 / -8.5 / -2.4	-15.56 / 1.3	8.42		ARAK 65 01-06-002

15

Name	Coordinates	Type	Dimensions	mag	code	Vo	W	N	ratio	D	SG coords	M	μ	P	Alt / Designation
N 755	153.9 −919 / 166.33 −66.30 −17.11 / 291.61	3B 0.24 / 52 −0 +7	3.6 0.30 / 2.8 79. / 4	10.53	0.67 0.02	1641. / 1669. / 1	256 / 222. / 225.	10	0.91 / 15 / 1.08	20.1 / 31.52	7.1 / −17.9 / −5.9	16.4	9.52 / 10.37	0.140	−2-06-005 2
0154−12	154.7 −1202 / 171.28 −68.25 −17.92 / 288.88	8B 0.27 / 52 −9 +7	2.9 0.23 / 2.1 86. / 2		0.67 0.00	1860. / 1877. / 1	173 / 140. / 144.	10	0.52 / 20 / 1.05	22.8 / 31.79	7.0 / −20.5 / −7.0	14.0	9.24 / 9.90	0.217	UGCA 22 / −2-06-006
N 746	154.8 4441 / 135.12 −16.37 0.35 / 343.52	10B 0.11 / 17 +1	2.2 0.75 / 2.3 45. / 2	12.5	3 0.10 0.53	712. / 915. / 1	127 / 141. / 145.	15	0.68 / 14 / 1.06	12.3 / 30.44	11.8 / −3.5 / 0.1	−17.94 / 8.3	9.37 / 8.86 / 9.68	0.31 / 0.152 / 2.04	1438 07-05-003
N 770	156.5 1843 / 144.39 −41.07 −9.45 / 319.36	1B 0.22 / 52−14	1.2 0.75 / 1.2 45. / 2	13.8	3 0.10 0.18	2451. / 2583.		71		32.3 / 32.55	24.2 / −20.8 / −5.3	−18.75 / 11.3	9.69		TURN 151A
N 772	156.6 1846 / 144.40 −41.01 −9.45 / 319.42	3A 0.23 / 52−14	7.1 0.66 / 6.7 53. / 2	10.77 / 7.96 / 2.81	1 0.15 0.18	2473. / 2605. / 2	489	25 / 19 / 54	1.34 / 211 9 / 1.09	32.6 / 32.57	24.5 / −20.9 / −5.4	−21.80 / 63.8	10.91 / 10.37	0.28	ARP 78, TURN 151B / 1466 03-06-011 2
N 779	157.2 −612 / 163.49 −63.32 −17.15 / 294.98	3X 0.15 / 52 −5	4.4 0.25 / 3.2 84. / 4	11.18 / 8.64 / 2.54	2 0.67 0.01	1397. / 1435. / 1	375 / 339. / 340.	15 / 12 / 211	0.72 / 211 13 / 1.11	17.3 / 31.19	7.0 / −15.0 / −5.1	−20.01 / 16.2	10.20 / 9.20 / 10.73	0.10 / 0.029 / 3.43	−1-06-016 2
N 784	158.4 2835 / 140.91 −31.60 −6.33 / 328.82	8 0.15 / 17 +6	6.1 0.30 / 4.9 79. / 2	11.39 / 10.88 / 0.51	1 0.67 0.20	201. / 362. / 1	119 / 96. / 102.	5	1.23 / 6 / 1.24	4.7 / 28.35	4.0 / −2.4 / −0.5	−16.96 / 6.7	8.98 / 8.57 / 9.25	0.40 / 0.209 / 1.90	1501 05-05-045 2
U 1546	200.6 1824 / 145.75 −41.02 −10.47 / 319.40	5 0.19 / 52−14	1.3 1.00 / 1.4 / 2	14.3	3 0.00 0.20	2372. / 2499.		9		31.2 / 32.47	23.3 / −20.0 / −5.7	−18.17 / 12.8	9.46	1.86	SCI 26
U 1547	200.6 2148 / 144.19 −37.84 −9.26 / 322.63	10B 0.26 / 52 −0+14	2.2 0.92 / 2.3 26. / 2	14.09	7 0.03 0.28	2649. / 2788. / 6	165	10	0.83 / 11 / 1.09	35.1 / 32.73	27.5 / −21.0 / −5.6	−18.64 / 23.6	9.65 / 9.92		1546 03-06-000 2
U 1551	200.8 2350 / 143.37 −35.92 −8.57 / 324.57	8B 0.21 / 52 −0+14	2.9 0.53 / 2.7 63. / 2	12.7	3 0.25 0.32	2672. / 2817. / 1	133 / 119. / 123.	10	0.56 / 16 / 1.06	35.6 / 32.75	28.6 / −20.4 / −5.3	−20.05 / 28.1	10.21 / 9.66 / 10.06	0.28 / 0.399 / 0.71	DDO 17 / 1551 04-05-045
N 803	201.0 1548 / 147.16 −43.40 −11.46 / 316.95	5 0.07 / 52 −0	3.3 0.50 / 2.9 65. / 2	12.59 / 10.34 / 2.25	1 0.29 0.13	2096. / 2214. / 1	267 / 252. / 255.	15 / 8 / 211	0.87 / 211 11 / 1.06	27.6 / 32.20	19.8 / −18.5 / −5.5	−19.61 / 23.4	10.04 / 9.75 / 10.63	0.52 / 0.131 / 3.96	1554 03-06-028 2
U 1561	201.3 2358 / 143.45 −35.75 −8.63 / 324.73	10 P 0.12 / 17 −0	1.7 0.70 / 1.6 50. / 2	14.0	3 0.13 0.29	595. / 740. / 1	80	10	30 / 1.02	9.3 / 29.83	7.5 / −5.3 / −1.4	−15.83 / 4.3	8.52 / 7.94	0.26	V ZW 173 / 1561 04-05-048
0202−06	202.0 −625 / 165.89 −62.78 −18.36 / 295.07	8B 0.19 / 52 −5	3.1 0.51 / 2.7 64. / 2	12.42	0.28 0.02	1364. / 1397. / 6	188 / 170. / 173.	10	0.77 / 12 / 1.11	16.8 / 31.13	6.8 / −14.5 / −5.3	13.2	9.22 / 10.05	0.148	UGCA 24 / −1-06-039
N 821	205.7 1046 / 151.56 −47.54 −14.23 / 312.49	−5 0.08 / 52 −0	3.5 0.68 / 3.4 51. / 2	11.61	2 0.00 0.19	1778. / 1874.		100		23.2 / 31.82	15.2 / −16.6 / −5.7	−20.21 / 23.0	10.28		1631 02-06-034 2

16

(1) Name	(2) α / ℓ / SGL	(3) δ / b / SGB	(4) Type / Q_{xyz} / Group	(5) D_{25} / $D^{b,i}_{25}$ / source	(6) d/D / i	(7) $B^{b,i}_T$ / $H_{-0.5}$ / $B{-}H$	(8) source / A^{i-o}_B / A^b_B	(9) V_h / V_o / Tel	(10) W_{20} / W_R / W_D	(11) e_v / e_w / source	(12) F_c / source e_f / f_h	(13) R / μ	(14) SGX / SGY / SGZ	(15) $M^{b,i}_B$ / $\Delta^{b,i}_{25}$	(16) $\log L_B$ / $\log M_H$ / $\log M_T$	(17) M_H/L_B / M_H/M_T / M_T/L_B	(18) Alternate Names (UGC/ESO · MCG · AGN? RCII)
U 1670	208.2 / 155.37 / 308.53	631 / -51.05 / -16.19	9 / 0.16 / 52-10	2.5 / 2.5 / 2	1.00 /	14.81 / /	7 / 0.00 / 0.14	1611. / 1690. / 1	118 / /	10 / /	0.38 / 18 / 1.07	20.7 / 31.58	12.4 / -15.5 / -5.8	-16.77 / 15.1	8.90 / 9.01 /	1.29 / /	DDO 18 / 1670 · 01-06-052 · 2
N 864	212.8 / 157.55 / 308.16	546 / -51.14 / -17.51	5X / 0.17 / 52-10	4.4 / 4.3 / 2	0.75 / 45.	11.31 / /	2 / 0.10 / 0.14	1561. / 1633. /	244 / 290. / 292.	8 / 7 / 217	1.36 / 217 10 / 1.13	20.0 / 31.51	11.8 / -15.0 / -6.0	-20.20 / 25.1	10.27 / 9.96 / 10.79	0.49 / 0.150 / 3.27	1736 · 01-06-061 · 2
U 1807	218.0 / 139.98 / 343.25	4232 / -17.12 / -4.36	10 / 0.39 / 17-1	2.5 / 2.6 / 2	1.00 /	/ /	/ 0.00 / 0.27	631. / 815. / 1	80 / /	10 / /	0.40 / 17 / 1.09	10.9 / 30.19	10.4 / -3.1 / -0.8	/ 8.3	8.47 / /	/ /	1807 · 07-05-040
N 895	219.2 / 171.81 / 296.90	-545 / -59.53 / -22.33	6A / 0.07 / 51-0	3.9 / 3.6 / 4	0.72 / 49.	12.16 / /	2 / 0.12 / 0.02	2287. / 2309. / 1	263 / 301. / 303.	10 / /	1.06 / 217 8 / 1.13	28.5 / 32.27	11.9 / -23.5 / -10.8	-20.11 / 30.0	10.24 / 9.97 / 10.90	0.54 / 0.118 / 4.57	NGC 894 / -1-07-002 · 2
N 891	219.3 / 140.37 / 342.98	4207 / -17.43 / -4.76	3A / 0.55 / 17-1	12.2 / 9.5 / 2	0.25 / 84.	9.95 / 6.78 / 3.17	1 / 0.67 / 0.29	529. / 712. / 2	490 / 455. / 456.	10 / /	1.69 / 211 5 / 1.17	9.6 / 29.91	9.1 / -2.8 / -0.8	-19.96 / 26.6	10.18 / 9.65 / 11.20	0.30 / 0.028 / 10.66	1831 · 07-05-046 · 2
N 899	219.6 / 200.79 / 280.52	-2102 / -68.42 / -25.42	10 / 0.31 / 52-6	1.5 / 1.5 / 6	0.78 / 43.	/ /	/ 0.09 / 0.03	1563. / 1524. / 1	143 / 167. / 171.	10 / /	1.19 / 7 / 1.01	18.5 / 31.34	3.1 / -16.5 / -8.0	/ 8.1	9.72 / 9.82 /	0.809 / /	UGCA 26 / 545-07 · -4-06-030 · 2
0219-20	219.7 / 200.65 / 280.59	-2058 / -68.37 / -25.43	9 P / 0.31 / 52-6	1.2 / 1.1 / 6	0.62 / 57.	/ /	/ 0.18 / 0.01	1585. / 1546. /	/ /	86 / /	/ /	18.8 / 31.38	3.1 / -16.7 / -8.1	/ 6.0	/ /	/ /	IC 223 / 545-08 · -4-06-031 · 2
U 1839	220.0 / 166.25 / 302.04	-51 / -55.63 / -21.21	8 / 0.30 / 52-3 +1	2.9 / 2.0 / 2	0.16 / 90.	14.5 / /	3 / 0.67 / 0.05	1527. / 1567. / 6	155 / /	20 / /	0.28 / 37 / 1.08	19.1 / 31.41	9.5 / -15.1 / -6.9	-16.91 / 11.2	8.96 / 8.84 /	0.77 / /	1839 · 00-07-08A
N 907	220.7 / 200.80 / 280.67	-2056 / -68.14 / -25.66	7BP / 0.17 / 52-6	1.8 / 1.5 / 6	0.36 / 75.	/ /	/ 0.51 / 0.04	1723. / 1683. / 2	209 / /	20 / /	0.37 / 34 / 1.01	20.5 / 31.56	3.4 / -18.2 / -8.9	/ 9.0	8.99 / /	/ /	UGCA 28 / 545-10 · -4-06-034 · 2
N 908	220.8 / 202.12 / 280.10	-2127 / -68.31 / -25.75	5A / 0.31 / 52-6	6.1 / 5.1 / 6	0.44 / 70.	10.46 / 8.29 / 2.17	1 / 0.37 / 0.02	1508. / 1466. / 2	389 / 374. / 376.	10 / /	1.08 / 14 / 1.06	17.8 / 31.26	2.8 / -15.8 / -7.8	-20.80 / 26.5	10.51 / 9.58 / 11.03	0.12 / 0.035 / 3.31	UGCA 29 / 545-11 · -4-06-035 · 2
U 1865	221.9 / 143.35 / 337.51	3549 / -23.10 / -7.97	9 / 0.65 / 17-1	3.3 / 3.3 / 2	0.82 / 39.	13.73 / /	7 / 0.07 / 0.22	577. / 742. / 1	87 / 109. / 114.	15 / /	0.53 / 5 11 / 1.12	9.8 / 29.96	9.0 / -3.7 / -1.4	-16.23 / 9.4	8.68 / 8.51 / 9.51	0.67 / 0.100 / 6.74	DDO 19 / 1865 · 06-06-028 · 2
N 918	223.1 / 152.18 / 321.29	1816 / -38.96 / -15.48	5X / 0.06 / 52-0	3.4 / 3.4 / 2	0.62 / 56.	12.1 / 9.96 / 2.14	3 / 0.18 / 0.45	1516. / 1625. / 1	257 / 263. / 265.	10 / /	0.73 / 15 / 1.15	20.3 / 31.54	15.3 / -12.3 / -5.4	-19.44 / 20.2	9.97 / 9.34 / 10.61	0.24 / 0.055 / 4.35	SCI 31 / 1888 · 03-07-011 · 2
0223-21	223.7 / 203.24 / 279.98	-2139 / -67.74 / -26.44	9 / 0.31 / 52-6	1.8 / 1.8 / 6	0.86 / 34.	13.94 / /	7 / 0.05 / 0.01	1555. / 1510. / 2	160 / /	15 / /	0.52 / 22 / 1.01	18.4 / 31.33	2.9 / -16.2 / -8.2	-17.39 / 9.7	9.15 / 9.05 /	0.80 / /	UGCA 31, DDO 21 / 545-16 · -4-06-040 · 2

ID	Coordinates	Type	Dim	Ratio/PA	mag		Velocity	n		Dist/mod	Vel. comp		mag		Names
0223-10	223.9 -1003 / 179.74 -61.71 / 292.64 -24.48	9X / 0.08 / 51 -0	3.3 / 2.6 / 4	0.33 / 78.		0.60 / 0.04	2090. / 2091.	100		25.7 / 32.05	9.0 / -21.6 / -10.7	19.5			DDO 20 / -2-07-007 2
0224-24	224.2 -2431 / 210.71 -68.44 / 276.80 -26.88	8B / 0.27 / 52 -6	2.9 / 2.5 / 6	0.50 / 65.		0.29 / 0.02	1515. / 1459. / 2	25	0.72 / 18 / 1.01	17.8 / 31.25	1.9 / -15.7 / -8.0	13.0	9.22		UGCA 32 / 479-04 / -4-06-041 2
N 925	224.3 3322 / 144.89 -25.16 / 335.50 -9.49	7X / 0.48 / 17 -1	11.3 / 10.8 / 2	0.65 / 54.	10.20 / 8.72 / 1.48	1 / 0.16 / 0.24	554. / 710. / 4	8	1.88 / 25 3 / 1.17	9.4 / 29.86	8.4 / -3.8 / -1.5	-19.66 / 29.6	10.06 / 9.83 / 10.66	0.59 / 0.146 / 4.03	1913 / 05-06-045 2
N 936	225.1 -123 / 168.59 -55.26 / 301.88 -22.59	-2B / 0.24 / 52 -3 +1	6.2 / 6.0 / 2	0.81 / 40.	11.06	1 / 0.00 / 0.04	1358. / 1392.	26		16.9 / 31.14	8.3 / -13.3 / -6.5	-20.08 / 29.6	10.22		1929 / 00-07-017 2
N 941	226.0 -122 / 168.88 -55.10 / 301.96 -22.80	6X / 0.26 / 52 -3 +1	3.3 / 3.1 / 2	0.77 / 44.	12.87	1 / 0.09 / 0.04	1627. / 1660. / 1	10	0.41 / 28 / 1.07	20.4 / 31.54	9.9 / -15.9 / -7.9	-18.67 / 18.5	9.66 / 9.03 / 10.24	0.23 / 0.061 / 3.84	1954 / 00-07-022 2
N 949	227.8 3655 / 144.06 -21.62 / 339.04 -8.55	7A / 0.66 / 17 -1	3.5 / 3.3 / 2	0.67 / 52.	12.21 / 9.71 / 2.50	2 / 0.15 / 0.19	610. / 774. / 1	10	0.64 / 18 / 1.11	10.3 / 30.06	9.5 / -3.6 / -1.5	-17.85 / 9.9	9.33 / 8.67 / 10.05	0.22 / 0.041 / 5.20	1983 / 06-06-048 2
N 955	228.0 -120 / 169.51 -54.76 / 302.15 -23.27	3 / 0.35 / 52 -3 +1	3.2 / 2.5 / 2	0.34 / 77.	12.34	2 / 0.57 / 0.04	1534. / 1566.	90		19.2 / 31.41	9.4 / -14.9 / -7.6	-19.07 / 14.0	9.82		1986 / 00-07-27A 2
0228-04	228.2 -401 / 172.73 -56.74 / 299.35 -24.05	9 / 0.31 / 52 -3 +1	3.3 / 3.3	1.00		0.00 / 0.02	1627. / 1648. / 1	5	0.65 / 12 / 1.12	20.2 / 31.53	9.0 / -16.1 / -8.2	19.5	9.26		UGCA 33 / -1-07-020
N 959	229.3 3517 / 145.08 -22.99 / 337.70 -9.56	9 / 0.57 / 17 -1	2.6 / 2.4 / 2	0.61 / 57.	12.54 / 11.03 / 1.51	2 / 0.19 / 0.18	609. / 767. / 1	10	0.45 / 17 / 1.06	10.1 / 30.03	9.2 / -3.8 / -1.7	-17.49 / 7.1	9.19 / 8.46 / 9.76	0.19 / 0.050 / 3.76	2002 / 06-06-051 2
U 2014	229.8 3828 / 143.79 -20.04 / 340.61 -8.19	10 / 0.61 / 17 -1	2.4 / 2.2 / 2	0.57 / 60.		0.22 / 0.17	570. / 737. / 1	10	0.37 / 15 / 1.05	9.9 / 29.97	9.2 / -3.2 / -1.4	6.4	8.36 / 8.84	0.335	DDO 22 / 2014 / 06-06-054 2
U 2017	229.9 2837 / 148.32 -28.99 / 331.68 -12.64	10 / 0.17 / 17 -3	2.4 / 2.5 / 2	0.82 / 39.		0.07 / 0.31	1019. / 1157. / 1	10	0.56 / 15 / 1.06	14.8 / 30.86	12.7 / -6.9 / -3.2	10.8	8.90 / 9.73	0.148	2017 / 05-07-000
U 2023	230.3 3317 / 146.19 -24.73 / 335.99 -10.65	10 / 0.54 / 17 -1	2.9 / 3.0 / 2	0.94 / 23.	12.8	3 / 0.02 / 0.28	611. / 763. / 6	5	0.62 / 10 / 1.16	10.1 / 30.02	9.0 / -4.0 / -1.9	-17.22 / 8.8	9.08 / 8.63	0.36	DDO 25 / 2023 / 05-07-007 2
U 2034	230.6 4019 / 143.15 -18.28 / 342.34 -7.47	10 / 0.63 / 17 -1	3.3 / 3.3 / 2	0.87 / 33.	12.9	3 / 0.05 / 0.20	581. / 752. / 1	5	0.93 / 7 / 1.14	10.2 / 30.04	9.6 / -3.1 / -1.3	-17.14 / 9.8	9.05 / 8.95 / 9.35	0.79 / 0.392 / 2.02	DDO 24 / 2034 / 07-06-032 2
N 972	231.3 2906 / 148.40 -28.42 / 332.26 -12.71	2 P / 0.06 / 52 -0	3.9 / 3.6 / 9	0.50 / 65.	11.46	2 / 0.29 / 0.35	1539. / 1677.	18	0.36 / 217 30 / 1.14	21.4 / 31.65	18.5 / -9.7 / -4.7	-20.19 / 22.5	10.27 / 9.02	0.06	2045 / 05-07-010 2

(1) Name	(2) α / l / SGL	(3) δ / b / SGB	(4) Type / Q_{xyz} / Group	(5) D_{25} / $D_{25}^{b,i}$ / source	(6) d/D / i	(7) $B_T^{b,i}$ / $H_{-0.5}$ / B–H	(8) source / A_B^{i-0} / A_B^b	(9) V_h / V_o / Tel	(10) W_{20} / W_R / W_D	(11) e_v / e_w / source	(12) F_c / source e_F / f_h	(13) R / μ	(14) SGX / SGY / SGZ	(15) $M_B^{b,i}$ / $\Delta_{25}^{b,i}$	(16) log L_B / log M_H / log M_T	(17) M_H/L_B / M_H/M_T / M_T/L_B	(18) Alternate Names / UGC/ESO MCG RCII / AGN?
U 2053	231.5	2932	10	2.3	0.60	14.64	7	1034.	64	5	0.60	15.1	13.1	-16.25	8.69	1.84	DDO 26
	148.23	-28.01	0.18	2.2	58.		0.20	1173.	55.		5 9	30.89	-6.8	9.7	8.96	1.067	2053 05-07-015 2
	332.68	-12.55	17 -3	2			0.36	1	64.		1.04		-3.3		8.93	1.73	
N 986	231.6	-3915	2B	3.3	0.79	11.66	2	1983.	119	30	0.39	23.2	-3.5	-20.17	10.26	0.07	299-07 -7-06-015 2
	248.65	-65.63	0.06	3.2	42.		0.08	1868.			23	31.83	-20.0	21.7	9.12		
	260.18	-28.83	51 -0	6			0.05	2			1.03		-11.2				
0232+59	232.6	5926	-5	0.8	0.84			15.		30		3.6	3.6				UGCA 34, MAFFEI 1
	135.84	-0.57	0.18	2.4	36.		0.00	224.				27.79	0.0	2.5			10-04-000 2
	359.30	1.44	14-11	2			5.74						0.1				
N 988	232.9	-935	5B	4.0	0.46			1509.	290	15	0.93	18.3	6.6				UGCA 35
	181.94	-59.72	0.39	3.4	68.		0.34	1504.			11	31.32	-15.0	18.2	9.45		-2-07-037 2
	293.69	-26.54	52 -1	2			0.05	6			1.17		-8.2				
N 991	233.1	-722	5X	3.1	0.90	12.2	5	1538.	85	8	0.68	18.8	7.4	-19.17	9.86	0.23	-1-07-023 2
	178.75	-58.23	0.43	3.1	30.		0.04	1541.			13	31.37	-15.2	17.0	9.23		
	296.11	-26.07	52 -1	4			0.03	1			1.10		-8.3				
U 2080	233.3	3845	6X	5.4	0.97	11.71	1	909.	140	8	1.42	14.2	13.3	-19.05	9.81	0.82	IC 239
	144.33	-19.50	0.08	5.6	17.		0.01	1074.			217 5	30.76	-4.5	23.2	9.72		2080 06-06-065 2
	341.19	-8.66	17 +1	2			0.21	1			1.25		-2.1				
U 2082	233.4	2513	5	5.8	0.21	11.6	3	710.	217	10	1.08	10.7	8.9	-18.55	9.61	0.34	2082 04-07-016
	150.89	-31.68	0.18	4.4	89.	10.72	0.67	834.	180.		7	30.15	-5.4	13.7	9.14	0.107	
	328.87	-14.82	17 -0	2		0.88	0.35	1	184.		1.18		-2.7		10.11	3.16	
N 1015	235.6	-132	1B	3.1	1.00	12.3	3	2631.	188	20	0.64	33.1	16.1	-20.30	10.31	0.23	2124 00-07-066
	172.18	-53.66	0.11	3.1			0.00	2655.			31	32.60	-25.2	30.0	9.68		
	302.53	-25.15	51+15	2			0.06	1			1.13		-14.1				
N 1022	236.1	-653	1B	2.5	0.91	12.13	2	1503.		23		18.5	7.4	-19.20	9.87		-1-07-025 2
	179.02	-57.36	0.48	2.5	28.		0.03	1506.				31.33	-14.7	13.5			
	296.83	-26.68	52 -1	4			0.04						-8.3				
N 1003	236.1	4040	6A	6.3	0.47	11.54	1	626.	230	8	1.59	10.7	10.1	-18.61	9.64	1.03	2137 07-06-051 2
	144.00	-17.54	0.54	5.6	67.	9.98	0.33	794.	209.		217 5	30.15	-3.1	17.5	9.65	0.201	
	343.15	-8.22	17 -1	2		1.56	0.24	1	212.		1.22		-1.5		10.35	5.12	
N 1012	236.3	2956	13 P	3.1	0.55	12.2	3	986.	222	10	1.09	14.4	12.6	-18.59	9.63	0.60	2141 05-07-027
	149.08	-27.19	0.18	3.0	61.		0.00	1123.	211.		9	30.79	-6.3	12.6	9.41	0.157	
	333.53	-13.30	17 -3	2			0.43	1	214.		1.08		-3.3		10.21	3.84	
0236-61	236.5	-6133	10	6.5	0.16			511.		20		4.5	-2.3				115-21
	282.79	-51.43	0.15	4.6	90.		0.67	329.				28.28	-3.3	6.0			
	235.21	-25.94	14-14	6			0.02	7					-2.0				
0236-27	236.9	-2740	-2B	2.1	0.74	13.20	2	1423.		22	0.24	16.4	0.9	-17.88	9.34	0.21	IC 1826 = IC 1830
	219.95	-66.12	0.19	1.9	46.		0.00	1345.			10 35	31.08	-14.2	9.1	8.67		UGCA 37, HARO 18
	273.48	-29.96	51 -3 +1	6			0.00						-8.2				416-06 -5-07-012 2

Name	Coordinates	Type	Dimensions	logR/PA	Mag	Flags	Velocity	C9	C10	C11	C12	C13	C14	C15	C16	Other designations
N 1035	237.1 -820 / 181.37 -58.12 / 295.32 -27.26	5A / 0.24 / 52 -1	2.3 / 1.7 / 4	0.25 / 84.	12.2 / 10.18 / 2.02	5 / 0.67 / 0.05	1236. / 1232. / 1	280	20	0.58 / 217 11 / 1.03	15.0 / 30.88	5.7 / -12.1 / -6.9	-18.68 / 7.4	9.66 / 8.93	0.19	-1-07-027 2 / ARP 135
N 1023	237.3 3851 / 145.03 -19.08 / 341.66 -9.30	-2B / 0.57 / 17 -1	7.5 / 6.5 / 9	0.40 / 72.	9.95	1 / 0.00 / 0.25	624. / 787.	315	30 / 40 / 56	1.03 / 216 24	10.5 / 30.11	9.8 / -3.3 / -1.7	-20.16 / 19.9	10.26 / 9.07	0.07	2154 06-06-073 2
0237-34	237.5 -3444 / 237.30 -65.72 / 265.31 -30.18	-5 / 0.50 / 14-12	20.1 / 18.8 / 9	0.70 / 50.	8.99	2 / 0.00 / 0.05	53. / -51.		9		0.1 / 20.00	0.0 / -0.1 / -0.1	-11.01 / 0.5	6.60		FOR / 356-04 -6-07-001
N 1036	237.7 1905 / 155.52 -36.58 / 323.52 -18.32	13 P / 0.10 / 17 -0	1.7 / 1.7 / 2	0.72 / 48.	13.2	3 / 0.00 / 0.25	789. / 889.		12	0.43 / 12 20	11.2 / 30.25	8.6 / -6.3 / -3.5	-17.05 / 5.6	9.01 / 8.53	0.33	MARK 370 / 2160 03-07-041 2
U 2162	237.8 101 / 170.08 -51.38 / 305.40 -24.92	10 / 0.30 / 52 -2 +1	2.0 / 2.0 / 2	1.00	11.20	/ 0.00 / 0.05	1194. / 1226. / 1	65	25	0.21 / 20 / 1.05	15.1 / 30.89	7.9 / -11.1 / -6.3	8.8	8.57		DDO 27 / 2162 00-07-077 2
N 1042	237.9 -839 / 182.07 -58.17 / 295.02 -27.53	6X / 0.47 / 52 -1	4.4 / 4.3 / 1	0.88 / 32.		1 / 0.04 / 0.06	1377. / 1371. / 1	119 / 178. / 181.	10	1.07 / 217 6 / 1.20	16.7 / 31.11	6.3 / -13.4 / -7.7	-19.91 / 21.0	10.16 / 9.52 / 10.28	0.23 / 0.170 / 1.35	-2-07-054 2
0238-06	238.1 -619 / 178.86 -56.61 / 297.58 -27.02	10 / 0.46 / 52 -1	1.9 / 1.9 / 2	0.90 / 29.		/ 0.04 / 0.05	1333. / 1336. / 1	110	10	0.46 / 17 / 1.04	16.3 / 31.06	6.7 / -12.9 / -7.4	9.0	8.88		UGCA 38 / -1-07-031
0238+59	238.1 5923 / 136.50 -0.33 / 359.60 0.80	4 / 0.18 / 14-11	1.1 / 4.0 / 2	1.00		/ 0.00 / 6.18	2. / 209.	365	5 / 40 / 51	1.89 / 51 25	3.4 / 27.67	3.4 / 0.0 / 0.0	4.0	8.95	0.049	UGCA 39, MAFFEI 2 / 10-04-000 2
N 1051	238.6 -709 / 180.15 -57.07 / 296.71 -27.35	9B / 0.41 / 52 -1	2.1 / 2.1 / 2	0.87 / 33.	11.44	/ 0.05 / 0.03	1300. / 1300. / 1	205 / 310. / 312.	10	0.72 / 19 / 1.05	15.9 / 31.01	6.3 / -12.6 / -7.3	9.7	9.12 / 10.43		IC 249, UGCA 40 / -1-07-033
N 1052	238.6 -828 / 182.01 -57.92 / 295.27 -27.66	-5 / 0.49 / 52 -1	2.7 / 2.5 / 1	0.59 / 59.	11.03 / 7.95 / 3.08	.1 / 0.00 / 0.06	1462. / 1456.		12		17.8 / 31.25	6.7 / -14.2 / -8.2	-19.81 / 13.0	10.12		L2 / -1-07-034 2
N 1055	239.2 13 / 171.35 -51.75 / 304.68 -25.50	3 / 0.25 / 52 -2 +1	7.1 / 6.1 / 2	0.48 / 67.	9.41	1 / 0.32 / 0.06	1002. / 1030. / 1	410 / 405. / 406.	10	1.48 / 211 6 / 1.37	12.6 / 30.51	6.5 / -9.4 / -5.4	-19.48 / 22.4	9.98 / 9.68 / 11.03	0.50 / 0.045 / 11.08	2173 00-07-081 2
N 1068	240.1 -14 / 172.11 -51.94 / 304.27 -25.84	3A / 0.34 / 52 -2 +1	8.0 / 7.9 / 2	0.90 / 29.		/ 0.04 / 0.06	1151. / 1176. / 2	307	20	0.90 / 211 16 / 1.10	14.4 / 30.80	7.3 / -10.7 / -6.3	-21.39 / 33.2	10.75 / 9.22	0.03	M 77, ARP 37, 3C 71 / S2 / 2188 00-07-083 2
N 1058	240.2 3708 / 146.37 -20.38 / 340.41 -10.63	5A / 0.50 / 17 -1	3.6 / 3.7 / 2	0.98 / 16.	11.93	1 / 0.01 / 0.21	520. / 676. / 3	42	5	1.27 / 211 4 / 1.21	9.1 / 29.80	8.4 / -3.0 / -1.7	-17.87 / 9.8	9.34 / 9.19	0.70	2193 06-07-001 2
N 1073	241.1 110 / 170.92 -50.73 / 305.84 -25.66	5B / 0.30 / 52 -2 +1	5.0 / 5.1 / 2	1.00	11.47	1 / 0.00 / 0.08	1212. / 1242. / 1	92	8	1.23 / 7 / 1.17	15.2 / 30.91	8.0 / -11.1 / -6.6	-19.44 / 22.6	9.97 / 9.59	0.42	2210 00-08-001 2

(1) Name	(2) α / ℓ / SGL	(3) δ / b / SGB	(4) Type / ρ_{xyz} / Group	(5) D_{25} / D_{25}^i / source	(6) d/D / i	(7) $B_T^{b,i}$ / $H_{-0.5}$ / $B-H$	(8) source / A_B^{i-0} / A_B^b	(9) V_h / V_o / Tel	(10) W_{20} / W_R / W_D	(11) e_v / e_w / source	(12) F_c / source,e_F / f_h	(13) R / μ	(14) SGX / SGY / SGZ	(15) $M_B^{b,i}$ / Δ_{25}	(16) $\log L_B$ / $\log M_H$ / $\log M_T$	(17) M_H/L_B / M_H/M_T / M_T/L_B	(18) Alternate Names; UGC/ESO MCG / AGN? RCII
N 1079	241.6 / 223.89 / 271.72	−2913 / −65.21 / −31.05	0XP / 0.20 / 51 −3 +1	5.4 / 5.0 / 6	0.71 / 49.	12.30	2 / 0.00 / 0.00	1464 / 1377.		20		16.9 / 31.13	0.4 / −14.4 / −8.7	−18.83 / 24.7	9.72		416-13 −5-07-017 / 2
N 1084	243.5 / 182.47 / 296.34	−747 / −56.55 / −28.68	5A / 0.49 / 52 −1	3.4 / 3.0 / 9	0.54 / 63.	10.91	2 / 0.25 / 0.06	1408 / 1401.	328 / 327. / 329.	6 / 20 / 217	1.15 / 217 15	17.1 / 31.17	6.7 / −13.5 / −8.2	−20.26 / 15.0	10.30 / 9.62 / 10.67	0.21 / 0.089 / 2.35	−1-08-007 / 2
N 1087	243.9 / 173.77 / 304.07	−43 / −51.65 / −26.89	5X / 0.33 / 52 −4 +1	3.7 / 3.3 / 9	0.53 / 63.	11.18	1 / 0.26 / 0.11	1525 / 1545.		10	0.87 / 211 25	19.0 / 31.39	9.5 / −14.0 / −8.6	−20.21 / 18.3	10.28 / 9.43	0.14	2245 00-08-009 / 2
N 1090	244.0 / 173.50 / 304.37	−27 / −51.44 / −26.84	5B / 0.14 / 51+15	3.3 / 2.9 / 9	0.50 / 65.	12.20 / 9.90 / 2.30	1 / 0.29 / 0.11	2765 / 2786. / 1	338 / 331. / 332.	10	0.87 / 12 / 1.17	34.9 / 32.71	17.6 / −25.7 / −15.7	−20.51 / 29.6	10.40 / 9.96 / 10.97	0.36 / 0.096 / 3.77	2247 00-08-011 / 2
N 1097	244.3 / 226.91 / 270.25	−3029 / −64.65 / −31.65	3B / 0.13 / 51 −3 +1	9.8 / 8.9 / 6	0.62 / 57.	10.02	2 / 0.18 / 0.05	1275 / 1181. / 4	402 / 436. / 438.	10	1.55 / 7 / 1.16	14.5 / 30.81	0.1 / −12.4 / −7.6	−20.79 / 37.7	10.51 / 9.87 / 11.32	0.23 / 0.036 / 6.45	UGCA 41, ARP 77; 416-20 −5-07-024 / 2
U 2259	244.8 / 147.15 / 341.05	3720 / −19.79 / −11.33	8B / 0.55 / 17 −1	2.8 / 2.8 / 2	0.80 / 41.	13.1	3 / 0.08 / 0.24	589 / 742. / 1	137 / 167. / 170.	20	0.76 / 12 / 1.09	10.0 / 29.99	9.2 / −3.2 / −2.0	−16.89 / 8.2	8.95 / 8.76 / 9.82	0.65 / 0.087 / 7.45	2259 06-07-009 / 2
U 2275	245.4 / 169.64 / 308.87	341 / −48.12 / −25.87	9 / 0.24 / 52 −2 +1	5.5 / 5.7	1.00	13.3	3 / 0.00 / 0.13	1025 / 1061. / 4	109	7	1.19 / 8 / 1.08	13.1 / 30.58	7.4 / −9.2 / −5.7	−17.28 / 21.8	9.10 / 9.42	2.09	DDO 28; 2275 01-08-000 / 2
U 2302	246.6 / 171.73 / 307.13	156 / −49.24 / −26.72	9 / 0.32 / 52 −2 +1	5.5 / 5.8 / 2	1.00	13.90	7 / 0.00 / 0.20	1105 / 1133. / 4	71	5	1.15 / 6 / 1.08	14.0 / 30.73	7.5 / −10.0 / −6.3	−16.83 / 23.7	8.92 / 9.44	3.30	DDO 29; 2302 00-08-023 / 2
N 1110	246.8 / 183.77 / 296.26	−803 / −56.09 / −29.53	10 / 0.40 / 52 −1	2.9 / 2.1 / 2	0.23 / 86.		0.67 / 0.08	1336 / 1325. / 1	193 / 158. / 162.	10	0.88 / 11 / 1.05	16.3 / 31.06	6.3 / −12.7 / −8.0	10.0	9.30 / 9.86	0.279	NGC ID UNCERTAIN, UGCA 43 −1-08-010
0246-02	246.8 / 177.06 / 302.00	−251 / −52.65 / −28.20	10 / 0.28 / 52 −2 +1	1.5 / 1.5 / 2	1.00		0.00 / 0.11	1094 / 1103. / 1	118	20	0.64 / 16 / 1.03	13.5 / 30.66	6.3 / −10.1 / −6.4	5.9	8.90		UGCA 44 −1-08-000
0248+04	248.5 / 169.97 / 309.75	415 / −47.19 / −26.41	12 P / 0.19 / 52 −2 +1	0.5 / 0.5 / 5	0.60 / 58.	14.5	3 / 0.19 / 0.18	990 / 1026.	116 / 141. / 145.	100		12.6 / 30.51	7.2 / −8.7 / −5.6	−16.01 / 1.8	8.60		MARK 600; 01-08-008 / 2
U 2345	249.3 / 176.03 / 303.84	−121 / −51.16 / −28.37	9B / 0.30 / 52 −4 +1	4.1 / 4.1 / 2	0.80 / 41.		0.08 / 0.16	1508 / 1521. / 1		10	0.81 / 11 / 1.15	18.8 / 31.37	9.2 / −13.7 / −8.9	22.5	9.36 / 10.11	0.176	DDO 30; 2345 00-08-035 / 2
N 1140	252.2 / 188.35 / 294.14	−1014 / −56.32 / −31.32	10 P / 0.35 / 52 −1	1.7 / 1.5 / 4	0.57 / 60.	12.54	2 / 0.22 / 0.10	1508 / 1484. / 1	218 / 209. / 212.	15	0.94 / 217 8 / 1.02	18.2 / 31.31	6.4 / −14.2 / −9.5	−18.77 / 8.0	9.70 / 9.46 / 10.00	0.58 / 0.286 / 2.01	−2-08-019 / 2

Name														Name
N 1145	252.3 -1850 / 202.78 -60.47 / 284.13 -32.76	5 / 0.11 / 51 -0 +4	3.3 0.17 90. / 2.4 / 6	/ 0.67 / 0.06	1968. 1911. / 1	20		0.59 / 16 / 1.06	23.6 / 31.87 / 4.8 -19.3 -12.8	16.5	9.34		UGCA 45 / 546-29 -3-08-042
U 2411	253.4 7533 / 130.66 14.83 / 14.18 8.04	9 / 0.14 / 42-19+18	4.3 0.09 90. / 4.1 / 6	/ 0.67 / 1.43	2547. 2767. / 1	15	238 201. 204.	0.95 / 16 / 1.14	37.8 / 32.89 / 36.3 9.2 5.3	45.3	10.10 / 10.72	0.240	2411 13-03-000
0255-02	255.4 -233 / 179.08 -50.88 / 303.03 -30.18	9B / 0.29 / 52 -4 +1	3.1 0.66 53. / 2.9 / 2	/ 0.15 / 0.19	1539. 1542. / 1	10	217	1.05 / 11 / 1.08	19.1 / 31.41 / 9.0 -13.9 -9.6	16.2	9.61		IC 1870, UGCA 46 / -1-08-020
0255-54	255.4 -5446 / 271.79 -54.30 / 241.81 -30.22	7B / 0.14 / 14+14	6.8 0.17 90. / 4.8 / 6	11.0 11.82 -0.82 / 3 0.67 0.00	578. 403. / 7	10 / 5 / 301	143 114. 119.		5.4 / 28.64 / -2.2 -4.1 -2.7	-17.64 / 7.6	9.25 / 9.46	1.61	A 0255 / 154-23 2
N 1156	256.8 2502 / 156.32 -29.20 / 331.24 -19.62	10B / 0.10 / 17 -0	3.5 0.83 38. / 3.8 / 2	11.56 / 1 0.06 0.63	372. 477. / 1	6 / 7 / 211	116 150. 153.	1.17 / 217 8	6.4 / 29.02 / 5.3 -2.9 -2.1	-17.46 / 7.1	9.18 / 8.78 / 9.67	0.40 / 0.131 / 3.09	2455 04-08-006 2
U 2459	257.1 4852 / 143.66 -8.54 / 352.17 -7.15	8 / 0.12 / 18 -1	2.6 0.22 87. / 2.7 / 2	/ 0.67 / 1.82	2464. 2640. / 1	10	341 303. 305.	0.96 / 15 / 1.16	34.7 / 32.70 / 34.1 -4.7 -4.3	27.4	10.04 / 10.86	0.151	2459 08-06-000
U 2460	257.2 234 / 173.99 -46.98 / 308.78 -29.02	4B / 0.11 / 51 -0	1.7 1.00 / 1.8 / 2	13.1 / 3 0.00 0.29	2700. 2721.	200			34.2 / 32.67 / 18.7 -23.3 -16.6	-19.57 / 18.0	10.02		IC 277, MARK 602 / 2460 00-08-064 2
U 2463	257.4 4003 / 148.07 -16.23 / 344.72 -12.02	9X / 0.07 / 18 -0	2.9 0.70 50. / 2.9 / 2	13.3 / 3 0.13 0.43	1899. 2051. / 1	25	220	0.62 / 20 / 1.10	26.8 / 32.14 / 25.3 -6.9 -5.6	-18.84 / 22.7	9.73 / 9.48	0.56	2463 07-07-013
N 1172	259.3 -1502 / 197.50 -57.34 / 288.91 -33.90	-5 / 0.28 / 51 -0 +4	2.2 0.91 28. / 2.1 / 5	12.81 / 2 0.00 0.09	1534. 1485.	10			18.3 / 31.31 / 4.9 -14.3 -10.2	-18.50 / 11.2	9.59		-3-08-059 2
N 1169	300.1 4612 / 145.41 -10.63 / 350.22 -9.05	3X / 0.13 / 18 -1	4.9 0.61 57. / 5.4 / 5	11.31 8.44 2.87 / 2 0.19 1.01	2397. 2564. / 1	15	482 528. 529.	1.09 / 14 / 1.41	33.7 / 32.64 / 32.8 -5.7 -5.3	-21.33 / 53.1	10.72 / 10.15 / 11.63	0.26 / 0.033 / 8.11	2503 08-06-025 2
N 1179	300.3 -1906 / 204.68 -58.82 / 284.07 -34.67	6X / 0.31 / 51 -7 +4	5.4 0.71 49. / 5.0 / 6	12.0 / 5 0.12 0.05	1781. 1716. / 6	10	207 228. 230.	1.06 / 9 / 1.25	21.2 / 31.64 / 4.2 -16.9 -12.1	-19.64 / 31.0	10.05 / 9.71 / 10.67	0.46 / 0.111 / 4.18	UGCA 48 / 547-01 -3-08-060 2
N 1187	300.4 -2304 / 212.10 -60.05 / 279.25 -35.02	5B / 0.35 / 51 -6 +4	5.4 0.86 35. / 5.2 / 6	11.29 / 6 0.05 0.00	1394. 1314. / 2	10	284 432. 434.	1.14 / 7 10 / 1.04	16.3 / 31.05 / 2.1 -13.1 -9.3	-19.76 / 24.7	10.10 / 9.56 / 11.13	0.29 / 0.027 / 10.74	UGCA 49 / 480-23 -4-08-016 2
N 1171	300.7 4312 / 147.02 -13.19 / 347.75 -10.82	5 / 0.14 / 18 -2 +1	3.1 0.48 67. / 3.1 / 2	11.9 10.07 1.83 / 3 0.31 0.81	2747. 2906. / 1	15	287 271. 274.	0.66 / 23 / 1.11	37.9 / 32.89 / 36.3 -7.9 -7.1	-20.99 / 34.3	10.59 / 9.82 / 10.86	0.17 / 0.090 / 1.89	2510 07-07-018
N 1199	301.3 -1549 / 199.19 -57.27 / 288.06 -34.49	-5 / 0.12 / 51-14	2.2 0.86 34. / 2.1 / 5	12.41 / 2 0.00 0.09	2581. 2528.	50			31.6 / 32.50 / 8.1 -24.7 -17.9	-20.09 / 19.4	10.23		IC 1188 / -3-08-067 2

(1) Name	(2) α ℓ SGL	(3) δ b SGB	(4) Type ϱ_{xyz} Group	(5) $D_{25}^{b,i}$ $D_{25}^{b,i}$ source	(6) d/D i	(7) $B_T^{b,i}$ $H_{-0.5}$ $B-H$	(8) source A_i^{-o} A_B^b	(9) V_h V_o Tel	(10) W_{20} W_R W_D	(11) e_v e_w source	(12) F_c source/e_F f_h	(13) R μ	(14) SGX SGY SGZ	(15) $M_B^{b,i}$ $\Delta_{25}^{b,i}$	(16) $\log L_B$ $\log M_H$ $\log M_T$	(17) M_H/L_B M_H/M_T M_T/L_B	(18) Alternate Names Alternate Names UGC/ESO MCG AGN? RCII
N 1201	302.0 218.55 275.35	-2616 -60.39 -35.51	-2A 0.36 51 -5 +4	3.0 2.7 6	0.61 57.	11.58	2 0.00 0.00	1722. 1630.		50		20.2 31.53	1.5 -16.4 -11.8	-19.95 15.9	10.17		480-28 -4-08-023 2
U 2531	302.9 147.90 347.12	4211 -13.87 -11.72	8A 0.14 18 -2 +1	4.6 4.6 2	0.53 63.	11.7	3 0.25 0.68	2717. 2872. 1	358	25	1.07 12 1.23	37.4 32.87	35.7 -8.2 -7.6	-21.17 50.2	10.66 10.22	0.36	IC 284 2531 07-07-023 2
N 1209	303.7 199.64 288.18	-1548 -56.74 -35.06	-5 0.13 51 -14	2.6 2.3 5	0.58 59.	12.22	2 0.00 0.11	2679. 2624.		77		32.9 32.58	8.4 -25.6 -18.9	-20.36 22.1	10.34		-3-08-073 2
0304-14	304.7 197.30 290.14	-1413 -55.80 -35.06	10 P 0.31 51 -0 +4	1.3 1.3 2	1.00		 0.00 0.14	1530. 1480. 1	85	15	0.22 28 1.02	18.3 31.31	5.2 -14.1 -10.5	6.9	8.74		UGCA 52 -2-08-053
0306-23	306.3 213.14 279.12	-2315 -58.79 -36.39	12 0.07 51 -0	1.8 1.6 6	0.57 60.		 0.22 0.01	2888. 2803. 6	187 176. 179.	25	0.51 27 1.05	35.2 32.73	4.5 -28.0 -20.9	16.4	9.60 10.17	0.269	IC 1892, UGCA 55 ARP 332, VV 260 480-36 -4-08-030
N 1222	306.4 182.60 303.27	-309 -49.21 -32.98	13 P 0.08 51 -0	1.1 1.1 4	0.88 33.		 0.00 0.13	2482. 2473.		66		31.0 32.46	14.3 -21.8 -16.9	10.0			-1-09-005 2
N 1232	307.5 208.77 282.22	-2046 -57.81 -36.51	5X 0.50 51 -7 +4	7.2 6.9 6	0.79 42.	10.39	1 0.08 0.03	1684. 1607. 2	250 319. 321.	10	1.48 6 1.10	20.0 31.50	3.4 -15.7 -11.9	-21.11 40.3	10.64 10.08 11.08	0.28 0.102 2.75	ARP 41 547-14 -4-08-032 2
0307-20	307.8 208.84 282.21	-2047 -57.74 -36.58	9B 0.44 51 -7 +4	1.0 1.0 6	0.91 28.		 0.03 0.02	1779. 1702.		73		21.1 31.62	3.6 -16.6 -12.6	6.2			NGC 1232A
0307-41	307.8 248.01 256.93	-4112 -58.46 -35.45	7 0.27 53 -9 +7	4.6 4.0 6	0.50 65.	11.81	 0.29 0.00	950. 804. 2	149 131. 135.	10	0.77 13 1.03	10.1 30.02	-1.9 -8.0 -5.9	11.8	8.78 9.77	0.102	300-14 -7-07-007
0308-22	308.1 212.12 279.97	-2235 -58.21 -36.77	5 0.21 51 -6 +4	3.3 2.3 6	0.17 90.	10.74	 0.67 0.00	1328. 1244. 2	260 222. 225.	15	0.62 45 1.02	15.4 30.93	2.1 -12.1 -9.2	10.3	9.00 10.17	0.067	IC 1898, UGCA 56 481-02 -4-08-036
0308-33	308.5 232.84 266.55	-3320 -59.47 -36.67	10 0.17 51 -0 +1	2.3 1.6 6	0.14 90.		 0.67 0.00	1130. 1009. 2	143 114. 119.	15	0.45 26	12.6 30.50	-0.6 -10.1 -7.5	5.9	8.65 9.35	0.202	357-07 -6-08-000
N 1249	308.6 268.22 242.50	-5332 -53.40 -32.43	6B 0.27 53 -0 +7	5.4 4.7 6	0.57 60.	11.6 10.21 1.39	5 0.22 0.00	1007. 828. 7	235 229. 231.	90		10.7 30.15	-4.2 -8.0 -5.7	-18.55 14.7	9.61 10.35	5.46	155-06 2
N 1241	308.8 190.72 296.49	-907 -52.32 -35.07	3B 0.11 51 -0	3.6 3.4 4	0.67 52.	12.3	5 0.15 0.22	2168. 2134.		65		26.6 32.13	9.7 -19.5 -15.3	-19.83 26.4	10.12		ARP 304, VV 334 -2-09-011 2

23

ID													Name
0310-05 310.4 -526 186.31 -49.86 300.96 -34.57	13 P 0.10 51-13				0.15 0.17	2252. 2231.		67		27.9 32.23	11.8 -19.7 -15.8		MARK 604 -1-09-000
N 1255 311.4 -2555 218.61 -58.25 275.81 -37.61	4X 0.47 51 -5 +4	3.6 0.61 3.2 57. 6	11.41 9.82 1.59	2 0.19 0.00	1699. 1600. 2	260 265. 267.	15	0.95 14 1.03	19.9 31.50	1.6 -15.7 -12.2	-20.09 18.6	10.23 9.55 10.58	0.21 0.093 2.24	UGCA 60 481-13 -4-08-050 2
N 1253 311.6 -300 183.68 -48.12 303.89 -34.18	6B 0.08 52 -0	5.2 0.49 4.6 66. 2	11.91 9.97 1.94	2 0.30 0.19	1709. 1696. 1	328 318. 320.	15	1.46 56 7 1.17	21.1 31.62	9.7 -14.5 -11.9	-19.71 28.3	10.08 10.11 10.92	1.08 0.154 6.98	UGCA 62, ARP 279 -1-09-018 2
0311-25 311.6 -2522 217.61 -58.10 276.51 -37.66	10 0.49 51 -5 +4	4.1 0.30 3.2 80. 6	12.48	0.67 0.00	1735. 1638. 2	182 150. 154.	10	0.94 11 1.02	20.4 31.54	1.8 -16.0 -12.4	19.1	9.56 10.09	0.291	UGCA 61
0311-57 311.8 -5733 273.28 -50.82 237.88 -31.45	12B 0.31 53 -0 +1	3.5 0.36 2.8 75. 6	10.63	0.51 0.00	1140. 951. 7	232 202. 205.	20		12.3 30.46	-5.6 -8.9 -6.4	10.1	10.08		481-14 -4-08-051 2
0312-04 312.2 -458 186.15 -49.22 301.65 -34.87	10 0.09 51-13	2.1 0.83 2.1 38. 4	13.87	7 0.06 0.17	2218. 2197. 1	164 215. 218.	12	0.58 15 1.05	27.6 32.20	11.9 -19.2 -15.8	-18.33 16.9	9.52 9.46 10.36	0.87 0.128 6.79	116-12 DDO 32 -1-09-021 2
U 2620 313.0 8004 129.17 19.23 18.53 9.77	9X 0.10 42 -0+18	2.3 0.80 2.6 41. 2	13.3	3 0.08 0.75	2244. 2464. 1	151	10	0.68 26 1.06	34.2 32.67	31.9 10.7 5.8	-19.37 26.0	9.94 9.75	0.64	2620 13-03-004
0314-24 314.0 -2423 216.03 -57.36 277.77 -38.19	10 0.10 51 -0 +4	1.7 0.85 1.6 36. 6		0.05 0.00	2077. 1982. 2	74 97. 103.	15	0.03 29 1.01	24.8 31.97	2.6 -19.3 -15.3	11.6	8.82 9.50	0.208	UGCA 63 481-16 -4-08-000
0314+03 314.7 326 177.50 -43.27 311.49 -32.82	-3 0.03 52 -0	0.7 0.58 0.6 60. 2	14.6	3 0.00 0.29	890. 900.		220		11.3 30.26	6.3 -7.1 -6.1	-15.66 2.0	8.46		ARAK 99 00-09-043
0314-35 314.9 -3542 237.26 -58.02 263.41 -37.71	7 1.01 51 -1	3.2 0.58 2.8 60. 6	12.62	0.21 0.00	1572. 1438. 2	160 149. 153.	20	0.81 13 1.01	18.0 31.28	-1.6 -14.2 -11.0	14.7	9.32 9.98	0.221	357-12 -6-08-002
N 1291 315.5 -4119 247.57 -57.03 256.43 -36.85	 0.14 53 -9 +7	11.1 0.91 10.8 28. 9	9.42	2 0.00 0.00	829. 678.	84	7 15 203	1.25 203 10	8.6 29.68	-1.6 -6.7 -5.2	-20.26 27.1	10.30 9.12	0.07	301-02 -7-07-008 2
N 1292 316.0 -2748 222.43 -57.55 273.40 -38.61	5A 0.52 51 -0 +1	2.8 0.43 2.3 70. 6	12.2 10.41 1.79	5 0.39 0.00	1370. 1261. 2	263 239. 242.	20	0.83 18 1.01	15.7 30.99	0.7 -12.3 -9.8	-18.79 10.5	9.71 9.22 10.24	0.33 0.095 3.42	418-01 -5-08-026 2
0316-26 316.4 -2601 219.19 -57.17 275.68 -38.74	4 0.33 51 -5 +4	2.9 0.36 2.3 75. 6	11.27	0.51 0.00	1802. 1699. 6	208	15	0.41 26 1.07	21.2 31.64	1.6 -16.5 -13.3	14.2	9.06		UGCA 64 481-18 -4-08-057 2
0316-23 316.6 -2358 215.54 -56.68 278.31 -38.77	10 0.66 51 -5 +4	1.9 1.00 1.9 6		0.00 0.00	1536. 1440. 2	100	10	0.51 17 1.01	17.9 31.27	2.0 -13.8 -11.2	9.9	9.02		UGCA 65 481-19 -4-08-000

Table columns (1)–(18). Each object spans three stacked data lines; within each cell the stacked values are separated by line breaks.

(1) Name	(2) α / ℓ / SGL	(3) δ / b / SGB	(4) Type / Q_{xyz} / Group	(5) D_{25} / $D_{25}^{b,i}$ / source	(6) d/D / i	(7) $B_T^{b,i}$ / $H_{-0.5}$ / B-H	(8) source / A_B^{i-o} / A_B^b	(9) V_h / V_o / Tel	(10) W_{20} / W_R / W_D	(11) e_v / e_w / source	(12) F_c / source/e_F / f_h	(13) R / μ	(14) SGX / SGY / SGZ	(15) $M_B^{b,i}$ / $\Delta_{25}^{b,i}$	(16) log L_B / log M_H / log M_T	(17) M_H/L_B / M_H/M_T / M_T/L_B	(18) Alternate Names / UGC-ESO / MCG / AGN? / RCII
N 1282	316.9 150.73 347.83	4111 -13.34 -14.43	-5 0.07 18 -0 +1	1.6 1.7 2	0.80 41.	13.0	3 0.00 0.70	2200. 2343.		150		30.7 32.44	29.1 -6.3 -7.7	-19.44 15.2	9.97		2675 07-07-068 2
N 1297	317.0 207.60 284.31	-1917 -55.21 -38.61	-5 0.71 51 -7 +4	2.0 1.9 6	0.91 28.	12.6	5 0.00 0.06	1554. 1475.		27		18.3 31.31	3.5 -13.9 -11.4	-18.71 10.2	9.68		547-30 -3-09-017 2
N 1300	317.5 208.16 283.94	-1935 -55.20 -38.76	4B 0.71 51 -7 +4	6.8 6.5 6	0.78 43.	10.95 8.79 2.16	1 0.09 0.06	1583. 1502. 2	292 373. 375.	15	1.00 11 12 1.05	18.8 31.37	3.5 -14.2 -11.7	-20.42 35.7	10.36 9.55 11.16	0.15 0.025 6.28	UGCA 66 2
U 2684	317.6 166.33 326.22	1707 -32.74 -27.68	10 0.04 17 -0	2.0 1.9 2	0.53 63.		 0.25 0.41	357. 417. 1	88 77. 84.	8	0.35 226 15 1.03	5.5 28.71	4.1 -2.7 -2.6	3.1	7.83 8.72	0.129	2684 03-09-000
0317-32	317.6 231.50 267.18	-3239 -57.57 -38.63	3B 0.40 51 +2 +1	2.0 1.4 6	0.13 90.			1287. 1161.		146		14.5 30.81	-0.6 -11.3 -9.1	5.9			357-16 -5-08-027 2
N 1313	317.7 283.36 228.00	-6641 -44.64 -28.20	7B 0.14 14 -14	9.1 8.8 9	0.82 38.	9.26	2 0.07 0.05	462. 254.	203 272. 274.	6 10 301		3.7 27.86	-2.2 -2.4 -1.8	-18.60 9.5	9.63 10.31	4.76	082-11 2
N 1302	317.7 219.68 275.40	-2614 -56.92 -39.03	 0.48 51 -5 +4	3.8 3.6 6	0.87 33.	11.4	5 0.00 0.00	1704. 1599.		7	0.63 205 16	20.0 31.51	1.5 -15.5 -12.6	-20.11 21.0	10.24 9.23	0.10	481-20 -4-08-058 2
0317-49	317.9 261.49 246.21	-4947 -53.89 -35.00	5 0.49 53 -7	4.3 3.7 6	0.55 62.		 0.24 0.00	1031. 856. 7		20		11.1 30.22	-3.7 -8.3 -6.3	12.0			IC 1914 200-03
N 1311	318.6 265.30 243.21	-5222 -52.66 -34.24	9 0.13 14 +14	2.8 2.1 6	0.29 81.		 0.67 0.00	570. 389. 7		20		5.2 28.58	-1.9 -3.8 -2.9	3.2			200-07
N 1310	319.1 240.10 261.22	-3719 -57.03 -38.30	5A 1.15 51 -1	1.8 1.6 6	0.50 65.		 0.29 0.00	1715. 1573.		146		16.9 31.14	-2.1 -13.3 -10.6	7.9			357-19 -6-08-004 2
N 1309	319.8 202.21 289.12	-1535 -53.15 -38.87	4A 0.18 51 -12 +11	2.8 2.9 4	1.00	11.90	2 0.00 0.10	2138. 2070. 1	161	10	0.84 217 10 1.09	26.0 32.07	6.6 -19.1 -16.3	-20.17 22.0	10.26 9.67	0.26	-3-09-028 2
0320-11	320.5 196.27 294.46	-1123 -51.06 -38.36	7B 0.08 51 -0	2.5 2.4 2	0.85 35.	9.67	 0.05 0.15	2808. 2755. 1	239 348. 350.	15	0.64 17 1.06	34.7 32.70	11.3 -24.8 -21.5	24.3	9.72 10.93	0.062	UGCA 67 -2-09-036
N 1316	320.8 240.16 261.06	-3723 -56.68 -38.62	-2XP 1.15 51 -1	5.7 5.2 9	0.68 51.		2 0.00 0.00	1774. 1631.		25		16.9 31.14	-21.47 25.7	-21.47 25.7	10.78		ARP 154 357-22 -6-08-005 2

N 1317	320.9	-3717		3.1	0.83	11.94	2	1943.		51		16.9	-2.1	-19.20	9.87		
	239.97	-56.68	1.30	3.0	38.		0.00	1800.				31.14	-13.5	14.8			
	261.18	-38.66	51 -1	9			0.00						-10.9				357-23 -6-08-006 2
N 1315	320.9	-2133	-2B	1.7	0.94			1730.		73		17.7	2.7				
	211.87	-55.08	0.88	1.6	23.		0.00	1639.				31.24	-13.6	8.3			
	281.48	-39.68	51 -4	6			0.03						-11.5				548-03 -4-09-002 2
0321-19	321.5	-1957	6 P	1.9	0.78			1838.	125	15	0.44	21.9	4.0				UGCA 68
	209.32	-54.44	0.49	1.8	42.		0.08	1753.	147.		23	31.71	-16.4	11.5	9.12	0.183	
	283.56	-39.72	51 -7 +4	6			0.04	2	151.		1.01		-14.0		9.86		548-05 -3-09-033
0321+72	321.7	7240	10	2.1	0.28			2074.	269	15	0.70	31.5	30.6				UGCA 69
	133.91	13.38	0.09	2.2	81.		0.67	2284.	234.		15	32.49	7.0	20.2	9.70	0.155	
	12.99	4.78	42 -0+18	2			1.54	1	236.		1.05		2.6		10.51		12-04-000
N 1326	322.0	-3638	-2B	3.8	0.67	11.30	2	1363.	280	20	0.80	16.9	-1.8	-19.84	10.13	0.13	
	238.76	-56.52	1.36	3.5	53.	8.16	0.00	1221.	305.		203 14	31.14	-12.9	17.3	9.26	0.039	
	261.96	-38.99	51 -1	6		3.14	0.00	2	307.		1.04		-10.5		10.67	3.47	357-26 -6-08-011 2
U 2729	322.2	6824	12	3.3	0.74			1936.	260	10	0.86	29.4	29.0				
	136.39	9.87	0.09	5.0	46.		0.00	2140.	307.		15	32.34	5.0	42.9	9.80	0.053	
	9.74	2.01	12-20+19	2			2.28	1	309.		1.31		1.0		11.07		2729 11-05-000
N 1320	322.3	-312	2	1.6	0.31			3000.		200		37.7	17.1				MARK 607
	186.34	-46.14	0.07	1.2	79.		0.66	2977.				32.88	-24.8	13.2			
	304.62	-36.79	51 -0	4			0.08						-22.6				-1-09-036 2
N 1325	322.3	-2141	4A	5.4	0.36	11.78	1	1595.	346	15	0.83	17.7	2.7	-19.46	9.98	0.22	UGCA 70
	212.26	-54.81	0.88	4.3	75.	9.54	0.52	1503.	318.		18	31.24	-13.4	22.2	9.33	0.032	
	281.32	-40.01	51 -4	6		2.24	0.03	6	320.		1.17		-11.5		10.81	6.89	548-07 -4-09-004 2
0322-21	322.6	-2131	3B	2.2	1.00	13.39	2	1333.		10		17.7	2.7	-17.85	9.33		NGC 1325A
	212.02	-54.69	0.89	2.2			0.00	1241.				31.24	-13.1	11.4			
	281.54	-40.07	51 -4	6			0.04						-11.3				548-10 -4-09-006 2
0323-37	323.0	-3712		1.1	0.50			2066.		146		16.9	-2.1				NGC 1316C
	239.75	-56.27	1.30	0.9	65.		0.00	1922.				31.14	-13.5	4.4			
	261.19	-39.08	51 -1	6			0.00						-11.1				357-27 -6-08-012 2
0323-362	323.1	-3631	9B	2.1	0.96			1836.	95	15	0.68	16.9	-1.9				NGC 1326A
	238.52	-56.31	1.30	2.1	19.		0.01	1694.			18	31.14	-13.3	10.4	9.14		
	262.05	-39.22	51 -1	6			0.00	2			1.00		-11.0				357-28 -6-08-013 2
0323-16	323.2	-1624	9	2.5	0.19			1878.	247	15	0.64	22.6	5.4				UGCA 71
	204.01	-52.75	0.42	1.8	90.		0.67	1804.			19	31.77	-16.5	11.9	9.35		
	288.21	-39.78	51-12+11	2			0.12	1			1.04		-14.5				-3-09-041
0323-363	323.4	-3632	7B	3.5	0.33			1012.	185	10	1.21	16.9	-1.8				NGC 1326B
	238.54	-56.25	1.41	2.8	77.		0.58	870.	154.		8	31.14	-12.5	13.8	9.67	0.487	
	262.02	-39.28	51 -1	6			0.00	2	158.		1.01		-10.3		9.98		357-29 -6-08-014 2
N 1332	324.1	-2131	-2	1.7	0.29	11.17	2	1564.		42	0.70	17.7	2.7	-20.07	10.22	0.09	UGCA 72
	212.20	-54.36	0.89	1.3	80.		0.00	1471.			11 76	31.24	-13.2	6.7	9.20		
	281.56	-40.42	51 -4	6			0.03						-11.5				548-18 -4-09-011 2

(1) Name	(2) α / ℓ / SGL	(3) δ / b / SGB	(4) Type / Q_{xyz} / Group	(5) D_{25} / $D_{25}^{b,i}$ / source	(6) d/D / i	(7) $B_T^{b,i}$ / $H_{-0.5}$ / $B-H$	(8) source / A_B^{i-o} / A_B^b	(9) V_h / V_o / Tel	(10) W_{20} / W_R / W_D	(11) e_v / e_w / source	(12) F_c / source,e_F / f_h	(13) R / μ	(14) SGX / SGY / SGZ	(15) $M_B^{b,i}$ / Δ_{25}	(16) $\log L_B$ / $\log M_H$ / $\log M_T$	(17) M_H/L_B / M_H/M_T / M_T/L_B	(18) UGC/ESO MCG / AGN? / RCII
N 1331	324.3	-2132	-5	1.2	1.00	14.17	6	1326.		84		17.7	2.6	-17.07	9.02		IC 324
	212.25	-54.32	0.89	1.2			0.00	1233.				31.24	-13.0	6.2			
	281.55	-40.47	51 -4	6			0.03						-11.3				
0324-52	324.3	-5257	6	2.1	0.56	12.5	5	1068.		17		11.5	-4.4	-17.80	9.31		548-19 -4-09-012 2
	265.51	-51.63	0.45	1.8	61.		0.23	882.				30.30	-8.3	6.0			
	242.17	-34.83	53 -8 +7	6			0.00						-6.6				
N 1337	325.7	-834	6A	4.7	0.36	11.52	1	1246.	269	8	1.34	15.0	5.5	-19.36	9.94	0.57	IC 1933
	193.56	-48.51	0.07	3.9	75.	9.91	0.52	1199.	239.		217 6	30.88	-10.2	17.1	9.69	0.174	155-25 2
	298.33	-39.02	52 -0	1		1.61	0.16	1	242.		1.14		-9.4		10.45	3.28	-2-09-042 2
0325-17	325.9	-1736	-5 P	1.5	0.40	14.77	2	1834.		11	-0.28	21.9	4.8	-16.94	8.97	0.27	UGCA 73, HARO 20
	206.25	-52.63	0.47	1.2	72.		0.00	1753.			10 75	31.71	-16.0	7.7	8.40		
	286.74	-40.56	51 -0 +4	6			0.13						-14.3				
N 1339	326.1	-3228	-5	2.0	0.77	12.6	5	1293.		68		14.6	-0.6	-18.22	9.48		548-23 -3-09-045 2
	231.24	-55.78	0.46	1.8	43.		0.00	1161.				30.82	-11.1	7.7			
	267.15	-40.43	51 +2 +1	6			0.00						-9.5				
N 1341	326.1	-3719	5B	1.4	1.00	13.0	5	1866.		68		16.9	-2.1	-18.14	9.45		418-04 -5-09-004 2
	239.85	-55.64	1.53	1.4			0.00	1720.				31.14	-13.2	6.9			
	260.89	-39.67	51 -1	6			0.00						-11.1				
N 1344	326.3	-3115	-5	3.9	0.70	11.25	2	1235.		53		13.9	-0.2	-19.46	9.98		358-08 -6-08-020 2
	229.09	-55.67	0.28	3.6	50.		0.00	1107.				30.71	-10.5	14.6			NGC 1340
	268.74	-40.61	51 +2 +1	6			0.00						-9.0				
N 1345	327.1	-1758	12 P	2.4	0.76	13.98	2	1527.		15	0.48	18.1	3.8	-17.30	9.11	0.76	418-05 -5-09-005 2
	206.98	-52.50	0.59	2.3	45.		0.10	1444.			10 20	31.28	-13.1	12.2	9.00		UGCA 74, HARO 21
	286.30	-40.88	51 -4	2			0.12						-11.8				
U 2765	327.2	6812	12	3.7	0.44			1679.	329	10	0.98	26.1	25.7		9.81		548-26 -3-09-046 2
	136.89	9.97	0.11	5.0	70.			1881.	310.		14	32.08	4.5	38.1		0.061	2765 11-05-000
	9.89	1.53	12-20+19	2			2.28	1	312.		1.28		0.7		11.03		
0327-15	327.2	-1524	9	2.2	0.92			1890.	136	10	0.84	22.7	5.8		9.55		
	203.17	-51.46	0.36	2.2	26.		0.03	1816.			16	31.78	-16.2	14.6			-3-09-047
	289.67	-40.61	51-12+11	2			0.15	1			1.05		-14.8				UGCA 75
N 1351	328.6	-3502	-5	2.5	0.50	12.8	5	1491.		94		16.9	-1.4	-18.34	9.53		358-12 -6-08-022 2
	235.78	-55.27	1.57	2.2	65.		0.00	1350.				31.14	-12.8	10.9			
	263.71	-40.57	51 -1	6			0.00						-11.0				
N 1350	329.1	-3347	2B	5.0	0.54	11.16	2	1786.		20		16.9	-1.1	-19.98	10.18		358-13 -6-08-023 2
	233.59	-55.18	1.40	4.4	62.	8.28	0.25	1648.				31.14	-13.0	21.7			
	265.32	-40.87	51 -1	6		2.88	0.00						-11.3				
N 1353	329.9	-2058	3X	3.6	0.43	11.83	2	1517.		77		17.8	2.8	-19.42	9.96		UGCA 76
	212.01	-52.90	0.86	3.0	70.		0.38	1421.				31.25	-13.0	15.6			548-31 -4-09-022 2
	282.37	-41.74	51 -4	6			0.04						-11.9				

Name	Coordinates (RA/Dec; l/b; SGL/SGB)	Type	Dimensions	Photometry	Velocity	W	N	HI	Δ / μ	SGX/Y/Z	M / a	B_T	Extra	Other names
0330-17	330.0 -1753 / 207.28 -51.83 / 286.49 -41.56	10 / 0.44 / 51 -0 +4	2.4 / 2.1 / 6 · 0.50 / 65.		1961. / 1876. / 1	160	20	0.42 / 29 / 1.03	23.6 / 31.86	5.0 / -16.9 / -15.6	14.5	9.17		UGCA 77; 548-32 -3-10-003
0330-52	330.2 -5205 / 263.65 -51.18 / 242.72 -36.00	5B / 0.54 / 53 -8 +7	2.8 / 2.4 / 6 · 0.51 / 64.	11.7 9.64 2.06 / 0.29 0.13 / 5 0.27 0.00	1079. / 892. / 7	229 / 213. / 216.	23		11.6 / 30.33	-4.3 / -8.4 / -6.8	-18.63 / 8.1	9.64		IC 1954; 200-36 2
N 1357	330.9 -1350 / 201.55 -49.97 / 291.89 -41.28	1A / 0.25 / 51 -11	2.3 / 2.1 / 4 · 0.65 / 54.	12.34 / — / 2 0.16 0.10	2035. / 1964.		58		24.7 / 31.96	6.9 / -17.2 / -16.3	-19.62 / 15.1	10.03	2.43	-2-10-001 2
N 1359	331.5 -1941 / 210.21 -52.13 / 284.12 -42.05	9BP / 0.40 / 51 -0 +4	2.4 / 2.3 / 6 · 0.67 / 53.	12.38 / — / 2 0.15 0.07	1976. / 1883. / 2	239 / 254. / 256.	15	0.97 / 12 / 1.01	23.7 / 31.87	4.3 / -17.1 / -15.9	-19.49 / 15.9	10.04	0.54 / 0.176 / 3.06	548-39 -3-10-007 2
0331-21	331.6 -2137 / 213.23 -52.72 / 281.52 -42.16	4B / 0.45 / 51 -0 +4	1.9 / 1.9 / 6 · 0.87 / 33.	12.2 / — / 5 0.05 0.03	1860. / 1760. / 6	220	15	0.53 / 21 / 1.12	22.1 / 31.73	3.3 / -16.1 / -14.9	-19.53 / 12.3	9.99 / 9.72 / 10.47	0.16	IC 1953, UGCA 78; 548-40 -4-09-026 2
0331-50	331.7 -5035 / 261.28 -51.54 / 244.33 -36.79	5 / 0.08 / 14 +14	3.0 / 2.1 / 6 · 0.21 / 89.		639. / 455. / 7		20	0.67 / — / 0.00	6.0 / 28.91	-2.1 / -4.4 / -3.6	3.7	10.00 / 9.22		IC 1959; 200-39
N 1365	331.8 -3618 / 237.94 -54.57 / 261.91 -40.98	3B / 1.58 / 51 -1	11.2 / 9.8 / 9 · 0.53 / 63.	9.88 7.18 2.70 / — / 2 0.26 0.00	1639. / 1492. / 4	403 / 410. / 411.	20	1.58 / 7 / 1.14	16.9 / 31.14	-1.8 / -12.7 / -11.2	-21.26 / 48.4	10.70 / 10.04 / 11.37	0.22 / 0.046 / 4.75	358-17 -6-08-026 2 S1.8
N 1366	331.9 -3122 / 229.43 -54.49 / 268.40 -41.79	-2 / 0.16 / 51 +2 +1	2.0 / 1.6 / 6 · 0.36 / 75.	12.87 / — / 6 0.00 0.00	1046. / 914.		68		11.6 / 30.32	-0.2 / -8.6 / -7.7	-17.45 / 5.4	9.17		418-10 -5-09-013 2
N 1371	332.8 -2505 / 218.92 -53.36 / 276.83 -42.45	1X / 0.75 / 51 -4	6.8 / 6.2 / 6 · 0.67 / 53.	11.43 / — / 6 0.15 0.00	1472. / 1360. / 6	412 / 471. / 472.	10	1.26 / 9 / 1.35	17.1 / 31.16	1.5 / -12.5 / -11.5	-19.73 / 31.0	10.08 / 9.73 / 11.30	0.44 / 0.027 / 16.41	UGCA 79; 482-10 -4-09-029 2
N 1375	333.3 -3526 / 236.43 -54.30 / 262.97 -41.45	-2 / 1.36 / 51 -1	2.0 / 1.6 / 6 · 0.36 / 75.	13.20 / — / 6 0.00 0.00	669. / 523.		104		16.9 / 31.14	-1.5 / -11.8 / -10.5	-17.94 / 7.9	9.37		358-24 -6-08-030 2
N 1374	333.4 -3524 / 236.37 -54.28 / 263.00 -41.47	-5 / 1.64 / 51 -1	2.1 / 2.0 / 6 · 0.87 / 33.	12.30 / — / 6 0.00 0.00	1251. / 1105.		82		16.9 / 31.14	-1.5 / -12.4 / -11.0	-18.84 / 9.9	9.73		358-23 -6-08-029 2
N 1379	334.2 -3537 / 236.74 -54.11 / 262.68 -41.59	-5 / 1.64 / 51 -1	2.4 / 2.4 / 6 · 1.00	12.07 / — / 6 0.00 0.00	1386. / 1239.		92		16.9 / 31.14	-1.6 / -12.5 / -11.2	-19.07 / 11.8	9.82		358-27 -6-09-001 2
N 1380	334.5 -3508 / 235.91 -54.06 / 263.30 -41.74	-2A / 1.54 / 51 -1	3.9 / 3.4 / 6 · 0.56 / 61.	11.10 / — / 2 0.00 0.00	1809. / 1663.		48		16.9 / 31.14	-1.5 / -12.8 / -11.5	-20.04 / 16.8	10.21		358-28 -6-09-002 2
N 1381	334.7 -3528 / 236.48 -54.02 / 262.85 -41.72	-2A / 1.54 / 51 -1	2.6 / 1.9 / 6 · 0.26 / 83.	12.34 / — / 6 0.00 0.00	1776. / 1629.		92		16.9 / 31.14	-1.6 / -12.8 / -11.5	-18.80 / 9.4	9.71		358-29 -6-09-003 2

(1) Name	(2) α / ℓ / SGL	(3) δ / b / SGB	(4) Type / Q_{xyz} / Group	(5) D_{25} / $D_{25}^{b,i}$ / source	(6) d/D / i	(7) $B_T^{b,i}$ / $H_{-0.5}$ / B-H	(8) source / A_B^{i-o} / A_B^b	(9) V_h / V_o / Tel	(10) W_{20} / W_R / W_D	(11) e_v / e_w / source	(12) F_c / source/e_F / f_h	(13) R / μ	(14) SGX / SGY / SGZ	(15) $M_B^{b,i}$ / $\Delta_{25}^{b,i}$	(16) log L_B / log M_H / log M_T	(17) M_H/L_B / M_H/M_T / M_T/L_B	(18) Alternate Names / UGC/ESO MCG / AGN? RCII
0334-44	334.8 / 250.93 / 251.81	-4407 / -52.95 / -39.54	3 / 0.50 / 53 -7	2.9 / 2.2 / 6	0.27 / 82.		/ 0.67 / 0.00	1074. / 903.		104		11.6 / 30.33	-2.8 / -8.5 / -7.4	7.5			IC 1970 / 249-07 -7-08-003 / 2
0334-34	334.8 / 235.49 / 263.61	-3453 / -54.00 / -41.85	2 / 1.59 / 51 -1	2.1 / 1.6 / 6	0.27 / 82.		/ 0.67 / 0.00	1616. / 1471.		146		16.9 / 31.14	-1.4 / -12.7 / -11.4	7.9			NGC 1380A / 2
U 2800	334.9 / 135.62 / 12.59	7114 / 12.81 / 3.07	10 / 0.14 / 12-19	2.9 / 3.4 / 2	0.53 / 63.		/ 0.25 / 1.48	1180. / 1385. / 1		15	0.84 / 226 9 / 1.11	20.0 / 31.50	19.5 / 4.3 / 1.1	19.9	9.44 / 10.45	0.098	2800 12-04-002
N 1386	335.0 / 237.67 / 261.91	-3610 / -53.94 / -41.64	-2A / 1.36 / 51 -1	3.0 / 2.4 / 6	0.38 / 74.	12.1	5 / 0.00 / 0.00	845. / 696.		57		16.9 / 31.14	-1.7 / -11.9 / -10.7	-19.04 / 11.8	9.81		358-35 -6-09-005 / S2 / 2
N 1387	335.1 / 236.84 / 262.54	-3541 / -53.93 / -41.76	-2 / 1.64 / 51 -1	2.5 / 2.5	1.00	12.0	5 / 0.00 / 0.00	1239. / 1091.		66		16.9 / 31.14	-1.6 / -12.3 / -11.1	-19.14 / 12.3	9.85		358-36 -6-09-007 / 2
0335-35	335.1 / 236.28 / 262.98	-3521 / -53.94 / -41.82	-2 / 1.54 / 51 -1	1.3 / 1.3 / 6	1.00	13.30	6 / 0.00 / 0.00	1925. / 1778.		146		16.9 / 31.14	-1.6 / -12.9 / -11.7	-17.84 / 6.4	9.33		NGC 1380B, NGC 1382 / ? / 2
N 1385	335.2 / 218.45 / 277.38	-2440 / -52.73 / -43.01	6B / 0.75 / 51 -4	4.6 / 4.1 / 6	0.58 / 59.	11.43 / 9.46 / 1.97	2 / 0.21 / 0.01	1503. / 1390. / 2	224. / 218. / 221.	15	0.87 / 14 / 1.02	17.5 / 31.21	1.6 / -12.7 / -11.9	-19.78 / 21.0	10.10 / 9.36 / 10.46	0.18 / 0.079 / 2.27	482-16 -4-09-036 / 2
N 1389	335.3 / 237.24 / 262.23	-3555 / -53.88 / -41.75	-5 / 1.50 / 51 -1	2.2 / 2.0 / 6	0.62 / 57.	12.39	6 / 0.00 / 0.00	1076. / 927.		117		16.9 / 31.14	-1.7 / -12.1 / -10.9	-18.75 / 9.9	9.69		358-38 -6-09-010 / 2
0335+66	335.5 / 138.52 / 9.21	6634 / 9.11 / -0.16	12 / 0.12 / 12+20+19	1.3 / 2.1 / 2	0.62 / 56.		/ 0.00 / 2.78	1501. / 1697. / 1	284. / 295. / 297.	10	1.25 / 9 / 1.07	23.8 / 31.88	23.5 / 3.8 / -0.1	14.6	10.00 / 10.57	0.273	UGCA 81 / 11-05-000
N 1395	336.3 / 216.20 / 279.41	-2311 / -52.12 / -43.28	-5 / 0.55 / 51 -4	3.9 / 3.8 / 6	0.90 / 29.	11.16	6 / 0.00 / 0.02	1695. / 1586.		21		20.0 / 31.50	2.4 / -14.3 / -13.7	-20.34 / 22.2	10.33		482-19 -4-09-039 / 2
N 1399	336.6 / 236.72 / 262.55	-3537 / -53.63 / -42.07	-5 / 1.59 / 51 -1	3.3 / 3.3	1.00	10.85	2 / 0.00 / 0.00	1443. / 1294.		59		16.9 / 31.14	-1.6 / -12.4 / -11.3	-20.29 / 16.3	10.31		358-45 -6-09-012 / 2
N 1398	336.8 / 221.54 / 274.85	-2630 / -52.77 / -43.29	2B / 0.46 / 51 -4	7.6 / 7.0 / 6	0.70 / 50.	10.47	2 / 0.13 / 0.00	1401. / 1281. / 2	458. / 550. / 551.	10	1.07 / 16 / 1.10	16.1 / 31.04	1.0 / -11.7 / -11.1	-20.57 / 32.9	10.42 / 9.48 / 11.46	0.12 / 0.011 / 10.98	482-22 -4-09-040 / 2
N 1404	337.0 / 236.94 / 262.35	-3545 / -53.54 / -42.12	-5 / 1.59 / 51 -1	2.7 / 2.7 / 6	0.94 / 24.	11.20	2 / 0.00 / 0.00	1908. / 1759.		65		16.9 / 31.14	-1.7 / -12.7 / -11.6	-19.94 / 13.3	10.17		358-46 -6-09-013 / 2

Name	Coordinates	Type / D_{25} / Group	Axial data	B mag / colour	Velocities	W	n	HI / W_{20}	Dist / μ	M / diam	mag	param	Other names
N 1411	337.1 −4415 / 251.00 −52.51 / 251.48 −39.88	−2A / 0.47 / 53 −7	2.2 2.1 0.72 48. 6	11.70 / 6 0.00 / 0.00	1040. / 867.		56		11.2 / 30.25	−18.55 / 6.9	9.61		249-11 −7-08-004 2
N 1401	337.2 −2253 / 215.82 −51.84 / 279.82 −43.48	−2 / 0.83 / 51 −4	2.4 1.9 0.29 81. 6	0.00 / 0.01	1569. / 1461.		104		18.4 / 31.32	10.2			482-26 −4-09-042 2
N 1400	337.3 −1851 / 209.71 −50.56 / 285.38 −43.36	−2A / 0.07 / 51 +8 +4	2.5 2.5 0.87 34. 6	11.94 / 2 0.00 / 0.14	483. / 389.		40		5.0 / 28.49	−16.55 / 3.7	8.81		548-62 −3-10-022 2
N 1406	337.5 −3130 / 229.82 −53.31 / 268.01 −42.96	5B / 0.20 / 51 +2 +1	3.4 2.4 0.19 90. 6	11.9 9.33 2.57 / 0.67 0.00	1068. 931. 2	352 314 316.	15	0.93 31 1.02	11.8 / 30.36	−18.46 / 8.3	9.58 9.07 10.37	0.31 0.050 6.28	UGCA 83 / 418-15 −5-09-020 2
U 2813	337.5 7109 / 135.85 12.87 / 12.67 2.86	10 / 0.12 / 12−19	2.3 2.7 0.48 67. 2	0.31 1.48	1407. 1611. 1		20	0.39 226 19 1.06	22.9 / 31.80	18.1	9.11 10.08	0.108	2813 12-04-000
N 1407	338.0 −1844 / 209.63 −50.36 / 285.55 −43.52	−5 / 0.42 / 51 −8 +4	4.3 4.3 0.91 28. 6	10.65 / 2 0.00 / 0.15	1811. 1716.	172 151 155	50		21.6 / 31.67	−21.02 / 27.1	10.60		548-67 −3-10-030 2
0338-35	338.3 −3547 / 236.99 −53.28 / 262.23 −42.37	10B / 1.26 / 51 −1	2.6 2.4 0.64 55. 6	0.17 0.00	2029. 1879. 2	96 92. 98.	25	0.79 10 1.01	16.9 / 31.14	11.8	9.25 9.46	0.606	NGC 1427A / 358-49 −6-09-016 2
N 1415	338.8 −2243 / 215.72 −51.44 / 280.06 −43.85	0.80 / 51 −4	3.9 3.4 0.50 65. 6	12.4 / 5 0.00 / 0.02	1508. 1399.		50		17.7 / 31.24	−18.84 / 17.6	9.73		IC 1983 / 482-33 −4-09-047 2
U 2824	338.8 7629 / 132.50 17.12 / 16.62 6.46	13 P / 0.21 / 42−19+18	5.4 6.4 1.00 2	11.4 / 3 0.00 / 0.80	2511. 2722. 1	255	40	0.15 66 1.39	37.3 / 32.86	−21.46 / 69.7	10.78 9.29	0.03	IC 334 / 2824 13-03-007
0338-45	338.9 −4531 / 252.89 −51.92 / 249.78 −39.76	10 / 0.24 / 53 −0	2.1 1.9 0.72 48. 6	0.12 0.00	1566. 1389. 7		20		17.8 / 31.25	9.9			IC 1986 / 249-13
N 1421	340.2 −1339 / 202.82 −47.87 / 292.60 −43.48	4X / 0.19 / 51−11	3.5 2.5 0.18 90. 4	10.91 9.56 1.35 / 2 0.67 0.09	2099. 2021. 1	383	15	0.98 217 15 1.08	25.5 / 32.03	−21.12 / 18.6	10.64 9.79	0.14	−2-10-008 2
N 1425	340.3 −3003 / 227.52 −52.56 / 269.88 −43.74	3A / 0.83 / 51 −0 +1	6.8 5.7 0.44 69. 6	11.34 8.85 2.49 / 2 0.36 0.00	1510. 1376. 2	368	15	0.94 8 1.03	17.4 / 31.20	−19.86 / 29.0	10.14 9.42	0.19	UGCA 84 / 419-04 −5-09-023 2
N 1427	340.4 −3534 / 236.62 −52.85 / 262.40 −42.84	−5 / 1.59 / 51 −1	2.2 2.0 0.64 55. 6	11.94 / 6 0.00 / 0.00	1567. 1416.		66		16.9 / 31.14	−19.20 / 9.9	9.87		358-52 −6-09-021 2
N 1433	340.5 −4723 / 255.69 −51.18 / 247.40 −39.34	1B / 0.55 / 53 −7	5.9 5.8 0.92 27. 9	10.64 / 2 0.03 / 0.00	1071. 889.		10		11.6 / 30.31	−19.67 / 19.6	10.06		249-14 2

(1) Name	(2) α / ℓ / SGL	(3) δ / b / SGB	(4) Type / ϱ_{xyz} / Group	(5) D_{25} / $D_{25}^{b,i}$ / source	(6) d/D / i	(7) $B_T^{b,i}$ / $H_{-0.5}$ / $B\text{-}H$	(8) source / A_B^{i-o} / A_B^b	(9) V_h / V_o / Tel	(10) W_{20} / W_R / W_D	(11) e_v / e_w / source	(12) F_c / source/e_f / f_h	(13) R / μ	(14) SGX / SGY / SGZ	(15) $M_B^{b,i}$ / $\Delta_{25}^{b,i}$	(16) $\log L_B$ / $\log M_H$ / $\log M_T$	(17) M_H/L_B / M_H/M_T / M_T/L_B	(18) Alternate Names UGC/ESO MCG — AGN? RCII
N 1428	340.5 236.21 262.73	-3519 -52.83 -42.91	-2 0.82 51 -1	1.5 1.2 6	0.47 68.		 0.00 0.00	228. 77.		146		16.9 31.14	-1.4 -11.2 -10.4	 5.9			358-53 -6-09-022 2
N 1426	340.6 215.22 280.68	-2216 -50.91 -44.26	-5 0.66 51 -4	2.4 2.2 6	0.61 57.	12.28	2 0.00 0.03	1441. 1332.		10		16.8 31.13	2.2 -11.9 -11.8	-18.85 10.8	9.73		549-01 -4-09-054 2
0341-36	341.2 238.05 261.20	-3626 -52.68 -42.80	9 1.15 51 -1	2.0 2.0 6	0.92 27.		 0.03 0.00	895. 741. 2	107	40	0.16 31 1.01	16.9 31.14	-1.8 -11.7 -10.9	 9.9	8.62		358-54 -6-09-024 2
N 1437	341.7 237.36 261.72	-3601 -52.59 -43.00	1B 1.38 51 -1	2.8 2.6 6	0.69 51.	12.19	6 0.13 0.00	1175. 1022.		68		16.9 31.14	-1.8 -12.0 -11.3	-18.95 12.8	9.77		358-58 -6-09-025 2
U 2847	342.0 138.18 10.63	6756 10.58 0.34	6X 0.19 14 -11	16.1 27.9 9	0.96 20.	6.40	2 0.02 2.70	32. 229.	195	6 15 72	2.87 7	3.9 27.96	3.8 0.7 0.0	-21.56 31.8	10.82 10.05	0.17	IC 342 2847 11-05-003
N 1439	342.6 215.15 280.95	-2205 -50.42 -44.73	-5 0.45 51 -4	2.2 2.3 6	0.96 19.	12.52	6 0.00 0.06	1670. 1560.		10		19.7 31.47	2.7 -13.8 -13.9	-18.95 13.2	9.77		549-09 -4-09-056 2
N 1440	342.8 209.80 286.10	-1825 -49.19 -44.63	-2X 0.52 51 -8 +4	2.2 2.1 6	0.69 50.	12.53	6 0.00 0.16	1534. 1437.		90		18.2 31.30	3.6 -12.4 -12.8	-18.77 11.2	9.70		NGC 1441 549-10 -3-10-043 2
N 1448	342.9 251.52 250.34	-4448 -51.39 -40.67	6A 0.32 53 -7	7.9 5.7 9	0.23 86.	10.63 8.32 2.31	2 0.67 0.00	1178. 1000. 2	416 379. 380.	15	1.50 19 1.07	12.9 30.55	-3.3 -9.2 -8.4	-19.92 21.5	10.16 9.72 10.95	0.36 0.059 6.19	NGC 1457 249-16 -7-08-005 2
N 1452	343.1 210.36 285.59	-1847 -49.25 -44.73	 0.40 51 -8 +4	3.0 2.8 6	0.61 57.	12.73	6 0.00 0.13	1904. 1805.		90		22.8 31.79	4.4 -15.6 -16.1	-19.06 18.6	9.82		549-12 -3-10-044 2
U 2855	343.3 136.97 12.17	6959 12.25 1.70	5X 0.14 12 -19	4.3 5.3 2	0.52 64.	10.9	3 0.27 1.74	1210. 1410. 1	440	10	1.39 9 1.23	20.3 31.53	19.8 4.3 0.6	-20.63 31.4	10.44 10.00	0.36	2855 12-04-004 2
0344-35	344.4 235.83 262.81	-3505 -52.04 -43.74	9 1.35 51 -1	4.6 3.4 6	0.25 84.	9.68	 0.67 0.00	1930. 1777. 2	298	20	0.83 20 1.02	16.9 31.14	-1.6 -12.6 -12.1	 16.8	9.29		358-63 -6-09-030 2
N 1461	346.2 207.64 288.83	-1633 -47.74 -45.28	-2A 0.30 51 -0 +4	3.2 2.3 4	0.20 90.	12.66	2 0.00 0.09	1450. 1356.		90		17.1 31.17	3.9 -11.4 -12.2	-18.51 11.5	9.60		-3-10-047 2
0346-80	346.6 294.78 214.38	-8016 -34.11 -22.17	7B 0.07 53 -0	2.1 1.8 6	0.44 69.		 0.37 0.30	1632. 1406. 7		20		19.1 31.40	-14.6 -10.0 -7.2	 10.0			015-01

Name	α (1950) / δ	l / b	SGL / SGB	Type	log D₂₅ / code	Dim / axis / PA	m_B / corr	V_hel / V_0 / src	V group	N	block	D (Mpc) / μ	components	M	log L	ratio	Other names
N 1473	347.2 -6822	282.84 -41.38	224.66 -29.68	7B 0.25 / 53 -5 +1	1.5 1.4 6	0.56 61.	0.23 / 0.21	1338. / 1118. / 7		20		15.0 / 30.89	-9.3 / -9.2 / -7.4	6.1			054-19
0347-27	347.4 -2708	223.28 -50.57	273.72 -45.61	5 0.61 / 51 -4	3.9 2.7 6	0.14 90.	11.45 / 0.67 0.00	1535. / 1404. / 2	235 198. 201.	20	0.51 30 1.01	17.8 / 31.25	0.8 / -12.4 / -12.7	14.0	9.01 / 10.20	0.064	UGCA 85 / 482-46 -5-10-004
0347-49	347.6 -4901	257.66 -49.61	244.84 -39.77	7 0.43 / 53 -7	3.9 2.7 6	0.20 90.	0.67 / 0.00	977. / 787. / 7		20		10.3 / 30.07	-3.4 / -7.2 / -6.6	8.1			IC 2000 / 201-03
0349-38	349.9 -3836	241.45 -50.88	257.71 -43.91	10 0.24 / 53 -7	1.8 1.8 6	0.91 28.	0.03 / 0.00	881. / 715. / 7		20		9.3 / 29.83	-1.4 / -6.5 / -6.4	4.9			302-14 -6-09-000
0350-71	350.1 -7147	286.33 -39.23	221.40 -27.78	7 0.13 / 53 -0	4.3 4.0 6	0.58 59.	10.73 / 0.21 0.22	1429. / 1206. / 7	209 201. 204.	20		16.3 / 31.06	-10.8 / -9.5 / -7.6	19.0	10.35		054-21
N 1483	351.2 -4738	255.37 -49.37	246.16 -40.90	4 0.53 / 53 -7	1.3 1.2 6	0.86 35.	0.05 / 0.00	1143. / 954.		14		12.5 / 30.48	-3.8 / -8.6 / -8.2	4.4			201-07 2
0352-49	352.1 -4908	257.54 -48.86	244.29 -40.38	10 0.12 / 53 -0	1.9 1.8 6	0.70 50.	0.13 / 0.00	1584. / 1391. / 7		20		18.0 / 31.27	-5.9 / -12.3 / -11.6	9.5			IC 2009
0352-36	352.2 -3608	237.54 -50.46	260.85 -45.03	-5 0.82 / 51 -1	1.8 1.7 6	0.84 36.	12.27 6 / 0.00 0.00	1384. / 1223.		57		15.7 / 30.98	-1.8 / -10.9 / -11.1	-18.71 / 7.8	9.68		201-08
N 1482	352.5 -2039	214.13 -47.79	283.04 -47.01	1 P 0.31 / 51 -4	2.1 1.9 6	0.60 58.	0.19 / 0.08	1655. / 1542.		104		19.6 / 31.46	3.0 / -13.0 / -14.3	10.9			IC 2006 2
N 1487	354.1 -4231	247.46 -49.76	252.23 -43.38	9 P 0.24 / 53 -7	2.6 2.3 9	0.62 56.	12.12 2 / 0.18 0.00	856. / 677.		10		8.9 / 29.74	-2.0 / -6.1 / -6.1	-17.62 / 6.0	9.24		359-07 -6-09-037 2
0355+66	355.0 6659	139.77 10.64	10.87 -1.21	9 0.16 / 14 -11	1.0 1.5 2	0.89 31.	0.00 / 2.42	72. / 262. / 1	145 227. 229.	10	1.90 6 1.03	4.4 / 28.19	4.3 / 0.8 / -0.1	1.9	9.19 / 9.46	0.534	549-33 -3-10-054 2
N 1493	355.9 -4621	253.19 -48.85	247.26 -42.17	6B 0.59 / 53 -7	3.9 3.9 6	1.00	11.8 5 / 0.00 0.00	1059. / 870. / 2	119	10	0.96 13 1.02	11.3 / 30.27	-3.3 / -7.8 / -7.6	-18.47 / 12.9	9.58 / 9.07	0.31	VV 78 / 249-31 -7-09-002 2
N 1494	356.2 -4903	257.17 -48.23	243.99 -41.02	4A 0.53 / 53 -7	3.2 2.8 6	0.55 61.	11.9 11.01 0.89 5 / 0.24 0.00	1123. / 928. / 7	205 192. 195.	14		12.2 / 30.43	-4.0 / -8.3 / -8.0	-18.53 / 10.0	9.60 / 10.03		UGCA 86 / 11-05-000 2
0357-46	357.7 -4601	252.61 -48.61	247.49 -42.59	10 0.37 / 53 -7	2.1 2.1 6	0.96 19.	0.01 / 0.00	901. / 712. / 7		20		9.4 / 29.87	-2.7 / -6.4 / -6.4	5.8			249-33 2 / 201-12 2 2.65 / 249-36

(1) Name	(2) α / ℓ / SGL	(3) δ / b / SGB	(4) Type / ε_{xyz} / Group	(5) D_{25} / $D_{25}^{b,i}$ / source	(6) d/D / i	(7) $B_T^{b,i}$ / $H_{-0.5}$ / $B-H$	(8) source / A_B^{i-o} / A_B^b	(9) V_h / V_o / Tel	(10) W_{20} / W_R / W_D	(11) e_v / e_w / source	(12) F_c / source,e_F / f_h	(13) R / μ	(14) SGX / SGY / SGZ	(15) $M_B^{b,i}$ / Δ_{25}	(16) $\log L_B$ / $\log M_H$ / $\log M_T$	(17) M_H/L_B / M_H/M_T / M_T/L_B	(18) Alternate Names / UGC/ESO MCG / AGN? RCII
N 1511	359.3 / 281.37 / 224.40	-6746 / -40.74 / -30.94	5AP / 0.29 / 53 -5 +1	3.9 / 3.3 / 6	0.40 / 72.	11.5	5 / 0.43 / 0.21	1350. / 1127.		14		15.1 / 30.90	-9.3 / -9.1 / -7.8	-19.40 / 14.6	9.95		055-04
N 1507	401.9 / 193.05 / 309.94	-219 / -37.56 / -45.89	9B / 0.07 / 53 -0	3.4 / 2.9 / 2	0.35 / 76.	11.8 / 10.75 / 1.05	3 / 0.54 / 0.36	864. / 810. / 1	205 / 174. / 177.	10	0.99 / 36 9 / 1.08	10.6 / 30.13	4.7 / -5.7 / -7.6	-18.33 / 9.0	9.52 / 9.04 / 9.90	0.33 / 0.139 / 2.36	2947 00-11-009 / 2
N 1510	401.9 / 248.77 / 250.15	-4333 / -48.22 / -44.30	-5 P / 0.44 / 53 -7	8.3 / 6.5 / 6	0.33 / 77.	13.48	2 / 0.00 / 0.00	968. / 782.		70		10.3 / 30.06	-2.5 / -6.9 / -7.2	-16.58 / 19.5	8.82		250-03 -7-09-006 / 2
N 1512	402.3 / 248.66 / 250.19	-4329 / -48.16 / -44.39	13 / 0.37 / 53 -7	13.4 / 10.5 / 6	0.33 / 77.	10.88	2 / 0.58 / 0.00	911. / 725. / 2	271 / 239. / 242.	10	1.57 / 6 / 1.04	9.5 / 29.90	-2.3 / -6.4 / -6.7	-19.02 / 29.1	9.80 / 9.53 / 10.68	0.53 / 0.069 / 7.65	250-04 -7-09-007 / 2
U 2953	402.5 / 138.46 / 13.18	6941 / 13.11 / 0.35	2AP / 0.08 / 12+15	5.8 / 6.9 / 9	0.72 / 48.	10.14	2 / 0.11 / 1.18	898. / 1092. / 1	489 / 607. / 608.	10	1.37 / 56 6 / 1.36	16.2 / 31.05	15.8 / 3.7 / 0.1	-20.91 / 32.6	10.56 / 9.79 / 11.54	0.17 / 0.018 / 9.69	IC 356, ARP 213 / 2953 12-04-011 / 2
N 1515	402.9 / 264.08 / 237.43	-5414 / -45.84 / -39.32	1X / 0.88 / 53 -1	5.7 / 4.0 / 9	0.21 / 89.	11.17	2 / 0.67 / 0.00	1169. / 961. / 7		20		13.4 / 30.64	-5.6 / -8.7 / -8.5	-19.47 / 15.7	9.98		156-36 / 2
0403-17	403.7 / 211.72 / 287.30	-1755 / -44.38 / -49.55	1C / 0.18 / 51 -9 +4	2.6 / 2.0 / 6	0.30 / 80.		0.67 / 0.00	1890. / 1777. / 1	146 / 118. / 123.	15	0.40 / 24 / 1.03	22.7 / 31.78	4.4 / -14.1 / -17.3	13.3	9.11 / 9.73	0.242	UGCA 87
N 1518	404.7 / 216.32 / 282.07	-2118 / -45.29 / -49.86	8B / 0.12 / 53 -0	3.3 / 2.8 / 6	0.48 / 67.	11.96	2 / 0.32 / 0.03	935. / 810.		10	1.24 / 11 20	10.5 / 30.11	1.4 / -6.6 / -8.0	-18.15 / 8.6	9.45 / 9.28	0.68	550-05 -3-11-011
0404-17	404.9 / 211.13 / 288.20	-1721 / -43.90 / -49.80	9A / 0.21 / 51 -9 +4	2.1 / 2.1 / 2	0.96 / 19.		0.01 / 0.01	1866. / 1754. / 1	98	10	0.33 / 26 / 1.05	22.4 / 31.75	4.5 / -13.8 / -17.1	13.7	9.03		550-07 -4-10-013 / 2
0405-55	405.9 / 265.55 / 235.82	-5527 / -45.06 / -39.01	10 / 0.97 / 53 -1	1.7 / 1.5 / 6	0.60 / 58.		0.19 / 0.00	1062. / 851. / 7		20		13.4 / 30.64	-5.8 / -8.4 / -8.3	5.9			UGCA 88 / -3-11-012
N 1527	406.9 / 255.13 / 244.08	-4801 / -46.71 / -43.07	-2 / 0.54 / 53 -7	3.5 / 2.9 / 6	0.42 / 71.	12.1	5 / 0.00 / 0.00	1031. / 832.		54		11.0 / 30.22	-3.5 / -7.3 / -7.5	-18.12 / 9.3	9.44		IC 2032 / 156-42
0407-45	407.6 / 251.66 / 246.91	-4538 / -46.96 / -44.31	-5 / 0.16 / 53 -0	1.2 / 1.1 / 6	0.73 / 47.	12.18	6 / 0.00 / 0.00	1458. / 1264.		90		16.5 / 31.08	-4.6 / -10.8 / -11.5	-18.90 / 5.3	9.75		201-20 / IC 2035 250-07 / 2
N 1533	408.8 / 266.45 / 234.67	-5615 / -44.43 / -38.88	-2B / 0.89 / 53 -1	2.4 / 2.4 / 6	0.86 / 35.	11.88	2 / 0.00 / 0.00	790. / 576.		12		13.4 / 30.64	-5.8 / -8.1 / -8.1	-18.76 / 9.4	9.70		157-03 / 2

Name	Pos (arc/RA/Dec)	Type	Size	B_T		V	V₂	N		Dist			M		Other names	
N 1536	410.0 / 266.86 −44.16 / 234.17 −38.80 / −5637	5B / 0.79 / 53 −1	2.0 / 2.0 / 9 · 0.90 / 29.	13.06	2 / 0.04 / 0.00	1592. / 1377. / 7		20		13.4 / 30.64	−6.3 / −8.8 / −8.7	−17.58 / 7.8	9.22		157-05	2
N 1531	410.1 / 233.14 −46.59 / 263.91 −49.41 / −3259	−5 P / 0.19 / 53 −13	1.5 / 1.3 / 6 · 0.67 / 53.	12.80	2 / 0.00 / 0.00	1163. / 998.		80		13.0 / 30.57	−0.9 / −8.4 / −9.9	−17.77 / 4.9	9.30		359-26 −5-11-001	2
N 1532	410.2 / 233.17 −46.57 / 263.87 −49.42 / −3300	2B / 0.21 / 53 −13	16.4 / 12.3 / 6 · 0.27 / 82.	10.65	6 / 0.67 / 0.00	1212. / 1047. / 2	538 / 505. / 506.	10	1.69 / 36 12 / 1.08	13.6 / 30.67	−0.9 / −8.8 / −10.3	−20.02 / 48.8	10.20 / 9.96 / 11.56	0.57 / 0.025 / 22.79	359-27 −5-11-002	2
N 1537	411.7 / 231.45 −46.10 / 265.59 −50.05 / −3146	−3 / 0.32 / 53 −13	3.4 / 3.1 / 6 · 0.67 / 53.	11.62	6 / 0.00 / 0.00	1344. / 1181.		68		15.2 / 30.91	−0.8 / −9.8 / −11.7	−19.29 / 13.8	9.91		420-12 −5-11-005	2
N 1543	411.8 / 268.40 −43.53 / 232.71 −38.24 / −5752	−2B / 0.95 / 53 −1	3.0 / 2.6 / 9 · 0.50 / 65.	11.57	2 / 0.00 / 0.00	1400. / 1183.		52		13.4 / 30.64	−6.4 / −8.5 / −8.4	−19.07 / 10.2	9.82		118-10	2
N 1546	413.5 / 266.09 −43.82 / 234.24 −39.45 / −5611	−2 / 0.97 / 53 −1	3.0 / 2.6 / 6 · 0.54 / 62.	12.50	2 / 0.00 / 0.00	1190. / 974.		90		13.4 / 30.64	−6.0 / −8.4 / −8.5	−18.14 / 10.2	9.45		157-12	2
N 1549	414.7 / 265.40 −43.79 / 234.60 −39.86 / −5543	−5 / 0.97 / 53 −1	2.8 / 2.6 / 6 · 0.75 / 45.	10.87	2 / 0.00 / 0.00	1153. / 937.		16		13.4 / 30.64	−5.9 / −8.3 / −8.6	−19.77 / 10.2	10.10		157-16	2
N 1553	415.1 / 265.62 −43.69 / 234.36 −39.80 / −5554	−2A / 0.97 / 53 −1	4.5 / 3.9 / 6 · 0.53 / 63.	10.47	2 / 0.00 / 0.00	1280. / 1064.		18		13.4 / 30.64	−6.0 / −8.4 / −8.6	−20.17 / 15.3	10.26		157-17	2
0415-60	415.7 / 271.36 −42.25 / 229.88 −37.08 / −6020	−2XP / 0.58 / 53 −2 +1	1.9 / 1.8 / 6 · 0.85 / 36.	12.4	5 / 0.00 / 0.00	1129. / 907. / 7		20		12.3 / 30.44	−6.3 / −7.5 / −7.4	−18.04 / 6.5	9.41		IC 2056	2
0416-56	416.8 / 265.73 −43.42 / 234.02 −39.89 / −5603	7 / 0.97 / 53 −1	2.6 / 1.9 / 6 · 0.15 / 90.	11.80	0.67 / 0.00	1369. / 1152. / 7	213 / 177. / 180.	21 / 21 / 301		13.4 / 30.64	−6.1 / −8.4 / −8.7	7.4	9.83	1.99	118-16 · IC 2058	2
N 1559	417.0 / 274.50 −41.20 / 227.36 −35.51 / −6254	6B / 0.56 / 53 −2 +1	3.3 / 3.0 / 9 · 0.61 / 57.	10.65 / 9.01 / 1.64	3 / 0.19 / 0.01	1293. / 1068. / 7	299 / 310. / 312.	13	0.96 / 16 / 1.20	14.3 / 30.78	−7.9 / −8.6 / −8.3	−20.13 / 12.5	10.24 / 10.54	0.23 / 0.074 / 3.16	157-18 · 084-10	2
N 1530	417.0 / 135.20 17.78 / 17.50 3.95 / 7512	3B / 0.20 / 42 −18	4.8 / 4.8 / 2 · 0.57 / 60.	11.5	2 / 0.22 / 0.61	2460. / 2661. / 1	326. / 334. / 335.	10		36.6 / 32.82	34.8 / 11.0 / 2.5	−21.32 / 51.3	10.72 / 10.09 / 11.22	3.59	VII ZW 12 · 3013 13-04-004	2
N 1566	418.9 / 264.30 −43.39 / 234.82 −40.74 / −5503	4X / 0.92 / 53 −1	8.4 / 8.0 / 9 · 0.81 / 40.	10.19	2 / 0.07 / 0.00	1479. / 1262.	233 / 305. / 307.	9 / 10 / 301		13.4 / 30.64	−6.0 / −8.5 / −9.0	−20.45 / 31.3	10.37 / 10.93		157-20	S1.5 / 2
0419-21	419.0 / 218.59 −42.32 / 280.85 −53.16 / −2157	7A / 0.13 / 53 −0	6.1 / 4.9 / 6 · 0.38 / 74.	10.61	0.48 / 0.03	906. / 768. / 2	191 / 162. / 166.	10	1.11 / 9 / 1.04	10.1 / 30.02	1.1 / −5.9 / −8.1	14.5	9.12 / 10.04	0.119	UGCA 90 · 550-24 −4-11-010	2

(1) Name	(2) α / ℓ / SGL	(3) δ / b / SGB	(4) Type / ϱ_{xyz} / Group	(5) D_{25} / $D_{25}^{b,i}$ / source	(6) d/D / i	(7) $B_T^{b,i}$ / $H_{-0.5}$ / B-H	(8) source / A_B^{i-o} / A_B^b	(9) V_h / V_o / Tel	(10) W_{20} / W_R / W_D	(11) e_v / e_w / source	(12) F_c source/e_f / f_h	(13) R / μ	(14) SGX / SGY / SGZ	(15) $M_B^{b,i}$ / $\Delta_{25}^{b,i}$	(16) logL_B / logM_H / logM_T	(17) M_H/L_B / M_H/M_T / M_T/L_B	(18) UGC/ESO MCG / Alt. Names / AGN? RCII
N 1574	421.0 / 266.88 / 232.50	-5705 / -42.58 / -39.70	-2A / 0.98 / 53 -1	3.5 / 3.5 / 6	0.89 / 31.	11.30	2 / 0.00 / 0.00	890. / 669.		40		13.4 / 30.64	-6.1 / -7.9 / -8.3	-19.34 / 13.7	9.93		
0421-63	421.7 / 275.27 / 226.17	-6344 / -40.40 / -35.33	9 / 0.56 / 53 -2 +1	1.8 / 1.4 / 6	0.27 / 82.		0.67 / 0.03	1300. / 1072. / 7		20		14.5 / 30.80	-8.2 / -8.5 / -8.4	5.9			157-22 / / 2
N 1569	426.0 / 143.67 / 11.94	6445 / 11.24 / -4.93	10B / 0.49 / 14-11	3.1 / 4.1 / 9	0.51 / 64.	9.69	1 / 0.28 / 1.99	-87. / 87.		8	1.53 / 60 20	1.6 / 26.06	1.6 / 0.3 / -0.1	-16.37 / 1.9	8.74 / 7.94	0.16	VII ZW 16 / 3056 11-06-001 / 2
N 1596	426.5 / 264.06 / 233.82	-5508 / -42.31 / -41.54	-2A / 0.92 / 53 -1	3.5 / 2.6 / 6	0.23 / 87.	11.90	2 / 0.00 / 0.00	1473. / 1252.		64		13.4 / 30.64	-6.1 / -8.4 / -9.1	-18.74 / 10.2	9.69		157-31 / / 2
N 1560	427.1 / 138.35 / 16.08	7148 / 16.04 / 0.78	7A / 0.35 / 14-11	8.8 / 7.1 / 2	0.22 / 88.	10.99 / 9.22 / 1.77	1 / 0.67 / 0.58	-28. / 164. / 1	153. / 122. / 127.	10	1.77 / 51 9 / 1.42	3.0 / 27.37	2.9 / 0.8 / 0.0	-16.38 / 6.2	8.74 / 8.72 / 9.43	0.96 / 0.197 / 4.84	IC 2062 / 3060 12-05-005 / 2
0427+63	427.4 / 144.71 / 11.35	6330 / 10.51 / -6.04	10 / 0.49 / 14-11	2.0 / 2.7 / 2	0.53 / 63.		0.00 / 2.01	-104. / 66. / 1		20	1.10	1.5 / 25.87	1.5 / 0.3 / -0.2	1.2			UGCA 92 / 11-06-000
N 1617	430.6 / 263.33 / 233.75	-5442 / -41.83 / -42.27	1 / 0.93 / 53 -1	4.0 / 3.4 / 9	0.50 / 65.	10.92	2 / 0.29 / 0.00	1040. / 818.		90		13.4 / 30.64	-5.8 / -7.8 / -8.9	-19.72 / 13.3	10.08		157-41 / / 2
0430-49	430.9 / 256.76 / 239.01	-4947 / -42.55 / -45.40	5 / 0.08 / 53 -0	2.5 / 1.8 / 6	0.16 / 90.		0.67 / 0.00	1856. / 1641. / 7		20		21.5 / 31.66	-7.8 / -12.9 / -15.3	11.3			202-35
N 1625	434.6 / 199.36 / 312.98	-323 / -31.17 / -53.89	3B / 0.07 / 34 -0	2.3 / 1.6 / 4	0.20 / 90.	12.40	2 / 0.67 / 0.11	3038. / 2953.		48		38.3 / 32.92	15.4 / -16.5 / -30.9	-20.52 / 17.9	10.40		-1-12-038 / / 2
0435-52	435.7 / 259.95 / 235.61	-5216 / -41.49 / -44.43	9B / C.10 / 53 -0	1.5 / 1.4 / 6	0.72 / 48.		0.11 / 0.00	1642. / 1421. / 7		20		18.8 / 31.37	-7.6 / -11.1 / -13.2	7.7			202-41
U 3130	437.8 / 135.82 / 18.78	7533 / 19.00 / 3.49	3X / C.18 / 42-18	2.9 / 2.9 / 2	0.61 / 57.	12.8 / 10.03 / 2.77	3 / 0.19 / 0.55	2485. / 2683. / 1	313. / 327. / 329.	10	0.79 / 14 / 1.08	37.0 / 32.84	35.0 / 11.9 / 2.3	-20.04 / 31.3	10.21 / 9.93 / 10.99	0.52 / 0.087 / 6.02	IC 381 / 3130 13-04-007 / 2
N 1637	438.9 / 199.53 / 314.40	-256 / -30.02 / -54.70	5X / 0.07 / 53-20	5.1 / 5.0 / 1	0.82 / 39.	11.41	1 / 0.07 / 0.11	726. / 639. / 1	205. / 270. / 272.	8	1.21 / 217 6 / 1.26	8.9 / 29.74	3.6 / -3.7 / -7.2	-18.33 / 13.0	9.52 / 9.11 / 10.44	0.38 / 0.047 / 8.22	UGCA 93 / 00-12-068 / 2
U 3137	439.4 / 135.22 / 19.28	7620 / 19.56 / 4.10	3 / 0.10 / 12+11	3.6 / 3.0 / 2	0.25 / 85.	13.0 / 11.21 / 1.79	3 / 0.67 / 0.50	994. / 1193. / 1	246. / 209. / 212.	10	1.02 / 10 / 1.09	17.9 / 31.27	16.9 / 5.9 / 1.3	-18.27 / 15.7	9.50 / 9.53 / 10.30	1.06 / 0.169 / 6.28	3137 13-04-008

N 1640	440.1 218.88 282.86	-2032 -37.21 -58.15	3B 0.07 51 -0	3.2 3.2 6	1.00	12.43	2 0.00 0.02	1602. 1452. 2	157	15	0.36 32 1.01	19.1 31.40	2.2 -9.8 -16.2	-18.97 17.8	9.78 8.92	0.14	 551-27 -3-12-018 2
0440-08	440.5 205.11 305.87	-810 -32.18 -56.85	9 0.08 34 -0	2.0 1.9 2	0.71 49.		 0.12 0.23	2522. 2414. 1	209 229. 231.	20	0.57 18 1.03	31.4 32.49	10.1 -13.9 -26.3	 17.4	 9.56 10.42	 0.138	UGCA 94 -1-12-047
U 3144	441.3 136.56 18.59	7450 18.74 2.76	10B 0.13 12+20+19	2.2 2.1 2	0.56 60.	13.8	3 0.22 0.53	1635. 1830. 1	158 145. 149.	10	0.78 14 1.04	26.1 32.08	24.7 8.3 1.3	-18.28 16.0	9.50 9.61 9.99	1.29 0.421 3.06	DDO 33 3144 12-05-018 2
N 1672	444.9 268.78 227.53	-5920 -38.99 -40.41	3B 0.64 53 -3 +1	5.9 5.7 9	0.84 37.	10.97	2 0.06 0.00	1310. 1077.		28		14.5 30.81	-7.5 -8.2 -9.4	-19.84 24.1	10.13		 118-43 2
0445-57	445.4 266.36 229.17	-5726 -39.29 -41.86	10 0.83 53 +3 +1	1.4 1.4 6	1.00		 0.00 0.00	1210. 978. 7		20		13.3 30.62	-6.5 -7.5 -8.9	 5.4			 158-03
U 3174	446.0 197.55 320.56	9 -26.92 -54.97	10 0.05 53-20	2.0 2.0 2	0.82 39.	13.95	7 0.07 0.28	669. 588. 1	131 166. 170.	20	0.64 16 1.04	8.2 29.58	3.6 -3.0 -6.7	-15.63 4.8	8.44 8.47 9.58	1.06 0.077 13.77	DDO 34 3174 00-13-014 2
0447-29	447.2 230.07 265.98	-2917 -38.13 -58.08	9 0.19 51-10	3.5 3.4 6	0.89 31.	13.02	7 0.04 0.00	1471. 1290. 2	155 242. 244.	10	0.63 18 1.02	17.0 31.15	-0.6 -9.0 -14.4	-18.13 16.9	9.44 9.09 10.46	0.44 0.043 10.31	UGCA 95, DDO 228 421-19 -5-12-003 2
N 1688	447.7 269.38 226.71	-5953 -38.53 -40.24	5B 0.76 53 -3 +1	2.4 2.3 6	0.77 44.	12.3	5 0.09 0.00	1230. 995.		13		13.5 30.65	-7.1 -7.5 -8.7	-18.35 9.1	9.53		 119-06 2
N 1679	448.1 233.59 260.98	-3204 -38.54 -57.28	10B 0.20 53 -0+10	3.2 3.0 6	0.75 45.		 0.10 0.00	1059. 870. 2	158 178. 181.	10	1.00 10 1.02	11.8 30.35	-1.0 -6.3 -9.9	 10.3	 9.14 9.98	 0.147	UGCA 96 422-01 -5-12-004
0449-02	449.1 200.70 316.71	-239 -27.67 -56.91	12 P 0.09 34 -0	0.8 0.7 4	0.40 72.		 0.43 0.15	2194. 2100.		220		27.5 32.20	10.9 -10.3 -23.1	 5.6			ARAK 113 00-13-018
U 3190	449.7 134.04 20.69	7807 21.06 5.34	5A 0.14 12 -0+19	1.9 2.0 2	0.95 21.	12.59	2 0.02 0.39	1565. 1766.	171	10 15 217	0.52 217 15	25.5 32.03	23.7 9.0 2.4	-19.44 14.9	9.97 9.33	0.23	IC 391 3190 13-04-011 2
0450-28	450.0 229.49 266.73	-2840 -37.39 -58.85	10 0.20 51-10	1.7 1.6 6	0.75 45.		 0.10 0.00	1460. 1279. 6	145	40	0.48 38 1.04	16.9 31.14	-0.5 -8.7 -14.4	 7.9	 8.94		UGCA 97 422-05 -5-12-005
0450-25	450.9 225.52 272.97	-2519 -36.32 -59.94	9X 0.14 51 -0	5.4 5.0 6	0.71 49.	13.12	7 0.12 0.00	1374. 1202. 2	98 103. 108.	7	0.95 8 1.03	16.0 31.02	0.4 -8.0 -13.8	-17.90 23.4	9.35 9.36 9.86	1.01 0.317 3.20	UGCA 98, DDO 229 485-21 -4-12-019 2
0450-61	450.9 271.58 224.78	-6144 -37.77 -39.07	7 0.15 53 -0 +1	2.6 2.2 6	0.44 70.		 0.37 0.01	980. 743. 7		20		10.4 30.08	-5.7 -5.7 -6.5	 6.7			 119-16

(1) Name	(2) α ℓ SGL	(3) δ b SGB	(4) Type $Q_{v.z}$ Group	(5) D_{25} $D_{25}^{b,i}$ source	(6) d/D i	(7) $B_T^{b,i}$ $H_{-0.5}$ B-H	(8) source A_B^{i-o} A_B^b	(9) V_h V_o Tel	(10) W_{20} W_R W_D	(11) e_v e_w source	(12) F_c source/e_F f_h	(13) R μ	(14) SGX SGY SGZ	(15) $M_B^{b,i}$ $\Delta_{25}^{b,i}$	(16) log L_B log M_H log M_T	(17) M_H/L_B M_H/M_T M_T/L_B	(18) Alternate Names Alternate Names UGC/ESO MCG AGN? RCII
N 1690	451.7 197.04 323.93	133 -24.97 -55.51	-5 0.09 34 -0	1.2 1.3 2	1.00	13.9	3 0.00 0.30	2180. 2100.		220		27.7 32.21	12.7 -9.2 -22.8	-18.31 10.5	9.52		ARAK 115
N 1703	452.1 269.16 226.21	-5949 -38.00 -40.64	3B 0.-6 53 -3 +1	3.5 3.5 6	1.00		 0.00 0.00	1526. 1290.		10		17.4 31.20	-9.1 -9.5 -11.3	17.8			3198 00-13-027
U 3203	452.8 142.66 16.06	6814 15.52 -3.40	12 P 0.08 12 +15	2.4 2.6 2	0.73 47.	12.0	3 0.11 0.64	762. 937.		140		14.4 30.79	13.8 4.0 -0.9	-18.79 10.9	9.71		119-19 2
N 1705	453.1 261.09 231.84	-5327 -38.74 -45.51	-2A? 0.10 53 +1	1.7 1.6 6	0.71 49.	12.64	2 0.00 0.16	640. 409. 7		20		6.0 28.87	-2.6 -3.3 -4.2	-16.23 2.8	8.68		IC 396 3203 11-07-002 2
0454-62	454.2 272.89 223.48	-6253 -37.15 -38.40	10 0.37 53 -3 +1	2.7 2.2 6	0.41 71.		 0.41 0.00	1409. 1170. 7		20		15.9 31.01	-9.0 -8.6 -9.9	10.2			158-13 2
0454-56	454.7 264.70 228.90	-5619 -38.20 -43.54	10 0.09 53 -0	1.4 1.3 6	0.62 56.		 0.18 0.00	1775. 1540. 7		20		20.4 31.55	-9.7 -11.2 -14.1	7.7			085-14
N 1744	457.9 226.99 270.63	-2606 -35.03 -61.28	7B 0.12 53 -0	12.0 10.3 6	0.50 65.	11.41 9.92 1.49	7 0.29 0.00	742. 562. 2	217 199. 202.	7	1.56 36 5 1.05	7.8 29.47	0.0 -3.8 -6.9	-18.06 23.5	9.42 9.34 10.43	0.85 0.082 10.34	158-15
U 3234	500.5 185.16 344.62	1620 -15.01 -47.72	10 0.06 30 -0	2.0 2.7 2	1.00		 0.00 1.50	1396. 1370. 1	145	10	0.79 10 1.07	18.7 31.36	12.1 -3.3 -13.8	14.7	9.33		486-05 -4-12-029 2
N 1796	502.1 270.61 223.82	-6112 -36.56 -40.30	53P 0.19 53 -0 +1	1.7 1.5 6	0.50 65.	12.6	5 0.29 0.00	1000. 759. 1		67		10.6 30.12	-5.8 -5.6 -6.8	-17.52 4.6	9.20		DDO 35 3234 03-13-000 2
0502+01	502.2 198.33 326.77	145 -22.62 -57.63	13 P 0.08 34 -0			14.4	3 0.29 0.33	3087. 3000.		220		39.3 32.97	17.6 -11.5 -33.2	-18.57	9.62		119-30 2 ARAK 118
N 1809	502.4 280.75 217.89	-6938 -34.75 -33.33	12 0.23 53 -0 +1	3.2 2.8 6	0.38 74.		 0.48 0.38	1302. 1060. 7		20		14.7 30.83	-9.7 -7.5 -8.1	12.0			00-13-069 056-48
N 1784	503.1 211.90 301.13	-1157 -28.85 -63.15	5B 0.13 34 +5 +4	3.6 3.6 1	0.64 55.	11.92 9.18 2.74	1 0.16 0.37	2321. 2182. 1	345 376. 378.	10	1.19 217 7 1.12	28.7 32.29	6.7 -11.1 -25.6	-20.37 30.2	10.34 10.11 11.09	0.58 0.103 5.65	-2-13-042 2
N 1792	503.6 241.70 248.46	-3803 -36.44 -57.13	4A 0.20 53 -10	6.1 5.3 6	0.50 65.	10.53 8.24 2.29	2 0.29 0.03	1216. 1003. 2	316 306. 308.	25	1.00 11 14 1.08	13.6 30.66	-2.7 -6.9 -11.4	-20.13 21.0	10.24 9.27 10.76	0.11 0.032 3.26	305-06 -6-12-004 2

Name	Positions	Type	Dimensions	mag	—	Velocities	—	—	—	—	—	mag	—	Other designations
N 1800	504.5 234.43 258.22 / -3201 -35.13 -60.45	-2XP 0.13 53-12+10	2.2 1.9 6 / 0.50 65.	13.10	2 0.00 / 0.00	722. / 522.	90		7.4 / 29.36	-0.7 -3.6 -6.5	-16.26 4.1	8.70		422-30 -5-13-004 2
0505-16	505.2 216.76 291.40 / -1621 -30.15 -64.16	9B 0.18 34 -3	2.5 2.2 4 / 0.54 62.		0.24 / 0.15	2044. / 1889. 1	7	54 / 40. / 53.	0.61 13 1.05 / 24.9 31.98	4.0 -10.1 -22.4	16.0	9.40 / 8.87	3.403	DDO 36 / -3-14-001 2
N 1808	506.0 241.22 248.65 / -3735 -35.90 -57.79	0.30 53-10	7.6 7.1 6 / 0.70 50.	10.65	2 0.00 / 0.05	995. / 782.	10		1.19 6 20 / 10.8 30.17	-2.1 -5.4 -9.2	-19.52 22.4	10.00 / 9.26	0.18	305-08 -6-12-005 2
N 1824	506.2 268.78 224.37 / -5947 -36.26 -41.75	9B 0.60 53 -3 +1	3.2 2.3 6 / 0.25 84.		0.67 / 0.01	1254. / 1012. 7	20		13.9 / 30.72	-7.4 -7.3 -9.3	9.3			119-36
0506-38	506.5 242.19 247.37 / -3822 -35.92 -57.40	10 0.27 53-10	4.3 4.1 6 / 0.82 39.		0.07 / 0.05	1022. / 807. 2	7	134 / 169. / 173.	1.21 7 1.00 / 11.1 30.23	-2.3 -5.5 -9.4	13.3	9.30 / 10.04	0.182	305-09 -6-12-006
0506-09	506.8 209.65 307.30 / -919 -26.89 -63.41	10 0.15 34 -0	1.6 1.5 2 / 0.50 65.		0.29 / 0.28	2693. / 2561. 1	25	215	0.44 24 1.02 / 33.6 32.63	9.1 -12.0 -30.1	14.7	9.49		UGCA 101 / -2-14-001
0507-63	507.3 272.75 221.80 / -6303 -35.66 -39.13	10 0.28 53 -0 +1	2.1 2.0 6 / 0.80 41.		0.08 / 0.02	1469. / 1225. 7	20		16.7 / 31.11	-9.6 -8.6 -10.5	9.8			085-47
0508-02	508.3 203.40 320.71 / -245 -23.53 -61.25	13 P 0.08 34 -0	0.9 0.9 2 / 0.74 46.		2 0.00 / 0.42	2838. / 2729.	17		0.23 209 40 / 35.8 32.77	13.3 -10.9 -31.4	9.4	9.34		UGCA 102, II ZW 33 / 00-14-010 2
N 1827	508.4 240.65 248.97 / -3702 -35.33 -58.50	6 0.29 53-10	2.8 2.0 6 / 0.20 90.		0.67 / 0.03	1049. / 836. 2	13	200 / 164. / 168.	0.72 18 1.02 / 11.6 30.32	-2.2 -5.6 -9.9	6.8	8.85 / 9.72	0.134	362-06 -6-12-008
0508-31	508.8 234.27 257.98 / -3140 -34.16 -61.42	9 0.26 53-12+10	2.8 2.7 6 / 0.91 27.	12.95	7 0.03 / 0.00	980. / 778. 2	10	145	0.88 11 1.02 / 10.8 30.16	-1.1 -5.0 -9.5	-17.21 8.5	9.08 / 8.95	0.74	UGCA 103, DDO 230
0509-14	509.4 215.66 294.97 / -1453 -28.65 -65.08	6A 0.17 34 -3	2.8 2.9 2 / 0.81 40.	12.0	5 0.07 / 0.30	1976. / 1823. 1	10	233 / 307. / 309.	0.70 16 1.08 / 24.1 31.91	4.3 -9.2 -21.9	-19.91 20.4	10.16 / 9.46 / 10.75	0.20 / 0.052 / 3.89	422-41 -5-13-011 2 / UGCA 104, A 0509 / -2-14-004 2
0509+62	509.6 148.53 15.03 / 6231 13.66 -9.29	10 0.17 14-11	5.1 6.2 2 / 0.67 52.	11.7	3 0.15 / 1.36	112. / 264. 1	7	139 / 139. / 144.	1.71 5 1.31 / 4.5 28.26	4.3 1.1 -0.7	-16.56 8.1	8.82 / 9.02 / 9.66	1.59 / 0.228 / 6.97	UGCA 105 / 10-08-000
N 1832	509.8 216.62 292.93 / -1545 -28.90 -65.24	4B 0.18 34 -3	2.6 2.4 4 / 0.57 60.	11.68	2 0.22 / 0.20	1929. / 1772.	8 15 217	292 / 292. / 294.	0.76 217 25 / 23.5 31.85	3.8 -9.1 -21.3	-20.17 16.5	10.26 / 9.50 / 10.61	0.17 / 0.078 / 2.24	-3-14-010 2
0510-33	510.1 235.96 255.23 / -3302 -34.20 -60.99	10 0.29 53-10	3.9 3.8 6 / 0.90 29.	13.16	2 0.04 / 0.00	935. / 729. 2	7	98	1.28 6 1.02 / 10.2 30.03	-1.3 -4.8 -8.9	-16.87 11.3	8.94 / 9.30	2.28	UGCA 106, DDO 231 / 362-09 -5-13-000 2

(1) Name	(2) α / ℓ / SGL	(3) δ / b / SGB	(4) Type / Q_xyz / Group	(5) D_25 / D_25^{b,i} / source	(6) d/D / i	(7) B_T^{b,i} / H_{-0.5} / B-H	(8) source / A_B^{i-o} / A_B^b	(9) V_h / V_o / Tel	(10) W_20 / W_R / W_D	(11) e_v / e_w / source	(12) F_c / source,e_t / f_h	(13) R / μ	(14) SGX / SGY / SGZ	(15) M_B^{b,i} / Δ_25^{b,i}	(16) log L_B / log M_H / log M_T	(17) M_H/L_B / M_T/M_H / M_T/L_B	(18) UGC/ESO / MCG / AGN? RCII
N 1853	511.4 / 265.82 / 225.45	-5727 / -35.82 / -44.05	7 / 0.29 / 53 -0 +1	1.9 / 1.5 / 6	0.35 / 76.		/ 0.55 / 0.00	1388. / 1145. / 7		20		15.6 / 30.96	-7.9 / -8.0 / -10.8	/ 6.8			158-22
N 1843	511.7 / 211.61 / 305.00	-1041 / -26.40 / -64.93	7 / 0.17 / 34 +5 +4	2.0 / 2.2 / 2	0.95 / 21.		/ 0.02 / 0.47	2611. / 2470. / 1	215	15	0.84 / 13 / 1.05	32.5 / 32.56	7.9 / -11.3 / -29.5	/ 20.9	/ 9.86 /		UGCA 107 / -2-14-008
0513-30	513.3 / 233.32 / 259.03	-3036 / -32.96 / -62.77	6B / 0.13 / 51 -0	3.0 / 2.3 / 6	0.29 / 81.	10.93	/ 0.67 / 0.00	1481. / 1279. / 2	245 / 210. / 213.	15	0.73 / 15 / 1.01	17.2 / 31.18	-1.5 / -7.7 / -15.3	/ 11.6	/ 9.20 / 10.17	/ 0.108 /	UGCA 108 / 423-02 / -5-13-013
U 3273	513.7 / 156.43 / 11.61	5330 / 9.01 / -17.69	9 / 0.06 / 15 -0	3.2 / 3.7 / 2	0.35 / 76.	11.6	3 / 0.54 / 1.81	616. / 734. / 1	203 / 172. / 176.	8	1.15 / 6 / 1.23	11.4 / 30.29	10.7 / 2.2 / -3.5	-18.69 / 12.3	9.67 / 9.26 / 10.02	0.39 / 0.174 / 2.27	3273 / 09-09-001
0515-37	515.0 / 241.09 / 247.22	-3710 / -34.06 / -59.47	5 / 0.27 / 53 -11+10	5.0 / 3.6 / 6	0.15 / 90.		/ 0.67 / 0.07	1348. / 1130. / 2	294 / 256. / 258.	10	1.23 / 10 / 1.03	15.4 / 30.93	-3.0 / -7.2 / -13.2	/ 16.2	/ 9.61 / 10.49	/ 0.131 /	362-11 / -6-12-012
0516-21	516.9 / 223.56 / 278.34	-2136 / -29.49 / -66.55	10 / 0.12 / 34 +2	2.1 / 2.0 / 6	0.72 / 48.		/ 0.12 / 0.09	1841. / 1661. / 2	143 / 153. / 157.	15	0.45 / 25 / 1.01	22.2 / 31.73	1.3 / -8.7 / -20.4	/ 13.0	/ 9.14 / 9.94	/ 0.158 /	UGCA 109, DDO 37 / 553-33 / -4-13-000 / 2
N 1892	516.9 / 274.92 / 219.34	-6501 / -34.32 / -38.00	3 / 0.29 / 53 -4 +1	2.6 / 2.1 / 6	0.30 / 80.	10.72	/ 0.67 / 0.12	1362. / 1115. / 7	233 / 199. / 202.	20		15.4 / 30.93	-9.4 / -7.7 / -9.5	/ 9.4	/ / 10.04		085-61
N 1879	517.9 / 235.45 / 254.93	-3212 / -32.41 / -62.82	10X / 0.19 / 53 -0+10	2.4 / 2.2 / 6	0.67 / 53.	13.00	2 / 0.15 / 0.00	1247. / 1038. / 1	145 / 146. / 149.	15	0.54 / 20 / 1.01	14.2 / 30.77	-1.7 / -6.3 / -12.7	-17.77 / 9.1	9.30 / 8.84 / 9.75	0.35 / 0.124 / 2.83	UGCA 110, DDO 232 / 423-06 / -5-13-016 / 2
0519-37	519.5 / 241.12 / 246.33	-3701 / -33.15 / -60.26	9 / 0.29 / 53 -11+10	2.3 / 1.8 / 6	0.29 / 81.		2 / 0.67 / 0.08	1299. / 1079. / 2	148 / 119. / 124.	15	0.42 / 27 / 1.01	14.8 / 30.85	-2.9 / -6.7 / -12.8	/ 7.8	/ 8.76 / 9.50	/ 0.180 /	362-19 / -6-12-019
N 1888	520.2 / 213.47 / 304.12	-1133 / -24.88 / -67.16	5BP / 0.14 / 34 -4	3.3 / 2.9 / 4	0.33 / 77.	11.90	2 / 0.58 / 0.43	2310. / 2160. / 1	111	20	0.04 / 46 / 1.05	28.6 / 32.28	6.2 / -9.2 / -26.4	-20.38 / 24.2	10.34 / 8.95 /	0.04	ARP 123 / -2-14-013 / 2
N 1889	520.3 / 213.48 / 304.14	-1133 / -24.86 / -67.18	-5 / 0.16 / 34 -4	0.7 / 0.8 / 5	0.88 / 33.		/ 0.00 / 0.43	2472. / 2322.		10		30.7 / 32.44	6.7 / -9.9 / -28.3	/ 7.2			ARP 123 / -2-14-014 / 2
U 3303	522.3 / 198.56 / 336.86	428 / -16.92 / -59.93	9 / 0.06 / 53 -0	3.8 / 4.1 / 2	0.77 / 44.		/ 0.09 / 0.64	528. / 437. / 1	179 / 210. / 213.	8	0.92 / 226 7 / 1.16	6.6 / 29.11	3.1 / -1.3 / -5.7	/ 7.9	/ 8.56 / 10.00	/ 0.036 /	3303 / 01-14-000
0524-69	524.0 / 280.46 / 215.82	-6948 / -32.89 / -34.10	93 / 0.50 / 14-12	645.7 / 660.4 / 9	0.85 / 35.	0.30	2 / 0.05 / 0.28	260. / 12.		5	5.40 / 102 28	0.1 / 18.49	0.0 / 0.0 / 0.0	-18.19 / 19.3	9.47 / 9.40 /	0.86	LMC, A 0524 / 56-115 / 2

0525-16	525.9 218.72 292.36	-1608 -25.49 -69.12	6B 0.11 34 -0	1.8 1.9 4	1.00	12.83	6 0.00 0.22	2171. 2002.		10		26.7 32.13	3.6 -8.8 -24.9	-19.30 14.8	9.91		SCI 58 -3-14-017 2
N 1947	526.5 273.32 218.91	-6348 -33.43 -39.57	-2 P 0.24 53 -4 +1	3.4 3.4 9	0.91 28.	11.70	2 0.00 0.13	1116. 866.		67		12.1 30.42	-7.3 -5.9 -7.7	-18.72 12.0	9.68		085-87 2
U 3317	527.2 139.35 20.88	7341 20.76 0.37	10 0.09 12 -0+19	2.3 2.5 2	1.00		 0.00 0.47	1241. 1424. 1	136	15	0.61 17 1.07	21.1 31.62	19.7 7.5 0.1	 15.4	 9.26		DDO 38 3317 12-06-006 2
N 1954	530.5 217.14 298.47	-1405 -23.65 -70.07	5B 0.08 34 -0	4.1 3.8 4	0.44 69.		 0.37 0.40	3143. 2977. 1	485 478. 479.	25 13 217	0.95 217 10 1.13	39.3 32.97	6.4 -11.8 -36.9	 43.6	 10.14 11.46	 0.048	 -2-15-003 2
N 1964	531.2 225.28 275.37	-2159 -26.51 -69.70	3X 0.14 34 -2	6.1 5.1 6	0.44 70.	11.07 8.11 2.96	1 0.37 0.05	1671. 1480. 2	427 415. 416.	15	1.21 11 20 1.05	20.0 31.51	0.6 -6.9 -18.8	-20.44 29.8	10.37 9.81 11.17	0.28 0.044 6.37	 554-10 -4-14-003 2
N 1963	531.6 241.07 243.84	-3627 -30.66 -62.44	5 0.27 53 -0+10	3.0 2.1 6	0.21 89.		 0.67 0.00	1324. 1098. 2	250 213. 215.	20	0.84 17 1.02	15.1 30.90	-3.1 -6.3 -13.4	 9.3	 9.20 10.09	 0.129	IC 2135, IC 2136 363-07 -6-13-004
0535-52	535.3 259.50 225.31	-5213 -32.38 -50.30	10 0.24 53 +6 +1	1.7 1.6 6	0.80 41.		 0.08 0.10	1279. 1030. 7		20		14.3 30.77	-6.4 -6.5 -11.0	 6.7			 204-22
0539-58	539.2 267.09 220.17	-5837 -32.11 -44.89	5B 0.32 53 -0 +1	2.0 1.9 6	0.75 45.		 0.10 0.18	1342. 1088. 7		20		15.1 30.90	-8.2 -6.9 -10.7	 8.4			 120-12
0539-35	539.3 240.70 242.52	-3544 -28.98 -64.05	9 0.21 53 -0+10	3.2 3.0 6	0.75 45.		 0.10 0.00	1273. 1044. 2	140 156. 160.	15	0.48 36 1.03	14.6 30.82	-2.9 -5.7 -13.1	 12.8	 8.81 9.96	 0.071	 363-15 -6-13-007
0539-22	539.8 227.08 270.79	-2258 -25.00 -71.30	10 0.17 34 -2	2.3 1.7 6	0.25 84.		 0.67 0.05	1731. 1532. 2	172 140. 144.	15	0.73 16 1.01	20.9 31.60	0.1 -6.7 -19.8	 10.4	 9.37 9.77	 0.398	UGCA 112 487-35 -4-14-011
N 2082	541.6 273.80 216.69	-6419 -31.75 -39.80	4B 0.20 53 -4 +1	1.4 1.4 6	0.94 24.	12.8	5 0.02 0.14	1104. 849. 7		20		12.1 30.41	-7.4 -5.5 -7.7	-17.61 4.9	9.24		 086-21 2
0541-19	541.8 223.54 281.84	-1919 -23.23 -72.69	5B 0.09 34 -0	1.9 1.6 6	0.35 76.		 0.55 0.17	2749. 2559. 2	280	40	0.77 20 1.01	34.1 32.66	2.1 -9.9 -32.5	 15.9	 9.84		UGCA 113 554-29 -3-15-011
0542-52	542.0 260.18 223.53	-5243 -31.41 -50.35	10 0.19 53 -6 +1	1.7 1.6 6	0.70 50.		 0.13 0.15	1084. 832. 7		20		11.8 30.36	-5.5 -5.2 -9.1	 5.5			 159-25
N 2090	545.2 239.43 242.88	-3415 -27.43 -65.96	5A 0.14 53+14	6.8 6.0 6	0.56 61.	11.5 8.61 2.89	 0.23 0.00	936. 707. 2	297 296. 298.	10	1.42 7 7 1.02	10.2 30.04	-1.9 -3.7 -9.3	-18.54 17.9	9.61 9.44 10.66	0.67 0.060 11.20	 363-23 -6-13-009 2

(1) Name	(2) α / l / SGL	(3) δ / b / SGB	(4) Type / Q_{vz} / Group	(5) D_{25}^i / $D_{25}^{b,i}$ / source	(6) d/D / i	(7) $B_T^{b,i}$ / $H_{-0.5}$ / $B-H$	(8) source / A_B^{i-o} / A_B^b	(9) V_h / V_o / Tel	(10) W_{20} / W_R / W_D	(11) e_v / e_w / source	(12) F_c / source/e_f / f_h	(13) R / μ	(14) SGX / SGY / SGZ	(15) $M_B^{b,i}$ / $\Delta_{25}^{b,i}$	(16) $\log L_B$ / $\log M_H$ / $\log M_T$	(17) M_H/L_B / M_H/M_T / M_T/L_B	(18) Alternate Names UGC/ESO / Alternate Names MCG / AGN? RCII
N 2101	545.3 259.52 223.26	-5206 -30.85 -51.13	10 P 0.25 53 -6 +1	2.0 1.8 6	0.58 59.		0.21 0.19	1192. 939. 7		20		13.3 30.61	-6.1 -5.7 -10.3	7.0			205-01
N 2104	545.9 258.92 223.51	-5134 -30.70 -51.65	4 0.24 53 -6 +1	2.1 1.8 6	0.44 69.		0.37 0.16	1151. 898. 7		20		12.7 30.53	-5.7 -5.4 -10.0	6.7			205-02
0548-14	548.7 219.77 297.72	-1447 -19.92 -74.53	8BP 0.08 53 -0	2.9 3.2 2	0.88 32.		0.04 0.58	908. 728. 1	133 200. 203.	10	0.84 22 1.10	10.7 30.15	1.3 -2.5 -10.3	10.0	8.90 10.06	0.068	UGCA 114 -2-15-011
U 3371	549.9 138.44 22.82	7518 22.81 1.44	10 0.11 12-11	4.6 4.9 2	0.82 39.		0.07 0.46	820. 1003. 1	149 190. 193.	8	0.90 9 1.21	15.6 30.97	14.4 6.1 0.4	22.3	9.29 10.37	0.083	DDO 39 3371 13-05-07A 2
0550-17	550.8 222.99 285.95	-1752 -20.70 -75.00	6P 0.09 34 -0	2.8 3.0 6	1.00		0.00 0.30	3125. 2934. 1	325	8	0.87 14 1.10	39.0 32.95	2.8 -9.7 -37.6	34.2	10.05		IC 438, UGCA 115 555-09 -3-15-025
0553+03	553.1 203.42 347.73	324 -10.76 -65.93	13 P 0.07 53 -0	1.3 2.1 2	0.90 29.	12.9	3 0.00 2.28	805. 688. 1	149	15	0.60 21 11 1.04	10.3 30.07	4.1 -0.9 -9.4	-17.17 6.3	9.06 8.63	0.37	UGCA 116, II ZW 40 01-16-000 2
0558-28	558.6 234.85 247.50	-2859 -23.10 -71.71	9B 0.09 51 -0	2.4 2.1 6	0.53 63.	14.08	7 0.25 0.01	1392. 1166. 6	189 173. 177.	15	0.63 21 1.06	16.4 31.07	-2.0 -4.8 -15.6	-16.99 10.1	8.99 9.06 9.94	1.19 0.132 8.97	UGCA 117, DDO 233 425-02 -5-15-002 2
U 3390	558.8 175.55 13.17	3607 6.63 -36.72	8X 0.06 20 -0	2.3 3.3 2	0.60 58.		0.20 2.37	1511. 1534. 1	220 216. 219.	15	0.72 16 1.16	22.0 31.71	17.2 4.0 -13.1	21.2	9.40 10.46	0.089	3390 06-14-001
N 2139	559.0 229.52 262.17	-2340 -21.14 -75.01	6X 0.22 34 +1	2.9 2.7 6	0.67 53.	11.84	2 0.15 0.06	1845. 1632. 2	235	15	1.10 11 9 1.01	22.4 31.75	-0.8 -5.7 -21.6	-19.91 17.7	10.16 9.80	0.44	IC 2154 488-54 -4-15-005 2
0601-20	601.4 226.76 272.42	-2039 -19.48 -76.81	7B 0.14 34 +6	2.1 2.1 6	0.96 20.		0.01 0.24	1982. 1776. 2	175	15	0.80 14 1.02	24.3 31.93	0.2 -5.5 -23.7	14.9	9.57		UGCA 118 555-27 -3-16-010
0602-26	602.8 232.30 252.69	-2607 -21.23 -74.32	7X 0.23 34 +1	3.9 3.2 6	0.40 72.		0.43 0.00	1814. 1593. 2	260 233. 236.	15	0.78 18 1.01	21.9 31.70	-1.8 -5.7 -21.1	20.5	9.46 10.51	0.090	UGCA 119 488-60 -4-15-008
0603-33	603.9 239.47 236.90	-3305 -23.37 -69.33	E C.10 55 -14	2.9 2.2 6	0.28 82.		0.67 0.02	790. 553. 7		20		8.3 29.58	-1.6 -2.4 -7.7	5.3			364-29 -6-14-000
N 2179	605.9 228.25 266.11	-2144 -18.94 -77.34	 0.17 34 -8	1.4 1.4 6	0.69 51.	13.13	2 0.00 0.27	2761. 2549.		90		34.2 32.67	-0.5 -7.5 -33.4	-19.54 14.0	10.01		555-38 -4-15-011 2

Name	Coordinates	Type	Dimensions	b/a, i	B_T		V_h / V_0		n		D / DM		M_B / diam		index	Alt. names
0607-61	607.0 -6148 / 270.97 -28.82 / 214.14 -43.10	5 / 0.21 / 53 -0 +1	3.9 / 2.8 / 6	0.20 / 90.		0.67 / 0.14	1205. / 942. / 7		20		13.5 / 30.65	-8.1 / -5.5 / -9.2	11.0	9.34 / 8.81 / 9.49		121-06
N 2188	608.4 -3406 / 240.85 -22.80 / 233.13 -68.99	9B / 0.10 / 53 -14	4.6 / 3.4 / 6	0.23 / 86.	11.62	2 / 0.67 / 0.01	757. / 515. / 2	145 / 116. / 121.	8	1.01 / 36 9 / 1.02	7.9 / 29.49	-1.7 / -2.3 / -7.4	-17.87 / 7.8	8.95 / 9.66	0.29 / 0.209 / 1.40	364-37 -6-14-008 2
0609-21	609.2 -2135 / 228.42 -18.17 / 265.06 -78.08	10 / 0.10 / 53 -0	2.4 / 2.5 / 6	0.83 / 37.		0.06 / 0.38	854. / 641. / 2	115 / 150. / 154.	10	0.98 / 9 / 1.01	9.7 / 29.94	-0.2 / -2.0 / -9.5	7.1	10.46 / 9.64	0.194	UGCA 120 / 556-02 -4-15-000 2
N 2196	610.1 -2147 / 228.70 -18.06 / 263.72 -78.17	2A / 0.10 / 34 +8	2.9 / 3.0 / 6	0.83 / 37.	11.62	2 / 0.06 / 0.42	2333. / 2119. / 2	407	30	0.72 / 56 / 1.02	28.8 / 32.30	-0.6 / -5.9 / -28.2	-20.68 / 25.2	10.29 / 9.69	0.15	UGCA 121 / 556-04 -4-15-014 2
N 2146	610.7 7822 / 135.66 24.90 / 24.65 4.17	2BP / 0.14 / 12 -11	5.3 / 5.5 / 2	0.85 / 36.	10.94	1 / 0.06 / 0.31	918. / 1108. / 1	494	15	1.22 / 6 8 / 1.24	17.2 / 31.18	15.6 / 7.2 / 1.3	-20.24 / 27.6	8.90	0.25	3429 13-05-022 2
0613-26	613.3 -2634 / 233.66 -19.21 / 245.13 -75.71	10 / 0.21 / 34 -1	1.8 / 1.8 / 6	0.86 / 34.		0.05 / 0.14	1799. / 1571. / 2	104	30	0.22 / 28 / 1.01	21.8 / 31.69	-2.3 / -4.9 / -21.1	11.5	10.97		UGCA 122, DDO 234 / KAR 48 / 489-22 -4-15-000 2
N 2207	614.2 -2121 / 228.68 -17.01 / 263.20 -79.21	4XP / 0.17 / 34 -8	4.6 / 4.7 / 2	0.67 / 53.	10.67	2 / 0.15 / 0.52	2680. / 2465.		60		33.3 / 32.61	-0.7 / -6.2 / -32.7	-21.94 / 45.7	9.00 / 10.57		UGCA 124 / 556-08 -4-15-020 2
0615-27	615.2 -2722 / 234.61 -19.11 / 241.76 -75.37	3 / 0.20 / 34 -1	3.9 / 2.8 / 6	0.14 / 90.		0.67 / 0.09	1696. / 1465. / 2	314 / 276. / 278.	15	0.38 / 25 / 1.01	20.5 / 31.56	-2.5 / -4.6 / -19.8	16.8	9.94 / 9.68 / 10.47	0.027	UGCA 126 / 489-29 -5-15-008
U 3439	615.8 7833 / 135.54 25.19 / 24.93 4.31	5 / 0.14 / 12 -0+19	3.3 / 2.9 / 2	0.41 / 71.	12.6 / 11.05 / 1.55	3 / 0.41 / 0.34	1495. / 1685. / 1	246 / 220. / 223.	10	0.89 / 217 9 / 1.08	24.7 / 31.97	22.4 / 10.4 / 1.9	-19.37 / 20.9	9.58 / 10.37	0.54 / 0.161 / 3.37	NGC 2146A / 3439 13-05-025 2
0616-70	616.4 -7053 / 281.37 -28.43 / 210.36 -34.46	7 / 0.10 / 53+17				0.67 / 0.37	1294. / 1035. / 7		20		14.7 / 30.84	-10.5 / -6.1 / -8.3				
0618-08	618.5 -826 / 217.07 -10.60 / 336.43 -78.97	5 / 0.07 / 53 -0	3.4 / 4.0 / 2	0.29 / 81.		0.67 / 2.04	739. / 561. / 1	315 / 281. / 283.	8	1.70 / 7 / 1.18	8.7 / 29.71	1.5 / -0.7 / -8.6	10.2	10.28	0.163	UGCA 127 / -1-17-000
0618-16	618.5 -1602 / 224.08 -13.90 / 292.70 -81.75	6A / 0.09 / 34 +8	1.9 / 2.3 / 4	0.62 / 56.	12.56	6 / 0.18 / 1.34	2845. / 2642.		13		35.7 / 32.77	2.0 / -4.7 / -35.4	-20.21 / 24.0			SCI 62 / -3-17-001 2
0618-20	618.9 -2002 / 227.88 -15.48 / 266.78 -80.81	12 P / 0.12 / 34 -6	3.5 / 3.9 / 6	0.82 / 39.		0.07 / 0.79	1981. / 1767. / 2	359 / 511. / 512.	20	1.12 / 12 / 1.02	24.4 / 31.94	-0.2 / -3.9 / -24.1	27.8	9.89 / 11.32	0.037	UGCA 128 / 556-15 -3-17-002 2
N 2217	619.7 -2713 / 234.86 -18.13 / 238.88 -76.09	-2B / 0.14 / 34 -1	4.9 / 5.0 / 6	0.92 / 26.	11.28	2 / 0.00 / 0.17	1615. / 1382.	307	10 / 20 / 205	0.80 / 205 20	19.5 / 31.45	-2.4 / -4.0 / -18.9	-20.17 / 28.5	10.26 / 9.38	0.13	489-42 -5-15-010 2

(1) Name	(2) α / ℓ / SGL	(3) δ / b / SGB	(4) Type / ρxyz / Group	(5) D25 / D25 / source	(6) d/D / i	(7) $B_T^{b,i}$ / $H_{-0.5}$ / B-H	(8) source / A_B^{i-o} / A_B^b	(9) Vh / Vo / Tel	(10) W20 / WR / WD	(11) ev / ew / source	(12) Fc / source/eF / fh	(13) R / μ	(14) SGX / SGY / SGZ	(15) $M_B^{b,i}$ / Δ_{25}	(16) log LB / log MH / log MT	(17) MH/LB / MH/MT / MT/LB	(18) Alternate Names UGC/ESO / MCG / AGN? RCII
0621-59	621.0 / 268.87 / 212.47	-5943 / -26.87 / -45.52	3 / 0.07 / 33 +5	3.4 / 3.2 / 6	0.70 / 50.		/ 0.13 / 0.15	2262 / 1994 / 7		/ / 20		27.0 / 32.16	-16.0 / -10.2 / -19.3	/ 25.2			121-26
0621-86	621.9 / 299.20 / 207.24	-8637 / -27.75 / -18.97	1 / 0.07 / 55 -0	3.9 / 3.1 / 6	0.20 / 90.		/ 0.67 / 0.62	1870. / 1638. / 7		/ / 20		22.4 / 31.75	-18.8 / -9.7 / -7.3	/ 20.3			005-04
N 2223	622.5 / 230.86 / 250.40	-2248 / -15.83 / -79.87	4X / 0:-5 / 34 -8	2.8 / 3.0 / 6	1.00 /	11.79	2 / 0.00 / 0.36	2716. / 2492. / 2	392	/ / 60	0.85 / 19 / 1.03	33.7 / 32.64	-2.0 / -5.6 / -33.2	-20.85 / 29.5	10.53 / 9.91	0.24	UGCA 129 / 489-49 -4-16-002 / 2
U 3463	622.6 / 155.87 / 22.77	5907 / 20.03 / -15.01	4X / 0.06 / 20 -0	2.6 / 2.6 / 2	0.72 / 48.	12.4	3 / 0.12 / 0.41	2696. / 2811. / 1	324 / 384. / 386.	/ / 10	0.70 / 24 / 1.07	38.9 / 32.95	34.7 / 14.6 / -10.1	-20.55 / 29.5	10.41 / 9.88 / 11.10	0.29 / 0.060 / 4.89	IC 2166 / 3463 10-10-001
N 2227	623.8 / 230.18 / 252.89	-2157 / -15.21 / -80.67	6B / 0.11 / 34 -6	2.5 / 2.4 / 6	0.55 / 62.	12.27	6 / 0.24 / 0.47	2221. / 1999. / 1		/ / 20		27.5 / 32.19	-1.3 / -4.3 / -27.1	-19.92 / 19.3	10.16		SCI 63 / 556-23 -4-16-004 / 2
U 3475	627.0 / 175.05 / 20.74	3931 / 13.13 / -34.54	9 / 0.09 / 15 -0	2.8 / 3.1 / 2	0.57 / 60.	12.8	3 / 0.22 / 1.13	487. / 512. / 1	182 / 171. / 175.	20 / 10 / 226	0.73 / 226 8 / 1.10	8.6 / 29.67	6.6 / 2.5 / -4.9	-16.87 / 7.8	8.94 / 8.60 / 9.82	0.46 / 0.060 / 7.58	3475 07-14-002
0630-52	630.0 / 261.19 / 212.52	-5223 / -24.17 / -52.96	10 / 0.20 / 53 -0	1.5 / 1.5 / 6	1.00 /		/ 0.00 / 0.16	1191. / 921. / 7		/ / 20		13.4 / 30.63	-6.8 / -4.3 / -10.7	/ 5.9			206-16
U 3496	632.0 / 127.61 / 26.38	8553 / 27.02 / 11.51	10 / 0.14 / 42+15	2.3 / 2.3 / 2	0.76 / 45.		/ 0.09 / 0.26	1583. / 1795. / 1	75	/ / 15	0.13 / 30 / 1.05	26.2 / 32.09	23.0 / 11.4 / 5.2	/ 17.6	8.97		3496 14-04-000
0635+75	635.0 / 138.92 / 25.67	7539 / 25.62 / 1.31	13 P / 0.09 / 12-11	0.8 / 0.8 / 2	0.85 / 35.		/ 0.00 / 0.34	794. / 972. / 1	80	/ / 10	0.09 / 17 28 / 1.01	15.3 / 30.93	13.8 / 6.6 / 0.4	/ 3.6	8.46		UGCA 130, MARK 5 / 13-05-000 / 2
U 3504	635.6 / 155.42 / 24.56	6007 / 21.91 / -14.19	6X / 0.18 / 24 -1	2.8 / 2.9 / 2	0.87 / 33.	11.9	3 / 0.05 / 0.31	2106. / 2222. / 1	233 / 359. / 361.	/ / 10	0.67 / 14 / 1.09	31.6 / 32.50	27.9 / 12.7 / -7.7	-20.60 / 26.8	10.43 / 9.67 / 11.00	0.17 / 0.047 / 3.70	3504 10-10-009
0639-58	639.9 / 268.05 / 209.23	-5829 / -24.25 / -47.07	3 P / 0.07 / 33 +5	3.5 / 2.9 / 6	0.27 / 83.	9.65	6 / 0.67 / 0.39	2598. / 2325. / 7	310 / 274. / 276.	/ / 20		31.4 / 32.48	-18.6 / -10.4 / -23.0	/ 26.6	10.76		122-01
U 3522	641.0 / 128.64 / 26.50	8459 / 27.11 / 10.60	13 P / 0.21 / 42-16	2.3 / 2.1 / 2	0.55 / 61.	14.3	3 / 0.00 / 0.23	2137. / 2346. / 1	208 / 196. / 199.	/ / 10	0.44 / 24 / 1.04	33.2 / 32.61	29.2 / 14.6 / 6.1	-18.31 / 20.4	9.52 / 9.48 / 10.36	0.93 / 0.134 / 6.91	VII ZW 92 / 3522 14-04-009
U 3530	642.0 / 155.37 / 25.40	6024 / 22.75 / -13.95	7X / 0.21 / 24 -1	2.5 / 2.4 / 2	0.64 / 55.	12.8 / 11.64 / 1.16	3 / 0.17 / 0.31	2109. / 2224. / 1	220 / 224. / 227.	/ / 15	0.69 / 13 / 1.05	31.6 / 32.50	27.7 / 13.2 / -7.6	-19.70 / 22.1	10.07 / 9.69 / 10.51	0.41 / 0.152 / 2.73	NGC 2273B / 3530 10-10-013 / 2

ID	Names															Coordinates
N 2280	UGCA 131 / 427-02 -5-16-020 2	0.43 / 0.088 / 4.91	10.60 / 10.24 / 11.29	-21.03 / 45.4	-3.9 / -3.0 / -22.7	23.2 / 31.83	1.51 / 11 10 / 1.09	10	400 / 387. / 389.	1896. / 1651. / 2	5 / 0.36 / 0.78	10.8 / 8.61 / 2.19	6.8 / 0.44 / 69.	6.8 / 6.7 / 6	6A / 0.09 / 34 +7	642.9 -2735 / 237.31 -13.53 / 217.15 -77.81
0644-47	255-19			7.2	-5.4 / -3.1 / -10.0	11.8 / 30.35		20		1055. / 782. / 7	0.04 / 0.24	11.5	2.1 / 0.88 / 32.	2.1 / 2.1 / 6	4 / 0.21 / 53 -0	644.4 -4729 / 256.81 -20.56 / 209.43 -58.09
N 2273	MARK 620 / 3546 10-10-015 2	0.09 / 0.020 / 4.68	10.50 / 9.46 / 11.17	-20.76 / 27.4	24.8 / 12.0 / -6.6	28.4 / 32.26	0.55 / 211 13 / 1.12	15	367 / 430. / 432.	1844. / 1961. / 1	3 / 0.13 / 0.31	11.5	3.4 / 0.70 / 50.	3.4 / 3.3 / 2	1B / 0.19 / 24 -1	645.6 6054 / 154.97 23.31 / 25.87 -13.47
U 3574	3574 10-10-017	0.64	9.96 / 9.77	-19.42 / 29.7	19.8 / 9.7 / -6.8	23.1 / 31.82	1.04 / 7 / 1.21	10	172	1445. / 1545. / 1	3 / 0.00 / 0.21	12.4	4.2 / 1.00	4.2 / 4.4 / 2	6A / 0.11 / 24 -2	648.9 5715 / 158.92 22.76 / 26.22 -17.13
0649-52	207-07			9.5	-6.3 / -3.3 / -9.7	12.0 / 30.40		20		1085. / 808. / 7	0.12 / 0.16		2.8 / 0.71 / 49.	2.8 / 2.7 / 6	6 / 0.24 / 53 -0	649.5 -5205 / 261.76 -21.21 / 207.60 -53.53
U 3580	3580 12-07-022 2	0.32 / 0.163 / 1.97	10.18 / 9.69 / 10.47	-19.97 / 21.7	18.4 / 9.2 / -1.7	20.6 / 31.57	1.06 / 11 / 1.13	10	236 / 218. / 220.	1204. / 1356. / 1	3 / 0.30 / 0.19	11.6 / 10.39 / 1.21	4.0 / 0.49 / 66.	4.0 / 3.6 / 2	1A / 0.08 / 12 -0	650.0 6939 / 145.69 25.65 / 26.52 -4.73
U 3587	3587 03-18-003	0.56 / 0.194 / 2.90	9.82 / 9.57 / 10.28	-19.07 / 14.7	9.3 / 4.5 / -14.7	18.0 / 31.27	1.06 / 13 / 1.20	10	239 / 212. / 215.	1254. / 1173. / 1	3 / 0.43 / 0.81	12.2	2.9 / 0.40 / 72.	2.9 / 2.8 / 2	12 / 0.09 / 20 +1	650.9 1922 / 195.87 9.18 / 26.01 -55.01
U 3598	3598 10-10-019	0.53 / 0.347 / 1.53	9.60 / 9.32 / 9.78	-18.51 / 18.6	26.3 / 13.2 / -7.2	30.3 / 32.41	0.36 / 17 / 1.04	20	115 / 106. / 112.	1996. / 2110. / 1	3 / 0.21 / 0.30	13.9	2.2 / 0.59 / 59.	2.2 / 2.1 / 2	10B / 0.22 / 24 -1	652.0 6043 / 155.38 24.02 / 26.67 -13.67
N 2310	309-07 -7-15-001 2		9.64	-18.63 / 10.3	-5.2 / -2.6 / -12.3	13.6 / 30.67		86		1188. / 917.	6 / 0.00 / 0.44	12.04	3.4 / 0.19 / 90.	3.4 / 2.6 / 6	-2 / 0.07 / 53 -0	652.7 -4048 / 250.73 -16.85 / 206.78 -64.81
N 2325	427-28 -5-17-005 2		10.58	-20.97 / 23.6	-5.6 / -1.9 / -25.4	26.1 / 32.08		58		2109. / 1854.	2 / 0.00 / 1.09	11.11	2.7 / 0.67 / 53.	2.7 / 3.1 / 6	-5 / 0.07 / 34 +7	700.7 -2837 / 239.95 -10.40 / 199.05 -76.87
U 3647	DDO 40 / 3647 09-12-027 2	1.12	9.13 / 9.18	-17.35 / 9.7	18.7 / 9.9 / -6.8	22.2 / 31.73	0.49 / 10 / 1.02	8	76	1382. / 1475. / 3	7 / 0.05 / 0.15	14.38	1.5 / 0.87 / 33.	1.5 / 1.5 / 6	10B / 0.11 / 24 -2	700.7 5637 / 160.03 24.16 / 27.91 -17.75
N 2268	3653 14-04-022 2	0.23 / 0.033 / 6.98	10.52 / 9.87 / 11.36	-20.81 / 34.2	30.2 / 15.4 / 6.0	34.4 / 32.68	0.80 / 217 13 / 1.02	7 / 10 / 211	403 / 481. / 482.	2226. / 2432.	1 / 0.12 / 0.21	11.87 / 9.21 / 2.66	3.5 / 0.70 / 49.	3.5 / 3.4 / 2	4X / 0.22 / 42 -16	700.8 8428 / 129.24 27.55 / 26.96 10.08
U 3658	3658 03-18-000	0.131	8.73 / 9.61	9.4	8.0 / 4.7 / -14.1	16.9 / 31.14	0.27 / 226 18 / 1.07	15 / 11 / 226	144 / 122. / 127.	1183. / 1089. / 1	0.37 / 0.52		2.0 / 0.44 / 70.	2.0 / 1.9 / 2	10 / 0.10 / 20 +1	701.8 1741 / 198.55 10.78 / 30.69 -56.62
U 3685	3685 10-11-002 2	0.50	10.32 / 10.02	-20.33 / 32.6	24.0 / 12.9 / -6.1	27.9 / 32.23	1.13 / 6 / 1.16	7	101	1797. / 1913. / 1	3 / 0.00 / 0.29	11.9	3.7 / 1.00	3.7 / 4.0 / 2	3B / 0.20 / 24 +1	704.5 6141 / 154.68 25.70 / 28.19 -12.67

(1) Name	(2) α ℓ SGL	(3) δ b SGB	(4) Type Q_xyz Group	(5) $D_{25}^{b,i}$ $D_{25}^{b,i}$ source	(6) d/D i	(7) $B_T^{b,i}$ $H_{-0.5}$ B-H	(8) source A_B^{i-o} A_B^b	(9) V_h V_o Tel	(10) W_{20} W_R W_D	(11) e_v e_w source	(12) F_c source/e_F f_h	(13) R μ	(14) SGX SGY SGZ	(15) $M_B^{b,i}$ $\Delta_{25}^{b,i}$	(16) $\log L_B$ $\log M_H$ $\log M_T$	(17) M_H/L_B M_H/M_T M_T/L_B	(18) Alternate Names UGC/ESO MCG AGN? RCII
U 3691	705.1 / 201.13 / 32.56	1515 / 10.44 / -58.98	6A / 0.10 / 30 +1	2.4 / 2.3 / 2	0.50 / 65.	12.3 / 11.36 / 0.94	3 / 0.29 / 0.58	2208 / 2102 / 1	265. / 251. / 253.	25	0.68 / 14 / 1.11	30.0 / 32.39	13.0 / 8.3 / -25.7	-20.09 / 20.1	10.23 / 9.63 / 10.57	0.25 / 0.117 / 2.18	3691 03-19-001
N 2339	705.4 / 197.83 / 32.04	1852 / 12.06 / -55.38	4X / 0.12 / 30 +1	2.6 / 2.7 / 2	0.75 / 45.	11.68	2 / 0.10 / 0.52	2256 / 2166	365. / 459. / 460.	9 / 15 / 217	0.88 / 217 10	30.9 / 32.45	14.9 / 9.3 / -25.5	-20.77 / 24.4	10.50 / 9.86 / 11.17	0.23 / 0.049 / 4.71	3693 03-19-002 2
N 2337	706.6 / 172.95 / 29.61	4432 / 21.80 / -29.77	10B / 0.10 / 15+12	2.6 / 2.7 / 2	0.83 / 38.	12.2	3 / 0.06 / 0.38	433. / 467. / 1	178. / 236. / 239.	15	0.98 / 10 / 1.07	8.2 / 29.58	6.2 / 3.5 / -4.1	-17.38 / 6.5	9.14 / 8.81 / 10.02	0.46 / 0.061 / 7.50	3711 07-15-010
U 3730	708.2 / 141.54 / 27.86	7334 / 27.54 / -0.78	13 P / 0.-3 / 12 -0	2.6 / 2.2 / 2	0.41 / 71.	14.03	2 / 0.00 / 0.19	2703. / 2868.	219	10 / 40 / 217	0.38 / 217 19	40.0 / 33.01	35.4 / 18.7 / -0.5	-18.98 / 25.7	9.78 / 9.58	0.63	A 0708, VV 123 / 3730 12-07-035 2
N 2344	708.8 / 170.26 / 29.82	4715 / 22.95 / -27.03	5B / 0.06 / 15 -0	2.2 / 2.3 / 2	0.92 / 26.	12.44	2 / 0.03 / 0.38	955. / 1002.		19		16.0 / 31.01	12.3 / 7.1 / -7.2	-18.57 / 10.7	9.62		3734 08-13-103 2
N 2276	710.5 / 127.67 / 27.09	8551 / 27.71 / 11.48	5X / 0.13 / 42-16	2.7 / 2.8 / 2	0.90 / 29.	11.64	1 / 0.03 / 0.23	2409. / 2619. / 1	169	20 / 12 / 211	0.79 / 56 8 / 1.08	36.8 / 32.83	32.1 / 16.4 / 7.3	-21.19 / 30.1	10.67 / 9.92	0.18	ARP 25, VII ZW 134 / 3740 14-04-028 2
N 2357	714.6 / 194.43 / 34.67	2327 / 15.90 / -50.57	5 / 0.13 / 30 +2 +1	3.5 / 2.6 / 2	0.16 / 90.	12.9 / 10.35 / 2.55	3 / 0.67 / 0.36	2273. / 2201. / 1	358. / 320. / 322.	15	0.74 / 23 / 1.10	31.5 / 32.49	16.5 / 11.4 / -24.3	-19.59 / 23.9	10.03 / 9.74 / 10.85	0.51 / 0.077 / 6.66	3782 04-17-014
0715-57	715.0 / 268.25 / 202.25	-5715 / -19.43 / -48.15	6 / 0.21 / 53 -0	2.1 / 1.8 / 6	0.32 / 78.		0.62 / 0.46	1095. / 812. / 7		20		12.2 / 30.43	-7.5 / -3.1 / -9.1	6.4			162-17
N 2300	715.8 / 127.70 / 27.19	8549 / 27.81 / 11.45	-5 / 0.14 / 42-16	4.8 / 4.5 / 9	0.61 / 57.	11.67	1 / 0.00 / 0.23	1958. / 2168.		25		31.0 / 32.46	27.0 / 13.9 / 6.1	-20.79 / 40.7	10.51		ARP 114 / 3798 14-04-031 2
N 2336	718.0 / 133.97 / 27.83	8016 / 28.20 / 5.94	4X / 0.20 / 42-17+16	5.7 / 5.2 / 9	0.59 / 59.	10.65 / 8.59 / 2.06	1 / 0.20 / 0.14	2196. / 2386. / 1	459. / 493. / 495.	15 / 8 / 217	1.33 / 217 8 / 1.36	33.9 / 32.65	29.8 / 15.7 / 3.5	-22.00 / 51.5	10.99 / 10.39 / 11.56	0.25 / 0.068 / 3.70	3809 13-06-006 2
U 3808	718.0 / 193.02 / 35.43	2515 / 17.32 / -48.67	73 / 0.11 / 30 +2 +1	2.0 / 1.9 / 2	0.53 / 63.	13.8	3 / 0.25 / 0.39	2382. / 2317. / 1	208. / 193. / 196.	10	0.49 / 26 / 1.04	33.1 / 32.60	17.8 / 12.7 / -24.9	-18.80 / 18.4	9.71 / 9.53 / 10.30	0.66 / 0.171 / 3.85	3808 04-18-006
U 3817	719.2 / 172.94 / 32.08	4512 / 24.13 / -28.89	10 / 0.11 / 15+12	2.0 / 1.9 / 2	0.53 / 63.		0.25 / 0.34	437. / 471. / 3	52. / 37. / 50.	5	0.41 / 226 6 / 1.03	8.3 / 29.60	6.2 / 3.9 / -4.0	4.6	8.25 / 8.26	0.969	3817 08-14-000
U 3826	720.0 / 154.87 / 30.05	6148 / 27.53 / -12.40	7X / C.16 / 24 +1	3.8 / 3.8 / 2	0.89 / 31.	12.9	3 / 0.04 / 0.19	1734. / 1847. / 3	70 / 103. / 109.	5	0.82 / 7 / 1.14	27.2 / 32.17	23.0 / 13.3 / -5.8	-19.27 / 30.2	9.90 / 9.69 / 9.97	0.62 / 0.526 / 1.17	3826 10-11-038

Name	Coordinates	Type	Size	Ax/PA	m₁	c	Vel	N	Q	r	d	(U V W)	M	m	L/S	Identifications
0720-72	720.6 -7238 / 284.15 -23.75 / 204.29 -32.84	10 / 0.10 / 53-17				0.62 / 0.69	1501. / 1235. / 7	20			17.6 / 31.22	-13.5 / -6.1 / -9.5	-19.17 / 9.6	9.86		058-30 2
N 2397	721.5 -6854 / 280.29 -22.59 / 203.55 -36.52	5B / 0.14 / 53-17	2.2 / 2.2 / 6	0.48 / 66.	11.7	5 / 0.31 / 0.72	1300. / 1027.	28			14.9 / 30.87	-11.0 / -4.8 / -8.9				IC 467
U 3834	721.6 7958 / 134.30 28.37 / 28.02 5.66	5X / 0.18 / 42-17+16	3.0 / 2.5 / 9	0.40 / 72.	11.9 / 10.67 / 1.23	3 / 0.43 / 0.15	2036. / 2225. / 1	15	316 / 292. / 294.	0.72 / 23 / 1.08	31.7 / 32.51	27.9 / 14.8 / 3.1	-20.61 / 23.1	10.44 / 9.72 / 10.76	0.19 / 0.092 / 2.10	3834 13-06-007 2
N 2377	722.6 -930 / 225.31 2.98 / 77.62 -80.49	3X / 0.06 / 30-0	1.9 / 3.4 / 2	0.90 / 29.		0.04 / 2.93	2455. / 2242. / 1	10	206	1.07 / 11 / 1.15	31.8 / 32.51	1.1 / 5.1 / -31.3	31.6	10.07		UGCA 132, 3C 178 / -2-19-000 2 L/S
N 2366	723.6 6918 / 146.44 28.54 / 29.49 -4.91	10B / 0.40 / 14-10	8.0 / 6.8 / 2	0.43 / 70.	10.66 / 10.78 / -0.12	1 / 0.38 / 0.16	102. / 248. / 4	5 / 4 / 211	114 / 96. / 102.	1.86 / 25 4 / 1.08	2.9 / 27.35	2.6 / 1.4 / -0.3	-16.69 / 5.8	8.87 / 8.78 / 9.19	0.83 / 0.396 / 2.09	DDO 42 / 3851 12-07-040 2
U 3860	724.8 4052 / 177.81 23.93 / 33.99 -33.03	10 / 0.05 / 15-0	1.9 / 1.9 / 2	0.70 / 50.	14.67	7 / 0.13 / 0.22	355. / 365. / 3	7	55	0.45 / 11 / 1.03	7.2 / 29.30	5.0 / 3.4 / -3.9	-14.63 / 4.0	8.04 / 8.16	1.32	DDO 43 / 3860 07-16-003 2
0726-45	726.1 -4535 / 257.81 -13.21 / 195.32 -59.22	10 / 0.21 / 53-16+15	2.4 / 2.7 / 6	0.83 / 37.		0.06 / 0.74	1007. / 722. / 7	20			11.3 / 30.26	-5.6 / -1.5 / -9.7	8.9			257-17
N 2417	729.5 -6209 / 273.80 -19.51 / 200.91 -43.01	4A / 0.15 / 33-4	3.5 / 3.7 / 6	0.67 / 53.	11.69	6 / 0.15 / 0.61	3173. / 2891.	14			39.0 / 32.96	-26.6 / -10.2 / -26.6	-21.27 / 42.1	10.70		SCI 203 / 123-15 2
0729-61	729.8 -6141 / 273.35 -19.30 / 200.72 -43.46	4 / 0.15 / 33-4	1.9 / 1.7 / 6	0.35 / 76.		0.55 / 0.66	3203. / 2920. / 7	20			39.4 / 32.98	-26.7 / -10.1 / -27.1	19.6			123-16
0730-74	730.0 -7457 / 286.78 -23.79 / 203.97 -30.45	10 / 0.10 / 53-17	2.1 / 2.1 / 6	0.48 / 67.		0.31 / 0.96	1144. / 881. / 7	20			12.9 / 30.55	-10.1 / -4.5 / -6.5	7.9			035-09
0731-68	731.3 -6805 / 279.78 -21.48 / 202.26 -37.17	9 / 0.04 / 14+20	1.8 / 2.0 / 6	0.95 / 21.		0.02 / 0.64	528. / 253. / 7	20			4.4 / 28.20	-3.2 / -1.3 / -2.6	2.6			059-01
U 3912	731.6 439 / 211.75 11.59 / 52.91 -67.57	10B / 0.11 / 31-24	2.1 / 2.1 / 2	0.61 / 57.	13.2	3 / 0.19 / 0.47	1236. / 1073. / 1	15	190	0.59 / 21 / 1.08	17.1 / 31.16	3.9 / 5.2 / -15.8	-17.96 / 10.5	9.38 / 9.06	0.48	3912 01-20-001
N 2403	732.1 6543 / 150.57 29.19 / 30.84 -8.33	6X / 0.30 / 14-10	23.8 / 21.4 / 2	0.54 / 62.	8.46 / 6.43 / 2.03	1 / 0.25 / 0.14	132. / 261. / 2	20 / 12 / 80	257 / 248. / 250.	2.54 / 3 / 1.61	4.2 / 28.14	3.6 / 2.2 / -0.6	-19.68 / 26.2	10.06 / 9.79 / 10.67	0.53 / 0.131 / 4.04	3918 11-10-007 2
0733-46	733.6 -4649 / 259.53 -12.59 / 193.65 -57.66	5 / 0.07 / 33-0	2.9 / 2.5 / 6	0.24 / 85.		0.67 / 0.74	2881. / 2593. / 7	20			35.5 / 32.75	-18.5 / -4.5 / -30.0	25.9			257-19

(1) Name	(2) α ℓ SGB→SGL	(3) δ b SGB	(4) Type Q_xyz Group	(5) D_{25} $D_{25}^{b,i}$ source	(6) d/D i	(7) $B_T^{b,i}$ $H_{-0.5}$ B-H	(8) source A_B^{i-o} A_B^b	(9) V_h V_o Tel	(10) W_{20} W_R W_D	(11) e_v e_w source	(12) F_c source/e_f f_h	(13) R μ	(14) SGX SGY SGZ	(15) $M_B^{b,i}$ Δ_{25}	(16) log L_B log M_H log M_T	(17) M_H/L_B M_H/M_T M_T/L_B	(18) Alternate Names UGC/ESO MCG AGN? RCII
N 2427	735.0	-4731	8X	5.0	0.40	11.18	2	977.	272	13		10.9	-5.8	-19.01	9.80	4.29	
	260.28	-12.70	0.22	4.8	72.	9.20	0.43	689.	246.			30.19	-1.4	15.3	10.43		
	193.65	-56.92	53-16+15	9		1.98	0.75	7	248.				-9.2				
N 2434	735.0	-6910	-5	1.6	0.94	11.60	2	1411.		54		16.4	-12.3	-19.48	9.98		208-27
	280.99	-21.54	0.19	1.8	24.		0.00	1137.				31.08	-5.0	8.6			2
	202.14	-36.04	53-17	6			0.70						-9.7				
U 3946	735.4	325	10	1.5	0.68	13.7	3	1198.	109	12	0.54	16.5	3.4	-17.39	9.15	0.67	059-05
	215.31	11.87	0.12	1.5	52.		0.14	1028.	110.		16	31.09	5.1	7.2	8.97	0.372	2
	56.65	-68.24	31-24	2			0.42	1	115.		1.04		-15.3		9.40	1.80	
N 2442	736.6	-6925	4B	5.5	0.91	10.50	2	1469.		25		17.1	-12.9	-20.66	10.46		NGC 2443
	281.30	-21.50	0.17	6.2	28.		0.03	1195.				31.16	-5.2	31.0			
	202.04	-35.76	53-7	9			0.69						-10.0				
0737-55	737.0	-5504	3	2.4	0.27			2818.		20		34.6	-21.4				163-11
	267.41	-15.76	0.11	2.0	83.		0.67	2530.				32.70	-6.5	20.2			
	196.99	-49.63	33 -0	6			0.50	7					-26.4				
0737-50	737.1	-5038	9B	2.4	0.67			1078.		20		12.1	-6.9				208-33
	263.30	-13.79	0.28	2.5	53.		0.15	789.				30.42	-1.8	8.8			
	194.83	-53.86	53-15	6			0.59	7					-9.8				
U 3964	737.8	-128	8X	2.0	0.77	13.8	3	1461.	178	15	0.45	19.6	2.5	-17.66	9.26	0.60	3964 00-20-004
	220.00	10.13	0.07	2.1	44.		0.09	1271.	209.		27	31.46	5.5	12.0	9.03	0.071	2
	65.98	-72.01	31 -0	2			0.56	1	212.		1.11		-18.7		10.18	8.44	
U 3966	738.0	4014	10	2.2	0.92	13.89	7	364.	91	10	0.66	7.4	5.0	-15.47	8.38	1.04	DDO 46
	179.23	26.17	0.06	2.3	26.		0.03	368.			12	29.36	3.8	5.0	8.40		3966 07-16-011
	37.07	-33.17	15 -0	2			0.23	1			1.06		-4.1				2
U 3974	739.0	1655	10	4.6	0.82	13.37	7	265.	88	5	1.22	2.1	0.8	-13.21	7.48	2.45	DDO 47
	203.10	18.53	0.16	4.5	39.		0.07	153.	110.		6	26.58	0.8	2.8	7.86	0.076	3974 03-20-010
	46.57	-55.52	14-19	2			0.10	1	115.		1.24		-1.7		8.99	32.36	2
U 3975	739.0	7255	8B	2.3	0.80	13.8	3	2480.	88	10	0.34	37.2	32.2	-19.05	9.81	0.47	3975 12-08-009
	142.29	29.76	0.19	2.2	41.		0.08	2639.	105.		17	32.85	18.7	23.9	9.48	0.396	
	30.14	-1.14	12 -0	2			0.09	1	111.		1.05		-0.7		9.88	1.18	
0743-58	743.7	-5802	5	2.3	0.14			2903.		20		35.6	-23.4				123-23
	270.61	-16.22	0.12	1.9	90.		0.67	2615.				32.76	-7.2	19.8			
	196.99	-46.53	33 +4	6			0.84	7					-25.9				
0748-54	748.1	-5421	10 P	2.6	0.48			1047.		20		11.8	-7.4				163-19
	267.49	-14.00	0.25	2.6	66.		0.31	756.				30.36	-1.9	9.0			
	194.29	-49.80	53 -0+15	6			0.69	7					-9.0				
U 4093	752.0	6026	3B	1.2	0.75	14.05	2	1545.		95		24.9	20.1	-17.93	9.36		IC 2209, MARK 13
	156.76	31.28	0.15	1.2	45.		0.10	1646.				31.98	13.7	8.7			4093 10-12-017
	34.20	-13.02	24 -3	2			0.15						-5.6				2

Name	Coords		Type	Diam	Ratio/PA	Mag	Vel				Dist		AbsMag	B-mags		final
N 2460	752.6 156.70 34.26	6029 31.35 -12.95	1A 0.14 24 -3	3.7 3.6 2	0.73 47.	12.35; 2 0.11 0.14	1452. 1553.	367 448. 449.	10 15 211	1.04 211 25	23.6 31.87	19.0 13.0 -5.3	-19.52 24.8	10.00 9.79 11.16		0.61 0.042 14.44
U 4115	754.2 207.01 54.26	1431 20.89 -56.26	10A 0.02 14+19	2.0 1.8 2	0.59 59.	14.0; 3 0.21 0.08	341. 214. 1	122 113. 118.	10	0.77 12 1.04	3.2 27.51	1.0 1.4 -2.6	-13.51 1.7	7.60 7.78 8.79		1.53 0.097 15.79
U 4121	754.8 159.40 35.16	5811 31.57 -15.10	9B 0.12 13 -0	2.5 2.0 2	0.36 75.	14.27; 7 0.52 0.11	1090. 1180. 1	165 137. 141.	10	0.60 5 17 1.04	18.9 31.38	14.9 10.5 -4.9	-17.11 11.0	9.04 9.15 9.78		1.31 0.237 5.53
0756-76	756.5 288.76 202.50	-7617 -22.70 -28.76	5 0.14 53-19	4.3 3.5 6	0.20 90.	; 0.67 0.81	1760. 1498. 7		20		21.0 31.61	-17.0 -7.0 -10.1	21.5			
0756-49	756.8 264.01 189.31	-4943 -10.58 -53.50	5 0.16 53-15	5.7 4.6 6	0.16 90.	9.02; 0.67 0.68	1119. 826. 7	334 296. 298.	20		12.8 30.54	-7.5 -1.2 -10.3	17.2	10.64		
U 4148	756.9 177.83 40.44	4219 30.10 -30.21	7 0.08 15 -0+10	2.5 1.8 2	0.15 90.	14.3; 3 0.67 0.16	737. 747. 1	148 118. 123.	10	0.47 23 1.04	12.8 30.54	8.4 7.2 -6.4	-16.24 6.7	8.69 8.68 9.43		0.99 0.178 5.57
N 2500	758.2 167.98 37.81	5054 31.58 -21.96	7B 0.23 15+11+10	2.7 2.8 2	1.00	12.05; 1 0.00 0.14	519. 572. 1	118	10 4 211	0.91 217 5 1.07	10.1 30.01	7.4 5.7 -3.8	-17.96 8.3	9.38 8.92		0.35
U 4173	758.6 133.73 29.53	8016 29.90 6.27	10 0.10 12 -0	3.2 2.4 2	0.25 85.	13.8; 3 0.67 0.09	866. 1054. 1	89 70. 78.	7	0.87 10 1.06	16.7 31.11	14.4 8.2 1.8	-17.31 11.7	9.12 9.32 9.22		1.58 1.243 1.27
N 2525	803.3 231.86 104.77	-1117 10.80 -72.31	5B 0.06 31 -0	2.5 2.6 1	0.76 45.	11.73; 1 0.09 0.38	1586. 1354. 1	232 278. 280.	15 11 217	0.55 217 13 1.18	21.1 31.62	-1.6 6.2 -20.1	-19.89 16.0	10.15 9.20 10.56		0.11 0.044 2.55
0804-76	804.9 289.62 202.22	-7656 -22.52 -27.99	10 0.14 53-19	1.4 1.5 6	0.75 45.	; 0.10 0.58	1747. 1486. 7		20		20.8 31.59	-17.0 -6.9 -9.8	9.1			
U 4238	805.0 137.83 30.89	7635 30.98 2.83	7B 0.17 12+10	2.8 2.5 2	0.61 57.	12.6 11.83 0.77; 3 0.19 0.11	1541. 1714. 1	187 181. 184.	10	0.84 14 1.06	25.5 32.03	21.9 13.1 1.3	-19.43 18.6	9.96 9.65 10.25		0.49 0.254 1.92
U 4260	807.6 173.20 40.90	4637 32.65 -25.53	10 0.11 21+19	1.8 1.9 2	0.95 21.	14.06; 7 0.02 0.14	2254. 2284. 1	154	10	0.44 17 1.04	33.3 32.61	22.7 19.7 -14.4	-18.55 18.5	9.61 9.48		0.75
N 2537	809.7 173.80 41.46	4609 32.96 -25.84	10BP 0.19 15+11+10	1.7 1.7 2	0.89 31.	12.15; 2 0.04 0.16	452. 479. 1	120 187. 190.	10 8 217	0.71 217 6 1.03	9.0 29.78	6.1 5.4 -3.9	-17.63 4.5	9.24 8.62 9.66		0.24 0.092 2.58
U 4278	810.4 174.12 41.68	4554 33.05 -26.03	7 0.24 15+11+10	4.3 3.1 2	0.13 90.	12.67; 1 0.67 0.16	565. 591. 1	206 170. 173.	8	1.10 9 1.11	10.6 30.12	7.1 6.3 -4.6	-17.45 9.6	9.17 9.15 9.91		0.95 0.176 5.41

Right-hand identification column:

| 4097 10-12-021 2 |
| 4115 02-21-000 |
| DDO 48 / 4121 10-12-046 2 |
| 035-18 |
| 209-09 |
| 4148 07-17-006 2 |
| 4165 09-13-110 2 |
| 4173 13-06-017 |
| UGCA 135 / -2-21-004 2 |
| 035-20 |
| 4238 13-06-018 2 |
| DDO 49 / 4260 08-15-047 2 |
| ARP 6, MARK 86, VV 138 / 4274 08-15-050 2 |
| IC 2233 / 4278 08-15-052 2 |

(1) Name	(2) α ℓ SGL	(3) δ b SGB	(4) Type ϱ_{xyz} Group	(5) D_{25} $D_{25}^{b,i}$ source	(6) d/D i	(7) $B_T^{b,i}$ $H_{-0.5}$ $B\text{-}H$	(8) source A_B^{i-o} A_B^b	(9) V_h V_o Tel	(10) W_{20} W_R W_D	(11) e_v e_w source	(12) F_c source/e_F f_h	(13) R μ	(14) SGX SGY SGZ	(15) $M_B^{b,i}$ $\Delta_{25}^{b,i}$	(16) $\log L_B$ $\log M_H$ $\log M_T$	(17) M_H/L_B M_H/M_T M_T/L_B	(18) Alternate Names Alternate Names AGN? UGC/ESO MCG RCII
0810-74	810.7 287.30 200.79	-7422 -21.08 -30.26	10 0.17 53+17	1.7 1.6 6	0.35 76.		 0.54 0.74	1247. 980. 7		20		14.4 30.80	-11.7 -4.4 -7.3	 6.7			 035-21
N 2541	811.0 170.18 40.50	4913 33.47 -22.89	6A 0.25 15-11+10	7.2 6.3 2	0.47 67.	11.76 10.26 1.50	1 0.32 0.16	553. 596. 1	209 188. 191.	8 5 211	1.54 217 4 1.22	10.6 30.13	7.4 6.3 -4.1	-18.37 19.5	9.54 9.59 10.30	1.12 0.195 5.76	 4284 08-15-054 2
U 4305	814.1 144.28 33.31	7052 32.71 -2.39	10 0.29 14-10	8.2 7.8 2	0.74 47.	10.92	1 0.11 0.09	158. 305. 1	73 76. 83.	5 4 211	1.94 25 3 1.49	4.5 28.26	3.7 2.5 -0.2	-17.34 10.2	9.13 9.25 9.23	1.31 1.027 1.28	A0814, HO II, ARP 268 DDO 50, 7 Zw 223 4305 12-08-033 2
N 2559	815.0 246.99 151.70	-2719 4.47 -67.62	5BP 0.16 31-21	4.6 5.5 6	0.50 65.		 0.29 1.59	1571. 1296. 6	419 420. 421.	30	1.02 13 1.34	20.0 31.50	-6.7 3.6 -18.5	 32.1	 9.62 11.22	 0.026	UGCA 136 494-41 -4-20-003
N 2549	815.0 159.66 37.86	5758 34.25 -14.48	-2A 0.13 13 -0	3.9 3.2 2	0.33 78.	11.92	2 0.00 0.12	1082. 1168.		75		18.8 31.38	14.4 11.2 -4.7	-19.46 17.6	9.98		 4313 10-12-124 2
0815-29	815.7 249.28 157.40	-2959 3.10 -66.16	7 0.15 31-21	3.5 6.2 6	0.89 31.		 0.04 2.86	1650. 1370. 2	272 456. 458.	99	1.16 9 1.04	20.9 31.60	-7.8 3.2 -19.1	 37.8	 9.80 11.36	 0.028	UGCA 137 431-02 -5-20-000
N 2552	815.7 169.10 40.90	5010 34.30 -21.73	9A 0.25 15-11+10	3.7 3.5 2	0.70 50.	12.54	2 0.13 0.13	527. 574. 1	152 159. 162.	8 6 211	0.82 211 7 1.11	10.0 30.01	7.1 6.1 -3.7	-17.47 10.2	9.18 8.82 9.87	0.44 0.088 4.95	 4325 08-15-062 2
N 2544	815.9 140.44 32.33	7409 32.17 0.75	-2 0.16 12+21	3.8 3.6 2	0.76 45.	12.3	3 0.00 0.05	2787. 2949.		39		41.3 33.08	34.9 22.1 0.5	-20.78 43.4	10.50		MARK 87 4312 12-08-034 2
N 2566	816.6 245.54 146.52	-2520 5.88 -68.13	2B 0.17 31-21	4.3 4.8 6	0.73 47.		 0.11 0.88	1649. 1377. 2	223 253. 255.	30	0.78 16 1.07	21.1 31.62	-6.6 4.3 -19.6	 29.6	 9.43 10.74	 0.049	UGCA 138 495-03 -4-20-008 2
0818+71	818.7 143.81 33.55	7112 33.01 -1.95	10 0.29 14-10				 0.11 0.08	113. 262.	38	10 7 225	-0.17 225 22	4.3 28.17	3.6 2.4 -0.1		 7.10		M81 DW A 12-08-000
N 2551	819.2 141.03 32.74	7335 32.53 0.29	 0.29 12-17	1.8 1.7 2	0.70 50.	12.98	2 0.00 0.07	2216. 2375.		74		34.1 32.66	28.7 18.4 0.2	-19.68 16.9	10.06		 4362 12-08-038 2
N 2577	819.8 201.12 56.34	2243 29.59 -46.14	-3 0.08 32 -0	2.0 1.9 2	0.63 55.	13.3	3 0.00 0.15	2148. 2054.		70		30.8 32.44	11.8 17.8 -22.2	-19.14 17.1	9.85		 4367 04-20-042 2
U 4390	822.2 140.86 32.90	7341 32.71 0.46	7B 0.28 12-17	2.2 2.2 2	0.92 26.	14.1	3 0.03 0.06	2167. 2326. 1	205	30	0.53 23 1.05	33.5 32.62	28.1 18.2 0.3	-18.52 21.5	9.60 9.58	0.96	 4390 12-08-042

Name	(pos)	Type				flux/etc									notes
U 4393	822.6 4608 / 174.09 35.18 / 43.75 -25.00	12B / 0.13 / 21+19	2.5 / 2.3 / 2	0.71 / 49.	12.7	3 / 0.12 / 0.15	2129. / 2154. / 1	127 / 133. / 138.	10	0.68 / 11 / 1.05	31.9 / 32.52	20.9 / 20.0 / -13.5	-19.82 / 21.4	10.12 / 9.69 / 10.04	0.37 / 0.443 / 0.83	4393 08-16-003
0822-60	822.7 -6043 / 275.54 -13.22 / 192.16 -42.07	5 / 0.09 / 33 -0	1.7 / 1.7 / 6	0.50 / 65.		0.29 / 0.83	2582. / 2291. / 7		20		31.8 / 32.51	-23.1 / -5.0 / -21.3	15.8			124-15
0824-53	824.9 -5352 / 269.93 -9.14 / 186.70 -47.73	12 / 0.18 / 53 -0+15	1.5 / 2.1 / 6	1.00 /		0.00 / 1.60	1054. / 758. / 7		20		12.0 / 30.39	-8.0 / -0.9 / -8.8	7.4			164-10
U 4426	825.2 4202 / 179.14 35.23 / 46.15 -28.53	10 / 0.04 / 15 -0+10	2.0 / 1.8 / 2	0.53 / 63.		0.25 / 0.13	392. / 396. / 1	96 / 85. / 91.	10	0.42 / 19 / 1.03	6.3 / 28.99	3.8 / 4.0 / -3.0	3.3	8.02 / 8.84	0.150	DDO 52 / 4426 07-18-004 2
N 2601	825.2 -6757 / 282.00 -16.90 / 196.32 -35.59	-2A / 0.08 / 33 -0	1.8 / 1.8 / 6	0.68 / 51.		0.00 / 0.53	3234. / 2953.		49		40.0 / 33.01	-31.2 / -9.1 / -23.3	21.0			060-05 2
0825+52	825.9 5215 / 166.60 35.94 / 41.61 -19.19	13 P / 0.07 / 21 -0	0.7 / 0.7 / 2	0.53 / 63.		0.00 / 0.13	1717. / 1773.		25	0.30 / 17 75	27.1 / 32.16	19.1 / 17.0 / -8.9	5.5	9.17		UGCA 140, MARK 89 / 09-14-000 2
U 4459	829.5 6620 / 149.32 34.95 / 36.29 -6.09	10 / 0.51 / 14-10	1.9 / 1.8 / 2	0.85 / 35.	14.2	3 / 0.05 / 0.11	19. / 144. / 3	277 / 241. / 243.	20	1.04	2.0 / 26.53	1.6 / 1.2 / -0.2	-12.33 / 1.1	7.12	0.17 / 0.075 / 2.19	VII ZW 238, DDO 53 / 4459 11-11-013 2
N 2591	830.6 7812 / 135.57 31.89 / 31.64 4.83	5 / 0.20 / 12-10	3.1 / 2.3 / 2	0.26 / 84.	12.1 / 10.33 / 1.77	3 / 0.67 / 0.09	1331. / 1509. / 1	637 / 602. / 602.	10	0.57 / 20 / 1.06	22.9 / 31.80	19.4 / 11.9 / 1.9	-19.70 / 15.4	10.07 / 9.29 / 10.41		4472 13-07-001 2
N 2613	831.1 -2248 / 245.35 10.04 / 137.80 -65.73	3A / 0.15 / 31-22+21	6.8 / 5.3 / 6	0.24 / 85.	10.47 / 7.55 / 2.92	2 / 0.67 / 0.23	1679. / 1411. / 2	224	25	1.32 / 11 21 / 1.17	21.9 / 31.71	-6.7 / 6.1 / -20.0	-21.24 / 33.9	10.69 / 10.00 / 11.55	0.21 / 0.028 / 7.31	UGCA 141 / 495-18 -4-21-003 2
0831-21	831.7 -2141 / 244.50 10.80 / 135.07 -65.82	10 / 0.13 / 31-22+21	2.4 / 1.9 / 6	0.23 / 86.		0.67 / 0.30	1777. / 1511. / 2		15	0.53 / 29 / 1.02	23.3 / 31.83	-6.7 / 6.7 / -21.2	12.9	9.26		UGCA 142 / 562-19 -4-21-004
N 2608	832.3 2839 / 195.46 34.06 / 55.41 -39.61	3B / 0.13 / 32 +2	2.0 / 1.9 / 9	0.64 / 55.	12.51	2 / 0.17 / 0.12	2129. / 2063.		20	0.05 / 210 25	31.0 / 32.46	13.6 / 19.7 / -19.8	-19.95 / 17.2	10.17 / 9.03	0.07	ARP 12 / 4484 05-20-027 2
0833-31	833.6 -3159 / 255.15 5.06 / 156.80 -61.84	8 / 0.12 / 31+21	3.9 / 3.7 / 6	0.20 / 90.		0.67 / 1.47	1552. / 1266. / 2	248	20	0.68 / 39 / 1.04	19.7 / 31.47	-8.5 / 3.7 / -17.3	21.3	9.27		UGCA 143 / 431-18 -5-21-000
U 4499	833.9 5149 / 167.11 37.18 / 42.99 -19.07	8X / 0.22 / 15-10	2.9 / 2.8 / 2	0.82 / 39.	13.0	3 / 0.07 / 0.12	696. / 749. / 1	142 / 181. / 184.	10	0.84 / 11 / 1.08	12.8 / 30.54	8.9 / 8.3 / -4.2	-17.54 / 10.5	9.21 / 9.05 / 10.00	0.70 / 0.114 / 6.16	4499 09-14-078 2
0834-26	834.1 -2614 / 248.57 8.58 / 145.15 -64.13	10 P / 0.02 / 54 -0 +1	1.5 / 1.6 / 6	0.88 / 32.		0.04 / 0.58	880. / 604.		20	0.61 / 219	10.5 / 30.10	-3.8 / 2.6 / -9.4	4.9	8.65		495-21 -4-21-005

Table columns: (1) Name · (2) α, l, SGL · (3) δ, b, SGB · (4) Type, Q_{xyz}, Group · (5) D_{25}, D^b_{25}, source · (6) d/D, i · (7) $B^{b,i}_T$, $H_{-0.5}$, B–H · (8) source, A^{i-o}, A^b_B · (9) V_h, V_o, Tel · (10) W_{20}, W_R, W_D · (11) e_v, e_w, source · (12) F_c, source/e_f, f_h · (13) R, μ · (14) SGX, SGY, SGZ · (15) $M^{b,i}_B$, Δ_{25} · (16) $\log L_B$, $\log M_H$, $\log M_T$ · (17) M_H/L_B, M_H/M_T, M_T/L_B · (18) Alternate Names — UGC/ESO, MCG, AGN? RCII

(1)	(2)	(3)	(4)	(5)	(6)	(7)	(8)	(9)	(10)	(11)	(12)	(13)	(14)	(15)	(16)	(17)	(18)
U 4508	835.4	−217	13 P	0.6	1.00	14.2	3	1943.		45		26.7	−0.3	−17.93	9.36		4508 00-22-025 2
	228.11	22.27	0.09	0.6			0.00	1736.				32.13	12.8	4.7			
	91.52	−61.41	31 -0	2			0.06						−23.4				
U 4514	835.9	5338	5B	2.3	0.52	13.2	3	697.	172	10	0.69	13.0	9.1	−17.36	9.14	0.61	4514 09-14-081 2
	164.81	37.40	0.22	2.0	64.	12.18	0.27	759.	155.		16	30.56	8.4	7.6	8.92	0.156	
	42.44	−17.30	15-10	2		1.02	0.11	1	159.		1.04		−3.9		9.72	3.87	
0839-74	839.4	−7459	7	2.4	0.50	13.6		1355.		20		15.9	−13.2				036-06
	288.88	−19.75	0.17	2.4	65.		0.29	1088.				31.01	−4.6	11.1			
	199.13	−28.88	53+17	6			0.59	7					−7.7				
U 4543	839.9	4555	8A	3.4	0.60	13.6	3	1961.	137	15	0.79	30.0	18.7	−18.78	9.70	1.10	4543 08-16-000
	174.56	38.17	0.15	3.1	58.		0.20	1983.	128.		10	32.38	20.0	27.2	9.74	0.430	
	46.82	−23.88	21 -0	2			0.12	1	133.		1.09		−12.1		10.11	2.55	
U 4559	841.1	3018	4	2.9	0.23	13.5	3	2086.	357	15	0.60	30.7	13.6	−18.94	9.77	0.64	4559 05-21-004
	194.08	36.33	0.15	2.1	86.	10.54	0.67	2027.	320.		18	32.44	20.4	18.8	9.57	0.067	
	56.34	−37.19	32 +2	2		2.96	0.14	1	322.		1.05		−18.6		10.75	9.54	
U 4576	842.6	7343	3B	1.7	0.22	13.30	2	2585.		64		38.8	32.1	−19.65	10.05		IC 2389
	140.32	34.07	0.22	1.2	87.		0.67	2744.				32.95	21.8	13.6			4576 12-09-011 2
	34.20	1.08	12-21	2			0.04						0.7				
N 2633	842.6	7417	33	2.7	0.64	12.62	2	2141.		43		33.2	27.5	−19.99	10.19		ARP 80
	139.68	33.88	0.30	2.5	55.		0.17	2302.				32.61	18.5	24.2			4574 12-09-013 2
	33.94	1.58	12+17	2			0.06						0.9				
N 2663	843.1	−3337	−5	3.3	0.63			2309.		33		29.6	−14.0	37.2			371-14 -6-20-001 2
	255.67	5.64	0.06	4.3	56.		0.00	2020.				32.35	5.7				
	157.70	−59.30	31 -0	6			1.85						−25.4				
N 2654	845.2	6024	1	3.9	0.20	11.94	2	1382.		10	0.78	23.3	17.4	−19.89	10.15	0.23	4605 10-13-017 2
	156.13	37.81	C.26	2.8	90.		0.67	1478.			211 66	31.83	14.8	19.1	9.51		
	40.47	−10.69	13 -6 +5	2			0.14						−4.3				
0848+73	848.0	7323	13 P	0.9	0.87			2399.		39		36.5	30.0	9.6			UGCA 146, MARK 16
	140.54	34.54	0.27	0.9	33.		0.00	2556.				32.81	20.8				12-10-000 2
	34.69	0.95	12-21	2			0.06						0.6				
N 2655	849.1	7825		5.9	0.90	10.92	1	1445.		23	1.13	24.4	20.5	−21.02	10.60	0.20	ARP 225
	134.92	32.69	0.23	5.8	29.		0.00	1624.			51 32	31.94	13.0	41.3	9.90		4637 13-07-010 2
	32.39	5.43	12-10	2			0.03						2.3				
N 2683	849.6	3338	3A	9.1	0.30	9.82	1	415.	446.	10	1.28	5.7	2.7	−18.96	9.78	0.10	4641 06-20-011 2
	190.43	38.77	0.09	7.1	79.	6.90	0.67	373.	415.		60 7	28.78	3.9	11.8	8.79	0.010	
	55.90	−33.42	15 -6	9		2.92	0.07	2	416.		1.09		−3.1		10.77	9.89	
N 2681	850.0	5130		3.6	1.00	11.03	1	710.		27	0.87	13.3	8.9	−19.59	10.03	0.12	4645 09-15-041 2 L/S
	167.33	39.69	0.20	3.6			0.00	761.			6 74	30.62	9.0	14.0	9.12		
	45.48	−18.19	15-10	2			0.07						−4.1				

51

U 4646 — IC 512; 4646 14-05-002

851.0 8542	6X	3.4	11.9
127.26 29.52	0.14	3.4 0.77	3 0.09
28.93 11.86	42+15	44. 2	0.23

1614. 1822. 1 · 149 · 15 · 0.56 17 1.11 · 26.7 32.14 · 22.9 12.7 5.5 · −20.24 26.5 · 10.29 9.41 · 0.13

N 2685 — ARP 336 L/S; 4666 10-13-039 2

851.7 5856	−2BP	5.1 0.55	11.75
157.77 38.90	0.13	4.6 62.	1 0.00
41.91 −11.59	13 −4	2	0.15

877. 965. · 304 303. 305. · 7 7 216 · 0.88 216 13 · 16.2 31.05 · 11.8 10.6 −3.3 · −19.30 21.8 · 9.91 9.30 10.76 · 0.24 0.034 7.09

N 2714 — 125-07

852.3 −5902	−2	1.4 1.00	
276.41 −9.24	0.15	1.8	0.00
186.80 −41.33	33 −0 +1	6	1.38

2715. 2421. · 60 · 33.7 32.64 · −25.1 −3.0 −22.2 · 17.7

N 2708 — 00-23-015 2

853.6 −310	3A	2.9 0.48	
231.55 25.67	0.09	2.4 67.	0.32
97.15 −57.76	31 −0	4	0.02

2007. 1795. 1 · 476 · 40 21 217 · 0.62 217 17 1.07 · 27.8 32.22 · −1.8 14.7 −23.5 · 19.5 · 9.51

U 4683 — 4683 10-13-046

854.0 5916	10	2.0 0.59	
157.27 39.12	0.13	1.8 59.	0.21
42.01 −11.16	13 −4	2	0.15

909. 999. 1 · 79 71. 79. · 15 · 1.03 · 16.9 31.14 · 12.3 11.1 −3.3 · 8.9 · 9.11

N 2701 — 4695 09-15-063 2

855.3 5357	4XP	2.2 0.84	12.69 10.95 1.74
164.04 40.22	0.11	2.1 36.	2 0.06
44.91 −15.67	21+20	9	0.05

2328. 2391. 1 · 254 364. 366. · 10 · 0.73 18 1.03 · 35.1 32.73 · 23.9 23.9 −9.5 · −20.04 21.5 · 10.21 9.82 10.92 · 0.41 0.080 5.12

U 4704 — 4704 07-19-011

855.8 3924	8	3.9 0.13	13.7
183.19 40.74	0.13	2.8 90.	3 0.67
53.34 −27.95	15 +6	2	0.05

599. 586. 1 · 137 109. 114. · 10 · 0.73 14 1.08 · 8.7 29.71 · 4.6 6.2 −4.1 · −16.01 7.1 · 8.60 8.61 9.39 · 1.03 0.166 6.21

N 2712 — 4708 08-17-003 2

856.2 4507	3B	2.7 0.56	12.43
175.64 41.01	0.17	2.4 61.	2 0.23
49.94 −23.12	21 −0	9	0.05

1840. 1857. · 200 · 28.6 32.28 · 16.9 20.1 −11.2 · −19.85 20.0 · 10.13

0857-68 — 060-19

857.0 −6852	5B	3.2 0.38	10.77
284.54 −15.05	0.17	2.7 74.	0.48
193.91 −33.27	53−22+21	6	0.36

1440. 1159. 7 · 234 205. 207. · 20 · 17.1 31.17 · −13.9 −3.4 −9.4 · 13.5 · 10.22

0859-26 — 497-02 -4-22-000

859.3 −2607	9	2.2 0.63	
252.03 13.10	0.07	2.3 55.	0.17
141.62 −58.73	31 −0+21	6	0.65

1960. 1682. 2 · 185 · 15 · 0.22 49 · 25.6 32.05 · −10.4 8.3 −21.9 · 17.2 · 9.04

N 2735 — ARP 287, VV 40; ZW COMPACT; 4744 04-22-002 2

859.7 2608	13 P	1.2 0.40	13.9
200.47 39.27	0.15	1.0 72.	3 0.00
63.63 −38.08	32 −1	2	0.10

2615. 2534. · 185 · 37.5 32.87 · 13.1 26.4 −23.1 · −18.97 10.9 · 9.78

0859+26 — NGC 2735A, ARP 287; VV 40; 04-22-003 2

859.7 2608	10 P		
200.47 39.27	0.11		0.00
63.63 −38.08	32 −1		0.10

2775. 2694. · 185 · 39.4 32.98 · 13.8 27.8 −24.3

N 2726 — 4750 10-13-054 2

901.0 6008	1	1.9 0.31	12.4
155.92 39.78	0.32	1.5 79.	3 0.64
42.33 −9.96	13 −6 +5	2	0.15

1432. 1526. · 105 · 24.0 31.90 · 17.5 15.9 −4.1 · −19.50 10.5 · 9.99

N 2715 — 4759 13-07-015 2

901.8 7817	5X	4.6 0.38	11.47
134.73 33.32	0.16	3.7 74.	1 0.47
33.02 5.63	12−10	2	0.01

1126. 1304. · 74 · 20.4 31.55 · 17.0 11.1 2.0 · −20.08 22.0 · 10.22

(1) Name	(2) α ℓ SGL	(3) δ b SGB	(4) Type Q_xyz Group	(5) D_25 D_25^{b,i} source	(6) d/D i	(7) B_T^{b,i} H_{-0.5} B-H	(8) source A_B^{i-o} A_B^b	(9) V_h V_o Tel	(10) W_20 W_R W_D	(11) e_v e_w source	(12) F_c source/e_F f_h	(13) R μ	(14) SGX SGY SGZ	(15) M_B^{b,i} Δ_25^{b,i}	(16) log L_B log M_H log M_T	(17) M_H/L_B M_H/M_T M_T/L_B	(18) Alternate Names UGC/ESO — Alternate Names MCG — AGN? RCII
N 2750	902.9 201.34 64.74	2538 39.84 -37.99	5X 0-15 32 -1	2.2 2.2 2	0.92 26.	12.3	3 0.03 0.12	2684. 2600. 1	203	30	0.73 14 1.05	38.4 32.92	12.9 27.3 -23.6	-20.62 24.7	10.44 9.90	0.29	4769 04-22-012 2
U 4777	903.6 189.47 57.85	3449 41.81 -30.77	10 0.27 21-18	2.4 1.7 2	0.20 90.	14.2	3 0.67 0.02	2056. 2019. 1	212 176. 179.	15	0.28 24 1.03	30.9 32.45	14.1 22.5 -15.8	-18.25 15.3	9.49 9.26 10.14	0.59 0.132 4.44	4777 06-20-024
N 2742	903.6 155.12 42.32	6041 39.95 -9.33	5A 0.18 13 -6 +5	3.2 2.9 2	0.55 61.	11.90	2 0.23 0.17	1296. 1393. 1	326	10 50 211	0.74 211 25	22.2 31.73	16.2 14.7 -3.6	-19.83 18.8	10.12 9.43	0.20	4779 10-13-057 2
U 4787	904.4 191.30 58.97	3328 41.79 -31.74	8 0.11 15 -6	2.3 1.7 2	0.26 84.	13.4	3 0.67 0.03	553. 509. 1	148 119. 124.	15	0.27 21 1.03	7.8 29.47	3.4 5.7 -4.1	-16.07 3.9	8.62 8.05 9.20	0.27 0.071 3.81	4787 06-20-028
N 2763	904.5 244.01 120.68	-1517 20.86 -58.28	6 0.22 31-14+12	2.0 2.1 1	1.00	12.52	1 0.00 0.18	1892. 1640. 1	210	15 11 79	0.64 217 9 1.07	25.7 32.05	-6.9 11.6 -21.8	-19.53 15.8	10.00 9.46	0.29	-2-23-010 2
N 2764	905.4 206.50 68.76	2139 39.23 -40.54	-2 0.14 32 +1	1.6 1.5 2	0.66 53.	13.27	2 0.00 0.13	2627. 2523.		90		37.4 32.86	10.3 26.5 -24.3	-19.59 16.4	10.03		4794 04-22-017 2
U 4797	905.5 224.12 85.77	608 32.98 -50.59	9 0.07 31-23	2.5 2.6 2	1.00	14.54	7 0.00 0.11	1315. 1140. 1	101	10	0.25 41 1.07	19.6 31.46	0.9 12.4 -15.1	-16.92 14.9	8.96 8.83	0.75	DDO 54
N 2770	906.4 191.60 59.48	3318 42.17 -31.61	5A 0.26 21-18	3.4 2.7 2	0.31 79.	11.4 10.45 0.95	3 0.64 0.02	1953. 1908. 1	342 310. 312.	15	0.90 17 1.07	29.6 32.36	12.8 21.7 -15.5	-20.96 23.3	10.58 9.84 10.81	0.18 0.107 1.73	4806 06-20-038 2
N 2732	906.9 133.42 32.64	7924 33.01 6.71	-2 0.18 42 -0+16	1.9 1.6 2	0.46 68.	12.83	2 0.00 0.02	2023. 2206.		87		31.8 32.51	26.6 17.1 3.7	-19.68 14.9	10.06		4818 13-07-016 2
0907-22	907.5 250.64 134.82	-2248 16.66 -57.46	10 0.06 54 -3 +1	1.7 1.9 6	0.75 45.		0.10 0.75	724. 453. 2	84	10	-0.02 74 1.01	7.0 29.21	-2.6 2.7 -5.9	3.9	7.67		UGCA 148, DDO 56
0907-33	907.5 258.64 152.90	-3309 9.83 -54.90	10 0.07 54 -0 +1	1.0 1.2 6	0.64 55.		0.17 1.28	1136. 846. 1	46 30. 45.	7		14.2 30.77	-7.3 3.7 -11.6	5.0	8.11		372-07 -5-22-000
N 2775	907.7 223.26 84.81	715 34.00 -49.51	2A 0.07 31-23	4.6 4.5 2	0.82 39.	11.02	2 0.07 0.11	1135. 965.		75		17.0 31.15	1.0 11.0 -12.9	-20.13 22.3	10.24		4820 01-24-005 2
N 2768	907.8 155.48 43.00	6015 40.57 -9.42	-5 0.31 13 -6 +5	7.3 6.4 2	0.48 67.	10.74	2 0.00 0.16	1408. 1503.		175		23.7 31.87	17.1 15.9 -3.9	-21.13 44.3	10.64		4821 10-13-065 2 L2

Name	Coordinates	Type	Size	Ratio/PA	Mag		Velocity		Index	Distance		Vectors	M			Alt. Name
0908-32	908.0 -3257 / 258.56 10.04 / 152.51 -54.87	10 / 0.07 / 31 -0+21	2.2 / 2.6 / 6	0.74 / 46.		0.10 / 1.05	1541. / 1251. / 2	154	25	0.45 / 25 / 1.04	19.9 / 31.50	-10.2 / 5.3 / -16.3	15.1	9.05		UGCA 149 / 372-08 -5-22-001
N 2748	908.0 7641 / 136.25 34.36 / 34.15 4.45	4A / 0.23 / 12-10	2.9 / 2.4 / 2	0.43 / 71.	12.00	0.39 / 0.01	1391. / 1562.		61		23.8 / 31.88	19.6 / 13.3 / 1.8	-19.88 / 16.7	10.14		4825 13-07-019 2
N 2788	908.3 -6744 / 284.34 -13.52 / 192.08 -33.51	12 / 0.14 / 53-22+21	1.5 / 1.2 / 6	0.22 / 87.		0.67 / 0.49	1538. / 1255.		47		18.4 / 31.33	-15.0 / -3.2 / -10.2	6.4			061-02 2
N 2776	908.9 4510 / 175.48 43.25 / 51.92 -21.83	5X / 0.10 / 21 -0	3.2 / 3.2 / 2	1.00	12.17	2 / 0.00 / 0.03	2626. / 2643.	229	9 / 40 / 54	0.91 / 54 11	38.7 / 32.94	22.1 / 28.2 / -14.4	-20.77 / 36.2	10.50 / 10.09	0.38	4838 08-17-056 2
U 4837	909.0 3543 / 188.44 43.01 / 58.22 -29.38	9 / 0.30 / 21-18	2.1 / 1.9 / 2	0.56 / 60.	14.59	7 / 0.22 / 0.01	1880. / 1848. / 1	149 / 137. / 141.	10	0.40 / 17 / 1.03	28.9 / 32.30	13.3 / 21.4 / -14.2	-17.71 / 16.0	9.28 / 9.32 / 9.94	1.11 / 0.240 / 4.62	DDO 55 / 4837 06-20-042 2
0909-14	909.0 -1450 / 244.35 21.99 / 120.04 -57.16	10 / 0.35 / 31-14+12	1.9 / 1.9 / 4	1.00	14.37	7 / 0.00 / 0.15	2054. / 1803. / 1	134	15	0.53 / 16 / 1.05	27.9 / 32.23	-7.6 / 13.1 / -23.4	-17.86 / 15.5	9.34 / 9.42	1.22	DDO 57 / -2-24-001 2
N 2781	909.1 -1436 / 244.17 22.16 / 119.62 -57.11	1X / 0.39 / 31-14+12	3.9 / 3.5 / 4	0.51 / 64.	11.97	2 / 0.28 / 0.15	2028. / 1778. / 1	418	40	0.59 / 26 / 1.14	27.5 / 32.20	-7.4 / 13.0 / -23.1	-20.23 / 28.1	10.28 / 9.47	0.15	-2-24-002 2
N 2778	909.2 3513 / 189.12 43.00 / 58.62 -29.75	-5 / 0.28 / 21-18	1.6 / 1.5 / 2	0.73 / 48.	13.0	3 / 0.00 / 0.02	2052. / 2017.		35		31.0 / 32.46	14.0 / 23.0 / -15.4	-19.46 / 13.6	9.98		TURN 3A / 4840 06-20-043
U 4841	909.4 7426 / 138.64 35.47 / 35.44 2.60	7X / 0.23 / 12+10	4.0 / 3.8 / 2	0.77 / 44.	12.95	1 / 0.09 / 0.03	1128. / 1289. / 1	200 / 237. / 240.	10	0.87 / 14 / 1.08	20.4 / 31.55	16.6 / 11.8 / 0.9	-18.60 / 22.6	9.63 / 9.49 / 10.57	0.72 / 0.084 / 8.60	A 0909, HOLMBERG III / 4841 12-09-032 2
N 2780	909.6 3507 / 189.27 43.07 / 58.76 -29.78	12BP / 0.16 / 21-18	1.1 / 1.1 / 2	0.73 / 47.	13.9	3 / 0.11 / 0.02	2208. / 2173.		35		32.9 / 32.59	14.8 / 24.4 / -16.3	-18.69 / 10.6	9.67		TURN 3B / 4843 06-20-047
0909-19	909.8 -1956 / 248.70 18.92 / 129.45 -57.14	7 / 0.21 / 31 -0+12	5.4 / 4.2 / 6	0.07 / 90.	10.54	0.67 / 0.49	2178. / 1914. / 2	356 / 318. / 320.	15	0.95 / 17 / 1.03	29.1 / 32.32	-10.0 / 12.2 / -24.4	35.7	9.88 / 11.02	0.072	UGCA 151 / 564-27 -3-24-001
N 2784	910.1 -2358 / 251.97 16.36 / 136.81 -56.72	-2A / 0.07 / 54 -3 +1	6.4 / 6.6 / 6	0.56 / 61.	12.07	2 / 0.00 / 0.71	708. / 434.		94		7.1 / 29.26	-2.8 / 2.7 / -5.9	-18.72 / 13.7	9.68		UGCA 152 / 497-23 -4-22-005 2
N 2782	910.9 4019 / 182.16 43.68 / 55.38 -25.51	1XP / 0.09 / 21 -0	3.9 / 3.7 / 2	0.78 / 43.		2 / 0.08 / 0.00	2537. / 2529.		11	0.61 / 6 74	37.3 / 32.86	19.1 / 27.7 / -16.1	-20.79 / 40.3	10.51 / 9.75	0.18	ARP 215 / 4862 07-19-036 2
0911-19	911.0 -1913 / 248.31 19.60 / 128.13 -56.88	10 / 0.05 / 54 -3 +1	2.3 / 2.2 / 6	0.61 / 57.		0.19 / 0.30	772. / 509. / 2	130	25	0.47 / 22 / 1.01	8.0 / 29.51	-2.7 / 3.4 / -6.7	5.1	8.28		UGCA 153 / 564-30 -3-24-000

(1) Name	(2) α / ℓ / SGL	(3) δ / b / SGB	(4) Type / Q_{xyz} / Group	(5) $D_{25}^{b,i}$ / D_{25} / source	(6) d/D / i	(7) $B_T^{b,i}$ / $H_{-0.5}$ / B-H	(8) source / A_B^{i-o} / A_B^b	(9) V_h / V_o / Tel	(10) W_{20} / W_R / W_D	(11) e_v / e_w / source	(12) F_c / source/e_F / f_h	(13) R / μ	(14) SGX / SGY / SGZ	(15) $M_B^{b,i}$ / $\Delta_{25}^{b,i}$	(16) $\log L_B$ / $\log M_H$ / $\log M_T$	(17) M_H/L_B / M_H/M_T / M_T/L_B	(18) Alternate Names UGC/ESO / Alternate Names MCG / AGN? RCII
N 2836	913.1 / 285.72 / 192.77	-6908 / -14.12 / -32.16	12 / 0.12 / 53-22+21	2.8 / 2.8 / 6	0.71 / 49.		/ 0.12 / 0.41	1702. / 1422. / 7		20		20.6 / 31.57	-17.0 / -3.9 / -11.0	16.8			061-03 / / 2
0913+53	913.2 / 163.83 / 47.38	5339 / 42.87 / -14.49	13 P / 0.12 / 21-20	0.6 / 0.6 / 2	0.66 / 53.	14.8	3 / 0.00 / 0.02	2225. / 2286.		43		34.0 / 32.66	22.3 / 24.2 / -8.5	-17.86 / 6.0	9.34		UGCA 154, MARK 104 / 09-15-000 / 2
U 4888	913.3 / 139.00 / 35.92	7358 / 35.92 / 2.36	5A / 0.30 / 12 -0	3.6 / 3.1 / 2	0.50 / 65.	12.33 / 10.52 / 1.81	1 / 0.29 / 0.03	2258. / 2417. / 1	308. / 298. / 300.	10	0.86 / 18 / 1.09	34.8 / 32.71	28.1 / 20.4 / 1.4	-20.38 / 31.5	10.34 / 9.94 / 10.91	0.40 / 0.108 / 3.67	IC 529 / 4888 12-09-035 / 2
0913-601	913.4 / 279.09 / 185.27	-6013 / -8.02 / -38.66	5 / 0.25 / 33 -2 +1	1.7 / 2.1 / 6	0.75 / 45.		/ 0.10 / 1.32	2913. / 2620. / 7		20		36.3 / 32.80	-28.2 / -2.6 / -22.7	22.3			126-03
0913-603	913.4 / 279.32 / 185.56	-6032 / -8.24 / -38.44	5 / 0.23 / 33 -2 +1	1.8 / 2.4 / 6	1.00		/ 0.00 / 1.27	2994. / 2701. / 7		20		37.4 / 32.86	-29.1 / -2.8 / -23.2	26.2			126-04
N 2793	913.7 / 190.06 / 59.86	3438 / 43.85 / -29.62	9BP / 0.29 / 21-15+12	1.3 / 1.3 / 9	0.87 / 34.	13.8	3 / 0.05 / 0.00	1674. / 1636.		22		26.3 / 32.10	11.5 / 19.8 / -13.0	-18.30 / 10.0	9.51		4894 06-21-002 / / 2
N 2811	913.8 / 246.20 / 122.52	-1606 / 22.10 / -56.11	C.09 / / 31 -0+12	2.5 / 2.1 / 2	0.40 / 72.	12.11	2 / 0.00 / 0.14	2514. / 2260.		75		33.7 / 32.64	10.1 / 15.8 / -27.9	-20.53 / 20.7	10.40		UGCA 155 / -3-24-003 / 2
N 2815	914.1 / 252.18 / 135.65	-2326 / 17.40 / -55.88	3B / 0.21 / 31 -0+12	3.7 / 3.3 / 6	0.34 / 77.	11.73	6 / 0.56 / 0.58	2289. / 2016.		30		30.3 / 32.41	-12.2 / 11.9 / -25.1	-20.68 / 29.2	10.46		UGCA 156 / 497-32 -4-22-006 / 2
N 2798	914.2 / 179.52 / 54.66	4213 / 44.31 / -23.63	-BP / 0.25 / 21-16	2.7 / 2.1 / 2	0.36 / 75.	12.49	2 / 0.52 / 0.00	1708. / 1710.		75		27.1 / 32.16	14.3 / 20.2 / -10.8	-19.67 / 16.6	10.06		ARP 283, VV 50 / 4905 07-19-055 / 2
N 2799	914.3 / 179.54 / 54.69	4212 / 44.33 / -23.63	9BP / 0.28 / 21-16	2.1 / 1.6 / 2	0.28 / 81.	13.5	3 / 0.67 / 0.00	1737. / 1739.		40		27.4 / 32.19	14.5 / 20.5 / -11.0	-18.69 / 12.8	9.67		ARP 283, VV 50 / 4909 07-19-056 / 2
N 2842	914.5 / 281.11 / 187.49	-6251 / -9.74 / -36.70	-2B / 0.23 / 33 -1	1.9 / 2.2 / 6	0.80 / 41.		/ 0.00 / 1.06	2780. / 2490.		90		34.6 / 32.69	-27.5 / -3.6 / -20.7	22.2			091-04
N 2787	914.8 / 144.04 / 38.55	6925 / 38.04 / -1.35	-2B / 0.06 / 12 -0	3.8 / 3.6 / 2	0.67 / 52.	11.65	2 / 0.00 / 0.16	620. / 759.		35		13.0 / 30.58	10.2 / 8.1 / -0.3	-18.93 / 13.7	9.76		4914 12-09-039 / / L/S 2
U 4922	915.2 / 171.34 / 51.03	4805 / 44.08 / -18.83	9A / 0.11 / 21 -0	3.7 / 3.3 / 2	0.53 / 63.	12.9	3 / 0.25 / 0.01	1993. / 2025. / 1	254	15	0.53 / 17 / 1.10	30.8 / 32.45	18.4 / 22.7 / -10.0	-19.55 / 29.7	10.01 / 9.51	0.31	4922 08-17-074

N 2835
915.7 -2209 18.52 / 251.43 / 133.31 -55.64
5B 0.12 54 -3 +1
6.8 0.72 48. / 6.9 0.9 6
10.5 5 0.11 / 8.96 0.38 / 1.54
890. / 620.
220 247. 249. / 8
1.57 36 5 1.08
10.8 30.16
-4.2 4.4 -8.9
-19.66 21.8
10.06 9.64 10.59
0.38 0.113 3.38
UGCA 157 / 564-35 -4-22-008 2

N 2805
916.4 6419 40.20 / 150.01 / 41.56 -5.48
7X 0.19 13 -5
6.7 0.71 49. / 6.4 2
11.52 1 0.12 / 0.14
1736. 1851. 1
116 121. 126. / 8 5 211
1.34 217 5 1.24
28.0 32.24
20.9 18.5 -2.7
-20.72 52.3
10.48 10.23 10.35
0.57 0.772 0.74
4936 11-12-003 2

0916-62
916.4 -6240 -9.46 / 281.13 / 187.14 -36.68
5 0.13 33 -3
2.2 0.56 61. / 2.4 6
0.23 1.10
2121. 1831. 7
20
26.2 32.09
-20.8 -2.6 -15.6
18.4
091-07

N 2814
917.1 6428 40.22 / 149.79 / 41.54 -5.31
13 0.25 13 -5
1.7 0.22 87. / 1.3 2
14.15 2 0.00 / 0.15
1663. 1778.
95
27.1 32.17
20.2 17.9 -2.5
-18.02 10.3
9.40
4952 11-12-004 2

0917-12
917.4 -1201 25.34 / 243.33 / 115.57 -54.74
10 0.34 31-16+12
1.5 0.83 38. / 1.5 4
0.06 / 0.15
1945. 1703. 1
67 81. 88. / 8
0.54 15 1.03
26.7 32.14
-6.7 13.9 -21.8
11.7
9.39 9.35
1.111
DDO 60 / -2-24-011 2

0917+64
917.5 6427 40.26 / 149.79 / 41.58 -5.30
13 P 0.30 13 -5
0.7 0.44 70. / 0.6 2
14.6 3 0.00 / 0.15
1467. 1582.
20
24.7 31.96
18.4 16.3 -2.3
-17.36 4.3
9.14
IC 2458, UGCA 159 / MK 108, 7 ZW 276 / 11-12-005 2

N 2820
917.7 6429 40.27 / 149.73 / 41.58 -5.26
5BP 0.28 13 -5
4.1 0.12 90. / 2.9 2
12.46 1 0.67 / 0.15
1576. 1691. 1
373 335. 337. / 25
1.14 12 1.06
26.0 32.08
19.4 17.2 -2.4
-19.62 22.0
10.04 9.97 10.86
0.85 0.130 6.53
4961 11-12-006 2

N 2848
917.8 -1618 22.71 / 247.03 / 122.99 -55.17
5X 0.43 31+14+12
2.6 0.76 45. / 2.6 2
12.43 2 0.09 / 0.14
2044. 1789. 1
208 245. 247. / 10
0.89 15 1.07
27.7 32.21
-8.6 13.3 -22.7
-19.78 21.0
10.10 9.77 10.56
0.47 0.163 2.88
UGCA 160 / -3-24-007 2

0918-12
918.3 -1223 25.29 / 243.80 / 116.26 -54.58
10 0.33 31-16+12
1.3 0.79 41. / 1.2 4
0.08 / 0.14
1906. 1663. 1
96 114. 119. / 10
0.47 17 1.02
26.2 32.09
-6.7 13.6 -21.4
9.2
9.31 9.54
0.586
DDO 61 / -2-24-012 2

N 2844
918.6 4022 45.15 / 182.11 / 56.62 -24.57
1A 0.14 21-17+16
1.9 0.51 64. / 1.7 2
13.37 2 0.28 / 0.00
1485. 1477.
90
24.1 31.91
12.1 18.3 -10.0
-18.54 12.0
9.61
4971 07-19-064 2

N 2841
918.6 5112 44.15 / 166.93 / 49.56 -15.99
3A 0.13 15+10
6.6 0.51 64. / 5.7 2
9.81 1 0.28 / 6.88 0.01 / 2.93
637. 686. 2
611 636. 637. / 10 6 211
1.63 211 7 1.09
12.0 30.39
7.5 8.8 -3.3
-20.58 20.0
10.42 9.79 11.37
0.23 0.026 8.83
4966 09-16-005 L2

N 2855
919.0 -1142 25.85 / 243.32 / 115.15 -54.30
10 0.32 31-16+12
2.4 0.92 26. / 2.4 2
12.30 2 0.00 / 0.15
1901. 1660.
26
26.3 32.10
-6.5 13.9 -21.3
-19.80 18.4
10.11
UGCA 161 / -2-24-015 2

0919-22
919.2 -2217 19.04 / 252.10 / 133.43 -54.82
10 0.11 54 -3 +1
3.6 0.22 88. / 2.7 6
0.67 / 0.27
849. 579. 2
141 112. 117. / 10
0.88 11 1.01
9.9 29.99
-3.9 4.2 -8.1
7.8
8.87 9.45
0.262
UGCA 162, DDO 62 / 565-01 -4-22-009 2

0919-68
919.2 -6842 -13.43 / 285.78 / 191.94 -32.11
5 0.08 33 -0
1.6 0.37 75. / 1.4 6
0.50 / 0.45
2353. 2072. 7
20
28.9 32.30
-23.9 -5.1 -15.4
11.8
061-08

(1) Name	(2) α / ℓ / SGL	(3) δ / b / SGB	(4) Type / Q_{xyz} / Group	(5) D_{25} / $D^{b,i}_{25}$ / source	(6) d/D / i	(7) $B^{b,i}_T$ / $H_{-0.5}$ / B-H	(8) source / A^{i-0}_B / A^b_B	(9) V_h / V_o / Tel	(10) W_{20} / W_R / W_D	(11) e_v / e_w / source	(12) F_c source/e_f / f_h	(13) R / μ	(14) SGX / SGY / SGZ	(15) $M^{b,i}_B$ / $\Delta^{b,i}_{25}$	(16) log L_B / log M_H / log M_T	(17) M_H/L_B / M_H/M_T / M_T/L_B	(18) Alternate Names / UGC/ESO MCG / AGN? RCII
N 2852	920.0 / 182.09 / 56.84	4023 / 45.41 / -24.39	-2 / 0.26 / 21-17+16	1.2 / 1.2 / 2	1.00	13.6	3 / 0.00 / 0.00	1888. / 1880.		40		29.3 / 32.33	14.6 / 22.3 / -12.1	-18.73 / 10.3	9.68		TURN 6A / 4986 07-19-065 2
N 2853	920.1 / 182.04 / 56.83	4025 / 45.43 / -24.35	1 / 0.31 / 21-17+16	1.9 / 1.7 / 2	0.51 / 64.	13.6	3 / 0.28 / 0.00	1799. / 1792.		39		28.2 / 32.25	14.0 / 21.5 / -11.6	-18.65 / 14.0	9.65		TURN 6B / 4987 07-19-066 2
N 2865	921.2 / 252.95 / 134.51	-2257 / 18.93 / -54.31	-5 / 0.11 / 31+17+12	2.5 / 2.6 / 6	0.93 / 24.	12.07	2 / 0.00 / 0.28	2714. / 2443.		75		35.7 / 32.77	-14.6 / 14.9 / -29.0	-20.70 / 27.1	10.47		498-01 -4-22-011 2
N 2859	921.3 / 190.15 / 61.15	3444 / 45.41 / -28.53	-2B / 0.32 / 21-15+12	5.0 / 4.9 / 2	0.90 / 30.	11.65	2 / 0.00 / 0.00	1594. / 1557.	155	20 / 20 / 210		25.4 / 32.03	10.8 / 19.6 / -12.2	-20.38 / 36.3	10.34 / 8.48	0.01	5001 06-21-030 2
N 2887	922.2 / 282.27 / 187.39	-6336 / -9.64 / -35.55	-5 / 0.24 / 33 -1	1.5 / 1.7 / 6	0.67 / 53.		/ 0.00 / 1.08	2850. / 2561.		90		35.5 / 32.75	-28.6 / -3.7 / -20.6	/ 17.6			091-09
0922-24	922.4 / 254.66 / 137.76	-2454 / 17.81 / -53.82	4XP / 0.23 / 31+17+12	1.5 / 1.5 / 6	0.82 / 38.	13.64	6 / 0.07 / 0.38	2413. / 2138.		25		31.9 / 32.52	-13.9 / 12.7 / -25.7	-18.88 / 14.0	9.74		SCI 84 / 498-05 -4-23-001 2
N 2872	923.0 / 220.53 / 82.72	1139 / 39.36 / -43.89	-5 / 0.12 / 32 -0	1.9 / 1.9 / 2	0.95 / 22.	12.5	3 / 0.00 / 0.08	2976. / 2825.		95		41.4 / 33.09	3.8 / 29.6 / -28.7	-20.59 / 23.0	10.43		ARP 307
0923+19	923.2 / 210.85 / 74.42	1936 / 42.50 / -38.93	13 P / 0.09 / 32 -0	0.9 / 0.9 / 2	0.87 / 33.	14.1	3 / 0.00 / 0.08	2494. / 2381.		220		35.9 / 32.78	7.5 / 26.9 / -22.6	-18.68 / 9.4	9.66		5018 02-24-008 2 / UGCA 164, MARK 400 / 03-24-058 2
N 2888	924.1 / 257.14 / 142.49	-2749 / 16.08 / -52.97	-5 / 0.15 / 31 -0	1.7 / 1.7 / 6	0.71 / 49.	12.99	2 / 0.00 / 0.51	2233. / 1952.		90		29.4 / 32.34	-14.0 / 10.8 / -23.4	-19.35 / 14.6	9.93		434-02 -5-23-001 2
0925-31	925.3 / 260.28 / 148.73	-3148 / 13.48 / -51.78	10 / 0.22 / 54 -1	1.7 / 1.9 / 6	1.00		/ 0.00 / 0.67	1082. / 794. / 2	98	15	0.33 / 24 / 1.01	13.6 / 30.67	-7.2 / 4.4 / -10.7	/ 7.5	/ 8.60	UGCA 165 / 434-05 -5-23-000
N 2880	925.7 / 151.45 / 43.35	6243 / 41.77 / -6.17	-2B / 0.30 / 13 -5	2.7 / 2.5 / 2	0.64 / 55.	12.41	2 / 0.00 / 0.09	1514. / 1621.		50		25.2 / 32.01	18.2 / 17.2 / -2.7	-19.60 / 18.4	10.03		5051 10-14-015 2
0926-60	926.2 / 280.45 / 184.21	-6033 / -7.14 / -37.28	12 P / 0.12 / 33 -3	2.1 / 2.5 / 6	0.48 / 67.		/ 0.31 / 1.71	2211. / 1918. / 7		20		27.4 / 32.19	-21.7 / -1.6 / -16.6	/ 20.0			126-13
N 2915	926.5 / 291.97 / 197.41	-7625 / -18.36 / -26.03	-0B / 0.05 / 14+20	1.7 / 1.6 / 6	0.50 / 65.	12.36	2 / 0.29 / 0.55	455. / 191.	157 / 139. / 143.	9 / 10 / 301		3.3 / 27.61	-2.9 / -0.9 / -1.5	-15.25 / 1.5	8.29 / 8.94	4.41	037-03 2

ID	Position	Type	Dimensions	B mag	Vel	Width	n	HI	Dist	SGX SGY SGZ	M	Flux	Corr	Names
N 2893	927.3 2946 / 197.44 46.01 / 66.13 -31.29	0.24 / 21 -0+12	1.3 1.00 / 1.3 / 2	13.3 / 3 0.00 / 0.02	1712. / 1650.		29	0.15 / 12 35	26.8 / 32.14	9.3 / 20.9 / -13.9	-18.84 / 10.2	9.73 / 9.01	0.19	5060 05-23-005 2
N 2902	928.5 -1431 / 247.37 25.83 / 120.40 -52.43	-2A 0.38 / 31 -0+12	1.2 0.92 / 1.3 26. / 5	12.8 / 5 0.00 / 0.25	2065. / 1816.		90		28.3 / 32.26	-8.7 / 14.9 / -22.5	-19.46 / 10.7	9.98		IC 543 / -2-24-030 2
N 2907	929.3 -1631 / 249.18 24.66 / 123.70 -52.44	1A 0.39 / 31-13+12	1.8 0.67 / 1.7 53. / 4	12.7 / 5 0.15 / 0.22	2065. / 1810.		90		28.2 / 32.25	-9.5 / 14.3 / -22.3	-19.55 / 14.0	10.01		-3-25-002 2
N 2903	929.4 2144 / 208.70 44.56 / 73.58 -36.44	4X 0.12 / 15 -0 +1	11.6 0.49 / 10.0 66. / 2	9.13 6.53 2.60 / 1 0.30 / 0.07	554. / 451. / 2	395 / 391 / 393	15	1.65 5 / 60 / 1.16	6.3 / 28.98	1.4 / 4.8 / -3.7	-19.85 / 18.4	10.13 / 9.25 / 10.91	0.13 / 0.022 / 6.02	NGC 2905
0930+55	930.5 5527 / 160.54 44.85 / 48.34 -11.54	13 P 0.05 / 15+10	0.5 0.75 / 0.5 45. / 2	0.00 / 0.02	755. / 826.		24	-0.17 / 209 90	14.3 / 30.78	9.3 / 10.5 / -2.9	2.1	8.14		5079 04-23-009 2
0930-16	930.9 -1633 / 249.48 24.92 / 123.80 -52.06	2X 0.36 / 31-13+12	2.3 0.88 / 2.3 32. / 2	0.04 / 0.23	2123. / 1868. / 1	260 / 421 / 423	15	0.78 / 15 / 1.07	28.9 / 32.30	-9.9 / 14.8 / -22.8	19.4	9.70 / 11.00	0.050	UGCA 166, MARK 116 / I ZW 18 / 09-16-000 2
0931-32	931.2 -3249 / 261.92 13.63 / 149.74 -50.30	9 0.12 / 54 -1	5.4 0.16 / 4.3 90. / 6	0.67 / 0.65	929. / 640. / 2	243 / 206 / 208	10	1.24 / 8 / 1.08	11.1 / 30.22	-6.1 / 3.6 / -8.5	13.9	9.33 / 10.23	0.125	UGCA 167 / -3-25-004
0931+11	931.4 1114 / 222.21 41.02 / 84.79 -42.43	13 P 0.08 / 32 -0		14.2 / 3 0.67 / 0.05	2510. / 2358.		46		35.8 / 32.77	2.4 / 26.3 / -24.1	-18.57	9.62		UGCA 168
N 2935	934.5 -2054 / 253.60 22.59 / 130.86 -51.33	4X 0.30 / 31-12	5.4 0.71 / 5.1 49. / 6	10.69 / 6 0.12 / 0.12	2275. / 2009. / 2	307 / 359 / 360	15	1.04 / 13 / 1.03	30.6 / 32.43	-12.5 / 14.5 / -23.9	-21.74 / 45.6	10.89 / 10.01 / 11.23	0.13 / 0.060 / 2.20	373-08 -5-23-006 / 02-25-000 2
N 2940	935.4 950 / 224.46 41.25 / 87.10 -42.36	-3 0.15 / 32 -0	1.1 0.79 / 1.1 42. / 6	14.5 / 3 0.00 / 0.06	2982. / 2824.		65		41.6 / 33.09	1.6 / 30.7 / -28.0	-18.59 / 13.4	9.63		UGCA 169 / 565-23 -3-25-011 2
0935-22	935.6 -2210 / 254.78 21.89 / 132.87 -51.04	7X 0.30 / 31-12	2.5 0.65 / 2.3 54. / 6	12.98 / 0.16 / 0.14	2412. / 2143. / 2	193 / 194 / 197	20	0.49 / 48 / 1.01	32.2 / 32.54	-13.8 / 14.8 / -25.0	21.6	9.51 / 10.37	0.136	TURN 7B / 02-25-012
U 5139	936.0 7125 / 140.73 38.65 / 38.79 1.31	10 0.26 / 14-10	3.7 1.00 / 3.8 / 2	1 0.00 / 0.08	141. / 289. / 3	44	5 4 211	1.04 5 / 211 / 1.13	4.4 / 28.22	3.4 / 2.8 / 0.1	-15.24 / 4.9	8.29 / 8.33	1.09	UGCA 170 / 565-29 -4-23-011 / A 0936, HO I, DDO 63 / KAR 57 / 5139 12-09-059 2
0936-38	936.3 -3847 / 266.86 9.98 / 157.83 -47.40	7 0.07 / 31 -0 +2	2.1 1.00 / 2.6 / 6	0.00 / 1.19	2470. / 2175. / 7		20		31.9 / 32.52	-20.0 / 8.2 / -23.5	24.2			315-17 -6-21-011
U 5151	937.2 4834 / 169.79 47.61 / 53.70 -16.16	13 P 0.03 / 15+10	1.0 0.89 / 0.9 31. / 2	13.4 / 3 0.00 / 0.03	414. / 450.		220		6.9 / 29.18	3.9 / 5.3 / -1.9	-15.78 / 1.8	8.50		ARAK 209 / 5151 08-18-021

(1) Name	(2) α / ℓ / SGL	(3) δ / b / SGB	(4) Type / Q_{xyz} / Group	(5) D_{25} / $D_{25}^{b,i}$ / source	(6) d/D / i	(7) $B_T^{b,i}$ / $H_{-0.5}$ / B-H	(8) source / A_B^{i-0} / A_B^b	(9) V_h / V_o / Tel	(10) W_{20} / W_R / W_D	(11) e_v / e_w / source	(12) F_c source/e_F / f_h	(13) R / μ	(14) SGX / SGY / SGZ	(15) $M_B^{b,i}$ / $\Delta_{25}^{b,i}$	(16) $\log L_B$ / $\log M_H$ / $\log M_T$	(17) M_H/L_B / M_H/M_T / M_T/L_B	(18) Alternate Names UGC/ESO MCG; AGN? RCII
N 2955	938.2 / 188.44 / 62.91	3607 / 48.93 / -25.17	3AP / 0.29 / 21 -0+12	1.6 / 1.4 / 2	0.50 / 65.	13.17	2 / 0.29 / 0.00	1750. / 1721.		190		27.7 / 32.21	11.4 / 22.3 / -11.8	-19.04 / 11.3	9.81		5166 06-21-073 2
N 2962	938.3 / 229.99 / 92.86	524 / 39.68 / -43.95	-2X / 0.15 / 31 -0	3.4 / 3.3 / 2	0.75 / 46.	12.66	2 / 0.00 / 0.09	2114. / 1937.		19		30.6 / 32.43	-1.1 / 22.0 / -21.2	-19.77 / 29.5	10.10		5167 01-25-011 2
U 5172	938.6 / 169.23 / 53.65	4854 / 47.76 / -15.76	9A / 0.08 / 21 -0	2.0 / 2.0 / 2	1.00	14.2	3 / 0.00 / 0.02	2597. / 2635. / 1		20	0.06 / 22 / 1.05	38.7 / 32.94	22.1 / 30.0 / -10.5	-18.74 / 22.6	9.69 / 9.24	0.35	5172 08-18-027
N 2950	939.0 / 155.18 / 46.90	5905 / 44.67 / -8.01	-2B / 0.22 / 13 -0 +5	3.5 / 3.3 / 2	0.70 / 49.	11.82	2 / 0.00 / 0.03	1362. / 1452.		38		23.3 / 31.84	15.8 / 16.9 / -3.2	-20.02 / 22.5	10.20		5176 10-14-032 2
0939+76	939.1 / 135.19 / 35.66	7635 / 35.94 / 5.43	13 P / 0.20 / 12 -0	1.1 / 0.9 / 2	0.43 / 71.	14.8	3 / 0.00 / 0.03	2355. / 2526.		38		36.2 / 32.79	29.3 / 21.0 / 3.4	-17.99 / 9.5	9.39		UGCA 171, MARK 118 / 13-07-036 2
N 2967	939.5 / 235.37 / 99.20	34 / 37.28 / -45.77	5A / 0.17 / 31 -0+19	2.8 / 2.9 / 2	1.00	12.14	2 / 0.00 / 0.16	2159. / 1963.		64		30.9 / 32.45	-3.4 / 21.3 / -22.1	-20.31 / 26.2	10.32		5180 00-25-007 2
N 2964	939.9 / 194.60 / 66.40	3205 / 49.01 / -27.74	4X / 0.22 / 21-12	3.0 / 2.6 / 9	0.57 / 60.	11.80	1 / 0.22 / 0.02	1310. / 1260.		41		21.9 / 31.70	7.8 / 17.8 / -10.2	-19.90 / 16.6	10.15		5183 05-23-027 2
N 2974	940.0 / 239.50 / 104.72	-328 / 35.01 / -47.12	-5 / 0.26 / 31-18	3.5 / 3.2 / 2	0.58 / 60.	11.65	2 / 0.00 / 0.10	1998. / 1786.		26		28.5 / 32.28	-4.9 / 18.8 / -20.9	-20.63 / 26.6	10.44		UGCA 172 / 00-25-008 2
N 2968	940.2 / 194.49 / 66.38	3210 / 49.08 / -27.63	3 P / 0.55 / 21-12	2.0 / 1.9 / 9	0.73 / 47.	12.67	1 / 0.11 / 0.02	1608. / 1559.		57		25.9 / 32.06	9.2 / 21.0 / -12.0	-19.39 / 14.4	9.95		5190 05-23-029 2
N 2970	940.6 / 194.42 / 66.41	3213 / 49.17 / -27.54	-2 / 0.43 / 21-12	0.7 / 0.6 / 2	0.80 / 41.	14.4	3 / 0.00 / 0.02	1678. / 1629.		90		26.7 / 32.13	9.5 / 21.7 / -12.3	-17.73 / 4.7	9.28		05-23-030 2
0940-05	940.9 / -241.19 / 107.03	-503 / 34.22 / -47.41	8BP / 0.23 / 31-18	1.3 / 0.9 / 2	0.15 / 90.		0.67 / 0.09	1951. / 1734.		52		27.9 / 32.23	-5.5 / 18.0 / -20.5	7.3			UGCA 173, ARP 253 / VV 52 / -1-25-031 2
0941-05	941.1 / 241.83 / 107.95	-541 / 33.87 / -47.55	7A / 0.21 / 31-18	2.4 / 2.4 / 2	0.89 / 31.		0.04 / 0.11	2028. / 1808. / 1	102 / 158. / 162.	15	0.30 / 22 / 1.06	28.8 / 32.30	-6.0 / 18.5 / -21.2	20.2	9.22 / 10.16	0.113	UGCA 175 / -1-25-034
N 2983	941.4 / 254.32 / 129.83	-2015 / 24.20 / -49.72	-2B / 0.16 / 31-12	2.5 / 2.3 / 6	0.60 / 58.	12.47	2 / 0.00 / 0.13	2015. / 1752.		100		27.4 / 32.19	-11.4 / 13.6 / -20.9	-19.72 / 18.4	10.08		UGCA 176 / 566-03 -3-25-017 2

Astronomical galaxy catalog data table (values transcribed per object; multi-line field values separated by " / ").

ID	Coordinates	Type	Dimensions	ratio/PA	B mag		Velocity		N						names
0941-31	941.8 -3157 15.79 -48.35 / 263.00 / 147.59	10 / 0.17 / 54 -1	1.8 1.9 6	1.8 / 0.73 / 47.		0.11 0.57	1209. 922. 2	70 71. 79.	10	0.35 19 1.01	15.8 31.00	-8.9 5.6 -11.8	8.8	8.75 9.11	0.436	UGCA 177 / 434-27 -5-23-000
N 2986	942.0 -2103 23.73 -49.58 / 255.05 / 131.06	-5 / 0.27 / 31-12	2.5 2.6 6	2.5 / 1.00	11.84	2 0.00 0.11	2397. 2132.		100		32.3 32.54	-13.7 15.8 -24.6	-20.70 24.5	10.47		UGCA 178 / 566-05 -3-25-019 2
0942-31	942.6 -3136 16.17 -48.26 / 262.89 / 147.03	8X / 0.14 / 54 -1	2.3 2.5 6	2.3 / 0.86 / 35.	12.77	7 0.05 0.52	1256. 970. 2	164 231. 234.	10	1.07 12 1.02	16.4 31.08	-9.2 6.0 -12.3	-18.31 12.0	9.52 9.50 10.27	0.96 0.171 5.65	UGCA 180, DDO 235 / 434-33 -5-23-010 2
N 2976	943.2 6808 40.91 -0.81 / 143.93 / 41.37	5AP / 0.50 / 14-10	5.0 4.6 2	5.0 / 0.58 / 59.	10.50	1 0.21 0.14	9. 142. 1		20	1.10 15 1.30	2.1 26.63	1.6 1.4 0.0	-16.13 2.8	8.64 7.74	0.13	5221 11-12-025 2
U 5224	943.3 312 39.55 -43.82 / 233.27 / 96.37	8B / 0.34 / 31-20+19	1.7 1.5 2	1.7 / 0.51 / 64.	14.6	3 0.28 0.10	1941. 1756. 1	184 166. 169.	15	0.44 28 1.02	28.4 32.27	-2.3 20.4 -19.7	-17.67 12.4	9.26 9.35 10.00	1.22 0.223 5.46	5224 01-25-018
N 2992	943.3 -1406 28.78 -48.83 / 249.71 / 120.47	1 P / 0.21 / 31-15+12	4.4 3.2 4	4.4 / 0.20 / 90.	11.91	2 0.67 0.22	2212. 1965.		64		30.5 32.42	-10.2 17.3 -22.9	-20.51 28.5	10.40		ARP 245 / -2-25-014 2
0943-30	943.3 -3008 17.35 -48.41 / 261.97 / 144.82	10 / 0.28 / 54 -1	2.8 2.2 6	2.8 / 0.17 / 90.		2 0.25 0.05	1004. 720. 1	148 118. 123.	10	0.71 15 1.01	12.6 30.49	-6.8 4.8 -9.4	8.1	8.91 9.51	0.249	UGCA 182 / 434-34 -5-23-011
N 2997	943.3 -3057 16.75 -48.25 / 262.55 / 146.02	5X / 0.28 / 54 -1	10.2 10.3 9	10.2 / 0.66 / 53.	9.97	1 0.15 0.48	1083. 798. 2	279 301. 302.	8	1.78 11.5 1.27	13.8 30.71	-7.6 5.2 -10.3	-20.74 41.5	10.49 10.06 11.04	0.37 0.105 3.55	UGCA 181 / 434-35 -5-23-012 2
N 2993	943.4 -1408 28.77 -48.81 / 249.76 / 120.52	1 P / 0.21 / 31-15+12	1.5 1.5 4	1.5 / 0.83 / 37.	12.82	2 0.06 0.22	2217. 1970.		64		30.5 32.42	-10.2 17.3 -22.9	-19.60 13.4	10.03		ARP 245 / -2-25-015 2
N 2990	943.7 556 41.10 -42.51 / 230.32 / 93.07	5 / 0.00 / 32 -0	1.2 1.1 2	1.2 / 0.53 / 63.	12.91	2 0.25 0.05	3168. 2994.		83		43.7 33.20	-1.7 32.2 -29.6	-20.29 14.0	10.31		5229 01-25-021 2
N 3001	944.0 -3012 17.41 -48.24 / 262.14 / 144.87	4X / 0.23 / 31 +5 +2	3.4 3.5 6	3.4 / 0.74 / 46.	12.2	5 0.10 0.47	2474. 2190. 2	398 501. 502.	25	0.95 19 1.03	32.5 32.56	-17.7 12.5 -24.3	-20.36 33.2	10.34 9.97 11.38	0.43 0.039 11.15	UGCA 183 / 434-38 -5-23-014 2
U 5238	944.3 44 38.36 -44.60 / 236.07 / 99.65	7X / 0.28 / 31-19	2.5 2.1 2	2.5 / 0.40 / 72.	13.7	3 0.43 0.18	1787. 1592. 1	247 220. 223.	20	0.69 21 1.04	26.3 32.10	-3.1 18.4 -18.4	-18.40 16.1	9.55 9.53 10.35	0.95 0.150 6.35	5238 00-25-015
0944+39	944.7 3919 50.18 -21.99 / 183.48 / 61.44	13 P / 0.31 / 21 -0+12			14.3	3 0.43 0.02	1629. 1617.		71		26.4 32.11	11.7 21.5 -9.9	-17.81	9.32		MARK 407 / 07-20-000 2
0944-63	944.8 -6303 -7.52 -33.99 / 283.78 / 184.87	5 / 0.20 / 33 -0 +1	1.8 1.7 6	1.8 / 0.14 / 90.		0.67 1.26	2907. 2618. 7		20		36.4 32.80	-30.0 -2.6 -20.3	18.1			091-18

(1) Name	(2) α / ℓ / SGL	(3) δ / b / SGB	(4) Type / Q_{xyz} / Group	(5) $D_{25}^{b,i}$ / D_{25} / source	(6) d/D / i	(7) $B_T^{b,i}$ / $H_{-0.5}$ / B-H	(8) source / A_B^{i-o} / A_B^b	(9) V_h / V_o / Tel	(10) W_{20} / W_R / W_D	(11) e_v / e_w / source	(12) F_c source/e_F / f_h	(13) R / μ	(14) SGX / SGY / SGZ	(15) $M_B^{b,i}$ / $B_{25}^{b,i}$ / $Δ_{25}$	(16) log L_B / log M_H / log M_T	(17) M_H/L_B / M_H/M_T / M_T/L_B	(18) Alternate Names / UGC/ESO MCG / AGN? RCII
U 5245	945.0 / 238.79 / 103.07	-148 / 37.01 / -45.38	8 / 0.12 / 31 -0	2.8 / 2.0 / 2	0.21 / 89.	14.4	3 / 0.09	1425. / 1220. / 1	170	25	0.42 / 27 / 1.05	21.3 / 31.64	-3.4 / 14.6 / -15.2	-17.24 / 12.4	9.09 / 9.08	0.97	5245 00-25-017
0945+33	945.1 / 193.17 / 66.40	3307 / 50.19 / -26.23	13 P / 0.64 / 21-12			13.9	3 / 0.67 / 0.01	1487. / 1443.		71		24.4 / 31.94	8.8 / 20.1 / -10.8	-18.04	9.41		MARK 408 / 06-22-000 2
U 5249	945.2 / 233.99 / 97.08	251 / 39.75 / -43.53	6B / 0.35 / 31-20+19	2.4 / 2.0 / 2	0.37 / 74.	13.2	3 / 0.49 / 0.12	1882. / 1695. / 1	247	15	0.72 / 19 / 1.04	27.7 / 32.21	-2.5 / 19.9 / -19.1	-19.01 / 16.2	9.80 / 9.60	0.64	5249 01-25-025
N 3003	945.6 / 192.35 / 66.04	3339 / 50.34 / -25.79	4E / 0.64 / 21-12	6.0 / 4.2 / 9	0.21 / 90.	11.48 / 10.22 / 1.26	1 / 0.67 / 0.00	1480. / 1439. / 1	286 / 248. / 250.	7 / 12 / 217	1.34 / 217	24.4 / 31.93	8.9 / 20.0 / -10.6	-20.45 / 29.9	10.37 / 10.11 / 10.73	0.55 / 0.244 / 2.27	5251 06-22-013 2
N 2985	945.9 / 139.01 / 38.67	7231 / 38.68 / 2.64	2K / 0.23 / 12 -7	4.0 / 3.8 / 2	0.79 / 42.	11.10	1 / 0.08 / 0.07	1277. / 1431.		50		22.4 / 31.75	17.4 / 14.0 / 1.0	-20.65 / 24.9	10.45		L/S / 5253 12-10-006 2
0946+55	946.1 / 158.99 / 49.80	5549 / 46.80 / -9.84	13 P / 0.12 / 13 -0 +5	0.5 / 0.5 / 2	0.53 / 63.	15.3	3 / 0.00 / 0.00	1485. / 1559.		95		25.0 / 31.99	15.9 / 18.8 / -4.3	-16.69 / 3.6	8.87		UGCA 184, MARK 22 / 09-16-000 2
0946-74	946.1 / 291.42 / 194.82	-7422 / -16.03 / -26.62	7 / 0.12 / 53-18	2.8 / 2.4 / 6	0.26 / 84.		0.67 / 0.70	1165. / 896. / 7		20		13.5 / 30.66	-11.7 / -3.1 / -6.1	9.5			037-05
N 3011	946.7 / 194.25 / 67.20	3227 / 50.48 / -26.44	-2 / -0.67 / 21-12	1.1 / 1.1 / 2	0.90 / 30.	14.0	3 / 0.00 / 0.01	1464. / 1417.		71		24.1 / 31.91	8.3 / 19.9 / -10.7	-17.91 / 7.7	9.36		MARK 409 / 5259 05-23-038 2
N 3023	947.3 / 236.51 / 99.91	51 / 39.04 / -43.86	6XP / 0.35 / 31-19	3.1 / 2.8 / 2	0.54 / 62.	12.5	3 / 0.24 / 0.17	1877. / 1663. / 1	151 / 137. / 141.	10	0.83 / 13 / 1.07	27.6 / 32.20	-3.4 / 19.6 / -19.1	-19.70 / 22.6	10.07 / 9.71 / 10.09	0.44 / 0.419 / 1.04	5269 00-22-022 2
U 5272	947.4 / 195.42 / 67.92	3143 / 50.56 / -26.82	10 / 0.12 / 15 -0 +1	2.2 / 1.8 / 2	0.40 / 72.	13.94	7 / 0.43 / 0.03	520. / 469. / 1	164 / 138. / 143.	15	0.61 / 15 / 1.03	6.5 / 29.06	2.2 / 5.4 / -2.9	-15.12 / 3.4	8.24 / 8.24 / 9.28	0.99 / 0.091 / 10.86	DDO 64 / 5272 05-23-041 2
N 3020	947.4 / 222.36 / 85.57	1303 / 45.33 / -38.16	6B / 0.12 / 21 -0	3.1 / 2.8 / 2	0.60 / 58.	12.5	3 / 0.20 / 0.02	1447. / 1305. / 1	247 / 247. / 249.	15	1.04 / 11 / 1.07	22.9 / 31.80	1.4 / 17.9 / -14.1	-19.30 / 18.7	9.91 / 9.76 / 10.52	0.70 / 0.174 / 4.06	5271 02-25-045 2
N 3026	948.0 / 200.02 / 70.51	2847 / 50.30 / -28.65	1C / C.49 / 21-13+12	2.5 / 2.0 / 2	0.30 / 80.	12.8 / 11.92 / 0.88	3 / 0.67 / 0.02	1497. / 1431. / 1	225	15	0.46 / 19 / 1.04	24.5 / 31.94	7.2 / 20.2 / -11.7	-19.14 / 14.3	9.85 / 9.24	0.25	5279 05-23-043
N 3021	948.0 / 192.19 / 66.31	3347 / 50.84 / -25.34	5B / 0.65 / 21-12	1.6 / 1.5 / 2	0.64 / 55.	12.6	3 / 0.17 / 0.00	1530. / 1490.		30		25.1 / 32.00	9.1 / 20.8 / -10.7	-19.40 / 11.0	9.95		5280 06-22-019 2

Nearby galaxies data table (galaxies as rows; multi-valued fields separated by " / ").

ID	Designation	Common name	(a)	B mag	(c)	x / y / z	dist	(f)	(g)	(h)	velocities	(j)	(k)	dimensions	type	(n)	SGL coords
0948+08	01-25-028		0.241	8.21 / 8.83	1.8	-0.1 / 4.0 / -3.4	5.2 / 28.59	0.78 / 10 / 1.00	8	116 / 113. / 118.	561. / 397. / 2	0.16 / 0.03		0.65 / 54. / 1.3 / 1.2 / 2	10 P / 0.05 / 15 -0 +1	804 / 43.26	948.7 / 228.71 / 91.34 -40.41
0949+01	00-25-024 2	UGCA 186, DDO 65		9.04	12.0	-3.1 / 19.7 / -18.7	27.3 / 32.18	0.17 / 27 / 1.02	10	97	1853. / 1662. / 1	0.05 / 0.10		0.87 / 33. / 1.5 / 1.5 / 2	10 / 0.35 / 31 -19	141 / 39.89	949.1 / 235.96 / 99.09 -43.12
N 3038	374-02 -5-24-001 2		0.03	10.57	-20.95 / 27.0	-20.6 / 12.9 / -25.8	35.5 / 32.75		33		2713. / 2426.	5 / 0.27 / 0.56	11.8	0.52 / 64. / 2.6 / 2.6 / 6	1A / 0.32 / 31 -4 +2	-3231 / 16.39	949.1 / 264.59 / 147.92 -46.72
N 3032	5292 05-23-046 2			9.94 / 8.36	-19.36 / 15.0	7.4 / 20.3 / -11.5	24.5 / 31.94	-0.42 / 210 26	20 / 20 / 210	209	1501. / 1439.	2 / 0.00 / 0.02	12.58	0.85 / 35. / 2.2 / 2.1 / 2	-2X / 0.57 / 21 -13+12	2928 / 50.66	949.2 / 199.02 / 70.11 -28.01
N 3059	037-07 2		0.22	10.09	-19.75 / 19.4	-12.8 / -3.2 / -6.7	14.8 / 30.85	0.72 / 16 / 1.10	10 / 10 / 301	158	1260. / 990.	5 / 0.00 / 0.69	11.1	1.00 / 3.9 / 4.5 / 6	5B / 0.14 / 53 -18	-7341 / -15.35	949.7 / 291.15 / 194.07 -26.89
N 3041	5303 03-25-039 2			10.10 / 9.44	-19.77 / 21.3	2.6 / 18.4 / -13.2	22.8 / 31.79		30	313	1419. / 1295. / 1	2 / 0.15 / 0.08	12.02	0.66 / 53. / 3.5 / 3.2 / 9	5X / 0.16 / 21 -0	1655 / 47.58	950.4 / 217.67 / 82.03 -35.44
N 3044	5311 00-25-031 2			10.09	-19.74 / 18.6	-2.4 / 15.0 / -13.9	20.6 / 31.57		24		1335. / 1145.	2 / 0.67 / 0.06	11.83	0.20 / 90. / 4.4 / 3.1 / 2	5B / 0.19 / 21 +10	149 / 40.38	951.1 / 236.20 / 99.20 -42.61
N 3027	5316 12-10-009 2		1.04 / 0.262 / 3.99	9.91 / 9.93 / 10.51	-19.30 / 22.8	15.1 / 12.3 / 1.0	19.5 / 31.45	1.35 / 17 5 / 1.13	10 / 6 / 217	234 / 222. / 225.	1057. / 1211. / 1	1 / 0.25 / 0.05	12.15 / 11.00 / 1.15	0.54 / 62. / 4.6 / 4.0 / 2	7B / 0.23 / 12 -7	7227 / 39.04	951.2 / 138.78 / 39.01 2.86
N 3031	5318 12-10-010 2	M 81	0.14 / 0.009 / 16.72	9.51 / 8.66 / 10.73	-18.29 / 8.3	1.0 / 0.9 / 0.0	1.4 / 25.67	2.37 / 211 26 / 1.15	4 / 7 / 211	446 / 473. / 474.	-43. / 96.	1 / 0.21 / 0.17	7.38 / 4.35 / 3.03	0.57 / 60. / 22.1 / 20.3 / 6	2A / 0.50 / 14 -10	6918 / 40.91	951.5 / 142.09 / 41.16 0.56
N 3034	5322 12-10-011 2	M 82, ARP 337	0.18	9.96 / 9.21	-19.42 / 15.8	3.9 / 3.4 / 0.1	5.2 / 28.57	1.78 / 221	7		210. / 352.	1 / 0.00 / 0.15	9.15	0.49 / 66. / 11.7 / 10.4 / 9	13 P / 0.16 / 14 -10	6955 / 40.57	951.7 / 141.41 / 40.75 1.03
N 3054	499-18 -4-24-005 2	UGCA 187		10.41	-20.55 / 32.0	-14.9 / 13.6 / -21.7	29.6 / 32.35		42		2199. / 1925.	5 / 0.16 / 0.23	11.8	0.64 / 55. / 3.9 / 3.7 / 6	3X / 0.14 / 31 +5 +2	-2527 / 22.15	952.1 / 260.18 / 137.53 -47.06
N 3056	435-07 -5-24-003 2			9.40	-18.02 / 7.8	-7.2 / 5.7 / -9.8	13.4 / 30.64		90		1047. / 768.	6 / 0.00 / 0.21	12.62	0.63 / 56. / 2.1 / 2.0 / 6	-2A / 0.30 / 54 -1	-2804 / 20.20	952.3 / 262.06 / 141.34 -46.78
N 3055	5328 01-25-034 2			10.06	-19.67 / 15.7	-2.2 / 20.1 / -17.7	26.9 / 32.15		68		1794. / 1615.	2 / 0.14 / 0.08	12.48	0.67 / 52. / 2.2 / 2.0 / 9	5X / 0.22 / 31 -0	430 / 42.22	952.7 / 233.55 / 96.13 -41.14
0952+08	02-25-056 2	UGCA 188, A 0953		8.85	-16.65 / 6.0	-0.4 / 15.8 / -12.9	20.4 / 31.55		100		1283. / 1122.	3 / 0.00 / 0.06	14.9	0.60 / 58. / 1.1 / 1.0 / 2	13 P / 0.16 / 21 +10	838 / 44.42	952.8 / 228.74 / 91.32 -39.25

Note: entry N 3031 (M 81) also carries the notations "11.9" and "L1.9".

(1) Name	(2) α / ℓ / SGL	(3) δ / b / SGB	(4) Type / Q_{xyz} / Group	(5) D_{25} / $D_{25}^{b,i}$ / source	(6) d/D / i	(7) $B_T^{b,i}$ / $H_{-0.5}$ / B–H	(8) source / A_B^{i-o} / A_B^b	(9) V_h / V_o / Tel	(10) W_{20} / W_R / W_D	(11) e_v / e_w / source	(12) F_c source/e_f / f_h	(13) R / μ	(14) SGX / SGY / SGZ	(15) $M_B^{b,i}$ / $\Delta_{25}^{b,i}$	(16) log L_B / log M_H / log M_T	(17) M_H/L_B / M_H/M_T / M_T/L_B	(18) Alternate Names UGC/ESO MCG — AGN? RCII
0953-32	953.0	-3255	6X	3.5	0.80	12.0	5	3018.	331	25	1.02	39.4	-23.3	-20.98	10.58	0.42	IC 2522, UGCA 189
	265.51	16.62	0.23	3.8	41.		0.08	2731.	449.		22	32.98	14.5	43.7	10.21	0.064	374-10 -5-24-004 2
	148.23	-45.83	31 -4 +2	6			0.54	2	450.		1.03		-28.3		11.41	6.66	
U 5340	953.8	2904	10 P	2.5	0.47	14.43	7	504.	113	10	0.86	5.9	1.7	-14.43	7.96	2.74	DDO 68
	199.88	51.60	0.12	2.2	67.		0.33	441.	97.		9	28.86	5.0	3.8	8.40	0.244	5340 05-24-004 2
	71.19	-27.50	15 -0 +1	2			0.04	1	103.		1.04		-2.7		9.01	11.24	
0954+33	954.6	3351	13 P	1.0	0.71	14.5	3	1509.		71		24.9	8.8	-17.48	9.18		IC 2524, MARK 411
	192.18	52.21	0.61	0.9	49.		0.00	1470.				31.98	20.9	6.5			06-22-039 2
	67.26	-24.27	21-12	2			0.01						-10.2				
U 5349	955.1	3732	9	2.5	0.38	13.4	3	1381.	220	15	0.56	23.4	9.4	-18.44	9.57	0.54	5349 06-22-043
	186.16	52.31	0.25	2.1	74.		0.47	1361.	191.		15	31.84	19.6	14.3	9.30	0.131	
	64.34	-21.74	21-14+12	2			0.03	1	194.		1.04		-8.7		10.18	4.10	
N 3067	955.4	3237	2X	2.2	0.40	12.24	2	1459.		36		24.2	8.1	-19.68	10.06		5351 06-22-046 2
	194.21	52.32	0.65	1.8	72.		0.43	1414.				31.92	20.4	12.7			
	68.40	-24.95	21-12	2			0.04	1					-10.2				
0955+13	955.9	1329	13 P			14.0	3	2839.		220		40.4	2.0	-19.03	9.80		ARAK 225
	223.11	47.38	0.23				0.43	2700.				33.03	32.5				02-26-008
	86.48	-36.18	32 +3				0.04						-23.8				
N 3078	956.1	-2641	-5	2.1	0.83	11.99	0	2506.		22		33.4	-17.6	-20.63	10.44		499-27 -4-24-009 2
	261.77	21.80	0.35	2.1	37.		0.00	2230.				32.62	15.1	20.5			
	139.22	-46.09	31 -5 +2	6			0.11						-24.1				
U 5364	956.4	3059	10	5.0	0.67	12.49	1	28.		20	1.22	1.0	0.3	-12.51	7.20	1.06	LEO A, DDO 69
	196.90	52.39	0.39	4.6	52.		0.15	-25.			8	25.00	0.9	1.3	7.22		5364 05-24-008 2
	69.94	-25.84	14+12	2			0.06	1			1.16		-0.4				
N 3087	957.0	-3359	-5	1.9	0.90	12.02	6	2662.		90		34.9	-21.3	-20.69	10.47		374-15 -6-22-005 2
	266.89	16.33	0.31	2.0	29.		0.00	2375.				32.71	12.6	20.4			
	149.47	-44.79	31 -4 +2	6			0.58						-24.6				
N 3081	957.2	-2235	1X	2.2	0.85	12.51	6	2413.		90		32.5	-15.5	-20.05	10.21		IC 2529
	259.02	25.04	0.25	2.2	35.		0.05	2146.				32.56	16.4	20.9			499-31 -4-24-012 2 S2
	133.30	-46.04	31 +5 +2	6			0.12						-23.4				
U 5373	957.4	534	10B	5.4	0.70	11.69	1	302.	63	5	1.33	1.6	-0.1	-14.27	7.90	0.69	SEX B, DDO 70
	233.21	43.78	0.23	5.1	50.		0.13	129.	59.		6	25.96	1.2	2.4	7.74	0.228	5373 01-26-005 2
	95.53	-39.63	14+12	2			0.04	4	68.		1.06		-1.0		8.38	3.03	
N 3089	957.4	-2805	2X	1.9	0.57	12.7	5	2653.		90		35.1	-19.1	-20.03	10.20		435-24 -5-24-014 2
	262.99	20.93	0.36	1.8	60.		0.22	2375.				32.73	15.4	18.4			
	141.18	-45.66	31 -5 +2	6			0.21						-25.1				
N 3073	957.5	5552	-5 P	1.3	0.92	13.4	3	1057.		105		19.3	12.0	-18.03	9.40		MARK 131
	157.98	48.25	0.28	1.3	27.		0.00	1132.				31.43	14.8	7.3			5374 09-17-007 2
	50.95	-8.70	13 -1	2			0.00						-2.9				

Name	(1)	(2)	(3)	(4)	(5)	(6)	(7)	(8)	(9)	(10)	(11)	(12)	(13)	(14)	Alt. Names
N 3065	957.6 7225 / 138.43 39.45 / 39.38 3.17	-2A / 0.21 / 12-14	2.0 0.95 / 2.1 22. / 2	12.87	2 0.00 / 0.08	1963. / 2117.		26		31.3 / 32.48	24.1 / 19.8 / 1.7		10.04		VII ZW 303 / 5375 12-10-014 2
0957-29	957.7 -2923 / 263.93 19.97 / 143.01 -45.44	5 / 0.33 / 31 -5 +2	6.1 0.10 / 4.5 90. / 6		0.67 / 0.23	2477. / 2197. / 2	491 / 453. / 454.	15	1.10 / 30 / 1.06	32.9 / 32.59	-18.4 / 13.9 / -23.4	43.2	10.13 / 11.41	0.053	IC 2531, UGCA 191 / 435-25 -5-24-015
N 3066	957.9 7222 / 138.46 39.49 / 39.43 3.15	4XP / 0.22 / 12-14	1.3 1.00 / 1.3 / 2	13.47	2 0.00 / 0.08	2050. / 2204.		31		32.4 / 32.55	24.9 / 20.5 / 1.8	-19.08 / 12.3	9.82		MARK 133, ZW COMPACT / 5379 12-10-015 2
0957-34	957.9 -3400 / 267.06 16.44 / 149.44 -44.61	7 / 0.37 / 31 -4 +2	1.5 0.88 / 1.6 32. / 6		0.04 / 0.59	2900. / 2613.		35		37.8 / 32.89	-23.2 / 13.7 / -26.6	17.7			IC 2532 / 499-34 -6-22-007
N 3095	957.9 -3119 / 265.28 18.52 / 145.72 -45.11	5X / 0.39 / 31 -4 +2	4.6 0.50 / 4.3 65. / 6	11.8	3 0.29 / 0.34	2721. / 2438.		23		35.8 / 32.77	-20.9 / 14.2 / -25.3	-20.97 / 45.0	10.58		435-26 -5-24-016 2
N 3079	958.6 5557 / 157.77 48.35 / 51.00 -8.54	9B / 0.29 / 13 -1	7.7 0.22 / 5.5 88. / 2	10.53 / 8.07 / 2.46	1 0.67 / 0.00	1125. / 1200. / 2	479 / 441. / 443.	10 / 8 / 217	1.42 / 217 5 / 1.07	20.4 / 31.54	12.7 / 15.6 / -3.0	-21.01 / 32.8	10.60 / 10.04 / 11.27	0.28 / 0.059 / 4.68	5387 09-17-010 2 / L2
U 5391	958.7 3730 / 186.16 53.02 / 64.88 -21.23	9 / 0.36 / 21-14+12	2.3 0.44 / 1.9 70. / 2	13.7	3 0.37 / 0.01	1569. / 1550. / 1	223 / 199. / 201.	20	0.59 / 14 / 1.04	25.8 / 32.06	10.2 / 21.8 / -9.3	-18.36 / 14.3	9.54 / 9.41 / 10.22	0.75 / 0.158 / 4.78	5391 06-22-054
U 5393	958.8 3322 / 193.04 53.07 / 68.28 -23.92	9B / 0.65 / 21-12	2.3 0.60 / 2.0 58. / 2	13.9	3 0.20 / 0.02	1448. / 1407. / 1	173	25	0.48 / 21 / 1.04	24.2 / 31.92	8.2 / 20.5 / -9.8	-18.02 / 14.1	9.40 / 9.25	0.70	5393 06-22-055
N 3077	959.4 6859 / 141.89 41.66 / 41.89 0.82	13 P / 0.50 / 14-10	5.4 0.77 / 5.3 43. / 2	10.46	1 0.00 / 0.14	10. / 148.	103 / 118. / 123.	9 / 5 / 51	1.90 / 51 10	2.1 / 26.59	1.5 / 1.4 / 0.0	-16.13 / 3.2	8.64 / 8.54 / 9.12	0.80 / 0.267 / 2.98	5398 12-10-017 2
N 3098	959.5 2457 / 206.81 52.07 / 75.75 -29.04	-2 / 0.32 / 21 +6	2.5 0.28 / 2.0 81. / 2	12.76	2 0.00 / 0.09	1340. / 1257.		90		22.4 / 31.75	4.8 / 19.0 / -10.9	-18.99 / 13.1	9.79		5397 04-24-012 2
1000-05	1000.2 -546 / 245.66 37.46 / 109.98 -43.02	7 / 0.08 / 15 -5 +1	3.2 0.15 / 2.3 90. / 2		0.67 / 0.07	666. / 449. / 1	157 / 126. / 130.	15	0.70 / 17 / 1.06	6.2 / 28.97	-1.6 / 4.3 / -4.2	4.2	8.28 / 9.28	0.100	UGCA 193 / -1-26-012
N 3109	1000.8 -2555 / 262.10 23.07 / 138.01 -45.09	10 / 0.15 / 14+12	24.4 0.21 / 17.8 89. / 6	9.6	5 0.67 / 0.13	403. / 130. / 2	129 / 102. / 108.	7 / 6 / 25	2.43 / 25 3 / 1.22	1.8 / 26.22	-0.9 / 0.8 / -1.2	-16.62 / 9.4	8.84 / 8.94 / 9.45	1.26 / 0.309 / 4.08	UGCA 194, DDO 236 / 499-36 -4-24-013 2
1000-21	1000.9 -2112 / 258.71 26.65 / 131.32 -45.18	7X / 0.09 / 31 -0	1.4 1.00 / 1.5 / 6		0.00 / 0.18	3069. / 2806. / 2	67	20	-0.15 / 42 / 1.01	41.0 / 33.06	-19.1 / 21.7 / -29.1	18.0	9.08		UGCA 195 / 567-10 -3-26-015
N 3104	1000.9 4100 / 180.29 53.10 / 62.40 -18.59	10A / 0.15 / 15 +7	3.4 0.67 / 3.2 52. / 2	12.5	3 0.15 / 0.03	618. / 617. / 1	118. / 118. / 123.	7	0.77 / 7 14 / 1.09	9.4 / 29.86	4.1 / 7.9 / -3.0	-17.36 / 8.8	9.14 / 8.72 / 9.55	0.38 / 0.147 / 2.59	ARP 264, VV 119 / 5414 07-21-007 2

64

(1) Name	(2) α / ℓ / SGL	(3) δ / b / SGB	(4) Type / ϱ_{xyz} / Group	(5) D_{25} / $D_{25}^{b,i}$ / source	(6) d/D / i	(7) $B_T^{b,i}$ / $H_{-0.5}$ / $B-H$	(8) source / A_B^{i-o} / A_B^b	(9) V_h / V_o / Tel	(10) W_{20} / W_R / W_D	(11) e_v / e_w / source	(12) F_c / source-e_F / f_h	(13) R / μ	(14) SGX / SGY / SGZ	(15) $M_{25}^{b,i}$ / $\Delta_{25}^{b,i}$	(16) $\log L_B$ / $\log M_H$ / $\log M_T$	(17) M_H/L_B / M_H/M_T / M_T/L_B	(18) Alternate Names UGC/ESO MCG	AGN?	RCII
N 3057	1001.0	8031	8B	2.7	0.84	13.34	7	1529	149	15	0.40	25.7	21.1	-18.71	9.68	0.35	DDO 67		
	130.66	34.24	0.15	2.6	37.		0.06	1717	200.		22	32.05	14.1	19.5	9.22	0.073	5404 14-05-010		2
	33.78	9.06	12+10	2			0.02	1	203.		1.07		4.0		10.35	4.77			
1001-26	1001.3	-2647	7A	3.9	0.44			961	203	10	1.03	12.0	-6.4	11.9	9.19	0.139	UGCA 196		
	262.80	22.48	0.18	3.4	69.		0.37	686	179.		10	30.40	5.5				499-37 -4-24-014		
	139.22	-44.92	54 -2 +1	6			0.17	2	182.		1.02		-8.5		10.04				
U 5423	1001.4	7037	10	1.4	0.79	14.7	3	333	69	10	-0.37	9.0	6.8	-15.06	8.22	0.21	M81 DW B		
	140.03	40.80	0.00	1.4	42.		0.08	479	77.	15	225	29.76	5.9	3.7	7.54	0.055	5423 12-10-021		
	40.86	2.10	14-10	2			0.16	2	85.	225	35		0.3		8.80	3.85			
1001-27	1001.5	-2720	5X	2.6	0.72	12.58	2	2809	358	30	0.71	37.2	-20.2	-20.27	10.30	0.36	IC 2537, UGCA 197		
	263.21	22.08	0.27	2.5	48.		0.12	2533			27	32.85	17.0	27.2	9.85		499-39 -4-24-015		2
	139.99	-44.83	31 -5 +2	6			0.20	2			1.02		-26.2						
U 5425	1001.7	1352	5	1.0	0.89	13.5	3	2713		83		38.9	1.7	-19.45	9.97		5425 02-26-022		2
	223.52	48.82	0.21	1.0	31.		0.04	2577				32.95	31.9	11.4					
	86.96	-34.77	32 +3	2			0.08	2					-22.2						
1001-64	1001.9	-6443	1	2.0	0.54			2152		20		26.8	-22.7	17.2			092-06		
	286.32	-7.66	0.11	2.2	62.		0.25	1867				32.14	-2.1						
	185.17	-31.49	33 +3	6			1.14	7					-14.0						
N 3113	1002.3	-2812	8	3.7	0.36			1088	215	10	0.92	14.3	-7.9	12.9	9.23	0.132	UGCA 198		
	263.96	21.53	0.31	3.1	75.		0.51	810.	185.		14	30.77	6.4		10.11		435-35 -5-24-021		2
	141.18	-44.57	54 -2 +1	6			0.22	6	188.		1.14		-10.0						
N 3115	1002.7	-729	-2	6.9	0.49	9.94	2	697		6	0.67	6.7	-1.9	-19.18	9.86	0.03	-1-26-018		2
	247.78	36.77	C.08	6.1	66.		0.00	475.			36100	29.12	4.5	11.9	8.32				
	112.47	-42.87	15 -5 +1	9			0.11						-4.5						
1003+29	1003.4	2912	13 P	0.7	0.91	14.5	3	1402		57		23.6	6.4	-17.37	9.14		UGCA 201, HARO 23		
	200.14	53.69	0.54	0.7	28.		0.00	1341				31.87	20.3	4.8			05-24-011		2
	72.54	-25.78	21 -0+12	2			0.06						-10.3						
1003-75	1003.8	-7514	5	2.4	1.00			1783		20		21.7	-19.0	17.7			037-10		
	292.92	-15.96	0.10	2.8			0.00	1517				31.68	-5.0						
	194.73	-25.17	53 -0	6			0.77	7					-9.2						
N 3136	1004.5	-6708	-5	2.5	0.67	11.16	2	1672		59		20.5	-17.6	-20.40	10.35		092-08		2
	287.98	-9.45	0.11	2.8	53.		0.00	1391				31.56	-2.3	16.8					
	187.32	-29.92	53-21	6			0.89						-10.2						
U 5455	1004.6	7053	10	2.0	1.00			1291	72	15	0.38	22.7	17.1	13.9	9.09		5455 12-10-000		
	139.53	40.84	0.19	2.1			0.00	1438			19	31.78	14.8						
	40.86	2.47	12 -7	2			0.15	1			1.05		1.0						
U 5458	1004.8	1231	12	1.4	0.65	13.7	3	2842		220		40.5	0.7	-19.34	9.93		IC 591, ARAK 231		
	225.86	48.88	0.24	1.3	54.		0.16	2700				33.04	33.2	15.4			5458 02-26-025		
	88.84	-34.79	32 +3	2			0.08						-23.1						

Name	ID	col1	col2	col3	col4	col5	col6	col7	col8	col9	col10	col11	col12	col13	coords
U 5459	5459 09-17-027 2	0.97 / 0.168 / 5.78	9.69 / 9.67 / 10.45	-18.74 / 16.6	11.9 / 16.1 / -3.4	20.3 / 31.54	1.06 / 12 / 1.09	15	280. / 242. / 245.	1121. / 1184. / 1	3 / 0.67 / 0.00	12.8 / 10.50 / 2.30	0.15 / 90.	4.0 / 2.8 / 2	5 / 0.27 / 13 -1	1004.8 5320 / 160.84 50.26 / 53.53 -9.72
U 5460	5460 09-17-028 2	0.37	9.55 / 9.12	-18.39 / 14.5	11.4 / 15.9 / -3.6	19.9 / 31.49	0.52 / 17 / 1.07	15	110	1094. / 1151. / 1	3 / 0.02 / 0.00	13.1	0.93 / 25.	2.5 / 2.5 / 2	7B / 0.26 / 13 +1	1004.9 5205 / 162.63 50.77 / 54.46 -10.57
U 5478	DDO 73 / 5478 05-24-020 2	0.76	9.08 / 8.96	-17.21 / 13.0	6.6 / 20.3 / -9.7	23.4 / 31.85	0.22 / 19 / 1.04	10	80	1381. / 1326. / 1	7 / 0.04 / 0.07	14.64	0.90 / 29.	1.9 / 1.9 / 2	10 / 0.51 / 21 -0+12	1006.6 3024 / 198.24 54.52 / 71.96 -24.49
N 3137	UGCA 203 / 435-47 -5-24-024	0.145	9.78 / 10.62	24.6	-8.3 / 6.5 / -10.0	14.5 / 30.80	1.46 / 7 / 1.05	10	269. / 242. / 244.	1109. / 831. / 2	0.46 / 0.22		0.39 / 73.	6.8 / 5.8 / 6	7 / 0.29 / 54 -2 +1	1006.9 -2849 / 265.22 21.68 / 141.90 -43.51
1007-66	IC 2554 / 092-12 2			14.6	-14.4 / -1.7 / -8.3	16.7 / 31.11		20		1375. / 1093.	0.43 / 0.78		0.40 / 72.	3.2 / 3.0 / 6	5BP / 0.11 / 53-21	1007.5 -6648 / 288.03 -9.00 / 186.81 -29.86
1008-25	UGCA 204, DDO 237 / 500-06 -4-24-019 2	0.441	9.39 / 9.74	24.7	-18.1 / 16.6 / -23.2	33.8 / 32.64	0.33 / 20 / 1.01	15	74. / 88. / 95.	2514. / 2243. / 2	0.07 / 0.13		0.81 / 40.	2.6 / 2.5 / 6	9 / 0.29 / 31 +5 +2	1008.3 -2534 / 263.28 24.42 / 137.40 -43.43
1008-04	SEX A, UGCA 205 / DDO 75 -1-26-030 2	1.19 / 0.235 / 5.07	7.73 / 7.81 / 8.44	-13.85 / 1.6	-0.3 / 0.9 / -0.9	1.3 / 25.59	1.58 / 36 6 / 1.10	5	64. / 76. / 83.	325. / 114. / 4	1 / 0.06 / 0.07	11.74	0.83 / 38.	4.4 / 4.3 / 9	10B / 0.19 / 14+12	1008.5 -426 / 246.14 39.89 / 109.05 -40.66
1008+58	UGCA 206, MARK 27 / 10-15-000 2			5.0	21.8 / 25.8 / -3.2	34.0 / 32.65		39		2178. / 2270.	0.00 / 0.00		0.53 / 63.	0.5 / 0.5 / 2	13 P / 0.09 / 13 -0	1008.5 5859 / 152.81 48.09 / 49.74 -5.47
N 3156	5503 01-26-019 2		9.61	-18.54 / 9.8	-2.5 / 14.5 / -11.3	18.6 / 31.34		90		1174. / 994.	3 / 0.00 / 0.04	12.8	0.63 / 55.	1.9 / 1.8 / 2	-5 / 0.20 / 21-10	1010.1 323 / 238.25 45.14 / 99.73 -37.63
N 3153	5505 02-26-032 2	0.27 / 0.135 / 2.04	10.24 / 9.68 / 10.55	-20.12 / 23.5	0.5 / 33.5 / -22.1	40.2 / 33.02	0.47 / 21 / 1.04	15	247. / 228. / 231.	2811. / 2672. / 1	3 / 0.31 / 0.09	12.9 / 11.51 / 1.39	0.48 / 67.	2.3 / 2.0 / 2	5 / 0.22 / 32 +3	1010.2 1255 / 226.26 50.24 / 89.20 -33.45
1010-47	263-15			24.4	-24.2 / 6.4 / -20.0	32.1 / 32.53		20		2497. / 2202. / 7	0.67 / 0.73		0.13 / 90.	3.2 / 2.6 / 6	5 / 0.13 / 31-8 +6	1010.3 -4703 / 276.98 7.40 / 165.31 -38.62
N 3162	5510 04-24-019 2	0.28 / 0.083 / 3.36	10.08 / 9.52 / 10.60	-19.71 / 20.7	3.6 / 19.3 / -10.5	22.2 / 31.73	0.83 / 211 18	10 / 5 / 211	188. / 258. / 260.	1306. / 1215. / 1	2 / 0.06 / 0.07	12.02	0.84 / 37.	3.2 / 3.2 / 2	4X / 0.50 / 21 -6	1010.8 2259 / 211.05 54.09 / 79.28 -28.07
N 3166	5516 01-26-024 2		10.28	-20.22 / 26.3	-2.9 / 17.3 / -13.3	22.0 / 31.71		50		1381. / 1203.	1 / 0.00 / 0.01	11.49	0.60 / 58.	4.6 / 4.1 / 2	0.15 / 21-10	1011.2 341 / 238.14 45.54 / 99.52 -37.26
U 5518	5518 07-21-028		9.00	18.2	13.3 / 28.2 / -10.1	32.8 / 32.58	-0.03 / 226 33 / 1.04	20 / 15 / 226	78	2115. / 2109. / 1	0.19 / 0.02		0.61 / 57.	2.1 / 1.9 / 2	10 / 0.07 / 13 -0	1011.2 3943 / 181.98 55.23 / 64.78 -17.94

(1) Name	(2) α / ℓ / SGL	(3) δ / b / SGB	(4) Type / ρxyz / Group	(5) D25 / Db,i25 / source	(6) d/D / i	(7) Bb,iT / H-0.5 / B-H	(8) source / Ai-oB / AbB	(9) Vh / Vo / Tel	(10) W20 / WR / WD	(11) ev / ew / source	(12) Fc / source/ef / fh	(13) R / μ	(14) SGX / SGY / SGZ	(15) Mb,iB / Δb,i25	(16) logLB / logMH / logMT	(17) MH/LB / MH/MT / MT/LB	(18) Alternate Names / UGC/ESO MCG RCII
U 5522	1011.3	7.16	6	2.9	0.64	13.3	3	1228.	232.	15	1.00	19.9	-1.5	-18.20	9.47	1.34	
	233.88	47.60	0.23	2.6	55.		0.17	1065.	238.		12	31.50	16.1	15.1	9.60	0.160	5522 01-26-025
	95.46	-35.78	21-10	2			0.01	1	241.		1.07		-11.7		10.39	8.37	
N 3169	1011.6	343	1AP	5.0	0.58	11.03	1	1229.		35	0.77	19.7	-2.6	-20.44	10.37	0.10	
	238.19	45.64	0.30	4.5	59.		0.21	1051.			11 75	31.47	15.5	25.9	9.36		5525 01-26-026 2
	99.53	-37.15	21-10	2			0.01						-11.9				
N 3175	1012.5	-2837	3XP	5.4	0.24	11.32	2	1125.		90		14.9	-8.6	-19.55	10.01		UGCA 207
	266.13	22.59	0.26	4.1	85.		0.67	849.				30.87	6.9	17.8			436-03 -5-24-028 2
	141.47	-42.30	54 -2 +1	6			0.22						-10.0				
N 3147	1012.7	7339	4A	4.4	0.86	11.33	2	2721.		80		40.9	31.6	-21.73	10.88		
	136.29	39.47	0.14	4.3	34.		0.05	2881.				33.06	25.8	51.4			5532 12-10-025 2
	39.26	4.82	42 -0	2			0.07						3.4				
U 5539	1013.3	256	10B	2.2	0.54	14.0	3	1278.	165.	20	0.53	20.5	-3.0	-17.56	9.22	0.87	
	239.45	45.51	0.29	1.9	62.		0.24	1097.	150.	12	226 15	31.56	16.1	11.4	9.15	0.192	5539 01-26-029 2
	100.64	-37.06	21-10	2			0.08	1	154.	226	1.04		-12.3		9.87	4.51	
1013+45	1013.4	4534	13 P	0.6	0.66	15.4	3	1643.		34		27.1	13.0	-16.77	8.90		UGCA 208, MARK 140
	171.92	54.32	0.10	0.6	53.		0.00	1668.				32.17	22.9	4.7			08-19-000 2
	60.39	-13.87	13 -0	2			0.00						-6.5				
N 3177	1013.8	2122	3A	1.7	0.83	12.87	2	1220.		65		21.1	2.8	-18.75	9.69		
	214.04	54.29	0.44	1.6	38.		0.06	1122.				31.62	18.3	9.9			5544 04-24-023 2
	81.25	-28.38	21 -6	2			0.07						-10.0				
1014-03	1014.7	-315	8B	2.5	0.47			1314.	187	15	0.67	20.3	-4.9		9.28		IC 600, UGCA 209
	246.34	41.83	0.12	2.2	67.		0.33	1109.			23	31.54	15.0	13.0			00-26-034
	108.17	-38.83	21+10	2			0.07	1			1.05		-12.7				
N 3185	1014.9	2156	1B	2.0	0.63	12.71	1	1241.		65		21.3	3.0	-18.93	9.76		
	213.23	54.71	0.44	1.9	55.		0.17	1146.				31.64	18.6	11.8			5554 04-24-024 2
	80.86	-27.86	21 -6	2			0.07						-10.0				
1014-48	1014.9	-4838	5	3.6	0.70			2742.		20		35.1	-27.2				
	278.52	6.53	0.19	4.0	50.		0.13	2447.				32.73	6.4	41.0			213-11
	166.84	-37.34	31 -8 +6	6			0.94	7					-21.3				
N 3187	1015.0	2208	5BP	3.3	0.47	13.20	1	1594.		22		26.1	3.7	-18.88	9.74		ARP 316, VV 307
	212.91	54.79	0.15	2.8	67.		0.33	1500.				32.08	22.8	21.3			5556 04-24-025 2
	80.69	-27.73	21 -6	2			0.07						-12.1				
N 3184	1015.2	4140	6X	7.5	0.93	10.37	1	599.	142	8	1.48	8.7	3.7	-19.34	9.93	0.27	NGC 3180
	178.36	55.62	0.17	7.4	26.		0.03	604.		4	217 4	29.71	7.5	18.8	9.36		5557 07-21-037 2
	63.70	-16.12	15 +7	2			0.00	1		211	1.44		-2.4				
N 3190	1015.4	2205	1AP	4.2	0.42	11.47	1	1310.		38		22.4	3.2	-20.29	10.31		ARP 316, VV 307
	213.04	54.86	0.52	3.5	71.		0.40	1216.				31.76	19.6	22.9			5559 04-24-026 2
	80.79	-27.68	21 -6	2			0.07						-10.4				

ID	Position	Type	Axes	Incl.	B mag	Ext	Vel								
N 3193	1015.7 2209 / 212.97 54.95 / 80.77 -27.58	-5 / 0.51 / 21 -6	2.7 / 2.7 / 2	1.00	11.78	1 / 0.00 / 0.07	1371. / 1277.		50		23.2 / 31.83	3.3 / 20.3 / -10.7	-20.05 / 18.3	10.21		ARP 316
N 3198	1016.7 4549 / 171.21 54.80 / 60.57 -13.27	5B / 0.15 / 15 +7	8.8 / 7.3 / 2	0.42 / 71.	10.55 / 8.67 / 1.88	1 / 0.40 / 0.00	660. / 686. / 4	318 / 297. / 299.	7 / 4 / 211	1.68 / 211 5 / 1.09	10.8 / 30.17	5.2 / 9.2 / -2.5	-19.62 / 23.0	10.04 / 9.75 / 10.77	0.51 / 0.095 / 5.37	5562 04-24-027 2 / 5572 08-19-020 2
N 3203	1017.2 -2627 / 265.61 24.95 / 138.49 -41.39	-2A / 0.19 / 31 +5 +2	3.0 / 2.2 / 6	0.19 / 90.	12.56	6 / 0.00 / 0.27	2424. / 2153.		90		32.6 / 32.57	-18.3 / 16.2 / -21.6	-20.01 / 20.9	10.20		500-24 -4-25-002 2
N 3208	1017.4 -2536 / 265.08 25.66 / 137.35 -41.37	4A / 0.14 / 31 -0 +2	2.1 / 2.1 / 6	0.84 / 37.	13.21	6 / 0.06 / 0.23	3007. / 2738.		20		40.0 / 33.01	-22.1 / 20.4 / -26.5	-19.80 / 24.5	10.11		SCI 90 / 500-25 -4-25-003 2
U 5588	1018.1 2537 / 207.24 56.32 / 77.87 -25.22	13 P / 0.41 / 21 -6	0.7 / 0.7 / 2	0.67 / 52.	13.8	3 / 0.00 / 0.05	1426. / 1350.		220		24.2 / 31.92	4.6 / 21.4 / -10.3	-18.12 / 4.9	9.44		ARAK 238 / 5588 04-25-000 2
N 3206	1018.5 5712 / 154.07 50.15 / 51.96 -5.68	6B / 0.22 / 13 -1	2.9 / 2.7 / 2	0.70 / 50.	12.3	3 / 0.13 / 0.00	1161. / 1245. / 1	209 / 226. / 229.	15	0.95 / 217 7 / 1.07	21.0 / 31.61	12.9 / 16.5 / -2.1	-19.31 / 16.6	9.92 / 9.59 / 10.39	0.48 / 0.160 / 2.98	5589 10-15-069 2
N 3223	1019.4 -3400 / 270.80 19.11 / 148.37 -40.22	3A / 0.65 / 31 +2	3.6 / 3.6 / 6	0.65 / 54.	11.2	5 / 0.16 / 0.41	2900. / 2617. / 2	425 / 480. / 481.	25	0.85 / 24 / 1.03	38.1 / 32.91	-24.8 / 15.3 / -24.6	-21.71 / 40.1	10.88 / 10.01 / 11.43	0.14 / 0.038 / 3.56	IC 2571 / 375-12 -6-23-023 2
U 5612	1020.1 7108 / 138.17 41.61 / 41.52 3.57	8B / 0.20 / 12 -7	3.3 / 3.1 / 6	0.69 / 51.	12.48	7 / 0.13 / 0.14	1011. / 1161. / 1	172 / 179. / 182.	10	0.73 / 16 / 1.09	19.1 / 31.41	14.3 / 12.6 / 1.2	-18.93 / 17.3	9.76 / 9.29 / 10.21	0.34 / 0.122 / 2.77	DDO 77 / 5612 12-10-032 2
N 3226	1020.7 2009 / 216.93 55.44 / 83.38 -27.65	-5 P / 0.45 / 21 -6	2.8 / 2.7 / 2	0.88 / 31.	12.27	2 / 0.00 / 0.03	1372. / 1270.		12		23.4 / 31.84	2.4 / 20.6 / -10.8	-19.57 / 18.4	10.02	0.10	ARP 94, VV 209 / TURN 16A / 5617 03-27-015 2
N 3227	1020.8 2007 / 217.00 55.45 / 83.43 -27.65	1XP / 0.38 / 21 -6	5.9 / 5.5 / 2	0.72 / 48.	11.40	2 / 0.12 / 0.03	1183. / 1081.	293	12 / 50 / 51	0.64 / 51 25	20.6 / 31.57	2.1 / 18.2 / -9.6	-20.17 / 33.1	10.26 / 9.27		ARP 94, VV 209 / TURN 16B / 5620 03-27-016 2 S1.2
1021-47	1021.1 -4705 / 278.54 8.39 / 164.57 -36.88	5 / 0.23 / 31 -8 +6	2.1 / 2.0 / 6	0.40 / 72.		0.43 / 0.79	2674. / 2380. / 7		20		34.4 / 32.68	-26.5 / 7.3 / -20.6	20.1			263-31
N 3225	1021.8 5824 / 152.14 49.86 / 51.33 -4.57	5 / 0.14 / 13 -0	2.3 / 2.0 / 2	0.55 / 61.	12.9 / 11.47 / 1.43	3 / 0.23 / 0.00	2137. / 2227. / 1	259 / 253. / 255.	15	0.67 / 15 / 1.04	33.6 / 32.63	20.9 / 26.1 / -2.7	-19.73 / 19.6	10.08 / 9.72 / 10.56	0.44 / 0.145 / 3.00	5631 10-15-077
U 5633	1022.0 1500 / 225.34 53.73 / 88.66 -29.94	8B / 0.28 / 21 -0	2.5 / 2.4 / 2	0.68 / 52.	14.09	7 / 0.14 / 0.12	1390. / 1264. / 1	179 / 185. / 188.	15	0.65 / 5 18 / 1.05	23.4 / 31.84	0.5 / 20.3 / -11.7	-17.75 / 16.4	9.29 / 9.39 / 10.21	1.25 / 0.150 / 8.31	5633 03-27-020 2 / DDO 79
N 3241	1022.0 -3214 / 270.20 20.87 / 145.99 -39.91	2A / 0.52 / 31 +2	2.3 / 2.2 / 6	0.71 / 49.	13.25	2 / 0.12 / 0.24	2874. / 2594.		90		37.9 / 32.89	-24.1 / 16.3 / -24.3	-19.64 / 24.3	10.05		436-16 -5-25-002 2

(1) Name	(2) α / ℓ / SGL	(3) δ / b / SGB	(4) Type / $\varrho_{x,y,z}$ / Group	(5) D_{25} / $D_{25}^{b,i}$ / source	(6) d/D / i	(7) $B_T^{b,i}$ / $H_{-0.5}$ / B-H	(8) source / A_B^{i-o} / A_B^b	(9) V_h / V_o / Tel	(10) W_{20} / W_R / W_D	(11) e_v / e_w / source	(12) F_c / source·e_F / f_h	(13) R / μ	(14) SGX / SGY / SGZ	(15) $M_B^{b,i}$ / $\Delta_{25}^{b,i}$	(16) log L_B / log M_H / log M_T	(17) M_H/L_B / M_H/M_T / M_T/L_B	(18) Alternate Names / UGC/ESO MCG AGN? RCII
N 3239	1022.4 / 221.64 / 86.29	1725 / 54.82 / -28.68	9BP / 0.34 / 15 -0 +1	5.5 / 5.2 / 2	0.70 / 50.	12.2	3 / 0.13 / 0.09	754. / 639. / 1	203	7 / 6 / 217	1.30 / 217 5 / 1.23	8.1 / 29.53	0.5 / 7.1 / -3.9	-17.33 / 12.3	9.12 / 9.12	0.98	ARP 263, VV 95 / 5637 03-27-025 2
1023+56	1023.8 / 154.36 / 52.95	5631 / 51.13 / -5.57	13 P / 0.13 / 13 +1	0.6 / 0.5 / 2	0.61 / 57.		0.00 / 0.00	799. / 880. / 1	62	20	-0.06 / 32 / 1.00	15.6 / 30.96	9.3 / 12.4 / -1.5	2.3	8.33		UGCA 211, MARK 32 / 09-16-000 2
N 3246	1024.1 / 240.48 / 100.54	407 / 48.37 / -34.12	8X / 0.06 / 30 -0	2.3 / 2.1 / 2	0.64 / 55.	12.9	3 / 0.17 / 0.05	2150. / 1977. / 1	262 / 274. / 276.	15	0.70 / 22 / 1.04	31.9 / 32.52	-4.8 / 25.9 / -17.9	-19.62 / 19.6	10.04 / 9.71 / 10.63	0.47 / 0.120 / 3.88	5661 01-27-009 2
U 5662	1024.2 / 201.66 / 75.73	2853 / 58.17 / -22.28	3B / 0.45 / 21 -8	3.4 / 2.4 / 2	0.11 / 90.	14.3	3 / 0.67 / 0.04	1327. / 1269. / 1	235	40	0.43 / 24 / 1.10	23.1 / 31.82	5.3 / 20.8 / -8.8	-17.52 / 16.2	9.20 / 9.16	0.91	NGC 3245A / 5662 05-25-012 2
N 3250	1024.4 / 274.97 / 155.36	-3941 / 14.94 / -38.23	-5 / 0.55 / 31 -7 +6	1.9 / 1.9 / 6	0.75 / 45.	11.61	2 / 0.00 / 0.49	2871. / 2582.		90		37.4 / 32.87	-26.7 / 12.3 / -23.2	-21.26 / 20.7	10.70		317-26 -7-22-007 2
N 3245	1024.5 / 201.90 / 75.87	2846 / 58.22 / -22.29	-2A / 0.34 / 21 -8	3.2 / 2.9 / 2	0.67 / 52.	11.66	2 / 0.00 / 0.04	1261. / 1202.		30		22.2 / 31.73	5.0 / 19.9 / -8.4	-20.07 / 18.8	10.22		5663 05-25-013 2
U 5666	1024.8 / 140.20 / 43.67	6840 / 43.62 / 2.29	9X / 0.47 / 14 -10	13.0 / 11.9 / 2	0.63 / 55.	10.79 / 10.05 / 0.74	1 / 0.17 / 0.07	38. / 177. / 1	123 / 119. / 123.	10 / 9 / 25	2.05 / 25 3 / 1.71	2.7 / 27.16	2.0 / 1.9 / 0.1	-16.37 / 9.4	8.74 / 8.91 / 9.59	1.49 / 0.212 / 7.01	IC 2574, VII ZW 330 / DDO 81 / 5666 12-10-038 2
N 3256	1025.7 / 277.37 / 160.16	-4339 / 11.73 / -37.08	13 P / 0.31 / 31 -6	8.3 / 7.5 / 6	0.36 / 75.	11.44	2 / 0.00 / 0.56	2886. / 2595.		17		37.4 / 32.86	-28.1 / 10.1 / -22.5	-21.42 / 81.9	10.76		VV 65 / 263-38 -7-22-010 2
U 5675	1025.8 / 218.21 / 84.39	1949 / 56.46 / -26.79	9 / 0.30 / 21 -6	2.0 / 1.9 / 2	0.77 / 44.		0.09 / 0.03	1104. / 1001. / 1	88	10	0.06 / 226 20 / 1.04	19.5 / 31.45	1.7 / 17.3 / -8.8	/ 10.8	8.64		5675 03-27-047
1026-31	1026.0 / 270.36 / 144.61	-3115 / 22.16 / -39.16	43 / 0.30 / 31 +2	2.0 / 2.1 / 6	1.00	13.14	6 / 0.00 / 0.20	3137. / 2859.		20		41.4 / 33.08	-26.2 / 18.6 / -26.1	-19.94 / 25.4	10.17		IC 2580, SCI 94 / 436-25 -5-25-004 2
N 3257	1026.5 / 272.90 / 149.89	-3524 / 18.76 / -38.58	-2X / 0.73 / 31 -2	0.9 / 1.0 / 6	0.88 / 33.		0.00 / 0.25	3063. / 2779.		90		37.5 / 32.87	-25.6 / 14.9 / -23.5	/ 10.9			375-36 -6-23-031 2
N 3254	1026.5 / 200.75 / 75.26	2945 / 58.75 / -21.38	4A / 0.38 / 21 -8	4.9 / 3.9 / 2	0.33 / 78.	11.56 / 9.58 / 1.98	2 / 0.60 / 0.04	1362. / 1309. / 1		10	1.01 / 211 29	23.6 / 31.87	5.6 / 21.3 / -8.6	-20.31 / 26.9	10.32 / 9.76	0.28	5685 05-25-018 2
N 3258	1026.6 / 272.89 / 149.82	-3521 / 18.82 / -38.56	-E / C.72 / 31 -2	1.7 / 1.7 / 6	0.88 / 32.	12.55	2 / 0.00 / 0.25	2848. / 2564.		90		37.5 / 32.87	-25.4 / 14.7 / -23.4	-20.32 / 18.6	10.32		375-37 -6-23-032 2

U 5688	1026.6	7019	9B	3.4	0.62	13.56	7	1916.	68	8	0.52	30.8	22.7	-18.88	9.74	0.57	VII ZW 331, VV 294
	138.44	42.57	0.16	3.2	56.		0.18	2063.	61.		13	32.44	20.8	28.8	9.50	1.011	DDO 80
	42.50	3.46	12 -0	2			0.10	1	69.		1.09		1.9		9.49	0.56	5688 12-10-044 2
1026-39	1026.8	-3950	7B	2.4	0.80			2818.		14		36.8	-26.5				NGC 3250E
	275.46	15.07	0.48	2.5	41.		0.08	2530.				32.83	12.1	26.9			
	155.43	-37.75	31 -7 +6	6			0.41						-22.5				317-34 -7-22-013 2
N 3260	1026.9	-3520	-5	1.3	0.75			2453.		90		37.5	-25.0				
	272.94	18.86	0.69	1.2	45.		0.00	2169.				32.87	14.5	13.1			
	149.79	-38.51	31 -2	6			0.25						-23.1				375-40 -6-23-033 2
N 3261	1026.9	-4424	5B	3.9	0.80	11.5	5	2571.	403	30	1.06	33.4	-25.3	-21.12	10.64	0.29	
	277.96	11.20	0.19	4.2	41.		0.08	2280.	559.		26	32.62	8.7	41.0	10.11	0.035	
	161.00	-36.67	31 -0 +6	6			0.61	2	560.		1.05		-19.9		11.57	8.50	263-40 -7-22-015 2
N 3263	1027.1	-4351	6B	6.8	0.22			3015.		37		39.1	-29.5				
	277.70	11.69	0.26	5.5	87.		0.67	2724.				32.96	10.5	62.8			
	160.33	-36.78	31 -6	6			0.57						-23.4				263-43 2
1027-350	1027.3	-3500	-3	1.2	0.18			1892.		90		37.5	-24.3				
	272.82	19.19	0.61	0.9	90.		0.00	1609.				32.87	14.4	9.9			
	149.35	-38.47	31 -2	6			0.26						-22.5				375-41 -6-23-035 2
1027-351	1027.6	-3507	-5					1781.		90		37.5	-24.3				
	272.94	19.12	0.58				0.00	1498.				32.87	14.2				
	149.49	-38.39	31 -2				0.26						-22.3				6-23-000
N 3268	1027.8	-3504	-5	2.0	0.73	12.59	2	2801.		90		37.5	-25.2	-20.28	10.30		
	272.95	19.19	0.69	1.9	47.		0.00	2518.				32.87	14.9	20.8			
	149.42	-38.36	31 -2	6			0.26						-23.2				375-45 -6-23-041 2
N 3273	1028.2	-3521	-2A	1.9	0.40			2459.		90		37.5	-25.0				
	273.18	18.99	0.69	1.6	72.		0.00	2176.				32.87	14.6	17.5			
	149.76	-38.24	31 -2	6			0.26						-22.9				375-49 -6-23-045 2
U 5707	1028.3	4323	6X	2.5	0.75	13.6	3	2800.	130	15	0.47	41.6	17.8	-19.49	9.99	0.53	
	174.22	57.51	0.07	2.4	45.		0.10	2816.	145.		15	33.09	36.3	29.2	9.71	0.287	
	63.85	-13.16	20 -0	2			0.00	1	149.		1.06		-9.5		10.25	1.83	5707 07-22-016
U 5706	1028.3	3446	10	1.6	0.82			1494.	53	15	0.03	25.4	7.8				KAR 70
	190.37	59.16	0.26	1.6	39.		0.07	1466.	55.	9	226 16	32.02	22.8	11.9	8.84	0.664	
	71.09	-18.23	21 +9	2			0.02	1	64.	226	1.03		-7.9		9.02		5706 06-23-000
U 5708	1028.6	444	7X	3.2	0.21	13.2	3	1177.	205	10	0.96	19.4	-2.9	-18.24	9.49	1.12	
	240.79	49.63	0.18	2.3	89.		0.67	1007.	169.		9	31.44	16.1	13.0	9.54	0.318	
	100.36	-32.85	21+10	2			0.05	1	172.		1.06		-10.5		10.03	3.51	5708 01-27-014
N 3275	1028.7	-3629	3B	2.8	1.00	12.1	5	3242.		90		42.4	-29.3	-21.04	10.61		
	273.92	18.09	0.28	2.9			0.00	2957.				33.14	16.1	35.9			
	151.17	-37.98	31 +2	6			0.23						-26.1				375-50 -6-23-046 2
N 3259	1029.1	6518	4X	2.3	0.55	12.7	3	1743.		60		28.7	19.7	-19.59	10.03		
	143.26	46.23	0.20	2.0	61.		0.23	1867.				32.29	20.8	16.8			
	46.55	0.49	12 -9	2			0.00						0.2				5717 11-13-027 2

(1) Name	(2) α / ℓ / SGL	(3) δ / b / SGB	(4) Type / Q_{vz} / Group	(5) D_{25} / $D_{25}^{b,i}$ / source	(6) d/D / i	(7) $B_T^{b,i}$ / $H_{-0.5}$ / $B-H$	(8) source / A_B^{i-o} / A_B^b	(9) V_h / V_o / Tel	(10) W_{20} / W_R / W_D	(11) e_v / e_w / source	(12) F_c / source/e_f / f_h	(13) R / μ	(14) SGX / SGY / SGZ	(15) $M_B^{b,i}$ / $\Delta_{25}^{b,i}$	(16) log L_B / log M_H / log M_T	(17) M_H/L_B / M_H/M_T / M_T/L_B	(18) Alternate Names / UGC/ESO MCG / AGN? RCII
N 3264	1029.2 / 153.94 / 53.57	5620 / 51.85 / -5.10	9B / 0.29 / 13 +1	3.3 / 2.8 / 2	0.47 / 67.	13.1 / 12.05 / 1.05	3 / 0.33 / 0.00	942. / 1023. / 1	182 / 160. / 164.	10	0.68 / 16 / 1.07	17.9 / 31.26	10.6 / 14.3 / -1.6	-18.16 / 14.6	9.46 / 9.19 / 10.04	0.54 / 0.141 / 3.80	5719 09-17-069
U 5720	1029.4 / 156.20 / 54.91	5439 / 52.81 / -6.12	13 F / 0.19 / 13 +1	1.1 / 1.1 / 2	0.91 / 28.	13.50	2 / 0.00 / 0.00	1446. / 1519.		16	0.41 / 17 77	24.9 / 31.98	14.2 / 20.3 / -2.7	-18.48 / 8.0	9.58 / 9.20	0.42	ARP 233, MARK 33 / 5720 09-17-070 2
N 3274	1029.5 / 203.78 / 77.29	2755 / 59.21 / -21.83	6XP / 0.-9 / 15 -0 +1	2.1 / 1.9 / 2	0.56 / 60.	12.87	2 / 0.22 / 0.05	543. / 481. / 1	178 / 166. / 169.	8	1.13 / 8 / 1.03	5.9 / 28.84	1.2 / 5.3 / -2.2	-15.97 / 3.3	8.58 / 8.67 / 9.42	1.24 / 0.180 / 6.88	5721 05-25-020 2
N 3277	1030.1 / 202.15 / 76.60	2846 / 59.44 / -21.26	2A / 0.35 / 21 -8	2.4 / 2.4 / 2	0.92 / 26.	12.48	2 / 0.03 / 0.04	1460. / 1402.	275	75		25.0 / 31.99	5.4 / 22.6 / -9.1	-19.51 / 17.5	10.00		5731 05-25-022 2
N 3252	1030.4 / 134.80 / 39.78	7401 / 40.06 / 6.00	6B / 0.28 / 12 -8 +7	2.0 / 1.7 / 2	0.39 / 73.	13.5	3 / 0.45 / 0.12	1136. / 1299. / 1		30	0.41 / 26 / 1.03	20.8 / 31.59	15.9 / 13.2 / 2.2	-18.09 / 10.3	9.43 / 9.05	0.42	5732 12-10-049 2
U 5740	1031.7 / 161.18 / 58.02	5101 / 54.98 / -8.08	9X / 0.07 / 15 -0 +7	2.2 / 2.0 / 2	0.63 / 56.	14.4	3 / 0.17 / 0.00	651. / 707. / 1	170	25	0.50 / 20 / 1.04	11.3 / 30.27	5.9 / 9.5 / -1.6	-15.87 / 6.6	8.54 / 8.61	1.16	5740 09-17-075
N 3287	1032.1 / 215.40 / 83.21	2154 / 58.51 / -24.47	7B / 0.42 / 21 -6	2.1 / 1.9 / 2	0.51 / 64.	12.6	3 / 0.28 / 0.01	1149. / 1058.		67		20.6 / 31.57	2.2 / 18.6 / -8.5	-18.97 / 11.4	9.78		5742 04-25-032 2
1033-24	1033.0 / 267.58 / 135.92	-2430 / 28.62 / -37.84	8B / 0.13 / 54 -0 +1	5.0 / 4.9 / 6	0.77 / 44.	12.98	7 / 0.09 / 0.19	1048. / 785. / 2	86 / 97. / 103.	10	0.91 / 8 / 1.03	14.2 / 30.76	-8.0 / 7.8 / -8.7	-17.78 / 20.3	9.30 / 9.21 / 9.74	0.81 / 0.296 / 2.75	UGCA 212, DDO 238 / 501-23 -4-25-024 2
N 3294	1033.4 / 184.62 / 69.29	3735 / 59.84 / -15.77	5A / 0.19 / 21 -0	3.4 / 3.0 / 9	0.52 / 63.	11.94 / 9.45 / 2.49	2 / 0.26 / 0.00	1582. / 1570.	449	9 / 50 / 217	0.58 / 217 25	26.7 / 32.13	9.1 / 24.0 / -7.2	-20.19 / 23.4	10.27 / 9.43	0.15	5753 06-23-021 2
N 3299	1033.8 / 230.77 / 92.24	1257 / 55.30 / -28.33	8X / 0.12 / 15 -1	2.0 / 2.0 / 2	0.86 / 34.	13.2	3 / 0.05 / 0.05	597. / 465. / 1	158	30	40 / 1.04	5.4 / 28.66	-0.2 / 4.7 / -2.6	-15.46 / 3.2	8.38 / 7.46	0.12	5761 02-27-029 2
U 5764	1033.9 / 196.16 / 74.36	3149 / 60.43 / -18.91	10 / C.19 / 15 -0 +1	2.0 / 1.8 / 2	0.64 / 55.	14.3	3 / 0.17 / 0.03	584. / 543. / 1	161 / 158. / 162.	20	0.57 / 16 / 1.03	6.9 / 29.19	1.8 / 6.3 / -2.2	-14.89 / 3.6	8.15 / 8.25 / 9.42	1.26 / 0.067 / 18.67	ARP 267, DDO 83 / 5764 05-25-025 2
1033-36	1033.9 / 275.11 / 151.55	-3657 / 18.25 / -36.88	10 / 0.08 / 54 -0 +1	3.9 / 3.9 / 6	0.80 / 41.		/ 0.08 / 0.18	959. / 675. / 2	67 / 76. / 83.	8	0.91 / 8	11.6 / 30.32	-8.2 / 4.4 / -7.0	/ 13.2	9.04 / 9.34	0.494	375-71 -6-24-000
N 3300	1034.0 / 228.51 / 90.76	1426 / 56.06 / -27.62	-2X / 0.19 / 32 +4	2.0 / 1.8 / 2	0.51 / 64.	13.1	3 / 0.00 / 0.07	2992. / 2867.		90		42.9 / 33.16	-0.5 / 38.0 / -19.9	-20.06 / 22.5	10.22		5766 02-27-030 2

N 3301	1034.2	2208		3.2	0.36	12.23	2	1333.		75		23.3	2.5	-19.61	10.04		
	215.25	59.05	0.53	2.6	75.		0.00	1244.				31.84	21.2	17.7			
	83.26	-23.93	21 -6	2			0.01						-9.5				5767 04-25-035 2
U 5776	1034.6	6433	13 P	0.7	1.00	14.80	2	1625.		55		27.1	18.3	-17.37	9.14		MARK 149, VII ZW 339
	143.47	47.17	0.19	0.7			0.00	1746.				32.17	20.0	5.5			
	47.50	0.48	12 -9	2			0.00						0.2				5776 11-13-000 2
N 3312	1034.7	-2718	2AP	3.3	0.40	12.1	5	2710.		72		48.6	-28.4	-21.33	10.72		IC 629
	269.69	26.50	0.00	2.8	72.		0.43	2442.				33.43	24.3	39.7			
	139.46	-37.47	31 -1	6			0.22						-28.7				501-43 -4-25-039 2
N 3314	1034.9	-2726	0BP	2.1	0.65			3031.		140		48.6	-28.8				
	269.82	26.42	0.00	2.0	54.		0.00	2762.				33.43	24.5	28.4			
	139.62	-37.42	31 -1	6			0.20						-28.9				501-46 -4-25-041 2
N 3318	1035.1	-4122	3X	2.4	0.55	11.9	5	2910.		100		37.9	-28.2	-20.99	10.59		
	277.68	14.58	0.42	2.2	61.		0.24	2622.				32.89	12.0	24.3			
	156.91	-35.90	31 -6	6			0.38						-22.2				317-52 -7-22-026 2
N 3310	1035.7	5346	4XP	3.6	0.93	11.27	2	994.	248	7	1.15	18.7	10.4	-20.09	10.23	0.29	ARP 217
	156.60	54.07	0.27	3.5	25.		0.02	1064.		40	56 10	31.36	15.5	19.1	9.69		
	56.18	-5.93	13 +1	2			0.00			56			-1.9				5786 09-18-008 2
1036-37	1036.0	-3751	5	2.4	1.00			3050.		20		40.0	-28.6				
	275.97	17.69	0.53	2.5			0.00	2766.				33.01	14.8	29.2			
	152.59	-36.33	31 +2	6			0.22	7					-23.7				317-54 -6-24-000
N 3321	1036.3	-1123	5	2.9	0.50			2489.	277	20	0.78	35.0	-14.1				UGCA 214
	258.86	39.62	0.07	2.6	65.		0.29	2262.	264.		36	32.72	24.7	26.6	9.87	0.137	
	119.73	-35.63	22 -0	2			0.13	1	266.		1.06		-20.4		10.73		-2-27-010
N 3319	1036.4	4156	6B	6.7	0.61	11.59	1	743.	221	10	1.29	11.5	4.6	-18.71	9.68	0.54	
	175.99	59.37	0.17	6.0	57.	10.38	0.19	754.	219.	8	217 9	30.30	10.2	20.1	9.41	0.092	
	65.95	-12.81	15 +7	2		1.21	0.00	4	222.	211	1.06		-2.5		10.45	5.91	5789 07-22-036 2
N 3320	1036.6	4739	6	2.2	0.49	12.4	3	2328.		10		35.9	17.1	-20.38	10.34		
	165.86	57.26	0.12	1.9	66.	10.82	0.30	2367.				32.78	31.1	19.9			
	61.24	-9.44	13 -0	2		1.58	0.00						-5.9				5794 08-20-010 2
1038-48	1038.2	-4818	5B	4.6	0.83			1052.		20		13.1	-10.5				
	281.67	8.82	0.10	5.8	37.		0.06	762.				30.58	2.8	22.2			
	164.98	-33.77	54 +4	6			1.28	7					-7.3				214-17
1038-23	1038.9	-2307	10	2.9	0.61	13.67	7	1199.	155	15	0.68	16.8	-9.4	-17.46	9.18	0.90	UGCA 215, DDO 85
	267.97	30.52	0.12	2.7	57.		0.19	941.	148.		16	31.13	9.7	13.2	9.13	0.161	
	134.24	-36.45	54 -0	6			0.19	2	152.		1.01		-10.0		9.92	5.61	501-79 -4-25-058 2
1039+48	1039.3	4802	13 P	0.9	0.33	14.8	3	1543.		43		26.3	12.5	-17.30	9.11		UGCA 216, MARK 151
	164.85	57.49	0.13	0.7	78.		0.00	1585.				32.10	22.8	5.4			
	61.18	-8.86	13 -0	2			0.00						-4.0				08-20-016 2
N 3338	1039.5	1401	5A	5.7	0.54	10.99	1	1300.	353	7	1.44	22.8	-0.7	-20.80	10.51	0.44	
	230.33	57.03	0.44	5.1	62.	9.02	0.25	1174.	355.	7	217 10	31.79	20.4	34.0	10.16	0.115	
	91.84	-26.61	21 -5 +1	9		1.97	0.06		357.	211			-10.2		11.09	3.82	5826 02-27-041 2

(1) Name	(2) α / ℓ / SGL	(3) δ / b / SGB	(4) Type / Q_{xyz} / Group	(5) D_{25} / D^b_{25} / source	(6) d/D / i	(7) $B^{b,i}_T$ / $H_{-0.5}$ / B-H	(8) source / A^{i-o}_B / A^b_B	(9) V_h / V_o / Tel	(10) W_{20} / W_R / W_D	(11) e_v / e_w / source	(12) F_c source/e_f / f_h	(13) R / μ	(14) SGX / SGY / SGZ	(15) $M^{b,i}_B$ / $\Delta^{b,i}_{25}$	(16) log L_B / log M_H / log M_T	(17) M_H/L_B / M_H/M_T / M_T/L_B	(18) Alternate Names UGC/ESO MCG / AGN? RCII
U 5829	1039.8 / 190.06 / 72.51	3443 / 61.52 / -16.30	10B / 0.20 / 15 +8 +7	4.9 / 4.8 / 2	0.86 / 34.	13.57	7 / 0.05 / 0.03	633. / 608. / 1	97 / 135. / 140.	7	1.19 / 6 / 1.22	8.0 / 29.51	2.3 / 7.3 / -2.2	-15.94 / 11.2	8.57 / 9.00 / 9.77	2.68 / 0.167 / 16.02	DDO 84 / 5829 06-24-006 / 2
U 5833	1040.4 / 218.84 / 85.41	2041 / 59.97 / -23.38	-2 / 0.58 / 21 +6	1.8 / 1.4 / 2	0.33 / 78.	14.8	3 / 0.00 / 0.01	1308. / 1214.		71		23.2 / 31.82	1.7 / 21.2 / -9.2	-17.02 / 9.5	9.00		MARK 416 / 5833 04-25-044 / 2
N 3347	1040.5 / 275.85 / 150.28	-3605 / 19.67 / -35.67	3B / 0.65 / 31 -3 +2	3.9 / 3.6 / 9	0.59 / 58.	12.12	2 / 0.20 / 0.18	2923. / 2642.		90		38.5 / 32.93	-27.2 / 15.5 / -22.5	-20.81 / 40.5	10.52		376-13 -6-24-007 / 2
N 3329	1040.5 / 131.70 / 37.69	7704 / 38.08 / 8.32	1A / 0.20 / 42+13	2.2 / 2.0 / 2	0.63 / 56.	12.3	3 / 0.17 / 0.03	1880. / 2056.		77		30.3 / 32.41	23.7 / 18.3 / 4.4	-20.11 / 17.7	10.24		NGC 3397 / 5837 13-08-033 / 2
N 3344	1040.8 / 210.04 / 81.21	2511 / 61.26 / -21.10	4X / 0.19 / 15 -0 +1	6.7 / 6.6 / 2	0.94 / 23.	10.45	1 / 0.02 / 0.03	585. / 513.	174	8 / 15 / 51	1.68 / 51 10	6.1 / 28.92	0.9 / 5.6 / -2.2	-18.47 / 11.8	9.58 / 9.25	0.47	5840 04-25-046 / 2
N 3346	1041.0 / 228.82 / 90.90	1508 / 57.88 / -25.80	6B / 0.50 / 21 -5 +1	2.5 / 2.6 / 2	0.96 / 19.	12.4	3 / 0.01 / 0.06	1257. / 1137. / 1	175	20	0.69 / 12 / 1.07	22.4 / 31.75	-0.3 / 20.1 / -9.7	-19.35 / 17.0	9.93 / 9.39	0.29	5842 03-28-001 / 2
1041-09	1041.1 / 258.59 / 117.90	-936 / 41.75 / -34.11	7 / 0.07 / 22 -0	1.9 / 1.8 / 2	0.85 / 35.		/ 0.05 / 0.06	2077. / 1858. / 1	182	30	0.62 / 19 / 1.04	30.2 / 32.40	-11.7 / 22.1 / -16.9	15.9	9.58		-2-28-001
U 5846	1041.2 / 147.02 / 51.09	6038 / 50.40 / -1.25	10 / 0.24 / 12 -0 +1	2.0 / 2.0 / 2	1.00	15.32	7 / 0.00 / 0.00	1018. / 1122. / 1	69	8	0.61 / 11 / 1.05	19.3 / 31.42	12.1 / 15.0 / -0.4	-16.10 / 11.3	8.63 / 9.18	3.54	DDO 86 / 5846 10-16-002 / 2
N 3351	1041.3 / 233.95 / 94.13	1158 / 56.36 / -27.09	3B / 0.54 / 15 -1	7.5 / 6.9 / 2	0.62 / 56.	10.28	1 / 0.18 / 0.04	783. / 649. / 1	288 / 302. / 304.	9 / 15 / 211	1.10 / 211 15	8.1 / 29.54	-0.5 / 7.2 / -3.7	-19.26 / 16.3	9.90 / 8.92 / 10.64	0.10 / 0.019 / 5.48	M 95 / 5850 02-28-001 / 2
N 3358	1041.3 / 276.02 / 150.32	-3608 / 19.71 / -35.51	/ 0.61 / 31 -3 +2	3.3 / 3.0 / 9	0.58 / 60.	12.55	2 / 0.00 / 0.15	2910. / 2629.		90		38.3 / 32.92	-27.1 / 15.4 / -22.2	-20.37 / 33.6	10.34		376-17 -6-24-009 / 2
U 5848	1041.3 / 151.84 / 54.30	5641 / 52.98 / -3.56	9 / 0.21 / 13 -2	2.1 / 2.0 / 2	0.70 / 50.	14.2	3 / 0.13 / 0.00	829. / 914. / 1	144 / 150. / 154.	15 / 11 / 226	0.28 / 226 25 / 1.04	16.3 / 31.06	9.5 / 13.2 / -1.0	-16.86 / 9.5	8.94 / 8.70 / 9.79	0.59 / 0.082 / 7.20	5848 09-18-015
N 3353	1042.3 / 152.31 / 54.76	5613 / 53.38 / -3.72	13 P / 0.28 / 13 -2	1.6 / 1.5 / 2	0.76 / 45.	13.20	2 / 0.00 / 0.00	866. / 949.		10		16.8 / 31.13	9.7 / 13.7 / -1.1	-17.93 / 7.4	9.36		MARK 35, HARO 3 / 5860 09-18-022 / 2
N 3359	1043.4 / 143.58 / 48.91	6330 / 48.59 / 0.63	5B / 0.21 / 12 -0 +1	7.1 / 6.4 / 2	0.63 / 55.	10.83 / 9.25 / 1.58	1 / 0.17 / 0.00	1013. / 1131. / 1	263 / 274. / 277.	10	1.70 / 60 6 / 1.26	19.2 / 31.42	12.6 / 14.5 / 0.2	-20.59 / 35.9	10.43 / 10.27 / 10.89	0.69 / 0.237 / 2.92	5873 11-13-037 / 2

Name		Type	Size	Mag	Vel/Lum				Names
N 3365	1043.7 205 247.76 50.77 104.83 -30.26	5 0.12 15 -0 +1	4.3 3.1 2 0.21 89.	12.2 10.96 1.24 3 0.67 0.08	986. 810. 1 251 214. 216. 15 1.10 11 1.10	10.7 30.15 -2.4 8.9 -5.4	-17.95 9.7	9.37 9.16 10.11 0.61 0.112 5.46	5878 00-28-006 2
N 3367	1043.9 1401 231.30 57.96 92.36 -25.65	5B 0.00 32 -4	2.3 2.3 2 0.96 19.	11.99 2 0.01 0.05	3037. 2913. 13	43.6 33.20 -1.6 39.2 -18.9	-21.21 29.3	10.68	5880 02-28-005 2
N 3368	1044.2 1205 234.46 57.03 94.34 -26.40	2X 0.52 15 -1	6.7 6.2 2 0.70 50.	9.92 1 0.13 0.05	899. 766. 4 367 429. 430. 10 7 211 1.38 211 9 1.10	8.1 29.54 -0.5 7.3 -3.6	-19.62 14.7	10.04 9.20 10.89 0.14 0.020 7.14	M 96 5882 02-28-006 2
N 3370	1044.4 1732 225.36 59.67 88.93 -24.00	5A 0.52 21 -3 +1	2.8 2.6 2 0.62 56.	12.0 3 0.18 0.04	1320. 1212. 64	23.4 31.85 0.4 21.4 -9.5	-19.85 17.8	10.13	5887 03-28-008 2
U 5889	1044.7 1420 230.96 58.28 92.13 -25.34	9 0.13 15 -1	1.8 1.8 2 0.95 21.	14.13 7 0.02 0.05	571. 449. 1 77 8 0.21 19 1.03	5.1 28.53 -0.2 4.6 -2.2	-14.40 2.7	7.95 7.63 0.47	NGC 3377A, DDO 88 5889 02-28-007 2
N 3377	1045.1 1415 231.19 58.33 92.26 -25.29	-5 0.49 15 -1	5.3 4.9 9 0.65 54.	10.99 2 0.00 0.06	718. 595. 40	8.1 29.54 -0.2 7.3 -3.4	-18.55 11.6	9.61	5899 02-28-009 2
N 3379	1045.2 1251 233.49 57.64 93.68 -25.86	-5 0.52 15 -1	5.4 5.4 9 0.93 25.	10.15 2 0.00 0.05	881. 752. 27	8.1 29.54 -0.5 7.3 -3.6	-19.39 12.8	9.95	M105 5902 02-28-011 2
N 3384	1045.6 1254 233.50 57.75 93.67 -25.75	-2B 0.54 15 -1	5.5 4.8 9 0.50 65.	10.81 2 0.00 0.06	770. 641. 27	8.1 29.54 -0.4 7.2 -3.5	-18.73 11.4	9.68	5911 02-28-012 2
N 3390	1045.7 -3116 274.26 24.38 144.29 -34.96	3 0.35 31 -0 +2	3.5 2.7 6 0.22 87.	12.45 2 0.67 0.28	2850. 2578. 90	38.0 32.90 -25.3 18.2 -21.8	-20.45 30.0	10.37	437-62 -5-26-007 2
N 3389	1045.8 1248 233.71 57.74 93.80 -25.74	5A 0.30 21 -5 +1	3.0 2.5 9 0.45 69.	11.94 2 0.36 0.05	1267. 1138. 32	22.5 31.76 -1.3 20.3 -9.8	-19.82 16.4	10.12	5914 02-28-013 2
1046+52	1046.1 5236 156.72 56.01 58.05 -5.35	13 0.16 13 -0	0.7 0.7 9 0.59 59.	14.4 3 0.00 0.00	2344. 2410. 220	36.3 32.80 19.1 30.7 -3.4	-18.40 7.4	9.55	UGCA 219, MARK 153 09-18-032 2
U 5918	1046.2 6548 140.91 47.11 47.21 2.20	10 0.01 14 -0	2.9 2.9 2 1.00	15.19 7 0.00 0.01	336. 465. 1 83 8 0.67 11 1.09	7.1 29.25 4.8 5.2 0.3	-14.06 6.0	7.82 8.37 3.60	DDO 87, KAR 72 7 Z 747 5918 11-13-039 2
N 3395	1047.0 3315 192.92 63.13 74.63 -15.81	6XP 0.30 21 -9	1.9 1.7 2 0.60 58.	12.19 2 0.20 0.00	1625. 1595. 6 0.99 217 10	27.4 32.19 7.0 25.4 -7.5	-20.00 13.6	10.19 9.87 0.47	ARP 270, ARAK 257 VV 246, TURN 20A 5931 06-24-017 2
N 3396	1047.1 3315 192.91 63.15 74.64 -15.79	10BP 0.29 21 -9	3.5 2.9 2 0.41 71.	12.20 2 0.41 0.00	1679. 1649. 16	28.0 32.24 7.1 26.0 -7.6	-20.04 23.7	10.21	ARP 270, VV 246 TURN 20B 5935 06-24-018 2

Table columns: (1) Name; (2) α / ℓ / SGL; (3) δ / b / SGB; (4) Type / ρ_{xyz} / Group; (5) D_{25} / $D_{25}^{b,i}$ / source; (6) d/D / i; (7) $B_T^{b,i}$ / $H_{-0.5}$ / $B-H$; (8) source / A_B^{i-o} / A_B^b; (9) V_h / V_o / Tel; (10) W_{20} / W_R / W_D; (11) e_v / e_w / source; (12) F_c / source e_F / f_h; (13) R / μ; (14) SGX / SGY / SGZ; (15) $M_B^{b,i}$ / Δ_{25}; (16) $\log L_B$ / $\log M_H$ / $\log M_T$; (17) M_H/L_B / M_H/M_T / M_T/L_B; (18) Alternate Names UGC/ESO, MCG, AGN? RCII

(1)	(2)	(3)	(4)	(5)	(6)	(7)	(8)	(9)	(10)	(11)	(12)	(13)	(14)	(15)	(16)	(17)	(18)
U 5947	1047.8	1955	10 P	1.5	0.61	14.63	7	1253.	98	12	0.23	22.8	1.1	-17.16	9.06	0.78	DDO 89
	221.53	61.34	0.62	1.4	57.		0.19	1158.	92.		21	31.79	21.0	9.3	8.95	0.386	5947 03-28-017 2
	87.02	-22.20	21 -3 +1	2			0.03	1	98.		1.02		-8.6		9.36	2.01	
N 3412	1048.3	1341	-2B	3.5	0.59	11.40	2	861.		75		8.1	-0.4	-18.14	9.45		5952 02-28-016 2
	232.87	58.72	0.52	3.2	59.		0.00	737.				29.54	7.4	7.6			
	93.19	-24.82	15 -1	2			0.05						-3.4				
U 5953	1048.4	4450	13 P	0.7	0.60	13.60	2	1776.		220		29.2	12.3	-18.73	9.68		MARK 155
	169.02	60.33	0.09	0.6	58.		0.00	1804.				32.33	26.1	5.1			5953 08-20-033 2
	64.75	-9.37	13 -0	2			0.00						-4.8				
N 3414	1048.5	2815	-2BP	3.3	0.87	11.75	2	1412.		20	-0.41	24.9	4.4	-20.23	10.28	0.01	ARP 162
	204.06	63.41	0.28	3.2	33.		0.00	1357.			210 26	31.98	23.3	23.3	8.38		5959 05-26-021 2
	79.30	-18.08	21 +7	2			0.00						-7.7				
N 3423	1048.7	607	6A	4.0	0.85	11.50	1	1008.	179	10	1.06	10.9	-1.8	-18.68	9.66	0.30	5962 01-28-012 2
	244.17	54.39	0.13	3.9	35.		0.05	851.	252.	8	217 7	30.18	9.5	12.4	9.13	0.060	
	101.00	-27.69	15 -3 +1	2			0.07	1	254.	79	1.15		-5.1		10.36	4.96	
N 3419	1048.7	1413	-2X	0.9	0.86	13.4	3	3016.	268	19	-0.32	43.4	-1.9	-19.79	10.11	0.07	5964 02-28-018 2
	232.08	59.07	0.00	0.9	34.		0.00	2894.	412.	20	210 25	33.19	39.4	11.4	8.95	0.016	
	92.70	-24.51	32 -4	2			0.04		414.	210			-18.0		10.75	4.38	
1048-19	1048.9	-1938	6B	3.9	0.16			2054.	278	15	0.79	29.2	-15.6		9.72	0.132	UGCA 221
	268.02	34.71	0.08	2.8	90.		0.67	1808.	240.		21	32.32	18.5	23.9	10.60		569-14 -3-28-015
	130.21	-33.86	22 -0	6			0.12	6	243.		1.06		-16.2				
U 5979	1049.2	6815	10A	1.9	0.80	14.4	3	1121.	111	10	0.33	20.6	14.5	-17.17	9.06	0.79	5979 11-13-043
	138.29	45.44	0.20	1.8	41.		0.08	1261.	135.		20	31.57	14.6	10.8	8.96	0.159	
	45.35	3.83	12 -0 +1	2			0.02	1	139.		1.03		1.4		9.76	4.98	
N 3433	1049.4	1025	5A	3.7	0.91	12.4	3	2720.	262	15	0.77	39.5	-4.1	-20.58	10.42	0.35	5981 02-28-023 2
	238.31	57.16	0.10	3.7	28.		0.03	2582.			18	32.98	35.3	42.7	9.96		
	96.62	-25.90	32 +4	2			0.04	1			1.14		-17.3				
N 3430	1049.4	3313	5X	4.2	0.57	11.92	2	1563.	347	10	1.05	26.7	6.7	-20.21	10.28	0.42	5982 06-24-026 2
	192.90	63.64	0.30	3.7	60.	9.58	0.22	1533.	358.	7	211 15	32.13	24.8	28.8	9.90	0.075	
	74.93	-15.40	21 -9	2		2.34	0.00		360.	211			-7.1		11.03	5.68	
N 3432	1049.7	3654	9B	4.9	0.27	11.07	1	611.	265	10	1.46	7.8	2.4	-18.38	9.54	0.50	ARP 206, VV 11
	184.74	63.16	0.19	3.7	82.	10.13	0.67	600.	229.	6	211 8	29.45	7.2	8.4	9.24	0.137	5986 06-24-028 2
	71.71	-13.43	15 +8 +7	9		0.94	0.00	4	232.	211	1.04		-1.8		10.11	3.66	
N 3437	1049.9	2312	5X	2.8	0.36	12.2	3	1291.	341	30	0.67	23.4	2.2	-19.65	10.05	0.23	5995 04-26-016 2
	215.18	62.82	0.41	2.2	75.		0.52	1212.			16	31.85	21.8	15.0	9.41		
	84.14	-20.25	21 -0 +1	2			0.01	1			1.05		-8.1				
U 5998	1050.2	5033	13 P	1.1	0.43	14.3	3	1263.		52		22.8	11.3	-17.49	9.19		MARK 156
	159.11	57.71	0.19	0.9	71.		0.00	1320.				31.79	19.7	6.0			5998 08-20-037 2
	60.11	-5.97	13 -3	2			0.00						-2.4				

Name	Designations									
N 3403	1050.2 133.45 40.63	7357 40.96 7.06	4A 0.24 12 -8 +7	3.3 2.9 2	0.47 67.	12.3	3 0.33 0.13	1195. 1359.	66			21.6 31.67	16.2 13.9 2.6	-19.37 18.3	9.94		5997 12-10-089 2
N 3443	1050.3 225.99 89.32	1750 61.08 -22.61	7A 0.68 21 -3 +1	2.5 2.2 2	0.50 65.	13.6 12.43 1.17	3 0.29 0.04	1132. 1028. 1	183. 164. 167.	10	0.41 30 1.05	21.1 31.62	0.2 19.5 -8.1	-18.02 13.6	9.40 9.06 10.02	0.46 0.108 4.21	6000 03-28-025 MARK 418
N 3442	1050.4 190.69 74.18	3411 63.75 -14.72	13 P 0.21 21 -9	0.8 0.8 2	0.79 42.	13.2	3 0.00 0.00	1738. 1713.	71			28.8 32.30	7.6 26.8 -7.3	-19.10 6.7	9.83		6001 06-24-000 2
N 3449	1050.5 275.98 145.93	-3240 23.66 -33.88	0.21 31 -0 +2	3.8 2.9 6	0.25 84.	12.6	5 0.00 0.30	3267. 2994.	90			43.2 33.18	-29.7 20.1 -24.1	-20.58 36.6	10.42		376-25 -5-26-010 2
N 3447	1050.7 227.57 90.15	1702 60.82 -22.87	7X 0.57 21 -3 +1	4.0 3.5 2	0.59 59.	12.5	3 0.21 0.03	1073. 965. 1	148. 138. 142.	8	0.99 9 1.11	20.1 31.51	0.0 18.5 -7.8	-19.01 20.5	9.80 9.60 10.05	0.63 0.348 1.81	VV 252 6006 03-28-027 2
U 6007	1050.9 227.58 90.15	1703 60.87 -22.82	10BP 0.46 21 -3 +1	1.8 1.5 2	0.51 64.	12.76	3 0.28 0.03	1031. 923.	77			19.7 31.47	0.0 18.2 -7.6	8.6			NGC 3447A, VV 252 6007 03-28-027 2
N 3445	1051.5 149.61 54.63	5715 53.67 -2.10	9X 0.24 12 -13+12	1.7 1.7 9	0.90 29.	12.76	2 0.04 0.00	2026. 2116. 1	181	25	0.71 10 1.03	32.4 32.55	18.7 26.4 -1.2	-19.79 16.1	10.11 9.73	0.42	ARP 24, VV 14 6021 10-16-023 2
N 3448	1051.6 153.05 56.86	5435 55.44 -3.56	13 0.24 13 -0	4.9 3.8 2	0.33 78.	12.15	2 0.00 0.00	1391. 1468.		10	1.18 56 10	24.5 31.94	13.4 20.4 -1.5	-19.79 27.2	10.11 9.96	0.71	ARP 205 6024 09-18-055 2
U 6027	1051.7 141.79 48.82	6415 48.66 1.81	10 0.21 12 -0+12	1.5 1.3 2	0.61 57.		3 0.19 0.01	1701. 1824. 1	82	15	30 1.02	28.2 32.25	18.6 21.2 0.9	10.7	8.90		6027 11-13-000 2
N 3454	1051.8 226.71 89.70	1737 61.31 -22.38	5B 0.68 21 -3 +1	2.4 1.8 2	0.25 85.	13.3	3 0.67 0.01	1138. 1033.		95		21.3 31.64	0.1 19.7 -8.1	-18.34 11.2	9.53		TURN 25A 6026 03-28-030 2
N 3455	1051.9 226.86 89.78	1733 61.30 -22.39	3X 0.64 21 -3 +1	2.8 2.5 2	0.67 52.	12.5	3 0.15 0.02	1092. 987.		95		20.5 31.56	0.1 19.0 -7.8	-19.06 15.0	9.82		TURN 25B 6028 03-28-031 2
U 6029	1052.1 159.64 60.74	5000 58.27 -6.02	13 P 0.21 13 -3	1.1 1.1 2	0.82 39.	13.9	3 0.00 0.00	1387. 1442.		52		24.3 31.93	11.8 21.1 -2.6	-18.03 7.8	9.40		MARK 157 6029 08-20-046 2
U 6035	1052.8 227.34 90.02	1724 61.43 -22.26	10B 0.63 21 -3 +1	1.4 1.4 2	0.93 25.	14.8	3 0.02 0.03	1073. 968. 1	56	7	0.21 226 15 1.02	20.3 31.54	0.0 18.8 -7.7	-16.74 8.3	8.89 8.82	0.86	6035 03-28-035
N 3458	1053.0 149.21 54.63	5723 53.72 -1.85	-2X 0.24 12 -13+12	1.6 1.4 2	0.59 59.	13.2	3 0.00 0.00	1800. 1891.		90		29.5 32.35	17.1 24.1 -1.0	-19.15 12.1	9.85		6037 10-16-026 2

(1) Name	(2) α / ℓ / SGL	(3) δ / b / SGB	(4) Type / Q_{xyz} / Group	(5) D_{25} / $D^{b,i}_{25}$ / source	(6) d/D / i	(7) $B^{b,i}_T$ / $H_{-0.5}$ / $B-H$	(8) source / A^{i-o}_B / A^b_B	(9) V_h / V_o / Tel	(10) W_{20} / W_R / W_D	(11) e_v / e_w / source	(12) F_c / source·e_F / f_h	(13) R / μ	(14) SGX / SGY / SGZ	(15) $M^{b,i}_B$ / Δ_{25}	(16) $\log L_B$ / $\log M_H$ / $\log M_T$	(17) M_H/L_B / M_H/M_T / M_T/L_B	(18) Alternate Names / UGC/ESO MCG / AGN? RCII
1053+06	1053.5 / 245.10 / 101.15	626 / 55.50 / -26.46	13 P / 0.-1 / 15 -3 +1			14.5	3 / 0.00 / 0.06	1204. / 1050.		220		13.7 / 30.68	-2.4 / 12.0 / -6.1	-16.18	8.67		ARAK 268
N 3471	1056.0 / 143.73 / 51.13	6148 / 50.82 / 0.88	1 / 0.19 / 12 -0+12	1.9 / 1.7 / 2	0.51 / 64.	12.7	3 / 0.28 / 0.00	2076. / 2188.		34		33.0 / 32.59	20.7 / 25.7 / 0.5	-19.89 / 16.4	10.15		01-28-022 / MARK 158
1056-49	1056.7 / 285.05 / 165.74	-4942 / 8.96 / -30.49	7B / 0.15 / 31 +9 +6	2.1 / 2.5 / 6	0.62 / 57.		/ 0.18 / 1.26	2741. / 2455. / 7		20		35.4 / 32.74	-29.6 / 7.5 / -18.0	/ 25.8			6064 10-16-039 2 / 215-13
N 3489	1057.7 / 234.40 / 93.73	1410 / 60.92 / -22.55	-2X / 0.39 / 15 -1	3.7 / 3.3 / 9	0.62 / 56.	11.13	2 / 0.00 / 0.02	695. / 576.		27		6.4 / 29.05	-0.4 / 5.9 / -2.5	-17.92 / 6.2	9.36		6082 02-28-039 2
N 3486	1057.8 / 202.06 / 79.43	2915 / 65.52 / -15.81	5X / 0.27 / 15 -4 +1	6.4 / 6.0 / 2	0.73 / 47.	10.75	1 / 0.11 / 0.00	679. / 632.	245 / 285. / 287.	10 / 6 / 217	1.53 / 217 6 / 1.09	7.4 / 29.36	1.3 / 7.0 / -2.0	-18.61 / 13.0	9.64 / 9.27 / 10.48	0.43 / 0.061 / 7.06	6079 05-26-032 2
N 3495	1058.6 / 249.88 / 104.31	353 / 54.70 / -26.15	7 / 0.12 / 15 -0 +1	4.5 / 3.4 / 2	0.25 / 85.	11.8 / 10.03 / 1.77	3 / 0.67 / 0.17	1143. / 980.	328 / 291. / 293.	10 / 15 / 217	0.95 / 217 12 / 1.09	12.8 / 30.54	-2.8 / 11.1 / -5.6	-18.74 / 12.7	9.69 / 9.16 / 10.49	0.30 / 0.047 / 6.42	6098 01-28-027 2
U 6112	1059.9 / 229.71 / 91.19	1700 / 62.77 / -20.90	7 / 0.58 / 21 -3 +1	2.3 / 1.9 / 2	0.39 / 73.	13.4	3 / 0.45 / 0.00	1037. / 932. / 1	190	10	0.70 / 18 / 1.03	20.5 / 31.56	-0.4 / 19.0 / -7.3	-18.16 / 11.4	9.46 / 9.32	0.74	6112 03-28-050 2
N 3501	1100.2 / 227.31 / 90.01	1815 / 63.39 / -20.31	5 / 0.64 / 21 -2 +1	3.6 / 2.5 / 2	0.16 / 90.	12.8 / 10.21 / 2.59	3 / 0.67 / 0.00	1139. / 1040. / 1	319 / 281. / 283.	15	0.64 / 16 / 1.07	21.7 / 31.69	0.0 / 20.4 / -7.5	-18.89 / 15.8	9.75 / 9.31 / 10.56	0.37 / 0.057 / 6.48	6116 03-28-051 2
N 3504	1100.5 / 204.58 / 80.64	2815 / 66.05 / -15.76	2X / 0.25 / 21 -7	2.6 / 2.5 / 9	0.85 / 35.	11.74	2 / 0.05 / 0.01	1529. / 1478.		29		26.5 / 32.12	4.1 / 25.2 / -7.2	-20.38 / 19.3	10.34		6118 05-26-039 2
N 3507	1100.8 / 227.14 / 89.93	1824 / 63.58 / -20.11	3B / 0.42 / 21 -3 +1	3.2 / 3.1 / 2	0.84 / 37.	11.4	3 / 0.06 / 0.00	977. / 879. / 1	158 / 213. / 215.	10	0.74 / 15 / 1.10	19.8 / 31.48	0.0 / 18.6 / -6.7	-20.08 / 17.9	10.23 / 9.33 / 10.37	0.13 / 0.091 / 1.40	6123 03-28-053 2
N 3511	1100.8 / 272.87 / 134.31	-2250 / 33.38 / -31.40	5A / 0.12 / 54 -5	6.8 / 5.6 / 6	0.33 / 77.	10.8	5 / 0.58 / 0.21	1104. / 854. / 6	325 / 294. / 296.	10	1.32 / 10 / 1.29	15.5 / 30.95	-9.2 / 9.4 / -8.1	-20.15 / 25.3	10.25 / 9.70 / 10.80	0.28 / 0.079 / 3.56	UGCA 223 / 502-13 -4-26-020 2
N 3510	1101.0 / 202.38 / 79.87	2909 / 66.21 / -15.24	9B / 0.26 / 15 -4 +1	4.0 / 2.9 / 2	0.25 / 85.	12.66 / 11.52 / 1.14	2 / 0.67 / 0.00	710. / 664. / 1	200 / 165. / 169.	10	1.11 / 10 9 / 1.09	7.9 / 29.49	1.3 / 7.5 / -2.1	-16.83 / 6.7	8.92 / 8.91 / 9.72	0.96 / 0.152 / 6.29	6126 05-26-040 2
N 3513	1101.3 / 273.07 / 134.47	-2258 / 33.32 / -31.29	5B / 0.13 / 54 -5	3.2 / 3.2 / 6	0.83 / 38.	11.7	5 / 0.06 / 0.21	1195. / 945. / 6	110. / 140. / 144.	8	0.77 / 13 / 1.14	17.0 / 31.16	-10.2 / 10.4 / -8.8	-19.46 / 15.9	9.98 / 9.23 / 9.96	0.18 / 0.188 / 0.95	UGCA 224 / 502-14 -4-26-021 2

Name	Position	Type	Dims	m_B	Velocity	Dist.	SG X Y Z	M_B	log	col.	Identifications
N 3512	1101.3 2818; 204.49 66.23; 80.68 -15.59	5X; 0.22; 21 -7	1.5 1.5 9; 0.97 17.	12.98; 2 0.01; 0.01	1388. 1338.; 241; 10 40 211; 0.39 211 49	25.0 31.99	3.9 23.7 -6.7	-19.01 10.9	9.80 9.19	0.25	6128 05-26-041 2
1102+29	1102.2 2924; 201.78 66.48; 79.77 -14.90	13 P; 0.28; 15 -4 +1	0.5 0.5 2; 0.89 31.	; 0.00; 0.00	636. 592.; ; 42;	6.9 29.21	1.2 6.6 -1.8	1.0			UGCA 225, MARK 36; HARO 4 05-26-046 2
N 3521	1103.2 14; 255.52 52.82; 108.58 -26.23	4X; 0.19; 15 -0 +1	11.7 10.5 9; 0.56 61.	9.40 6.48 2.92; 1 0.23; 0.08	804. 628. 4; 466 489. 490.; 10; 1.82 14 1.13	7.2 29.28	-2.0 6.1 -3.2	-19.88 22.1	10.14 9.53 11.18	0.25 0.022 10.99	6150 00-28-030 2
U 6151	1103.2 2005; 224.11 64.77; 88.57 -18.88	9; 0.46; 21 -2 +1	1.9 1.9 2; 1.00	14.75; 7 0.00; 0.00	1330. 1241. 3; 41; 7; 0.16 14 1.04	24.3 31.93	0.6 23.0 -7.9	-17.18 13.5	9.06 8.93	0.74	6151 03-28-057 2; DDO 91
N 3516	1103.4 7250; 133.24 42.41; 42.09 7.28	-2B; 0.19; 42+14	2.3 2.3 2; 0.86 34.	12.36; 2 0.00; 0.09	2548. 2709.; ; 22; 0.20 6 32	38.9 32.95	28.6 25.9 4.9	-20.59 26.1	10.43 9.38	0.09	6153 12-11-009 S1 2
U 6157	1103.7 1746; 229.10 63.93; 90.85 -19.75	8A; 0.00; 32 -0	1.9 1.8 2; 0.90 29.	14.1; 3 0.04; 0.00	2959. 2859.; 160; 15; 0.46 17 1.04	43.0 33.17	-0.6 40.5 -14.5	-19.07 22.6	9.82 9.73	0.81	6157 03-28-058
U 6162	1103.9 5128; 155.23 58.86; 60.49 -3.66	6; 0.19; 13 -0	2.5 2.1 2; 0.52 64.	13.4; 3 0.27; 0.00	2203. 2268. 1; 223 208. 211.; 10; 0.62 21 1.04	34.7 32.70	17.1 30.1 -2.2	-19.30 21.3	9.91 9.70 10.43	0.61 0.188 3.27	6162 09-18-086
U 6161	1104.0 4400; 167.85 63.19; 66.95 -7.44	8B; 0.16; 15 -9	2.9 2.6 2; 0.57 60.	13.3; 3 0.22; 0.00	765. 793. 1; 133 122. 127.; 10; 0.75 12 1.06	12.3 30.44	4.8 11.2 -1.6	-17.14 9.3	9.05 8.93 9.61	0.76 0.211 3.61	6161 07-23-024
U 6171	1104.6 1850; 227.10 64.58; 89.92 -19.11	10 P; 0.61; 21 -2 +1	2.5 1.9 2; 0.28 81.	14.1; 3 0.67; 0.00	1212. 1118. 1; 173 141. 145.; 15; 0.49 16 1.04	23.0 31.81	0.0 21.7 -7.5	-17.71 12.8	9.28 9.21 9.87	0.87 0.222 3.90	6171 03-28-061
1105-46	1105.5 -4615; 284.98 12.71; 161.51 -29.66	7; 0.10; 54 +4	3.9 3.5 6; 0.36 75.	; 0.52; 0.63	1055. 772. 7; ; 20;	13.4 30.64	-11.1 3.7 -6.6	13.7			265-07
1106-28	1106.5 -2806; 277.06 29.30; 140.53 -30.44	10; 0.11; 22 -6	3.5 2.7 2; 0.22 87.	; 0.67; 0.25	1499. 1239. 2; 182 148. 152.; 15; 0.77 15 1.01	21.4 31.65	-14.2 11.7 -10.8	16.9	9.43 10.03	0.252	UGCA 226; 438-05 -5-27-000
1107-37	1107.2 -3705; 281.42 21.23; 150.94 -30.25	-5; -0.33; 31-10	2.7 2.2 6; 0.24 85.	12.78; 2 0.00; 0.42	2885. 2611.; ; 90;	38.2 32.91	-28.8 16.0 -19.2	-20.13 24.5	10.24		NGC 3557B; 377-15 -6-25-004 2
N 3547	1107.3 1100; 242.52 61.03; 97.87 -21.62	5 P; 0.13; 21 -0 +1	2.1 1.8 2; 0.47 67.	12.86; 2 0.33; 0.01	1543. 1414.; ; 90;	26.5 32.11	-3.4 24.4 -9.7	-19.25 13.9	9.89		6209 02-29-007 2
1107-23	1107.5 -2326; 274.84 33.57; 135.16 -29.92	4A; 0.10; 22 -0	3.2 3.2 6; 0.85 36.	12.30; 2 0.05; 0.25	2090. 1841. 6; 59 71. 79.; 8; 0.76 9 1.11	29.5 32.35	-18.2 18.1 -14.7	-20.05 27.6	10.21 9.70 9.61	0.30 1.233 0.25	IC 2627, UGCA 227; 502-21 -4-27-002 2

(1) Name	(2) α / ℓ / SGL	(3) δ / b / SGB	(4) Type / ϱ_{xyz} / Group	(5) D_{25} / D^b_{25} / source	(6) d/D / i	(7) $B^{b,i}_T$ / $H_{-0.5}$ / B-H	(8) source / A^{i-o}_B / A^b_B	(9) V_h / V_o / Tel	(10) W_{20} / W_R / W_D	(11) e_v / e_w / source	(12) F_c / source e_F / f_h	(13) R / μ	(14) SGX / SGY / SGZ	(15) $M^{b,i}_B$ / Δ_{25}	(16) log L_B / log M_H / log M_T	(17) M_H/L_B / M_T/M_H / M_T/L_B	(18) Alt. UGC/ESO / Alt. MCG / AGN? RCII
N 3557	1107.6 / 281.58 / 151.15	-3716 / 21.09 / -30.16	-5 / 0.28 / 31-10	3.5 / 3.7 / 6	0.78 / 43.	10.91	2 / 0.00 / 0.49	3151. / 2877.		90		41.5 / 33.09	-31.5 / 17.3 / -20.9	-22.18 / 44.8	11.06		377-16 / -6-25-005 / 2
N 3549	1108.1 / 151.32 / 58.91	5340 / 57.84 / -1.99	5A / 0.00 / 12+18	3.2 / 2.5 / 2	0.35 / 76.	11.9 / 10.06 / 1.84	3 / 0.54 / 0.00	2867. / 2944. / 1	419. / 392. / 394.	10	0.73 / 16 / 1.06	43.0 / 33.17	22.2 / 36.8 / -1.5	-21.27 / 31.4	10.70 / 10.00 / 11.15	0.20 / 0.071 / 2.79	6215 / 09-18-097 / 2
N 3564	1108.2 / 281.70 / 151.16	-3717 / 21.13 / -30.04	-2 / 0.31 / 31-10	2.0 / 1.9 / 6	0.57 / 60.	12.76	2 / 0.00 / 0.49	2835. / 2561.		64		37.6 / 32.88	-28.5 / 15.7 / -18.8	-20.12 / 20.9	10.24		377-18 / -6-25-006 / 2
1108-48	1108.3 / 286.45 / 164.35	-4850 / 10.51 / -28.79	2B / 0.13 / 31 +9 +6	2.4 / 2.5 / 6	0.67 / 53.		/ 0.15 / 0.65	2717. / 2434. / 7		20		35.2 / 32.73	-29.7 / 8.3 / -17.0	/ 25.7			215-31
N 3568	1108.4 / 281.70 / 151.04	-3711 / 21.23 / -30.01	3B / 0.07 / 31-10	2.6 / 2.3 / 6	0.37 / 74.		/ 0.48 / 0.49	2446. / 2172.		18		32.7 / 32.58	-24.8 / 13.7 / -16.4	/ 22.0			377-20 / -6-25-009 / 2
N 3556	1108.7 / 148.29 / 56.98	5557 / 56.26 / -0.77	6B / 0.15 / 12 -0+1	7.8 / 5.9 / 2	0.29 / 81.	9.97 / 8.04 / 1.93	1 / 0.67 / 0.00	697. / 785. / 1	328. / 294. / 296.	8	1.67 / 217 4 / 1.32	14.1 / 30.74	7.7 / 11.8 / -0.2	-20.77 / 24.3	10.50 / 9.97 / 10.78	0.29 / 0.153 / 1.93	M 108
U 6251	1110.5 / 150.58 / 58.92	5352 / 57.95 / -1.58	9X / 0.59 / 12 -2 +1	1.9 / 1.8 / 2	0.76 / 45.	14.5	3 / 0.09 / 0.00	928. / 1006. / 3	61. / 62. / 71.	7	0.44 / 12 / 1.03	18.1 / 31.28	9.3 / 15.5 / -0.5	-16.78 / 9.5	8.90 / 8.96 / 9.03	1.13 / 0.851 / 1.32	6251 / 09-19-006 / 2
N 3585	1110.8 / 277.24 / 138.72	-2629 / 31.17 / -29.40	-5 / 0.12 / 22 -6	4.5 / 4.2 / 6	0.58 / 59.	10.74	2 / 0.00 / 0.26	1491. / 1236.		75		21.6 / 31.67	-14.1 / 12.4 / -10.6	-20.93 / 26.5	10.56		502-25 / -4-27-004 / 2
U 6258	1111.2 / 221.81 / 87.79	2147 / 67.11 / -16.47	10 / 0.27 / 21 +1	2.1 / 1.6 / 2	0.28 / 81.	14.5	3 / 0.67 / 0.00	1460. / 1382. / 1	215	20	0.39 / 21 / 1.03	26.1 / 32.08	1.0 / 25.0 / -7.4	-17.58 / 12.2	9.22 / 9.22	1.00	6258 / 04-27-008
1111-69	1111.2 / 294.38 / 186.08	-6900 / -8.05 / -23.53	5 / 0.09 / 53 -0+21	3.2 / 2.7 / 6	0.18 / 90.		/ 0.67 / 0.87	1334. / 1063. / 7		20		16.3 / 31.06	-14.9 / -1.6 / -6.5	/ 12.9			063-11
N 3583	1111.4 / 158.09 / 63.58	4836 / 61.63 / -4.05	3B / 0.15 / 13 -0	2.6 / 2.5 / 2	0.76 / 45.	11.6	3 / 0.09 / 0.00	2135. / 2188. / 1	362	25	0.78 / 21 / 1.06	34.0 / 32.65	15.1 / 30.3 / -2.4	-21.05 / 24.8	10.61 / 9.84	0.17	08-21-008 / 2
N 3593	1112.0 / 240.44 / 96.26	1305 / 63.21 / -19.77	/ 0.19 / 15 -2 +1	4.8 / 4.0 / 2	0.45 / 69.	11.70	2 / 0.00 / 0.00	628. / 510.	254. / 232. / 234.	8 / 12 / 223	0.54 / 210 26	5.5 / 28.71	-0.6 / 5.2 / -1.9	-17.01 / 6.4	9.00 / 8.02 / 10.00	0.11 / 0.010 / 10.12	6272 / 02-29-014 / 2
N 3596	1112.4 / 236.86 / 94.36	1504 / 64.41 / -18.92	5X / 0.46 / 21 -1	4.1 / 4.1 / 2	1.00	11.5	3 / 0.00 / 0.00	1193. / 1084. / 1	130	10 / 4 / 211	0.92 / 211 8 / 1.18	23.0 / 31.81	-1.7 / 21.7 / -7.5	-20.31 / 27.5	10.32 / 9.64	0.21	6277 / 03-29-013 / 2

Name	Position (RA Dec l b / SGL / SGB)	Type	Dimensions	Magnitudes	Velocity	HI	N	V(grp)	col7	col8	col9	col10	col11	Cross-IDs
N 3600	1113.0 4152 65.66 -7.05 / 170.31 / 69.64	1 / 0.14 / 15 -9	3.9 3.0 0.28 81. 2	11.7 3 / 10.64 0.67 0.00 / 1.06	719. / 740. / 1	1.07 / 10 / 1.10	10	221 / 186. / 189.	10.5 / 30.11	3.6 9.8 -1.3	-18.41 9.2	9.56 9.11 9.97	0.36 0.140 2.57	6283 07-23-038 2
1113-33	1113.5 -3341 24.86 -29.05 / 281.21 / 146.99	3B / 0.29 / 31+10	2.4 2.1 0.45 69. 6	13.36 6 / 0.36 / 0.31	3005. / 2738.		75		40.1 / 33.01	-29.4 19.1 -19.5	-19.65 24.6	10.05		SCI 99; 377-31 -6-25-014 2
N 3605	1114.1 1818 66.37 -17.28 / 230.61 / 91.40	-5 / 0.16 / 21 -1	1.3 1.2 0.52 63. 2	13.1 3 / 0.00 0.00	693. / 600.		65		16.8 / 31.13	-0.5 16.0 -5.0	-18.03 5.9	9.40		6295 03-29-019 2
N 3607	1114.3 1820 66.43 -17.22 / 230.59 / 91.39	-2A / 0.34 / 21 -1	5.0 4.9 0.90 30. 2	10.95 2 / 0.00 0.00	933. / 840.		35		19.9 / 31.49	-0.4 19.0 -5.9	-20.54 28.5	10.41		6297 03-29-020 2
N 3608	1114.4 1826 66.50 -17.16 / 230.39 / 91.30	-5 / 0.56 / 21 -1	3.3 3.2 0.84 36. 2	11.90 2 / 0.00 0.00	1210. / 1118.		50		23.4 / 31.84	-0.5 22.3 -6.9	-19.94 21.9	10.17		6299 03-29-022 2
N 3611	1114.9 450 58.20 -22.01 / 253.99 / 104.80	1AP / 0.20 / 21-11	2.6 2.7 1.00 2	12.67 2 / 0.00 0.13	1642. / 1490.	0.63 / 223 / 23	47 12 223	296	27.3 / 32.18	-6.5 24.4 -10.2	-19.51 21.5	10.00 9.50	0.32	6305 01-29-026 2
N 3614	1115.5 4602 63.77 -4.68 / 161.54 / 66.17	5X / 0.12 / 13 -0	4.3 3.9 0.64 55. 2	11.9 3 / 0.17 0.00	2339. / 2381. / 1	0.88 / 10 / 1.14	10	310 / 333. / 334.	36.5 / 32.81	14.7 33.2 -3.0	-20.91 41.6	10.56 10.00 11.13	0.28 0.076 3.72	6318 08-21-015 2
N 3610	1115.5 5904 54.46 1.55 / 143.54 / 54.71	-5 / 0.30 / 12-12	2.6 2.6 0.96 20. 9	11.60 2 / 0.00 0.00	1765. / 1869.		50		29.2 / 32.33	16.9 23.9 0.8	-20.73 22.2	10.48		6319 10-16-107 2
U 6320	1115.6 1907 67.05 -16.63 / 229.15 / 90.77	13 P / 0.51 / 21 -1	1.3 1.2 0.92 26. 2	13.6 3 / 0.00 0.00	1138. / 1050.		220		22.6 / 31.77	-0.3 21.7 -6.5	-18.17 7.9	9.46		ARAK 286; 6320 03-29-028
N 3613	1115.7 5816 55.11 1.19 / 144.35 / 55.42	-5 / 0.24 / 12-12	3.7 3.2 0.54 62. 2	11.60 2 / 0.00 0.00	2054. / 2154.		75		32.9 / 32.59	18.7 27.1 0.7	-20.99 30.7	10.59		6323 10-16-109 2
N 3621	1115.9 -3234 26.08 -28.54 / 281.24 / 145.72	7A / 0.06 / 15 -0	14.9 13.8 0.50 65. 6	9.4 5 / 7.35 0.29 0.37 / 2.05	734. / 470. / 2	2.20 / 36 / 1.12	10	290 / 278. / 280.	7.1 / 29.27	-5.2 3.5 -3.4	-19.87 28.6	10.14 9.90 10.81	0.58 0.125 4.65	UGCA 232; 377-37 -5-27-008 2
N 3623	1116.3 1323 64.23 -18.69 / 241.29 / 96.38	1X / 0.44 / 15 -2 +1	8.4 6.3 0.28 81. 2	9.50 1 / 6.69 0.67 0.00 / 2.81	806. / 692. / 2	0.79 / 211 / 1.09	20 14 223	500 / 467. / 469.	7.3 / 29.31	-0.8 6.9 -2.3	-19.81 13.4	10.12 8.52 10.93	0.03 0.004 6.50	M 65, ARP 317; 6328 02-29-018 2
N 3619	1116.5 5802 55.35 1.17 / 144.45 / 55.68	0.30 / 12-12	3.8 3.6 0.77 44. 2	12.60 2 / 0.00 0.00	1649. / 1748.		75		27.9 / 32.23	15.7 23.0 0.6	-19.63 29.3	10.04		6330 10-16-115 2
N 3626	1117.4 1838 67.22 -16.43 / 230.75 / 91.40	-2A / 0.32 / 21 -1	3.0 2.8 0.73 48. 2	11.70 2 / 0.00 0.00	1473. / 1383.	0.38 / 223 / 21	12 25 223	366 / 444. / 446.	26.3 / 32.10	-0.6 25.2 -7.4	-20.40 21.5	10.35 9.22 11.09	0.07 0.013 5.47	6343 03-29-032 2

(1) Name	(2) α / ℓ / SGL	(3) δ / b / SGB	(4) Type / Q_{xyz} / Group	(5) D_{25} / $D^{b,i}_{25}$ / source	(6) d/D / i	(7) $B^{b,i}_T$ / $H_{-0.5}$ / B-H	(8) source / A^{i-o}_B / A^b_B	(9) V_h / V_o / Tel	(10) W_{20} / W_R / W_D	(11) e_v / e_w / source	(12) F_c source/e_F / f_h	(13) R / μ	(14) SGX / SGY / SGZ	(15) $M^{b,i}_B$ / Δ_{25}	(16) log L_B / log M_H / log M_T	(17) M_H/L_B / M_T/M_T / M_T/L_B	(18) Alternate Names / Alternate Names / UGC/ESO MCG AGN? RCII
U 6345	1117.6 / 257.43 / 107.11	248 / 57.10 / -22.01	10B / 0.21 / 21-11	2.5 / 2.3 / 2	0.64 / 55.	13.2	3 / 0.17 / 0.10	1609. / 1450. / 1	142	15	0.76 / 11 / 1.05	26.9 / 32.15	-7.3 / 23.8 / -10.1	-18.95 / 18.1	9.77 / 9.62	0.70	DDO 94 / / 01-29-030 2
N 3627	1117.6 / 241.92 / 96.60	1317 / 64.42 / -18.43	3X / 0.35 / 15 -2 +1	8.0 / 6.9 / 2	0.51 / 65.	9.43 / 6.67 / 2.76	1 / 0.28 / 0.00	737. / 623. / 2	378. / 376. / 378.	15 / 11 / 211	1.17 / 7 / 1.13	6.6 / 29.09	-0.7 / 6.2 / -2.1	-19.66 / 13.3	10.06 / 8.81 / 10.74	0.06 / 0.012 / 4.79	M 66, ARP 16, ARP 317 / VV 308 / 6346 02-29-019 2
N 3630	1117.7 / 256.96 / 106.67	314 / 57.45 / -21.85	-2 / 0.24 / 21-11	2.0 / 1.7 / 2	0.44 / 70.	12.7	3 / 0.00 / 0.11	1520. / 1363.		77		25.9 / 32.07	-6.9 / 23.0 / -9.6	-19.37 / 12.9	9.94		6349 01-29-031 2
N 3628	1117.7 / 240.82 / 96.03	1353 / 64.80 / -18.19	3 P / 0.39 / 15 -2 +1	14.8 / 10.6 / 9	0.22 / 87.	9.48 / 6.90 / 2.58	1 / 0.67 / 0.00	846. / 734. / 4	479 / 442. / 443.	8 / 5 / 211	1.81 / 6 / 1.19	7.7 / 29.44	-0.8 / 7.3 / -2.4	-19.96 / 23.8	10.18 / 9.58 / 11.13	0.26 / 0.028 / 9.00	ARP 317, VV 308 / / 6350 02-29-020 2
N 3629	1117.9 / 208.19 / 83.35	2714 / 69.79 / -12.78	6A / 0.17 / 21 -0	2.1 / 2.0 / 2	0.78 / 43.	12.5	3 / 0.08 / 0.00	1520. / 1471. / 1	240 / 299. / 301.	20 / 12 / 79	0.76 / 8 / 1.04	26.8 / 32.14	3.0 / 26.0 / -5.9	-19.64 / 15.7	10.05 / 9.62 / 10.61	0.37 / 0.102 / 3.63	6352 05-27-058 2
N 3637	1118.1 / 269.37 / 120.60	-959 / 46.61 / -25.30	-2B / 0.22 / 22 -7	1.6 / 1.6 / 5	0.94 / 23.	12.6	5 / 0.00 / 0.11	1855. / 1649.		90		28.2 / 32.25	-13.0 / 22.0 / -12.1	-19.65 / 13.2	10.05		-2-29-020 2
N 3631	1118.3 / 149.48 / 59.82	5328 / 59.03 / -0.76	5A / 0.33 / 12 -2 +1	5.4 / 5.2 / 2	0.85 / 36.	10.98	1 / 0.05 / 0.00	1161. / 1239. / 4	133 / 181. / 185.	8	1.12 / 60 / 1.06	21.6 / 31.67	10.9 / 18.7 / -0.3	-20.69 / 32.8	10.47 / 9.79 / 10.49	0.21 / 0.197 / 1.06	ARP 27 / / 6360 09-19-047 2
N 3640	1118.5 / 256.90 / 106.45	331 / 57.80 / -21.58	-5 / 0.18 / 21-11	5.0 / 5.0 / 2	0.90 / 30.	11.14	2 / 0.00 / 0.11	1354. / 1198.		40		24.2 / 31.92	-6.4 / 21.6 / -8.9	-20.78 / 35.3	10.50		6368 01-29-033 2
N 3642	1119.4 / 142.56 / 54.70	5921 / 54.53 / 2.13	4A / 0.30 / 12-12	5.6 / 5.4 / 2	0.83 / 38.	11.54	1 / 0.06 / 0.00	1623. / 1729.		50		27.5 / 32.20	15.9 / 22.4 / 1.0	-20.66 / 43.4	10.46		6385 10-16-128 L2 2
N 3652	1119.9 / 177.68 / 73.68	3802 / 68.54 / -7.65	5BP / 0.20 / 13 -8	2.8 / 2.2 / 2	0.31 / 79.	12.1	3 / 0.64 / 0.00	2096. / 2100.		220		33.5 / 32.62	9.3 / 31.8 / -4.5	-20.52 / 21.5	10.40		ARAK 291 / / 6392 06-25-055 2
N 3655	1120.3 / 235.59 / 93.38	1652 / 66.97 / -16.48	5A / 0.24 / 21 -1	1.7 / 1.5 / 2	0.66 / 53.	12.15	2 / 0.15 / 0.00	1481. / 1384. / 1	325	20	0.52 / 21 / 1.02	26.5 / 32.11	-1.5 / 25.3 / -7.5	-19.96 / 11.6	10.18 / 9.37	0.16	6396 03-29-039 2
U 6399	1120.6 / 152.05 / 62.01	5111 / 60.95 / -1.52	9 / 0.84 / 12 -1	3.2 / 2.4 / 2	0.28 / 81.	13.4 / 12.33 / 1.07	3 / 0.67 / 0.00	805. / 873. / 1	173	20	0.38 / 19 / 1.06	17.0 / 31.15	7.8 / 14.6 / -0.4	-17.75 / 11.9	9.29 / 8.84	0.35	6399 9-19-060
N 3656	1120.8 / 148.12 / 59.42	5407 / 58.78 / -0.13	-5 P / 0.14 / 12-18	1.8 / 1.8 / 2	1.00	12.9	3 / 0.00 / 0.00	2828. / 2910.		70		42.6 / 33.14	21.6 / 36.6 / -0.1	-20.24 / 22.4	10.29		ARP 155, VV 22 / / 6403 09-19-063 2

N 3659	1121.1	1805	9B	2.0	0.53	12.5	3	1291.	263	30	0.74	24.5	-0.9	-19.45	9.97	0.35	
	233.09	67.74	0.46	1.8	63.		0.25	1200.			31	31.95	23.6	12.9	9.52		
	92.29	-15.83	21 -1	2			0.00	1			1.03		-6.7				6405 03-29-040 2
N 3657	1121.1	5312	5XP	2.8	1.00	12.2	3	1215.	215	10	0.93	22.4	11.1	-19.55	10.01	0.42	
	149.21	59.51	0.41	2.8			0.00	1293.			9	31.75	19.5	18.3	9.63		
	60.25	-0.52	12 -2 +1	2			0.00	1			1.09		-0.2				6406 09-19-065 2
N 3664	1121.8	336	9BP	1.9	0.90	13.00	7	1370.	134	10	0.74	24.4	-6.5	-18.94	9.77	0.56	ARP 5, VV 251, DDO 95
	258.01	58.39	0.19	1.9	29.		0.04	1216.			11	31.94	21.9	13.5	9.51		
	106.64	-20.77	21-11	2			0.10	1			1.04		-8.7				6419 01-29-041 2
N 3666	1121.8	1137	5A	4.2	0.30	11.6	3	1067.	280	10	1.11	17.0	-2.4	-19.55	10.01	0.36	
	246.39	64.17	0.23	3.3	80.	9.77	0.67	947.	246.		11	31.15	16.0	16.4	9.57	0.130	
	98.63	-18.08	21 +4 +1	2		1.83	0.08	1	249.		1.10		-5.3		10.46	2.80	6420 02-29-025 2
N 3665	1122.0	3902	-2A	3.8	0.86	11.70	2	2002.		50		32.4	9.4	-20.86	10.54		
	174.73	68.49	0.20	3.7	34.		0.00	2012.				32.56	30.8	35.0			
	72.97	-6.84	13 -8	2			0.00						-3.9				6426 07-24-003 2
N 3672	1122.5	-931	5A	3.7	0.47	11.1	5	1856.	412	10	1.15	28.4	-13.1	-21.17	10.66	0.25	UGCA 235
	270.42	47.55	0.26	3.2	67.		0.33	1654.		30	211 28	32.27	22.4	26.5	10.06		
	120.37	-24.14	22 -7	2			0.08			211			-11.6				-2-29-028 2
N 3669	1122.6	5800	9	2.2	0.27	12.3	3	1940.	410	50	0.48	31.5	17.6	-20.19	10.27	0.16	
	143.34	55.86	0.25	1.7	82.		0.67	2041.			26	32.49	26.1	15.6	9.48		
	56.09	1.87	12-12	2			0.00	1			1.03		1.0				6431 10-16-135 2
N 3673	1122.7	-2627	3B	3.9	0.70	12.0	5	1946.	337	20	0.51	27.6	-18.6	-20.20	10.27	0.13	UGCA 236
	280.08	32.29	0.15	3.8	50.		0.13	1696.	392.		25	32.20	16.2	30.6	9.39	0.018	
	138.91	-26.75	22 +4	6			0.25	2	393.		1.02		-12.4		11.14	7.30	503-16 -4-27-010 2
N 3675	1123.5	4352	3A	5.8	0.58	10.69	2	771.	425	10	1.15	12.8	4.6	-19.84	10.13	0.17	
	163.63	66.20	0.13	5.1	59.	7.72	0.21	805.	452.	7	211 10	30.53	11.9	19.1	9.36	0.020	
	68.75	-4.42	15 -9	9		2.97	0.00	4	453.	211	1.06		-1.0		11.05	8.41	6439 07-24-004 2
N 3681	1123.8	1708	4X	2.9	0.66	12.25	2	1246.	195	10	0.96	24.2	-1.4	-19.67	10.06	0.47	
	236.12	67.83	0.45	2.6	53.		0.15	1152.	199.	15	217 10	31.92	23.2	18.4	9.73	0.253	
	93.46	-15.60	21 -1	9			0.00		202.	217			-6.5		10.32	1.84	6445 03-29-048 2
U 6446	1123.8	5401	7A	3.9	0.67	12.7	3	643.	157	10	1.03	17.0	7.8	-18.45	9.57	0.83	
	147.58	59.14	0.56	3.6	52.		0.15	725.	159.		9	31.15	13.3	17.9	9.49	0.236	
	59.71	0.21	12 -1	2			0.00	1	162.		1.12		0.0		10.12	3.51	6446 9-19-079
U 6447	1123.9	5926	13 P	1.0	0.66	14.1	3	1318.		34		23.7	13.6	-17.77	9.30		IC 691, MARK 169
	141.66	54.78	0.48	0.9	53.		0.00	1425.				31.87	19.3	6.2			
	54.88	2.67	12 -0 +1	2			0.00						1.1				6447 10-16-139 2
U 6448	1123.9	6425	13 P	2.0	0.53	14.1	3	992.		43		19.0	12.1	-17.30	9.11		MARK 170
	137.28	50.57	0.17	1.8	63.		0.00	1121.				31.40	14.6	10.0			
	50.42	4.91	12 -4 +1	2			0.00						1.6				6448 11-14-025 2
N 3684	1124.6	1718	4A	2.9	0.66	12.15	2	1168.	247	7	0.97	23.4	-1.3	-19.69	10.07	0.44	
	236.00	68.07	0.41	2.6	53.		0.15	1075.	262.	10	217 15	31.84	22.5	17.8	9.71	0.144	
	93.37	-15.36	21 -1	9			0.00		264.	217			-6.2		10.55	3.02	6453 03-29-050 2

(1) Name	(2) α / ℓ / SGL	(3) δ / b / SGB	(4) Type / Qxyz / Group	(5) D25 / D25^{b,i} / source	(6) d/D / i	(7) BT^{b,i} / H-0.5 / B-H	(8) source / A_B^{i-0} / A_B^b	(9) Vh / Vo / Tel	(10) W20 / WR / WD	(11) ev / ew / source	(12) Fc / source/eT / fh	(13) R / μ	(14) SGX / SGY / SGZ	(15) MB^{b,i} / Δ25	(16) log LB / log MH / log MT	(17) MH/LB / MH/MT / MT/LB	(18) Alternate Names / UGC/ESO MCG / AGN? RCII
U 6456	1124.6 / 127.84 / 36.91	7916 / 37.33 / 11.37	13 P / 0.46 / 14+10	1.7 / 1.5 / 2	0.60 / 58.	14.3	3 / 0.00 / 0.11	-92. / 96. / 1		20	0.15 / 20 / 1.02	1.4 / 25.75	1.1 / 0.8 / 0.3	-11.45 / 0.6	6.77 / 6.44	0.47	VII ZW 403 / 6456 13-08-058 / 2
N 3683	1124.7 / 143.82 / 56.97	5709 / 56.72 / 1.74	5B / 0.27 / 12-12	2.1 / 1.8 / 2	0.43 / 71.	12.4	3 / 0.39 / 0.00	1686. / 1783.		90		28.4 / 32.27	15.5 / 23.8 / 0.9	-19.87 / 14.9	10.14		6458 10-16-143 / 2
1124+54	1124.9 / 147.14 / 59.63	5411 / 59.11 / 0.43	13 P / 0.00 / 12-18				/ 0.39 / 0.00	2895. / 2979.		32		43.5 / 33.19	22.0 / 37.5 / 0.3				09-19-081
N 3686	1125.1 / 235.71 / 93.23	1730 / 68.28 / -15.18	4B / 0.41 / 21 -1	2.9 / 2.8 / 9	0.79 / 42.	11.92	2 / 0.08 / 0.00	1168. / 1076. / 1	212 / 262. / 264.	20 / 9 / 217	0.50 / 217 12 / 1.08	23.5 / 31.86	-1.3 / 22.7 / -6.2	-19.94 / 19.2	10.17 / 9.24 / 10.58	0.12 / 0.046 / 2.60	6460 03-29-051 / 2
N 3687	1125.3 / 200.64 / 81.70	2947 / 71.50 / -10.23	1X / 0.15 / 13 -0	1.9 / 1.9 / 2	1.00	12.6	3 / 0.00 / 0.04	2407. / 2373.		90		37.2 / 32.85	5.3 / 36.2 / -6.6	-20.25 / 20.6	10.29		6463 05-27-073 / 2
N 3689	1125.6 / 212.73 / 85.30	2556 / 71.33 / -11.76	5X / 0.11 / 13 -0	1.8 / 1.7 / 9	0.63 / 55.	12.83	2 / 0.17 / 0.00	2750. / 2698.		79		41.2 / 33.07	3.3 / 40.2 / -8.4	-20.24 / 20.5	10.29		6467 04-27-037 / 2
1125-36	1125.6 / 284.78 / 149.90	-3615 / 23.39 / -26.58	4B / 0.29 / 31-11+10	2.6 / 2.5 / 6	0.63 / 56.	13.39	6 / 0.18 / 0.35	2976. / 2710.		25		39.6 / 32.99	-30.7 / 17.8 / -17.7	-19.60 / 28.9	10.03		SCI 105 / 378-03 -6-25-020 / 2
N 3694	1126.2 / 182.64 / 76.37	3541 / 70.56 / -7.55	13 P / 0.20 / 13-10	1.3 / 1.2 / 2	0.83 / 38.	13.5	3 / 0.00 / 0.01	2255. / 2250.		220		35.4 / 32.75	8.3 / 34.1 / -4.7	-19.25 / 12.4	9.89		ARAK 296 / 6480 06-25-076
U 6491	1126.7 / 184.11 / 76.92	3508 / 70.84 / -7.70	8 / 0.16 / 13-10	2.0 / 1.8 / 2	0.53 / 63.	14.3	3 / 0.25 / 0.01	2530. / 2523. / 1	170	20	0.22 / 25 / 1.03	38.8 / 32.94	8.7 / 37.4 / -5.2	-18.64 / 20.4	9.65 / 9.40	0.56	6491 06-25-077
N 3706	1127.3 / 285.08 / 149.78	-3608 / 23.62 / -26.24	-5 / 0.27 / 31-11+10	3.1 / 2.9 / 6	0.53 / 63.	11.9	5 / 0.00 / 0.34	3045. / 2780.		90		40.5 / 33.04	-31.4 / 18.3 / -17.9	-21.14 / 34.3	10.65		378-06 -6-25-022 / 2
N 3705	1127.5 / 252.07 / 101.21	931 / 63.75 / -17.50	2X / 0.27 / 21 +4 +1	4.6 / 4.0 / 2	0.48 / 67.	11.0 / 8.51 / 2.49	3 / 0.31 / 0.10	1017. / 891. / 1	361 / 352. / 354.	10	1.09 / 14 / 1.14	17.0 / 31.15	-3.2 / 15.9 / -5.1	-20.15 / 19.9	10.25 / 9.55 / 10.85	0.20 / 0.050 / 3.99	6498 02-29-039 / 2
1127+37	1127.7 / 178.59 / 75.29	3701 / 70.36 / -6.71	12 P / 0.19 / 13 -8			14.5	3 / 0.31 / 0.00	2012. / 2014.		71		32.6 / 32.57	8.2 / 31.3 / -3.8	-18.07	9.42		MARK 424 / 06-25-000 / 2
N 3717	1129.1 / 283.14 / 143.02	-3002 / 29.48 / -25.62	2A / 0.17 / 22 +4	6.1 / 4.5 / 6	0.19 / 90.	11.2 / 7.95 / 3.25	5 / 0.67 / 0.27	1731. / 1477. / 2	433 / 395. / 396.	20	1.10 / 15 / 1.05	24.6 / 31.95	-17.7 / 13.3 / -10.6	-20.75 / 32.3	10.49 / 9.88 / 11.17	0.25 / 0.052 / 4.71	UGCA 238 / 439-15 -5-27-015 / 2

Name	(coord)																Other names
N 3718	1129.9 / 146.99 / 60.71	5321 / 60.22 / 0.72	1BP / 1.17 / 12 -1	9.6 / 8.3 / 2	0.49 / 66.	10.96	1 / 0.30 / 0.00	987. / 1068. / 4	473 / 478. / 479.	10 / 4 / 211	1.35 / 211 11 / 1.05	17.0 / 31.15	8.7 / 15.6 / 0.2	-20.19 / 41.2	10.27 / 9.81 / 11.44	0.35 / 0.024 / 14.73	ARP 214 L2 / 6524 09-19-114 2
U 6531	1130.1 / 171.87 / 73.37	3921 / 69.73 / -5.28	8B / 0.18 / 13 +7	2.0 / 2.0 / 2	1.00	14.3	3 / 0.00 / 0.01	1565. / 1580. / 1	97	10	0.22 / 20 / 1.05	27.3 / 32.18	7.8 / 26.1 / -2.5	-17.88 / 15.9	9.34 / 9.09	0.56	6531 07-24-022
U 6534	1130.4 / 136.96 / 51.50	6334 / 51.66 / 5.18	6 / 0.36 / 12 -4 +1	2.9 / 2.2 / 2	0.30 / 80.	12.4	3 / 0.67 / 0.03	1273. / 1400. / 1	149	20	0.50 / 22 / 1.05	23.0 / 31.81	14.3 / 17.9 / 2.1	-19.41 / 14.8	9.96 / 9.22	0.19	6534 11-14-030 2
1130+55	1130.6 / 144.51 / 58.95	5521 / 58.66 / 1.68	13 P / 0.21 / 12+12	0.6 / 0.5 / 2	0.74 / 46.		0.00 / 0.00	1724. / 1815.		220		28.9 / 32.30	14.9 / 24.7 / 0.8	4.2			UGCA 239, MARK 177 / 09-19-000 2
N 3726	1130.7 / 155.34 / 66.20	4719 / 64.88 / -1.77	5X / 1.21 / 12 -1	5.5 / 5.1 / 2	0.70 / 50.	10.80 / 8.87 / 1.93	1 / 0.13 / 0.01	861. / 914. / 4	285 / 324. / 326.	10 / 6 / 211	1.34 / 217 8 / 1.06	17.0 / 31.15	6.7 / 15.3 / -0.5	-20.35 / 25.3	10.33 / 9.80 / 10.89	0.29 / 0.082 / 3.59	6537 08-21-051 2
U 6541	1130.8 / 151.89 / 64.23	4931 / 63.28 / -0.81	10 / 0.44 / 14 -0	1.4 / 1.3 / 2	0.57 / 60.	13.6	3 / 0.22 / 0.00	250. / 314. / 1	58 / 46. / 58.	8	-0.01 / 17 22 / 1.02	4.0 / 28.01	1.7 / 3.6 / -0.1	-14.41 / 1.5	7.96 / 7.19 / 7.97	0.17 / 0.168 / 1.03	MARK 178 / 6541 08-21-053 2
N 3729	1131.1 / 146.62 / 60.73	5325 / 60.27 / 0.91	1BP / 1.07 / 12 -1	3.2 / 3.0 / 2	0.73 / 47.	11.90	1 / 0.11 / 0.00	1035. / 1117.		90		17.0 / 31.15	8.9 / 15.8 / 0.3	-19.25 / 14.9	9.89		6547 09-19-117 2
N 3732	1131.7 / 273.46 / 120.98	-934 / 48.56 / -21.94	1X / 0.36 / 22 -7	1.3 / 1.3 / 5	0.92 / 26.	13.1	5 / 0.03 / 0.04	1712. / 1514.		90		26.9 / 32.15	-12.8 / 21.4 / -10.1	-19.05 / 10.2	9.81		-2-30-005 2
N 3733	1132.2 / 144.39 / 59.24	5508 / 58.97 / 1.79	5 / 0.52 / 12 -3 +1	4.5 / 3.8 / 2	0.50 / 65.	12.0 / 11.09 / 0.91	3 / 0.29 / 0.00	1188. / 1278. / 6	255 / 240. / 242.	10	1.14 / 8 / 1.21	22.1 / 31.72	11.3 / 18.9 / 0.7	-19.72 / 24.5	10.08 / 9.83 / 10.61	0.56 / 0.165 / 3.41	6554 09-19-123 2
N 3738	1133.0 / 144.59 / 59.61	5447 / 59.32 / 1.75	10 P / 0.39 / 14 -0	3.3 / 3.2 / 2	0.87 / 33.	12.05	1 / 0.05 / 0.00	225. / 314. / 1	115	30	0.74 / 12 / 1.07	4.3 / 28.16	2.2 / 3.7 / 0.1	-16.11 / 4.0	8.64 / 8.01	0.23	ARP 234 / 6565 09-19-130 2
N 3735	1133.2 / 131.73 / 44.98	7048 / 45.30 / 8.42	5A / 0.19 / 42-14	4.0 / 3.0 / 2	0.25 / 85.	11.6 / 9.20 / 2.40	3 / 0.67 / 0.01	2696. / 2853. / 1	509 / 473. / 474.	15	0.88 / 14 / 1.09	41.0 / 33.06	28.7 / 28.6 / 6.0	-21.46 / 35.9	10.78 / 10.11 / 11.37	0.21 / 0.055 / 3.90	6567 12-11-036 2
N 3755	1133.9 / 177.97 / 76.12	3641 / 71.63 / -5.72	5XP / 0.25 / 13 -7	3.4 / 2.9 / 2	0.43 / 71.	12.8 / 11.18 / 1.62	3 / 0.39 / 0.00	1572. / 1575. / 1	286 / 263. / 265.	15	0.98 / 13 / 1.08	27.6 / 32.20	6.6 / 26.6 / -2.7	-19.40 / 23.4	9.95 / 9.86 / 10.67	0.81 / 0.155 / 5.24	6577 06-26-008 2
N 3756	1134.0 / 144.60 / 59.87	5434 / 59.58 / 1.79	5X / 0.44 / 12 -3 +1	4.6 / 4.0 / 2	0.54 / 62.	11.90 / 9.65 / 2.25	1 / 0.25 / 0.00	1294. / 1382. / 1	302 / 298. / 300.	10	0.80 / 12 / 1.09	23.5 / 31.86	11.8 / 20.3 / 0.7	-19.96 / 27.4	10.18 / 9.54 / 10.85	0.23 / 0.049 / 4.71	6579 09-19-134 2
N 3769	1135.0 / 152.74 / 65.74	4810 / 64.75 / -0.76	3B / 1.00 / 12 -1	3.2 / 2.5 / 2	0.35 / 76.	11.2	3 / 0.54 / 0.00	714. / 773.		95		17.0 / 31.15	6.5 / 14.3 / -0.3	-19.95 / 12.4	10.17		ARP 280, TURN 38A / 6595 08-21-076 2

(1) Name	(2) α / ℓ / SGL	(3) δ / b / SGB	(4) Type / ϱ_{xyz} / Group	(5) D_{25} / $D_{25}^{b,i}$ / source	(6) d/D / i	(7) $B_T^{b,i}$ / $H_{-0.5}$ / B-H	(8) source / A_B^{i-o} / A_B^b	(9) V_h / V_o / Tel	(10) W_{20} / W_R / W_D	(11) e_v / e_w / source	(12) F_c source/e_F / f_h	(13) R / μ	(14) SGX / SGY / SGZ	(15) $M_B^{b,i}$ / $\Delta_{25}^{b,i}$	(16) $\log L_B$ / $\log M_H$ / $\log M_T$	(17) M_H/L_B / M_H/M_T / M_T/L_B	(18) Alternate Names / UGC/ESO MCG / AGN? RCII
1135+48	1135.1 / 152.71 / 65.75	4810 / 64.76 / -0.74	9BP / 0.76 / 12 -1	1.1 / 0.9 / 2	0.37 / 74.	14.0	3 / 0.49 / 0.00	685. / 744.		95		17.0 / 31.15	6.4 / 14.2 / -0.3	-17.15 / 4.5	9.05		NGC 3769A, ARP 280 / TURN 38B / 08-21-077 2
N 3773	1135.6 / 250.54 / 99.11	1223 / 67.18 / -14.65	-2AP / 0.35 / 21 -4 +1	1.6 / 1.5 / 2	0.80 / 41.	12.9	3 / 0.00 / 0.06	1014. / 904.		90		17.0 / 31.15	-2.6 / 16.2 / -4.3	-18.25 / 7.4	9.49		6605 02-30-005 2
N 3783	1136.5 / 287.45 / 151.29	-3728 / 22.94 / -24.42	1B / 0.21 / 31+11+10	1.5 / 1.7 / 6	0.94 / 23.	12.4	5 / 0.02 / 0.46	2891. / 2627.		31		38.5 / 32.93	-30.7 / 16.8 / -15.9	-20.53 / 19.1	10.40		S1 / 378-14 -6-26-004 2
N 3782	1136.6 / 154.48 / 67.11	4647 / 65.96 / -1.08	6XP / 0.93 / 12 -1	1.5 / 1.3 / 2	0.54 / 62.	12.9 / 11.70 / 1.20	3 / 0.24 / 0.02	740. / 793. / 1	133	12	0.84 / 11 / 1.02	17.0 / 31.15	6.2 / 14.6 / -0.3	-18.25 / 6.5	9.49 / 9.30	0.64	6618 08-21-087 2
U 6616	1136.7 / 140.01 / 56.39	5833 / 56.40 / 3.78	7A / 0.51 / 12 -3 +1	2.9 / 2.8 / 2	0.82 / 39.	13.3	3 / 0.07 / 0.00	1154. / 1261. / 1	119 / 151. / 155.	10	0.58 / 17 / 1.08	21.6 / 31.67	11.9 / 18.0 / 1.4	-18.37 / 17.7	9.54 / 9.25 / 10.07	0.51 / 0.152 / 3.37	6616 10-17-035
N 3780	1136.7 / 141.91 / 58.24	5632 / 58.14 / 2.95	5A / 0.18 / 12-16	3.0 / 2.8 / 2	0.82 / 39.	11.7	3 / 0.07 / 0.00	2393. / 2491. / 1	317 / 447. / 448.	10	0.73 / 22 / 1.08	37.2 / 32.86	19.6 / 31.6 / 1.9	-21.16 / 30.4	10.66 / 9.87 / 11.25	0.16 / 0.042 / 3.89	6615 09-19-150 2
N 3786	1137.1 / 191.59 / 80.53	3211 / 73.71 / -6.94	1XP / 0.00 / 13 -0	2.2 / 1.9 / 2	0.54 / 62.	12.7	3 / 0.24 / 0.01	2761. / 2744.		95		41.6 / 33.10	6.8 / 40.7 / -5.0	-20.40 / 23.1	10.35		ARP 294, VV 228 / TURN 39A L/S / 6621 05-28-008 2
N 3788	1137.1 / 191.48 / 80.49	3213 / 73.70 / -6.93	2XP / 0.17 / 13 -0	1.9 / 1.5 / 2	0.37 / 74.	12.7	3 / 0.49 / 0.01	2339. / 2322.		95		36.5 / 32.81	6.0 / 35.7 / -4.4	-20.11 / 16.0	10.24		ARP 294, VV 228 / TURN 39B / 6623 05-28-009 2
U 6628	1137.4 / 155.21 / 67.70	4612 / 66.48 / -1.20	9A / 1.21 / 12 -1	3.3 / 3.3	1.00	12.9	3 / 0.00 / 0.01	851. / 901. / 1	56	5	0.76 / 8 / 1.12	17.0 / 31.15	6.3 / 15.5 / -0.3	-18.25 / 16.4	9.49 / 9.22	0.54	6628 8-21-089
U 6633	1137.7 / 256.75 / 102.31	917 / 65.28 / -15.21	-2 / 0.08 / 21 -0 +1	1.5 / 1.1 / 2	0.28 / 81.	13.5	3 / 0.00 / 0.08	1772. / 1650.		220		29.5 / 32.35	-6.1 / 27.9 / -7.8	-18.85 / 9.5	9.73		IC 719, ARAK 308 / 6633 02-30-008 2
U 6637	1137.8 / 204.22 / 83.85	2840 / 74.25 / -8.18	13 P / 0.00 / 13 -0	1.1 / 0.9 / 2	0.48 / 67.	14.3	3 / 0.00 / 0.00	2970. / 2937.		55		44.0 / 33.22	4.7 / 43.3 / -6.3	-18.92 / 11.6	9.76		HARO 27 / 6637 05-28-010 2
1138-09	1138.0 / 275.74 / 121.60	-948 / 49.02 / -20.48	6X / 0.39 / 22 -2 +1	2.5 / 2.5 / 2	0.96 / 19.		0.01 / 0.02	1736. / 1540. / 1	127	10	0.78 / 13 / 1.07	27.3 / 32.18	-13.4 / 21.8 / -9.6	19.9	9.65		UGCA 241 / -2-30-014
1138+35	1138.2 / 180.40 / 77.57	3529 / 72.91 / -5.41	10 P / 0.26 / 13 -7	0.8 / 0.7 / 2	0.53 / 63.	14.4	3 / 0.25 / 0.00	1511. / 1510.		71		26.9 / 32.15	5.8 / 26.2 / -2.5	-17.75 / 5.5	9.29		MARK 426 / 06-26-016 2

Name	Other names	Coordinates	Type		Dimensions	Mag		Velocities			Distances		Abs Mag		Extinction
N 3804	6640 09-19-153 2	1138.2 5628 58.30 / 141.64 3.12 / 58.38	5A 0.44 12 -3 +1		2.5 0.71 49. / 2.3 / 2	13.0	3 0.12 0.00	1381. 1479. 6 / 179 192. 195. / 10 / 0.62 16 1.08			24.7 31.96 / 12.9 21.0 1.3 / -18.96 16.6		9.78 9.41 10.25		0.43 0.144 2.97
N 3810	6644 02-30-010 2	1138.4 1145 67.22 / 252.94 -14.22 / 99.97	5A 0.46 21 -4 +1		3.8 0.71 49. / 3.6 / 2	11.03	1 0.12 0.10	993. 882. 1 / 266 303. 305. / 15 6 211 / 1.07 211 9 1.13			16.9 31.14 / -2.8 16.1 -4.1 / -20.11 17.8		10.24 9.53 10.67		0.19 0.071 2.75
N 3813	6651 06-26-019 2	1138.7 3649 72.43 / 176.20 -4.79 / 76.39	5A 0.25 13 -7		2.1 0.61 57. / 1.9 / 2	12.11 9.87 2.24	2 0.19 0.00	1468. 1474. 1 / 309 322. 324. / 25 9 211 / 0.85 211 20 1.04			26.4 32.10 / 6.2 25.5 -2.2 / -19.99 14.6		10.19 9.69 10.64		0.32 0.112 2.86
1139-06	UGCA 242 / -1-30-022	1139.0 -612 52.38 / 273.68 -19.38 / 117.94	10 P 0.36 22 -8		1.7 0.72 48. / 1.6 / 2		0.12 0.11	1698. 1515. 1 / 115 / 25 / 0.09 30 1.02			27.3 32.18 / -12.1 22.7 -9.0 / 12.8		8.96		
U 6655	ARAK 311	1139.2 1615 70.37 / 244.01 -12.49 / 95.69	13 P 0.30 21 +4 +1		0.7 0.64 55. / 0.7 / 2	14.3	3 0.00 0.03	841. 750. / / 220			17.9 31.26 / -1.7 17.3 -3.9 / -16.96 3.7		8.98		
N 3818	6655 03-30-047	1139.4 -553 52.71 / 273.60 -19.21 / 117.64	-5 0.20 22 -8		2.0 0.69 50. / 1.9 / 2	12.60	2 0.00 0.10	1498. 1317. / / 65			25.1 32.00 / -11.0 21.0 -8.3 / -19.40 13.9		9.95		
U 6667	UGCA 243 / -1-30-023 2	1139.7 5153 62.28 / 146.27 1.46 / 62.66	5 1.51 12 -1		3.3 0.17 90. / 2.3 / 2	13.4 12.07 1.33	3 0.67 0.00	974. 1051. 1 / 205 169. 172. / 20 / 0.44 19 1.06			17.0 31.15 / 8.2 15.8 0.5 / -17.75 11.4		9.29 8.90 9.98		0.41 0.084 4.83
U 6670	6667 09-19-157	1139.8 1836 71.83 / 238.31 -11.52 / 93.50	10B 0.32 21 -0		3.0 0.33 78. / 2.3 / 2	13.1	3 0.60 0.00	918. 838. 1 / 217 185. 188. / 15 / 0.61 23 1.05			17.2 31.18 / -1.0 16.8 -3.4 / -18.08 11.6		9.42 9.08 10.06		0.45 0.105 4.32
U 6682	6670 03-30-053 2	1140.4 5923 55.90 / 138.54 4.55 / 55.82	9 0.56 12 -5 +1		1.8 0.84 37. / 1.7 / 2		3 0.06 0.00	1328. 1439. 6 / 93 123. 127. / 10 / 0.42 17 1.05			23.8 31.89 / 13.3 19.7 1.9 / 11.8		9.17 9.71		0.286
U 6706	DDO 96	1141.5 5519 59.53 / 142.01 3.08 / 59.62	9 P 0.30 12 -3 +1		2.1 0.87 33. / 2.1 / 2	13.3	3 0.05 0.00	1431. 1525. 1 / 175 / 20 / 0.26 25 1.05			25.3 32.01 / 12.8 21.8 1.4 / -18.71 15.5		9.68 9.07		0.25
U 6711	6682 10-17-049 2	1141.7 7001 46.30 / 131.23 8.78 / 45.99	13 P 0.23 42 -14		1.0 0.60 58. / 0.9 / 2	13.7	3 0.00 0.00	2544. 2700. / / 220			39.1 32.96 / 26.8 27.8 6.0 / -19.26 10.3		9.90		
U 6713	NGC 3846A, VV 320	1141.8 4907 64.72 / 149.29 0.66 / 65.33	9 1.55 12 -1		2.0 0.86 34. / 1.9 / 2	14.1	3 0.05 0.00	901. 966. 1 / 103 144. 148. / 7 / 0.52 226 11 1.04			17.0 31.15 / 7.1 15.5 0.2 / -17.05 9.4		9.01 8.98 9.75		0.93 0.169 5.52
1142-09	6706 09-19-169 2	1142.8 -949 49.47 / 277.41 -19.33 / 121.90	4B 0.39 22 -2 +1		2.7 0.29 81. / 2.1 / 2		0.67 0.01	1729. 1536. 1 / 232 198. 201. / 15 / 0.75 19 1.05			27.2 32.18 / -13.6 21.8 -9.0 / 16.7		9.62 10.28		0.219
N 3850	ARAK 317 / 6711 12-11-039 / 6713 08-21-094 / UGCA 245 / -2-30-027 / 6733 09-19-174 2	1142.9 5609 58.90 / 140.86 3.59 / 58.93	5B 0.52 12 -3 +1		2.2 0.49 66. / 1.9 / 2	13.5	3 0.30 0.00	1166. 1264. 1 / 185 / 25 / 0.40 30 1.03			21.8 31.69 / 11.2 18.7 1.4 / -18.19 12.1		9.47 9.08		0.41

(1) Name / SGL	(2) α / ℓ / SGL	(3) δ / b / SGB	(4) Type / Q_{xyz} / Group	(5) D_{25} / $D_{25}^{b,i}$ / source	(6) d/D / i	(7) $B_T^{b,i}$ / $H_{-0.5}$ / B-H	(8) source / A_B^{i-o} / A_B^b	(9) V_h / V_o / Tel	(10) W_{20} / W_R / W_D	(11) e_v / e_w / source	(12) F_c / source/e_f / f_h	(13) R / μ	(14) SGX / SGY / SGZ	(15) $M_B^{b,i}$ / $\Delta_{25}^{b,i}$	(16) log L_B / log M_H / log M_T	(17) M_H/L_B / M_H/M_T / M_T/L_B	(18) Alternate Names / UGC/ESO MCG / AGN? RCII
N 3870	1143.3	5030	-2	1.1	0.79			750.	115	15	0.12	17.0	7.1	5.5	8.58		MARK 186
	146.98	63.74	1.39	1.1	42.		0.00	822.			17 27	31.15	14.6				
	64.16	1.43	12 -1	2			0.00	1			1.02		0.4				
N 3877	1143.5	4746	5A	5.1	0.25	10.9	3	894.	347	10	0.77	17.0	6.7	-20.25	10.29	0.09	6742 08-22-001 2
	150.72	65.96	1.53	3.8	85.	8.63	0.67	954.	310.		23	31.15	15.7	18.9	9.23	0.032	
	66.68	0.38	12 -1	2		2.27	0.01	1	312.		1.14		0.1		10.72	2.68	
N 3882	1143.6	-560?	5	2.4	0.60			1817.		20		23.5	-21.5	22.0			170-11
	293.93	5.33	0.07	3.2	58.		0.19	1544.				31.85	3.2				
	171.57	-22.52	23 -0	6			1.80	7					-9.0				
1143+35	1143.8	3508	13 P			14.0	3	1346.		71		25.1	5.1	-18.00	9.39		MARK 429
	179.89	74.10	0.17				0.67	1346.				32.00	24.6				06-26-000 2
	78.35	-4.50	13 -7				0.00	1					-2.0				
N 3885	1144.3	-2739	1A	2.9	0.33	11.98	6	1948.		90		27.8	-19.9	-20.24	10.29		440-07 -5-28-006 2
	285.93	32.80	0.37	2.4	77.		0.58	1706.				32.22	16.3	19.5			
	140.76	-22.08	22 -4	6			0.32						-10.4				
N 3887	1144.6	-1635	4B	3.4	0.90	11.52	2	1212.	263	10	1.11	19.3	11.4	-19.91	10.16	0.34	-3-30-012 2
	281.57	43.33	0.08	3.3	30.		0.04	998.			19	31.43	14.0	18.6	9.68		
	129.03	-20.32	22 -0	4			0.04	1			1.12		-6.7				
N 3888	1144.9	5614	4X	1.8	0.84	12.7	3	2408.	283	15	0.59	37.5	19.3	-20.17	10.26	0.30	6765 09-19-189 2
	140.31	58.97	0.19	1.7	37.		0.06	2507.			28	32.87	32.1	18.6	9.74		
	58.96	3.88	12-16	2			0.00	1			1.03		2.5				
N 3892	1145.5	-1041	-2B	2.6	0.65	12.4	5	1727.		90		27.2	-14.0	-19.77	10.10		-2-30-030 2
	278.85	48.92	0.37	2.3	54.		0.00	1532.				32.17	21.6	18.3			
	122.95	-18.88	22 -2 +1	4			0.01						-8.8				
N 3893	1146.1	4900	5X	4.3	0.57	10.88	2	977.	307	8	1.30	17.0	7.2	-20.27	10.30	0.29	6778 08-22-007 2
	148.11	65.23	1.65	3.8	60.	8.77	0.22	1043.	312.	6	211 6	31.15	16.1	18.9	9.76	0.108	
	65.72	1.26	12 -1	2		2.11	0.00	1	314.	211	1.13		0.4		10.73	2.67	
U 6782	1146.3	2407	10	2.0	1.00			528.	92	10	0.23	5.0	0.1	2.9	7.63		DDO 97
	222.59	75.52	0.18	2.0			0.00	477.			20	28.52	5.0				
	88.83	-8.12	14 -0	2			0.03	1			1.05		-0.7				
U 6780	1146.3	-145	7AP	3.3	0.33	13.0	3	1736.	239	10	0.90	28.4	-11.0	-19.26	9.90	0.81	6780 00-30-024
	273.12	57.18	0.26	2.6	78.		0.60	1574.	206.		15	32.26	24.9	21.6	9.81	0.241	
	113.92	-16.46	22 +9	2			0.03	1	209.		1.07		-8.0		10.42	3.37	
1146-28	1146.3	-2801	9B	2.8	0.94			1934.	125	10	0.79	27.6	-20.0	24.2	9.67		UGCA 247, DDO 239
	286.56	32.58	0.41	3.0	23.		0.02	1692.			12	32.20	16.0				440-11 -5-28-007 2
	141.20	-21.69	22 -4	6			0.34	2			1.02		-10.2				
N 3900	1146.6	2718	-2A	3.5	0.52	12.14	2	1702.		50		29.4	2.1	-20.20	10.27		6786 05-28-034 2
	209.81	76.15	0.13	3.1	63.		0.00	1666.				32.34	29.1	26.6			
	85.86	-6.90	13 -9	2			0.06						-3.5				

N 3904	1146.7 286.99 142.27	-2900 31.66 -21.71	-5 0.17 22 -4	2.2 2.2 6	0.72 48.	11.73	2 0.00 0.22	1613. 1369.		75		23.4 31.84	-17.2 13.3 -8.6	-20.11 15.0	10.24		440-13 -5-28-009 2
N 3898	1146.7 139.76 58.93	5622 58.97 4.16	2A 0.56 12 -3 +1	4.7 4.4 9	0.75 46.	11.61	2 0.10 0.00	1172. 1272.	493	10 30 211	0.98 211 35	21.9 31.71	11.3 18.7 1.6	-20.10 28.1	10.23 9.66	0.27	L2 6787 09-19-204 2
N 3912	1147.5 212.12 86.44	2646 76.31 -6.91	1 0.13 13 -9	1.9 1.7 2	0.60 58.	12.8	3 0.20 0.04	1762. 1724.		90		30.0 32.39	1.9 29.7 -3.6	-19.59 14.9	10.03		6801 05-28-037 2
N 3913	1148.0 140.11 59.68	5538 59.70 4.05	7A 1.01 12 -1	2.9 2.8 2	0.94 23.	12.7	3 0.02 0.00	956. 1053. 1	64	8	0.54 12 1.09	17.0 31.15	8.9 15.2 1.2	-18.45 13.9	9.57 9.00	0.27	IC 740 6813 09-20-001 2
1148-75	1148.0 299.02 191.60	-7506 -12.97 -19.22	10 0.08 53 -0	2.4 2.4 6	0.57 60.		0.22 0.61	1823. 1567. 7		20		22.5 31.76	-20.8 -4.3 -7.4	15.8			039-02
U 6816	1148.1 139.11 58.67	5644 58.73 4.47	10B 0.91 12 -1	1.7 1.6 2	0.83 38.	14.12	7 0.06 0.00	896. 998. 1	140 182. 186.	10	0.50 15 1.03	17.0 31.15	9.0 14.8 1.4	-17.03 7.9	9.00 8.96 9.88	0.91 0.120 7.56	VV 273, DDO 98 6816 10-17-086 2
N 3917	1148.2 143.61 62.96	5207 62.79 2.76	6A 1.55 12 -1	4.7 3.5 2	0.26 84.	11.6 9.98 1.62	3 0.67 0.00	975. 1056. 1	279 243. 245.	10	0.78 217 9 1.12	17.0 31.15	8.1 15.8 0.8	-19.55 17.4	10.01 9.24 10.47	0.17 0.059 2.89	6815 09-20-008 2
U 6817	1148.2 166.25 74.96	3909 72.74 -2.16	10 0.45 14 -7	4.4 3.7 2	0.47 67.	13.15	7 0.33 0.00	248. 269. 1	67 54. 64.	5	0.98 7 1.12	3.1 27.49	0.8 3.0 -0.1	-14.34 3.3	7.93 7.96 8.45	1.08 0.324 3.34	DDO 99 6817 07-24-035 2
U 6818	1148.2 151.75 68.55	4605 67.78 0.47	13 P 1.42 12 -1	2.1 1.8 2	0.47 67.	13.9 12.77 1.13	3 0.00 0.00	803. 857. 1	163	15	0.54 14 1.03	17.0 31.15	6.0 15.3 0.1	-17.25 8.9	9.09 9.00	0.81	6818 08-22-015
N 3923	1148.5 287.28 141.82	-2832 32.22 -21.26	-5 0.40 22 -4	3.9 3.8 6	0.70 50.	10.83	2 0.00 0.27	1788. 1546.		65		25.8 32.06	-18.9 14.8 -9.3	-21.23 28.6	10.68		440-17 -5-28-012 2
N 3930	1149.1 168.25 75.83	3817 73.41 -2.32	5X 0.38 12 +6 +1	4.2 4.0 2	0.80 41.	12.3	3 0.08 0.00	919. 936. 1	191	20	0.93 10 1.15	18.5 31.33	4.5 17.9 -0.7	-19.03 21.6	9.80 9.46	0.46	6833 06-26-045 2
N 3928	1149.2 147.14 65.94	4858 65.54 1.72	13 P 1.65 12 -1	1.6 1.6 2	1.00	13.0	3 0.00 0.00	990. 1057.		43		17.0 31.15	7.2 16.2 0.5	-18.15 7.9	9.45		MARK 190 6834 08-22-019 2
U 6840	1149.5 142.94 62.79	5223 62.66 3.04	9B 1.35 12 -1	2.9 2.8 2	0.85 35.	14.25	7 0.05 0.00	1019. 1102. 1	151 208. 211.	15	1.08	17.0 31.15	8.3 16.0 1.0	-16.90 13.9	8.95 10.24	19.48	DDO 100 6840 09-20-019 2
1149+35	1149.8 177.74 78.78	3510 75.18 -3.35	13 P 0.08 13 -0				0.05 0.00	2165. 2168.		70		34.7 32.70	6.7 33.9 -2.0				MARK 641 06-26-000

(1) Name	(2) α / ℓ / SGL	(3) δ / b / SGB	(4) Type / ρxyz / Group	(5) D25^{b,i} / D25^{b,i} / source	(6) d/D / i	(7) BT^{b,i} / H−0.5 / B−H	(8) source / A_B^{i−o} / A_B^{b}	(9) Vh / Vo / Tel	(10) W20 / WR / WD	(11) ev / ew / source	(12) Fc source/eF / fh	(13) R / μ	(14) SGX / SGY / SGZ	(15) MB^{b,i} / Δ25^{b,i}	(16) log LB / log MH / log MT	(17) MH/LB / MH/MT / MT/LB	(18) Alternate Names UGC/ESO MCG / AGN? RCII
N 3936	1149.9 287.01 139.84	−2638 34.13 −20.72	5 0.34 22 −4	4.3 3.2 6	0.16 90.	11.8 9.97 1.83	5 0.67 0.38	2026. 1789.	335. 297. 299.	30	0.57 / 22 1.02	29.0 32.31	−20.7 17.5 −10.2	−20.51 27.1	10.40 9.49 10.84	0.13 0.045 2.79	UGCA 248 504-20 −4-28-004
N 3924	1150.0 145.18 64.75	501E 64.45 2.34	9 1.63 12 −1	1.9 1.9 2	0.95 21.	14.6	3 0.02 0.00	1003. 1077. 6	103	15	0.18 / 28 1.06	17.0 31.15	7.6 16.2 0.7	−16.55 9.4	8.81 8.64	0.67	6849 08-22-026
1150-03	1150.0 275.90 115.82	−323 56.09 −16.01	10 0.32 22 −9	1.4 1.4 2	0.86 34.		0.05 0.06	1665. 1499. 1	122	40	0.18 / 25 1.02	27.5 32.20	−11.5 23.8 −7.6	11.2	9.06		IC 2969, UGCA 249 00-30-000
N 3938	1150.2 153.88 70.24	4424 69.32 0.16	5A 1.24 12 −1	4.9 4.9 2	0.95 22.	10.89	1 0.02 0.00	805. 852.	112	6 5 211	1.31 / 7	17.0 31.15	5.5 15.4 0.0	−20.26 24.3	10.30 9.77	0.30	6856 07-25-001 2
N 3941	1150.3 170.72 76.87	3716 74.19 −2.48	−23 0.29 12 +6 +1	3.9 3.6 2	0.70 49.	11.4	3 0.00 0.00	927. 940.	215. 235. 238.	18 20 210	0.31 / 25 210	18.9 31.38	4.3 18.4 −0.8	−19.98 19.9	10.18 8.86 10.50	0.05 0.023 2.08	6857 06-26-051 2
N 3945	1150.6 135.33 54.84	6057 55.03 6.32	−2B 0.50 12 −5 +1	6.5 5.9 2	0.63 55.	11.46	2 0.00 0.04	1220. 1341.		75		22.5 31.76	12.9 18.2 2.5	−20.30 38.8	10.31		6860 10-17-096 2
1150-28	1150.9 287.80 141.60	−2815 32.62 −20.71	8 0.32 22 −4	4.3 3.1 6	0.15 90.		0.67 0.24	1702. 1461. 6	287. 249. 251.	10	1.15 / 11 1.12	24.7 31.96	−18.1 14.4 −8.7	22.4	9.94 10.60	0.215	UGCA 250 440-27 −5-28-015
N 3952	1151.1 276.59 116.23	−343 55.89 −15.83	9 P 0.34 22 −9	1.2 0.9 4	0.33 77.	12.2	5 0.58 0.06	1625. 1458.		90		27.0 32.16	−11.5 23.3 −7.4	−19.96 7.1	10.18		−1-30-044 2
N 3949	1151.1 147.63 66.83	4808 66.41 1.70	4A 1.61 12 −1	2.8 2.5 2	0.64 55.	11.23 9.44 1.79	2 0.17 0.00	804. 868. 1	284. 301. 303.	10	1.00 / 13 1.06	17.0 31.15	6.4 15.1 0.5	−19.92 12.4	10.16 9.46 10.51	0.20 0.089 2.26	6869 08-22-029 2
N 3953	1151.2 142.21 62.67	5237 62.58 3.37	5B 1.20 12 −1	5.9 5.2 2	0.56 61.	10.52 7.93 2.59	1 0.23 0.00	1054. 1139. 1	418. 436. 437.	10	0.99 / 9 1.31	17.0 31.15	8.4 16.3 1.1	−20.63 25.8	10.44 9.45 11.15	0.10 0.020 5.12	6870 09-20-026 2
N 3956	1151.4 285.19 133.23	−2018 40.31 −19.41	5A 0.51 22 −1	4.4 3.4 6	0.30 80.	11.8 10.65 1.15	5 0.67 0.08	1660. 1439. 2	320. 286. 288.	25	0.92 / 15 1.02	25.1 32.00	−16.2 17.3 −8.4	−20.20 24.9	10.27 9.72 10.77	0.28 0.089 3.16	UGCA 251 572-13 −3-30-016 2
N 3955	1151.4 286.14 135.93	−2253 37.83 −19.84	13 0.08 22 +1	3.3 2.6 6	0.29 81.	12.39	2 0.00 0.16	1345. 1117.		58		20.6 31.57	−13.9 13.5 −7.0	−19.18 15.6	9.86		504-26 −4-28-005 2
N 3957	1151.5 284.84 132.19	−1918 41.27 −19.21	−2A 0.48 22 −1	3.5 2.5 6	0.18 90.	12.91	6 0.00 0.07	1838. 1620.		90		27.5 32.20	−17.4 19.2 −9.0	−19.29 20.1	9.91		572-14 −3-30-017 2

ID	Name		mag												Type		
U 6877	IC 745, ARAK 332 / 6877 00-30-034		9.13	-17.34 / 4.0	-5.6 / 13.8 / -3.9	15.4 / 30.94		220		1201. / 1050.	3 / 0.00 / 0.02	13.6	0.89 / 31.	1.0 / 0.9 / 2	13 P / 0.14 / 22+10	25 / 59.72 / -14.57	1151.7 / 273.62 / 112.15
N 3962	UGCA 253 / -2-30-040 2		10.48	-20.73 / 27.0	-15.8 / 21.4 / -8.6	28.0 / 32.23		27		1822. / 1621.	2 / 0.00 / 0.05	11.50	0.91 / 29.	3.3 / 3.3 / 2	-5 / -0.32 / 22 +1	-1342 / 46.65 / -17.97	1152.1 / 282.65 / 126.43
1152-16	UGCA 254 / -3-30-019	0.233	9.55 / 10.18	13.7	-16.6 / 20.2 / -8.7	27.5 / 32.20	0.67 / 17 / 1.03	10	215 / 196. / 199.	1811. / 1601. / 1	0.30 / 0.05		0.49 / 66.	2.0 / 1.7 / 2	8 / 0.45 / 22 -1	-1635 / 43.92 / -18.51	1152.3 / 283.99 / 129.41
U 6900	DDO 101, KAR 82 / 6900 05-28-049 2		7.90	-14.26 / 3.1	0.8 / 5.8 / -0.4	5.9 / 28.84		100		552. / 541.	7 / 0.17 / 0.05	14.58	0.64 / 55.	2.0 / 1.8 / 2	10 / 0.24 / 14 +1	3149 / 77.06 / -3.96	1153.0 / 189.68 / 82.15
N 3972	6904 09-20-032 2	0.12	9.89 / 8.98	-19.25 / 14.4	8.5 / 14.6 / 1.4	17.0 / 31.15	0.52 / 19 / 1.09	15	265	848. / 947. / 1	3 / 0.67 / 0.00	11.9 / 10.43 / 1.47	0.28 / 81.	3.9 / 2.9 / 2	4A / 0.93 / 12 -1	5536 / 60.05 / 4.71	1153.1 / 138.86 / 59.98
U 6903	6903 00-31-002	0.66 / 0.075 / 8.74	9.88 / 9.70 / 10.82	-19.22 / 21.4	-10.7 / 27.6 / -7.3	30.5 / 32.42	0.73 / 17 / 1.06	15	202 / 327. / 328.	1894. / 1748. / 1	3 / 0.04 / 0.02	13.2	0.89 / 31.	2.5 / 2.4 / 2	6X / 0.21 / 22+12	132 / 60.87 / -13.92	1153.1 / 273.27 / 111.15
N 3976	6906 01-31-001 2	0.55 / 0.137 / 4.00	10.58 / 10.32 / 11.19	-20.98 / 29.7	-10.0 / 35.5 / -8.0	37.7 / 32.88	1.17 / 13 / 1.07	20	448 / 422. / 424.	2504. / 2381. / 1	3 / 0.54 / 0.00	11.9 / 9.14 / 2.76	0.35 / 76.	3.4 / 2.7 / 2	5X / 0.18 / 11+24	702 / 65.74 / -12.23	1153.4 / 267.53 / 105.77
N 3981	UGCA 255, ARP 289 / VV 8 / 572-20 -3-31-001 2	1.00	10.14 / 10.14	-19.87 / 63.8	-16.6 / 18.1 / -8.4	26.0 / 32.07	1.31 / 9 / 1.04	15	329	1717. / 1499. / 2	5 / 0.15 / 0.07	12.2	0.67 / 53.	9.0 / 8.4 / 6	4 P / 0.59 / 22 -1	-1936 / 41.12 / -18.78	1153.6 / 285.58 / 132.60
U 6912	VII ZW 430, VV 57 / 6912 10-17-104	1.18	9.28 / 9.35	-17.73 / 15.0	13.1 / 20.4 / 2.5	24.4 / 31.93	0.58 / 14 / 1.07	10	134	1357. / 1468. / 6	3 / 0.00 / 0.01	14.2	0.50 / 65.	2.4 / 2.1 / 2	13 P / 0.46 / 12 -3 +1	5828 / 57.47 / 5.80	1153.7 / 136.44 / 57.32
U 6917	6917 09-20-038	0.48 / 0.081 / 5.91	9.65 / 9.33 / 10.42	-18.65 / 20.4	7.4 / 15.6 / 0.9	17.0 / 31.15	0.87 / 13 / 1.15	10	206 / 212. / 215.	919. / 996. / 1	3 / 0.15 / 0.01	12.5	0.66 / 53.	4.5 / 4.1 / 2	9B / 1.71 / 12 -1	5042 / 64.46 / 3.06	1153.9 / 143.47 / 64.61
N 3982	S2 / 6918 09-20-036 2	0.18	9.97 / 9.21	-19.45 / 11.9	9.3 / 16.4 / 1.6	17.0 / 31.15	0.75 / 14 / 1.10	15	236	1110. / 1208. / 6	3 / 0.03 / 0.00	11.7	0.92 / 26.	2.4 / 2.4 / 2	4X / 0.72 / 12 -1	5524 / 60.28 / 4.74	1153.9 / 138.82 / 60.21
N 3985	ARAK 334 / 6921 08-22-045 2		8.86	-16.68 / 2.9	3.3 / 7.6 / 0.3	8.3 / 29.58		220		532. / 600.	3 / 0.15 / 0.02	12.9	0.67 / 52.	1.3 / 1.2 / 2	9B / 0.44 / 14 -4	4836 / 66.29 / 2.33	1154.1 / 145.96 / 66.58
U 6923	6923 09-20-040	0.35 / 0.105 / 3.30	9.21 / 8.75 / 9.73	-17.55 / 8.9	8.7 / 16.5 / 1.3	17.0 / 31.15	0.29 / 36 / 1.03	15	170 / 144. / 148.	1083. / 1172. / 1	3 / 0.43 / 0.00	13.6 / 12.03 / 1.57	0.40 / 72.	2.2 / 1.8 / 2	10 / 1.10 / 12 -1	5327 / 62.05 / 4.09	1154.2 / 140.51 / 62.05
U 6930	6930 08-22-046 2	0.71	9.57 / 9.42	-18.45 / 20.4	6.7 / 14.9 / 0.8	17.0 / 31.15	0.96 / 9 / 1.18	8	140	779. / 851. / 1	3 / 0.01 / 0.01	12.7	0.96 / 19.	4.2 / 4.1 / 2	7X / 1.56 / 12 -1	4933 / 65.52 / 2.77	1154.7 / 144.56 / 65.73

(1) Name	(2) α ℓ SGL	(3) δ b SGB	(4) Type Q_{xyz} Group	(5) D_{25} $D_{25}^{b,i}$ source	(6) d/D i	(7) $B_T^{b,i}$ $H_{-0.5}$ $B-H$	(8) source A_B^{i-o} A_B^b	(9) V_h V_o Tel	(10) W_{20} W_R W_D	(11) e_v e_w source	(12) F_c source/e_f t_h	(13) R μ	(14) SGX SGY SGZ	(15) $M_B^{b,i}$ $\Delta_{25}^{b,i}$	(16) $\log L_B$ $\log M_H$ $\log M_T$	(17) M_H/L_B M_H/M_T M_T/L_B	(18) Alternate Names UCG/ESO MCG AGN? RCII
N 3992	1155.0 140.10 61.91	5339 61.93 4.27	4B 1.23 12 −1	7.4 6.6 2	0.59 59.	10.40 7.96 2.44	1 0.20 0.00	1051. 1142. 1	479 517. 518.	10 7 211	1.28 217 6 1.32	17.0 31.15	8.7 16.1 1.4	−20.75 32.8	10.49 9.74 11.40	0.18 0.022 8.18	6937 09-20-044 L2 2
N 3990	1155.0 138.26 59.95	5544 60.04 5.01	−2 0.82 12 −1	1.6 1.4 2	0.59 59.	13.40	2 0.00 0.00	720. 820.		43		17.0 31.15	8.0 13.9 1.4	−17.75 6.9	9.29		TURN 47A 6938 09-20-043 2
N 3998	1155.3 138.18 59.96	5544 60.06 5.05	−2P 0.49 12 −3 +1	3.3 3.2 2	0.84 36.	11.55	2 0.00 0.00	1138. 1238.		16		21.6 31.67	10.8 18.6 1.9	−20.12 20.2	10.24		TURN 47B 6946 09-20-046 L1.9 2
U 6955	1155.8 165.10 76.25	3821 74.35 −1.07	10 0.38 12 +6 +1	5.5 4.8 2	0.53 63.	12.9	3 0.25 0.00	917. 938. 1	160 144. 148.	10	0.90 11 1.19	18.7 31.36	4.4 18.2 −0.3	−18.46 26.2	9.58 9.44 10.20	0.74 0.176 4.18	DDO 105 6955 06-26-063 2
U 6956	1155.8 142.31 64.25	5112 64.16 3.52	9B 1.71 12 −1	2.6 2.6 2	0.93 25.		3 0.02 0.03	915. 995. 1	71	7	0.47 14 1.07	17.0 31.15	7.5 15.5 1.0	12.9	8.93		6956 09-20-050 2
1155−22	1155.9 287.17 135.38	−2211 38.78 −18.70	10 0.55 22 −1	3.2 3.0 6	0.65 54.		0.16 0.13	1784. 1561. 2	160 159. 162.	15	0.56 20 1.01	26.5 32.12	−17.9 17.6 −8.5	23.2	9.41 10.23	0.150	UGCA 257, DDO 106 572-30 −4-28-008 2
N 4024	1156.0 285.74 131.12	−1804 42.75 −17.93	−5 0.57 22 −1	1.9 1.8 6	0.90 29.	12.61	6 0.00 0.06	1694. 1482. 2				25.9 32.07	−16.2 18.6 −8.0	−19.46 13.6	9.98		572-31 −3-31-004 2
N 4010	1156.0 146.70 67.69	4732 67.35 2.25	9 1.59 12 −1	3.7 2.8 2	0.26 84.	12.0 10.41 1.59	3 0.67 0.00	905. 968. 1	276 240. 242.	10	0.95 9 1.08	17.0 31.15	6.5 15.8 0.7	−19.15 13.9	9.85 9.41 10.37	0.36 0.111 3.26	6964 08-22-049 2
N 4013	1156.0 151.83 70.77	4414 70.09 1.07	4 1.34 12 −1	4.7 3.5 2	0.26 84.	11.6 8.48 3.12	3 0.67 0.00	835. 883. 1	398 362. 364.	10	0.90 11 1.12	17.0 31.15	5.4 15.6 0.3	−19.55 17.4	10.01 9.36 10.82	0.22 0.035 6.42	6963 07-25-009 2
U 6962	1156.0 154.07 71.91	4301 71.05 0.63	6X 1.01 12 −1	2.5 2.4 2	0.85 35.	12.76	2 0.05 0.00	784. 827. 1	188	20	0.71 28 1.06	17.0 31.15	5.1 15.4 0.1	−18.39 11.9	9.55 9.17	0.42	IC 749 6962 07-25-008 2
N 4020	1156.3 193.95 83.46	3041 78.03 −3.70	7 0.57 14 +1	2.0 1.8 2	0.53 63.	12.8	3 0.25 0.00	757. 742. 6		15		8.0 29.53	0.9 8.0 −0.5	−16.73 4.2	8.88 8.30	0.25	6971 05-28-066 2
U 6973	1156.3 153.97 71.94	4300 71.10 0.68	2 P 0.96 12 −1	3.0 2.6 2	0.52 64.	12.53	2 0.27 0.00	713. 756.		90		17.0 31.15	4.9 14.9 0.2	−18.62 12.9	9.64		IC 750 6973 07-25-010 2
1156+46	1156.3 148.81 69.14	4600 68.68 1.75	10 0.78 12 −1	1.3 1.2 2	0.83 38.		0.06 0.00	1154. 1210. 1	350 512. 513.	20	0.15 21 1.01	17.0 31.15	6.7 17.6 0.6	6.0	8.61 10.66	0.009	UGCA 259 08-22-000

Name	Position (1950; l,b; SGL,SGB)	Type	Dim.	B	ext.	V(a)	V(b)	n	HI	D / μ	SGX Y Z	M / size	log L	HI par.	Other names
U 6983	1156.6 5259 / 140.26 62.63 / 62.62 4.26	6B / 1.10 / 12 -1	4.1 3.8 / 0.75 45. / 2	12.8 / 11.66 / 1.14	3 / 0.10 / 0.00	1078. / 1166. / 1	200. / 231. / 233.	10	0.93 / 12 / 1.14	17.0 / 31.15	8.5 / 16.4 / 1.3	-18.35 / 18.9	9.53 / 9.39 / 10.47	0.72 / 0.084 / 8.58	6983 09-20-051 2
N 4027	1156.9 -1859 / 286.35 41.93 / 132.11 -17.90	8BP / 0.61 / 22 -1	3.3 3.2 / 0.77 44. / 9	11.50	2 / 0.09 / 0.06	1675. / 1461.		7		25.6 / 32.04	-16.3 / 18.1 / -7.9	-20.54 / 23.9	10.41		UGCA 260, ARP 22 / VV 66 / 572-37 -3-31-008 2
N 4026	1156.9 5114 / 141.93 64.21 / 64.28 3.69	-2 / 1.71 / 12 -1	5.0 3.7 / 0.26 83. / 2	11.68	2 / 0.00 / 0.02	878. / 958.		75		17.0 / 31.15	7.4 / 15.4 / 1.1	-19.47 / 18.4	9.98		6985 09-20-052 2
N 4030	1157.8 -49 / 277.35 59.22 / 113.81 -13.44	4A / 0.33 / 22 -10	4.0 3.8 / 0.78 43. / 2	11.8	3 / 0.08 / 0.04	1463. / 1311. / 6	349. / 459. / 460.	15	1.19 / 6 / 1.22	25.9 / 32.07	-10.2 / 23.0 / -6.0	-20.27 / 28.7	10.30 / 10.02 / 11.24	0.53 / 0.059 / 8.80	6993 00-31-016 2
N 4033	1158.0 -1734 / 286.18 43.37 / 130.71 -17.37	-5 / 0.38 / 22 -1	2.5 2.2 / 0.48 66. / 6	12.40	6 / 0.00 / 0.07	1521. / 1312.		90		23.9 / 31.89	-14.9 / 17.3 / -7.1	-19.49 / 15.4	9.99		572-42 -3-31-011 2
U 6998	1158.3 15 / 276.76 60.25 / 112.79 -13.03	9 / 0.24 / 22+12	1.9 1.8 / 0.80 41. / 2		3 / 0.08 / 0.04	1937. / 1789. / 1	69. / 79. / 86.	15 / 11 / 226	0.20 / 16 / 1.03	31.0 / 32.45	-11.7 / 27.8 / -7.0	16.3	9.18 / 9.47	0.516	6998 00-31-000
1158-24	1158.6 -2417 / 288.60 36.91 / 137.68 -18.43	8 / 0.52 / 22 -4	2.3 1.7 / 0.18 90. / 6		0.67 / 0.31	1808. / 1581. / 2	211. / 175. / 178.	20	0.63 / 21 / 1.01	26.6 / 32.12	-18.6 / 17.0 / -8.4	13.2	9.48 / 10.07	0.257	UGCA 263
N 4037	1158.8 1340 / 260.04 71.95 / 99.78 -8.90	5BP / 1.76 / 11 -0 +1	2.8 2.7 / 0.90 29. / 2	12.4	3 / 0.04 / 0.02	936. / 844. / 1	97	25	-0.01 / 36 / 1.08	17.2 / 31.18	-2.9 / 16.7 / -2.7	-18.78 / 13.6	9.70 / 8.46	0.06	505-03 -4-29-001
U 7007	1158.9 3337 / 179.97 77.56 / 80.90 -2.16	9 / 0.67 / 14 +1	2.0 2.0 / 1.00 / 2		0.00 / 0.00	786. / 786. / 1	67	10	-0.05 / 34 / 1.05	8.9 / 29.74	1.4 / 8.7 / -0.3	5.2	7.85		7002 02-31-015 2
N 4036	1158.9 6210 / 132.98 54.26 / 54.03 7.66	-2 / 0.44 / 12 -5 +1	4.0 3.3 / 0.40 72. / 9	11.39	1 / 0.00 / 0.01	1382. / 1510.		50		24.6 / 31.95	14.3 / 19.7 / 3.3	-20.56 / 23.7	10.42		7007 06-26-068
U 7009	1159.2 6237 / 132.67 53.84 / 53.62 7.84	10 / 0.30 / 12 -1	1.8 1.4 / 0.34 77. / 2	13.7	3 / 0.57 / 0.00	1120. / 1250. / 6	187	15	0.26 / 33 / 1.03	21.2 / 31.63	12.5 / 16.9 / 2.9	-17.93 / 8.7	9.36 / 8.91	0.35	7005 10-17-125 L2
N 4038	1159.3 -1835 / 286.95 42.47 / 131.81 -17.26	10BP / 0.59 / 22 -1	11.2 9.9 / 0.53 63. / 6	10.98	2 / 0.25 / 0.07	1658. / 1446.		8		25.5 / 32.04	-16.3 / 18.2 / -7.6	-21.06 / 73.7	10.62		7009 11-15-010 2
N 4039	1159.3 -1836 / 286.96 42.45 / 131.83 -17.27	10 P / 0.56 / 22 -1	9.8 8.4 / 0.46 68. / 6		0.34 / 0.07	1641. / 1429.		9		25.3 / 32.01	-16.1 / 18.0 / -7.5	62.1			UGCA 264, ARP 244 / VV 245 / 572-47 -3-31-014 2
N 4041	1159.7 6225 / 132.69 54.05 / 53.83 7.83	4A / 0.46 / 12 -5 +1	2.7 2.7 / 1.00 / 2	11.70	1 / 0.00 / 0.00	1231. / 1360.	244	10 / 10 / 217	1.03 / 217 / 10	22.7 / 31.78	13.3 / 18.1 / 3.1	-20.08 / 17.9	10.22 / 9.74	0.33	UGCA 265, ARP 244 / VV 245 / 572-48 -3-31-015 2 / 7014 10-17-129 2

(1) Name	(2) α ℓ SGL	(3) δ b SGB	(4) Type Q_{xyz} Group	(5) D_{25} $D^{b,i}_{25}$ source	(6) d/D i	(7) $B^{b,i}_T$ $H_{-0.5}$ B-H	(8) source $A^{b,i-o}_B$ A^b_B	(9) V_h V_o Tel	(10) W_{20} W_R W_D	(11) e_v e_w source	(12) F_c source/e_F f_h	(13) R μ	(14) SGX SGY SGZ	(15) $M^{b,i}_B$ Δ_{25}	(16) log L_B log M_H log M_T	(17) M_H/L_B M_H/M_T M_T/L_B	(18) Alt Names UGC/ESO Alt Names MCG AGN? RCII
U 7020	1200.1 131.46 51.70	6439 51.94 8.59	-2 0.32 12 +5 +1	1.3 1.2 2	0.52 63.	14.1	3 0.00 0.02	1447. 1586.		28		25.4 32.02	15.6 19.7 3.8	-17.92 8.9	9.36		MARK 195 2
N 4045	1200.2 276.00 110.94	216 62.29 -12.01	4 0.25 22+12	2.9 2.7 2	0.70 50.	12.52	2 0.13 0.00	1966. 1827.		81		31.5 32.49	-11.0 28.7 -6.5	-19.97 24.8	10.18		7020 07-25-014 2
N 4050	1200.4 286.40 129.31	-16C5 44.95 -16.51	3B 0.24 22 -1	2.5 2.3 4	0.73 48.	11.8	5 0.11 0.09	1904. 1700.		90		28.8 32.29	-17.5 21.3 -8.2	-20.49 19.3	10.39		7021 00-31-022 / -3-31-016 2
N 4051	1200.6 148.90 70.53	4448 70.09 2.04	4X -.06 12 -1	5.4 5.2 2	0.85 36.	10.90	1 0.05 0.00	710. 763. 1	267 394. 396.	15 6 211	1.02 211 7 1.23	17.0 31.15	5.3 14.7 0.6	-20.25 25.8	10.29 9.48 11.07	0.15 0.026 5.93	7030 08-22-059 S1.2 2
N 4061	1201.5 242.92 93.47	2030 77.15 -6.07	-5 0.40 11 +5 +1	1.3 1.3 2	0.76 45.	14.2	3 0.00 0.05	1604. 1545.		83		14.6 30.82	-0.9 14.5 -1.5	-16.62 5.5	8.84		VV 179, TURN 52A 7044 04-29-006 2
N 4062	1201.5 185.23 82.44	32-1 78.65 -2.14	5A 1.05 14 -1	4.5 3.8 2	0.46 68.	11.66 9.27 2.39	2 0.34 0.00	769. 764. 1	308 291. 293.	10 7 211	0.75 211 15 1.13	9.7 29.93	1.2 9.4 -0.3	-18.27 10.8	9.50 8.72 10.42	0.17 0.020 8.36	7045 05-29-004 2
N 4068	1201.6 138.90 62.98	5252 63.05 4.91	10A 0.47 14 -0	2.9 2.6 2	0.64 55.	12.6	3 0.17 0.00	213. 302. 1	83 79. 86.	7	0.94 8 1.07	3.8 27.93	1.7 3.4 0.3	-15.33 2.9	8.32 8.10 8.72	0.60 0.241 2.48	7047 09-20-079 2
N 4065	1201.6 242.98 93.48	2030 77.17 -6.05	-5 0.73 11 +7 +1	1.3 1.3 2	0.92 27.	13.5	3 0.00 0.05	1181. 1122.		152		16.0 31.02	-1.0 15.9 -1.7	-17.52 6.1	9.20		VV 179, TURN 52B 7050 04-29-007 2
N 4064	1201.7 249.08 95.18	1843 76.08 -6.63	1BP 0.98 11 +7 +1	3.9 3.4 2	0.47 67.	11.82	2 0.33 0.05	1026. 959.		60		16.9 31.14	-1.5 16.8 -2.0	-19.32 16.8	9.92		7054 03-31-033 2
U 7056	1201.7 132.12 53.55	6247 53.77 8.17	6B 0.45 12 -5 +1	2.0 2.0 2	0.95 21.	13.5	3 0.02 0.00	1275. 1406. 6	116	10	0.14 33 1.07	23.3 31.83	13.7 18.5 3.3	-18.33 13.6	9.52 8.87	0.23	IC 758 7056 11-15-014 2
U 7053	1201.8 279.48 114.51	-115 59.18 -12.60	10 0.33 22-10	1.9 1.8 2	0.85 35.		3 0.05 0.05	1463. 1311. 1	140	25	0.40 23 1.04	25.9 32.07	-10.5 23.0 -5.7	13.6	9.23		DDO 108 7053 00-31-000 2
1202-27	1202.2 290.57 141.52	-2751 33.62 -18.18	9 0.42 22 -4	2.4 2.5 6	0.83 37.		 0.06 0.35	1787. 1553. 2	242	60	0.62 21 1.01	25.9 32.06	-19.2 15.3 -8.1	18.9	9.45		UGCA 267, DDO 240 440-46 -5-29-005 2
N 4085	1202.8 140.60 65.16	5C38 65.16 4.36	5X 1.31 -2 -1	2.6 2.1 2	0.34 77.	12.31 10.01 2.30	1 0.57 0.02	714. 794.		90		17.0 31.15	6.7 14.5 1.2	-18.84 10.4	9.73		7075 09-20-086 2

Name	Coordinates	Type	Dimensions	mag (16)	(15)	Velocity (13)	(12)	N	(10)	Dist (8)	(7)	(6)	mag (5)	(3)	Alternate names
N 4088	1203.0 5049 / 140.35 65.01 / 65.00 4.45	5XP / 1.19 / 12 -1	5.3 0.41 71. / 4.4 / 2	10.68 / 8.42 / 2.26	1 / 0.41 / 0.02	763. / 844. / 1	376 / 357. / 358.	8	1.41 / 6 / 1.12	17.0 / 31.15	6.9 / 14.7 / 1.3	-20.47 / 21.8	10.38 / 9.87 / 10.91	0.31 / 0.092 / 3.37	ARP 18 / 7081 09-20-089 2
1203-27	1203.2 -2740 / 290.78 33.85 / 141.36 -17.94	5B / 0.40 / 22 -4	3.2 0.25 84. / 2.5 / 6	11.7	5 / 0.67 / 0.34	1851. / 1618. / 2	248 / 212. / 214.	15	0.63 / 25 / 1.01	26.7 / 32.13	-19.9 / 15.9 / -8.2	-20.43 / 19.5	10.36 / 9.48 / 10.40	0.13 / 0.120 / 1.10	IC 2995, UGCA 268 / 440-50 -5-29-008 2
N 4094	1203.3 -1416 / 286.69 46.88 / 127.62 -15.44	6 / 0.18 / 22 +1	4.1 0.41 71. / 3.4 / 2	12.0 / 10.19 / 1.81	5 / 0.41 / 0.10	1428. / 1232. / 1	271 / 246. / 249.	10	0.82 / 33 / 1.10	23.3 / 31.84	-13.7 / 17.8 / -6.2	-19.84 / 23.1	10.13 / 9.55 / 10.61	0.27 / 0.088 / 3.02	UGCA 269 / -2-31-016 2
U 7089	1203.4 4325 / 149.94 71.52 / 72.00 2.04	9 / 1.09 / 12 -1	3.3 0.29 81. / 2.5 / 2	13.1 / 11.88 / 1.22	3 / 0.67 / 0.00	778. / 826. / 1	157	10	0.62 / 12 / 1.07	17.0 / 31.15	5.0 / 15.4 / 0.5	-18.05 / 12.4	9.41 / 9.08	0.47	7089 07-25-020
1203-22	1203.5 -2234 / 289.48 38.84 / 136.10 -17.03	10B / 0.52 / 22 -1	3.3 0.86 35. / 3.3 / 6		0.05 / 0.19	1718. / 1498. / 6	136 / 190. / 193.	8	1.07 / 9 / 1.12	25.7 / 32.05	-17.7 / 17.1 / -7.5	24.8	9.89 / 10.41	0.300	UGCA 270, VV 46 / 505-13 -4-29-006
N 4096	1203.5 4745 / 143.54 67.80 / 67.92 3.51	5X / 0.40 / 14 -4	6.4 0.28 82. / 4.9 / 2	10.33 / 8.79 / 1.54	1 / 0.67 / 0.01	577. / 644. / 1	335 / 300. / 302.	15 11 211	1.23 / 211 8 / 1.21	8.8 / 29.72	3.3 / 8.1 / 0.5	-19.39 / 12.6	9.95 / 9.12 / 10.52	0.15 / 0.040 / 3.70	7090 08-22-067 L2 / 2
N 4100	1203.6 4952 / 141.11 65.91 / 65.93 4.23	5A / 1.02 / 12 -1	5.1 0.38 74. / 4.2 / 2	11.0 / 8.82 / 2.18	3 / 0.47 / 0.04	1076. / 1153. / 1	415	20	1.11 / 14 / 1.15	17.0 / 31.15	7.5 / 16.8 / 1.4	-20.15 / 20.8	10.25 / 9.57	0.21	7095 08-22-068 2
N 4102	1203.8 5259 / 138.10 63.08 / 62.98 5.28	3X / 1.42 / 12 -1	3.1 0.63 56. / 2.8 / 2	12.13	2 / 0.17 / 0.00	862. / 953. / 1	298	30	0.48 / 21 / 1.07	17.0 / 31.15	7.7 / 15.1 / 1.6	-19.02 / 13.9	9.80 / 8.94	0.14	7096 09-20-094 2
N 4105	1204.1 -2929 / 291.46 32.11 / 143.28 -18.01	-5 / 0.37 / 22 -5 +4	2.1 0.74 46. / 2.0 / 6	11.8	5 / 0.00 / 0.22	1890. / 1654.		25	-0.17 / 202 34	27.1 / 32.16	-20.7 / 15.4 / -8.4	-20.36 / 15.8	10.34 / 8.70	0.02	440-54 -5-29-013 2
N 4106	1204.2 -2929 / 291.49 32.12 / 143.29 -17.98	-2B / 0.18 / 22 -5 +4	1.9 0.65 54. / 1.8 / 6	12.13	2 / 0.00 / 0.22	2180. / 1944.		44		30.6 / 32.43	-23.3 / 17.4 / -9.5	-20.30 / 16.1	10.31		440-56 -5-29-014 2
N 4111	1204.5 4321 / 149.52 71.69 / 72.13 2.20	-2A / 1.09 / 12 -1	4.7 0.23 87. / 3.4 / 9	11.50	1 / 0.00 / 0.00	794. / 842.		14		17.0 / 31.15	5.0 / 15.5 / 0.6	-19.65 / 16.9	10.05		7103 07-25-026 L2 / 2
1204+40	1204.8 4005 / 155.97 74.37 / 75.23 1.16	10 / 0.58 / 12 -6 +1	1.4 0.86 34. / 1.4 / 2		0.05 / 0.00	881. / 914. / 1	67 / 88. / 94.	10	0.11 / 25 / 1.02	17.5 / 31.22	4.5 / 16.9 / 0.4	7.2	8.60 / 9.21	0.246	UGCA 271, DDO 109 / 07-25-000 2
N 4116	1205.1 258 / 277.87 63.41 / 110.61 -10.64	8B / 0.37 / 22-11	3.6 0.69 51. / 3.3 / 2	12.26 / 10.64 / 1.62	1 / 0.13 / 0.01	1310. / 1177. / 1	235 / 255. / 257.	15	1.09 / 67 11 / 1.10	25.0 / 31.99	-8.7 / 23.1 / -4.6	-19.73 / 24.1	10.08 / 9.89 / 10.66	0.63 / 0.169 / 3.74	7111 01-31-022 2
1205+67	1205.3 6740 / 129.33 49.20 / 48.95 10.01	13 P / 0.24 / 42 -0+12	0.8 1.00 / 0.8 / 2	14.4	3 / 0.00 / 0.02	2239. / 2391.		43		35.4 / 32.74	22.9 / 26.3 / 6.1	-18.34 / 8.3	9.53		UGCA 272, MARK 197 / 11-15-000 2

(1) Name	(2) α / ℓ / SGL	(3) δ / b / SGB	(4) Type / ϱ_{xyz} / Group	(5) D_{25} / $D^{b,i}_{25}$ / source	(6) d/D / i	(7) $B^{b,i}_T$ / $H_{-0.5}$ / B-H	(8) source / A^{i-o}_B / A^b_B	(9) V_h / V_o / Tel	(10) W_{20} / W_R / W_D	(11) e_v / e_w / source	(12) F_c / source/e_f / f_h	(13) R / μ	(14) SGX / SGY / SGZ	(15) $M^{b,i}_B$ / $\Delta^{b,i}_{25}$	(16) $\log L_B$ / $\log M_H$ / $\log M_T$	(17) M_H/L_B / M_H/M_T / M_T/L_B	(18) Alternate Names; UGC/ESO MCG; AGN? RCII
N 4121	1205.4 / 130.24 / 51.15	6524 / 51.38 / 9.35	-5 / 0.27 / 12 +5 +1	0.6 / 0.5 / 2	0.88 / 31.	14.3	3 / 0.00 / 0.03	1457. / 1600.		35		25.5 / 32.04	15.8 / 19.6 / 4.1	-17.74 / 3.7	9.29		TURN 56A; 11-15-026 2
N 4124	1205.6 / 269.56 / 103.20	1039 / 70.39 / -8.26	-2A / 1.93 / 11 -1	4.5 / 3.8 / 2	0.46 / 68.	12.0 / 9.35 / 2.65	3 / 0.00 / 0.00	1652. / 1551.		90		16.8 / 31.13	-4.0 / 17.1 / -2.5	-19.13 / 18.6	9.84		IC 3011
N 4123	1205.6 / 277.96 / 110.45	310 / 63.64 / -10.45	4B / 0.36 / 22 -11	4.6 / 4.4 / 2	0.82 / 39.	11.77	1 / 0.07 / 0.01	1339. / 1207. / 1	211	10	1.14 / 12 / 1.14	25.3 / 32.02	-8.7 / 23.3 / -4.6	-20.25 / 32.5	10.29 / 9.95	0.45	7117 02-31-036 2
N 4125	1205.6 / 130.19 / 51.11	6527 / 51.34 / 9.39	-5 / 0.34 / 12 +5 +1	6.6 / 6.4 / 2	0.84 / 36.	10.57	2 / 0.00 / 0.03	1352. / 1495.		20		24.2 / 31.92	15.0 / 18.6 / 3.9	-21.35 / 45.2	10.73		7116 01-31-023 2; TURN 56B
N 4127	1206.0 / 126.12 / 39.73	7705 / 40.06 / 12.64	4 / 0.32 / 42 -13	3.1 / 2.8 / 2	0.54 / 62.	12.5 / 10.49 / 2.01	3 / 0.24 / 0.15	1821. / 2006. / 6	277 / 271. / 273.	15	0.68 / 25 / 1.11	29.8 / 32.37	22.4 / 18.6 / 6.5	-19.87 / 24.4	10.14 / 9.63 / 10.72	0.30 / 0.081 / 3.76	7118 11-15-027 2 L2
N 4128	1206.1 / 128.69 / 47.63	6903 / 47.88 / 10.47	-2A / 0.27 / 42 +12	2.8 / 2.3 / 2	0.38 / 74.	12.7	3 / 0.00 / 0.02	2324. / 2481.		62		36.4 / 32.80	24.1 / 26.4 / 6.6	-20.10 / 24.4	10.23		7122 13-09-012 2; 7120 12-12-02A 2
U 7125	1206.2 / 163.31 / 78.15	3705 / 76.76 / 0.42	9 / 0.29 / 12 -6 +1	4.4 / 3.1 / 2	0.20 / 90.	13.0	3 / 0.67 / 0.00	1078. / 1098. / 1	187 / 152. / 156.	10	1.08 / 9 / 1.10	22.0 / 31.71	4.5 / 21.6 / 0.2	-18.71 / 19.9	9.68 / 9.76 / 10.13	1.23 / 0.436 / 2.81	7125 06-27-014
N 4129	1206.3 / 285.55 / 122.26	-846 / 52.38 / -13.45	5 / 0.19 / 22 -0	2.3 / 1.7 / 4	0.25 / 84.	12.38	2 / 0.67 / 0.15	1210. / 1034.		90		21.2 / 31.63	-11.0 / 17.4 / -4.8	-19.25 / 10.5	9.89		-1-31-006 2
N 4136	1206.7 / 193.68 / 84.68	3012 / 80.32 / -1.76	5X / 1.05 / 14 -1	4.0 / 4.0 / 2	1.00	11.7	3 / 0.00 / 0.01	618. / 606. / 1	112	8	1.07 / 7 / 1.17	9.7 / 29.93	0.9 / 9.4 / -0.3	-18.23 / 11.3	9.48 / 9.04	0.36	7134 05-29-025 2
N 4138	1207.0 / 147.28 / 71.70	4358 / 71.40 / 2.63	-2A / 0.84 / 12 -1	3.2 / 2.8 / 2	0.60 / 58.	12.30	2 / 0.00 / 0.00	1039. / 1091.		100		17.0 / 31.15	5.7 / 17.2 / 0.9	-18.85 / 13.9	9.73		7139 07-25-035 2
N 4142	1207.0 / 136.79 / 62.76	5323 / 62.88 / 5.86	6B / 0.26 / 12 -1	2.2 / 1.9 / 2	0.54 / 62.	13.5 / 12.15 / 1.35	3 / 0.24 / 0.00	1163. / 1257. / 1	187 / 173. / 176.	10	0.53 / 41 / 1.04	22.1 / 31.73	10.1 / 19.6 / 2.3	-18.23 / 12.3	9.48 / 9.22 / 10.03	0.54 / 0.156 / 3.49	7140 09-20-102 2
N 4143	1207.1 / 149.16 / 72.79	4249 / 72.40 / 2.47	-2X / 0.82 / 12 -1	3.2 / 2.9 / 2	0.63 / 55.	12.10	2 / 0.00 / 0.00	784. / 831.		100		17.0 / 31.15	4.7 / 15.4 / 0.7	-19.05 / 14.4	9.81		7142 07-25-036 2
N 4145	1207.5 / 154.25 / 75.32	40:0 / 74.62 / 1.68	7X / 0.36 / 12 -6 +1	5.9 / 5.4 / 2	0.69 / 51.	11.37 / 9.70 / 1.67	2 / 0.13 / 0.00	1013. / 1048. / 1	232 / 252. / 254.	10 / 4 / 211	1.21 / 206 8 / 1.25	20.7 / 31.58	5.2 / 20.0 / 0.6	-20.21 / 32.6	10.28 / 9.84 / 10.78	0.37 / 0.116 / 3.18	7154 07-25-040 2

Name	(1)	(2)	Type					Vel	Width								Other names
N 4144	1207.5 / 143.16 / 69.11	4644 / 69.01 / 3.81	6X / 0.42 / 14 -7	6.3 / 4.8 / 2	0.28 / 81.	11.2 / 10.25 / 0.95	3 / 0.67 / 0.00	267. / 331. / 1	179 / 147. / 151.	10	1.12 / 217 7 / 1.20	4.1 / 28.07	1.5 / 3.8 / 0.3	-16.87 / 5.7	8.94 / 8.35 / 9.56	0.25 / 0.062 / 4.13	7151 08-22-077 2
1207-29	1207.7 / 292.36 / 143.38	-2928 / 32.28 / -17.23	5A / 0.21 / 22 -5 +4	4.6 / 3.8 / 6	0.33 / 77.	11.5 / 10.46 / 1.04	5 / 0.58 / 0.24	2142. / 1907. / 2	308 / 277. / 279.	15	1.16 / 17 / 1.04	30.2 / 32.40	-23.1 / 17.2 / -8.9	-20.90 / 33.5	10.55 / 10.12 / 10.87	0.37 / 0.177 / 2.09	IC 764, UGCA 273 / 441-13 -5-29-025 2
N 4150	1208.0 / 190.44 / 84.32	3041 / 80.46 / -1.34	-2A / 0.75 / 14 -1	2.2 / 2.0 / 9	0.67 / 52.	12.45	2 / 0.00 / 0.00	244. / 235.		50		9.7 / 29.93	1.0 / 8.6 / -0.3	-17.48 / 5.7	9.18		7165 05-29-029 2
N 4151	1208.0 / 155.09 / 75.81	3941 / 75.06 / 1.61	2X / 0.36 / 12 -6 +1	6.3 / 6.1 / 2	0.87 / 33.	11.08	2 / 0.05 / 0.00	989. / 1022. / 4	139 / 202. / 205.	10 / 6 / 211	1.20 / 206 8 / 1.08	20.3 / 31.54	5.0 / 19.7 / 0.6	-20.46 / 36.2	10.38 / 9.81 / 10.63	0.27 / 0.153 / 1.80	TURN 57A / 7166 07-25-044 2 S1.5
U 7164	1208.0 / 127.86 / 45.98	7047 / 46.23 / 11.11	9X / 0.22 / 42-12	1.9 / 1.8 / 2	0.85 / 35.		0.05 / 0.04	2124. / 2288. / 1	101	15	0.16 / 34 / 1.03	33.9 / 32.65	23.1 / 23.9 / 6.5	17.8	9.22		7164 12-12-000
N 4152	1208.1 / 260.40 / 97.96	1619 / 75.42 / -5.93	5X / 0.11 / 13+12+11	2.2 / 2.2 / 2	0.88 / 32.	12.34	2 / 0.04 / 0.11	2160. / 2086. / 1	233 / 371. / 373.	10 / 15 / 201	0.74 / 201 24 / 1.05	34.5 / 32.69	-4.8 / 34.0 / -3.6	-20.35 / 22.2	10.33 / 9.82 / 10.95	0.30 / 0.074 / 4.12	7169 03-31-052 2
U 7170	1208.1 / 252.34 / 95.31	1906 / 77.47 / -5.05	5 / 0.21 / 13-11	2.9 / 2.0 / 2	0.09 / 90.	14.2	3 / 0.67 / 0.04	2444. / 2382. / 1	232	15	0.78 / 28 / 1.05	37.8 / 32.89	-3.5 / 37.5 / -3.3	-18.69 / 22.1	9.67 / 9.93	1.85	7170 03-31-055
U 7168	1208.1 / 127.89 / 46.11	7039 / 46.36 / 11.08	13 P / 0.24 / 42-12	1.4 / 1.0 / 2	0.22 / 87.	15.0	3 / 0.00 / 0.03	2181. / 2344.		105		34.5 / 32.69	23.5 / 24.4 / 6.6	-17.69 / 10.1	9.27		MARK 199 / 7168 12-12-000 2
N 4156	1208.3 / 154.75 / 75.77	3945 / 75.04 / 1.69	3B / 0.53 / 12 -6 +1	1.5 / 1.5 / 2	0.94 / 23.	13.93	2 / 0.02 / 0.00	726. / 759.		63		15.5 / 30.95	3.9 / 15.0 / 0.5	-17.02 / 6.8	9.00		TURN 57B / 7173 07-25-045 2
U 7175	1208.4 / 154.04 / 75.51	4002 / 74.83 / 1.80	9 / 0.25 / 12 -6 +1	2.1 / 1.6 / 2	0.28 / 81.	14.3	3 / 0.67 / 0.00	1168. / 1202. / 6	147 / 118. / 123.	30 / 13 / 206	0.33 / 206 9 / 1.05	22.9 / 31.80	5.7 / 22.2 / 0.7	-17.50 / 10.7	9.19 / 9.05 / 9.64	0.71 / 0.259 / 2.78	7175 07-25-046 2
U 7178	1208.5 / 280.17 / 111.52	217 / 63.08 / -10.02	10 / 0.39 / 22+11	1.6 / 1.5 / 2	0.70 / 50.		0.13 / 0.01	1339. / 1205. / 1	95 / 98. / 103.	15	0.42 / 26 / 1.02	25.3 / 32.02	-9.2 / 23.2 / -4.4	11.1	9.23 / 9.49	0.546	DDO 110 / 7178 00-31-036 2
N 4158	1208.6 / 247.83 / 94.06	2027 / 78.45 / -4.51	0.20 / 13-11	1.7 / 1.7 / 2	0.95 / 21.	12.6	3 / 0.00 / 0.03	2536. / 2480.		90		38.9 / 32.95	-2.7 / 38.7 / -3.1	-20.35 / 19.3	10.33		7182 03-31-060 2
N 4157	1208.6 / 138.46 / 65.33	5046 / 65.41 / 5.27	3X / 1.19 / 12 -1	7.0 / 4.9 / 2	0.20 / 90.	11.0 / 8.28 / 2.72	3 / 0.67 / 0.02	771. / 854. / 1	431 / 393. / 394.	10	1.46 / 7 8 / 1.23	17.0 / 31.15	6.8 / 14.7 / 1.5	-20.15 / 24.3	10.25 / 9.92 / 11.04	0.47 / 0.077 / 6.10	7183 09-20-106 2
1209+16	1209.1 / 260.00 / 97.61	1646 / 75.92 / -5.56	13 P / 0.18 / 13+12+11			13.8	3 / 0.67 / 0.12	2472. / 2400.		220		38.0 / 32.90	-5.0 / 37.5 / -3.7	-19.10	9.83		ARAK 350 / 03-31-064

(1) Name	(2) α ℓ SGL	(3) δ b SGB	(4) Type Q_xyz Group	(5) D₂₅ᵇⁱ D₂₅ᵇⁱ source	(6) d/D i	(7) B_Tᵇⁱ H₋₀.₅ B-H	(8) source A_Bⁱ⁻ᵒ A_Bᵇ	(9) V_h V_o Tel	(10) W₂₀ W_R W_D	(11) e_v e_w source	(12) F_c source/e_f f_h	(13) R μ	(14) SGX SGY SGZ	(15) M_Bᵇⁱ Δ₂₅ᵇⁱ	(16) log L_B log M_H log M_T	(17) M_H/L_B M_H/M_T M_T/L_B	(18) Alternate Names UGC/ESO MCG AGN? RCII
N 4162	1209.3 229.36 90.36	2424 80.58 -3.10	4A 0.15 13 -0	2.5 2.2 2	0.60 58.	11.99	2 0.20 0.06	2491. 2454.		69		38.5 32.93	-0.2 38.4 -2.1	-20.94 24.7	10.57		7193 04-29-046 2
N 4163	1209.6 163.21 78.97	3627 77.69 0.86	10A 0.35 14 -7	2.1 2.1 2	0.96 19.	13.0	3 0.01 0.00	170. 188. 6	55	10	0.32 225 13 1.08	2.1 26.62	0.4 2.1 0.0	-13.62 1.3	7.64 6.96	0.22	7199 06-27-026 2
N 4168	1209.7 267.66 100.79	1329 73.33 -6.44	-5 1.56 11 -1	3.1 3.0 9	0.78 43.	12.20	1 0.00 0.05	2428. 2342.		90		16.8 31.13	-3.5 18.6 -2.1	-18.93 14.7	9.76		7203 02-31-046 2
U 7207	1209.8 160.37 78.18	3718 77.11 1.17	10 0.34 12 -6 +1	2.5 2.1 2	0.52 64.		0.27 0.00	1055. 1077. 6	85 74. 81.	20 12 226	0.19 226 18 1.07	21.7 31.68	4.4 21.2 0.4	13.3	8.86 9.32	0.347	7207 06-27-027
N 4173	1209.8 197.21 85.59	2928 81.14 -1.36	7B 0.98 14 -1	4.7 3.3 2	0.19 90.	12.4	3 0.67 0.02	1127. 1113. 1	205 169. 172.	20	1.00 10 1.12	9.7 29.93	0.8 9.9 -0.3	-17.53 9.3	9.20 8.97 9.89	0.59 0.122 4.84	7204 05-29-033
U 7209	1210.2 269.76 101.85	1224 72.44 -6.70	4A 2.04 11 -1	2.5 2.3 2	0.71 49.	13.1	3 0.12 0.02	2235. 2144. 1	283	50	0.62 48 1.05	16.8 31.13	-3.8 18.2 -2.2	-18.03 11.3	9.40 9.07	0.46	IC 769 7209 02-31-047
N 4178	1210.4 271.84 103.07	1109 71.37 -7.03	8B 2.04 11 -1	5.0 4.0 2	0.35 76.	11.36 10.09 1.27	1 0.54 0.00	381. 285. 1	279 249. 251.	15 8 300	1.27 24 8 1.15	16.8 31.13	-3.5 15.1 -1.9	-19.77 19.6	10.10 9.72 10.55	0.42 0.149 2.80	7215 02-31-050 2
N 4179	1210.3 281.61 112.33	135 62.57 -9.78	-2 0.37 22-11	4.2 3.2 2	0.28 81.	11.78	2 0.00 0.02	1279. 1144. 1		50		24.5 31.95	-9.1 22.3 -4.1	-20.17 22.9	10.26		7214 00-31-038 2
1210-20	1210.4 290.86 133.93	-2008 41.56 -15.00	10 0.24 22 -1	1.7 1.5 6	0.60 58.		0.19 0.11	1550. 1340. 6	127	15	0.38 26 1.04	24.0 31.90	-16.1 16.7 -6.2	10.5	9.14		UGCA 274 573-03 -3-31-024
U 7218	1210.4 136.37 63.71	5233 63.84 6.09	10 P 0.93 12 -1	1.5 1.3 2	0.54 62.	14.4	3 0.24 0.00	791. 882. 1	112	15	0.08 38 1.02	17.0 31.15	7.3 14.7 1.8	-16.75 6.5	8.89 8.54	0.45	7218 09-20-113
N 4183	1210.8 145.40 71.92	4358 71.74 3.48	6A 1.05 12 -1	5.1 3.6 2	0.14 90.	12.3 10.60 1.70	3 0.67 0.00	934. 987. 1	251 213. 216.	10 8 79	1.06 217 7 1.13	17.0 31.15	5.4 16.4 1.0	-18.85 17.9	9.73 9.52 10.37	0.62 0.141 4.36	7222 07-25-051 2
N 4192	1211.2 265.42 99.29	1510 74.94 -5.58	2X 1.44 11 -1	8.7 6.6 2	0.26 83.	10.05 7.69 2.36	1 0.67 0.12	-142. -220. 1	464 429. 430.	10 8 300	1.30 24 10 1.32	16.8 31.13	-2.4 14.6 -1.4	-21.08 32.4	10.62 9.75 11.24	0.13 0.033 4.11	M 98 7231 03-31-079 2
N 4190	1211.2 160.62 78.63	3655 77.58 1.31	10 0.37 14 -7	1.9 1.9 2	0.95 21.	13.2	3 0.02 0.00	234. 255. 1	80	7	0.67 225 8 1.04	2.8 27.24	0.6 2.7 0.1	-14.04 1.6	7.81 7.56	0.57	VV 104 7232 06-27-030 2

Name	Pos 1	Pos 2	Type	Dim	ratio/PA	m_B	flags	Vel	N / disp	small	col	dist	sup	abs	mag	ext	Identifier
N 4189	1211.2 / 268.32 / 100.68	1343 / 73.72 / -6.02	6X / 1.99 / 11 -1	2.6 / 2.5 2	0.83 / 37.	12.41	1 / 0.06 / 0.05	2123. / 2039. / 6	295	30 / 24 / 79	0.64 / 217 / 9 / 1.08	16.8 / 31.13	-3.4 / 18.1 / -1.9	-18.72 / 12.3	9.68 / 9.09	0.26	IC 3050 / 7235 02-31-054 2
U 7239	1211.6 / 276.69 / 106.16	803 / 68.72 / -7.62	10 / 0.26 / 11 -0 +1	2.5 / 2.4 2	0.93 / 25.	13.8	3 / 0.02 / 0.00	1234. / 1126. / 1	128	15 / 14 / 226	0.11 / 226 / 25 / 1.06	24.9 / 31.98	-6.9 / 23.7 / -3.2	-18.18 / 17.4	9.46 / 8.90	0.27	7239 01-31-028
N 4194	1211.7 / 134.39 / 61.61	5448 / 61.77 / 6.95	-2BP / 0.18 / 42+11	2.5 / 2.3 2	0.70 / 49.	12.95	2 / 0.00 / 0.00	2528. / 2629.	112	15 / 20 / 56	0.05 / 56 / 23 / 1.06	39.1 / 32.96	18.5 / 34.2 / 4.7	-20.01 / 26.3	10.20 / 9.23	0.11	ARP 160, MARK 201 / I ZW 33, VV 261 / 7241 09-20-119 2
U 7249	1212.1 / 270.12 / 101.35	1305 / 73.29 / -6.00	10 / 2.83 / 11 -1	1.5 / 1.2 2	0.40 / 72.	14.5	3 / 0.43 / 0.06	627. / 541. / 6	130	15	0.08 / 39 / 1.02	16.8 / 31.13	-3.1 / 15.6 / -1.7	-16.63 / 5.9	8.84 / 8.53	0.49	DDO 114 / 7249 02-31-056 2
1212-35	1212.1 / 294.57 / 149.51	-3515 / 26.75 / -17.10	5B / 0.19 / 23 -5	2.9 / 2.8 6	0.65 / 54.		0.16 / 0.29	2689. / 2445. / 2	317	25	0.71 / 24 / 1.02	36.5 / 32.81	-30.1 / 17.7 / -10.7	29.8	9.83		380-01 -6-27-009
U 7257	1212.5 / 161.95 / 79.37	3614 / 78.26 / 1.34	9 / 0.26 / 12 -6 +1	1.7 / 1.5 2	0.66 / 53.	13.9	3 / 0.15 / 0.00	948. / 967. / 1	100 / 98. / 104.	20	0.57 / 14 / 1.02	20.2 / 31.52	3.7 / 19.8 / 0.5	-17.62 / 8.8	9.24 / 9.18 / 9.39	0.87 / 0.615 / 1.42	KAR 91 / 7257 06-27-039 2
N 4203	1212.6 / 172.97 / 81.98	3329 / 80.08 / 0.49	-2X / 0.94 / 14 -1	3.5 / 3.4 9	0.93 / 26.	11.55	2 / 0.00 / 0.00	1088. / 1094.	282	9 / 20 / 205	0.48 / 210 / 25	9.7 / 29.93	1.4 / 9.8 / 0.1	-18.38 / 9.6	9.54 / 8.45	0.08	7256 06-27-040 2
N 4204	1212.7 / 249.06 / 93.90	2056 / 79.50 / -3.45	8B / 0.31 / 14 -0	4.4 / 4.4 2	1.00	12.7	3 / 0.00 / 0.05	861. / 810. / 1	110	8	0.91 / 9 / 1.20	7.9 / 29.48	-0.5 / 7.9 / -0.5	-16.78 / 10.1	8.90 / 8.71	0.63	7261 04-29-051 2
N 4206	1212.7 / 270.18 / 101.19	1318 / 73.55 / -5.80	4A / 3.12 / 11 -1	5.3 / 3.7 2	0.19 / 90.	12.08 / 10.27 / 1.81	1 / 0.67 / 0.04	701. / 616. / 1	307 / 269. / 271.	25	1.01 / 14 / 1.11	16.8 / 31.13	-3.1 / 15.8 / -1.6	-19.05 / 18.2	9.81 / 9.46 / 10.58	0.45 / 0.076 / 5.87	IC 3064 / 7260 02-31-066 2
1212+06	1212.8 / 279.44 / 108.19	602 / 66.97 / -7.92	-5 P / 0.76 / 11-24	0.5 / 0.5 5	0.67 / 53.	15.00	2 / 0.00 / 0.00	1960. / 1844.		52		35.1 / 32.73	-10.7 / 32.5 / -4.8	-17.73 / 5.1	9.28		HARO 6 / 01-31-030 2
U 7267	1212.9 / 136.22 / 64.69	5139 / 64.82 / 6.18	9 / 0.44 / 14 -4	2.0 / 1.7 2	0.44 / 70.	14.0	3 / 0.37 / 0.01	473. / 561. / 6	134 / 113. / 118.	15	0.43 / 24 / 1.04	7.8 / 29.47	3.3 / 7.0 / 0.8	-15.47 / 3.9	8.38 / 8.21 / 9.16	0.69 / 0.115 / 5.98	7267 09-20-128
U 7271	1213.0 / 144.66 / 72.30	4342 / 72.16 / 3.77	7B / 0.68 / 14 -4	2.3 / 1.9 2	0.39 / 73.	13.9	3 / 0.45 / 0.00	546. / 599. / 6		20	0.24 / 33 / 1.05	7.5 / 29.37	2.3 / 7.1 / 0.5	-15.47 / 4.2	8.38 / 7.99	0.41	7271 07-25-052
N 4212	1213.1 / 268.88 / 100.37	1411 / 74.36 / -5.44	4A / 1.44 / 11 -1	3.1 / 2.8 9	0.59 / 58.	11.57	1 / 0.20 / 0.10	-82. / -163. / 7		20		16.8 / 31.13	-2.7 / 14.5 / -1.4	-19.56 / 13.7	10.02		7275 02-31-070 2
N 4214	1213.1 / 160.30 / 79.06	3636 / 78.07 / 1.57	10X / 0.40 / 14 -7	9.6 / 9.3 2	0.84 / 37.	10.14	1 / 0.06 / 0.00	288. / 309. / 4	95 / 124. / 128.	10	1.95 / 96 / 16 / 1.13	3.5 / 27.71	0.7 / 3.4 / 0.1	-17.57 / 9.5	9.22 / 9.04 / 9.63	0.66 / 0.258 / 2.55	7278 06-27-042 2

(1) Name	(2) α ℓ SGL	(3) δ b SGB	(4) Type ρ_{xyz} Group	(5) D_{25}^{i} D_{25} source	(6) d/D i	(7) $B_T^{b,i}$ $H_{-0.5}$ $B-H$	(8) source A_B^{i-o} A_B^{b}	(9) V_h V_o Tel	(10) W_{20} W_R W_D	(11) e_v e_w source	(12) F_c source e_F f_h	(13) R μ	(14) SGX SGY SGZ	(15) $M_B^{b,i}$ $\Delta_{25}^{b,i}$	(16) $\log L_B$ $\log M_H$ $\log M_T$	(17) M_H/L_B M_H/M_T M_T/L_B	(18) Alternate Names UGC/ESO Alternate Names MCG AGN? RCII
U 7279	1213.2 274.17 103.47	1058 71.54 -6.38	9 2.44 11 -1	2.2 1.6 2	0.11 90.	14.1	3 0.67 0.01	1978. 1883. 6	248 211. 213.	20	0.65 23 1.05	16.8 31.13	-4.2 17.7 -2.0	-17.03 7.8	9.00 9.10 10.01	1.24 0.124 10.04	IC 3074 7279 02-31-071 2
N 4217	1213.3 139.93 68.82	4722 68.85 4.95	3 0.95 12 -1	5.1 4.0 2	0.33 78.	11.3 8.59 2.71	3 0.60 0.00	1028. 1098. 1	421 392. 393.	20	1.09 7 14 1.14	17.0 31.15	6.5 16.8 1.6	-19.85 19.9	10.13 9.55 10.95	0.26 0.040 6.53	7282 08-22-087 2
N 4218	1213.3 138.86 67.81	4825 67.87 5.27	-2 0.07 12 -1	1.1 1.0 2	0.57 60.	13.2	3 0.00 0.00	1388. 1462. 1		55		25.3 32.02	9.5 23.4 2.3	-18.82 7.4	9.72		7283 08-22-088 2
N 4215	1213.4 279.19 107.61	641 67.63 -7.59	-2A 0.81 11-24	1.8 1.5 9	0.43 70.	13.1	3 0.00 0.00	2093. 1980.		90		35.1 32.73	-10.5 32.9 -4.6	-19.63 15.4	10.04 8.49	 0.03	7281 01-31-031 2
N 4216	1213.4 270.48 101.11	1326 73.75 -5.60	3X 2.35 11 -1	7.9 5.7 9	0.21 89.	10.17 7.26 2.91	1 0.67 0.05	139. 55. 1	528. 490. 491.	25 10 300	1.09 24 14 1.27	16.8 31.13	-2.9 14.8 -1.5	-20.96 28.0	10.58 9.54 11.29	0.09 0.018 5.17	7284 02-31-072 2
N 4220	1213.7 138.94 68.07	4810 68.13 5.26	-2A 1.19 12 -1	4.2 3.4 2	0.39 73.	12.20	2 0.00 0.00	979. 1052.		50		17.0 31.15	6.6 16.4 1.6	-18.95 16.9	9.77		7290 08-22-089 2
1213-11	1213.8 289.39 125.21	-1116 50.40 -12.26	10 0.26 22 -3	1.5 1.1 4	0.17 90.		 0.67 0.10	1165. 985. 6	140	15	0.28 26 1.02	20.6 31.57	-11.6 16.4 -4.4	6.6	8.91		DDO 116 -2-31-026 2
N 4219	1213.8 296.21 157.68	-4303 19.09 -17.62	4A 0.07 23 -0	4.1 3.6 6	0.32 78.	11.1	5 0.62 0.57	1993. 1739.	385. 355. 356.	15	1.32 10 1.04	27.1 32.17	-23.9 9.8 -8.2	-21.07 28.5	10.62 10.19 11.02	0.37 0.147 2.50	267-37 -7-25-005 2
N 4224	1214.0 278.53 106.64	744 68.66 -7.14	1A 0.69 11-24	2.0 1.7 2	0.49 66.	12.65	1 0.30 0.00	2651. 2543.		90		35.1 32.73	-10.2 34.0 -4.4	-20.08 17.4	10.22		7292 01-31-034 2
U 7300	1214.1 199.20 86.34	2900 82.15 -0.62	10 1.20 14 -1	1.8 1.8 2	1.00		 0.00 0.07	1215. 1202. 1	97	15	0.49 16 1.04	9.7 29.93	0.7 10.0 -0.1	5.1	8.46		DDO 117 7300 05-29-048 2
N 4236	1214.3 127.42 47.13	6945 47.35 11.35	8B 0.46 14+10	19.6 16.1 2	0.39 73.	9.47 9.03 0.44	1 0.45 0.05	2. 163. 1	176. 149. 153.	30 5 211	2.29 80 2 2.40	2.2 26.74	1.5 1.6 0.4	-17.27 10.3	9.10 8.97 9.82	0.75 0.142 5.29	7306 12-12-004 2
N 4235	1214.6 279.18 106.94	728 68.46 -7.08	1A 0.74 11-24	3.8 2.8 9	0.25 84.	11.98	1 0.67 0.00	2596. 2487.		90		35.1 32.73	-10.3 33.8 -4.4	-20.75 28.7	10.49		IC 3098 7310 01-31-036 2 L1.9
N 4233	1214.6 278.74 106.52	754 68.87 -6.95	-2 0.73 11-24	2.6 2.2 9	0.50 65.	13.08	1 0.00 0.00	2224. 2117.		90		35.1 32.73	-9.9 33.4 -4.3	-19.65 22.5	10.05		7311 01-31-037 2

Name	Coords (RA/Dec; SGL/SGB; l/b)	Type	Dim / ratio / PA	m / corr	Vel	W₂₀	N	H	Distance block	M_B / size	Flux block	Alt names
N 4234	1214.6 358 / 282.19 65.17 / 110.32 -8.08	10 / 0.52 / 11+24	1.3 1.3 2 / 0.92 26.	13.47 / 2 0.03 / 0.00	2075. 1952.		66		32.9 32.59 / -11.3 30.6 -4.6	-19.12 12.5	9.84		7309 01-31-035 2
1214-11	1214.6 -1125 / 289.74 50.30 / 125.41 -12.10	10 / 0.24 / 22 -3	1.1 1.0 2 / 0.49 66.	/ 0.30 / 0.08	1278. 1098. 6	115	15	0.16 29 1.01	22.1 31.72 / -12.5 17.6 -4.6	6.5	8.85		UGCA 277, DDO 118 / -2-31-028 2
U 7307	1214.6 1017 / 276.04 71.06 / 104.22 -6.25	10 / 2.87 / 11 -1	2.0 2.0 2 / 1.00	/ 0.00 / 0.00	1188. 1091. 1	86	10	0.16 21 1.05	16.8 31.13 / -4.1 16.3 -1.8	9.8	8.61		7307 02-31-000
N 4237	1214.7 1536 / 267.26 75.77 / 99.14 -4.64	5X / 2.56 / 11 -1	2.2 2.1 2 / 0.67 52.	12.26 / 2 0.15 / 0.09	945. 871.		90		16.8 31.13 / -2.6 16.3 -1.3	-18.87 10.3	9.74		7315 03-31-091 2
N 4242	1214.9 4554 / 140.82 70.31 / 70.30 4.77	8X / 0.68 / 14 -4	5.2 4.8 2 / 0.70 50.	11.47 / 1 0.13 / 0.00	516. 580. 1	142 147. 151.	10	1.06 9 1.26	7.5 29.38 / 2.5 7.1 0.6	-17.91 10.5	9.36 8.81 9.82	0.28 0.098 2.90	7323 08-22-098 2
U 7321	1215.0 2249 / 241.92 81.04 / 92.28 -2.36	6 / 0.28 / 14 -0	5.1 3.6 2 / 0.07 90.	13.1 / 3 0.67 / 0.09	403. 362. 1	234 197. 200.	10	1.02 11 1.13	3.8 27.89 / -0.2 3.8 -0.2	-14.79 4.0	8.11 8.18 9.65	1.18 0.034 35.06	7321 04-29-060
N 4244	1215.0 3805 / 154.56 77.16 / 77.76 2.39	6A / 0.39 / 14 -7	15.8 11.1 9 / 0.11 90.	9.83 8.68 1.15 / 1 0.67 / 0.00	247. 276. 4	220 183. 186.	5	1.99 4 1.18	3.1 27.43 / 0.6 3.0 0.1	-17.60 10.0	9.23 8.97 9.99	0.55 0.096 5.72	7322 06-27-045 2
N 4245	1215.1 2953 / 192.54 82.16 / 85.56 -0.14	1.25 / 14 -1	3.3 3.3 2 / 0.95 21.	12.19 / 2 0.00 / 0.06	890. 881.		65		9.7 29.93 / 0.7 9.7 0.0	-17.74 9.3	9.29		7328 05-29-049 2
N 4248	1215.4 4741 / 138.69 68.68 / 68.62 5.39	13 / 0.68 / 14 -4	3.0 2.6 2 / 0.45 69.	13.19 / 1 0.00 / 0.00	484. 556.		10	220 11	7.3 29.32 / 2.7 6.8 0.7	-16.13 5.5	8.64 7.73	0.12	7335 08-22-099 2
U 7332	1215.4 43 / 284.80 62.13 / 113.53 -8.79	10 / 0.47 / 11 -0 +1	2.3 2.1 2 / 0.68 52.	13.9 / 3 0.14 / 0.03	941. 805. 1	100. 100. 106.	15	0.79 11 1.04	17.6 31.23 / -6.9 15.9 -2.7	-17.33 10.8	9.12 9.28 9.50	1.44 0.610 2.35	7332 00-31-045
N 4251	1215.6 2827 / 202.96 82.55 / 86.96 -0.48	-2B / 1.20 / 14 -1	3.9 3.5 2 / 0.58 60.	11.53 / 2 0.00 / 0.07	1014. 999.		75		9.7 29.93 / 0.6 9.8 -0.1	-18.40 9.9	9.55		7338 05-29-050 2
1215-79	1215.8 -7927 / 301.42 -16.95 / 195.87 -17.03	6 / 0.06 / 14+20	2.4 2.4 6 / 0.57 60.	/ 0.22 / 0.51	430. 185. 7		20		3.1 27.46 / -2.9 -0.8 -0.9	2.2			IC 3104 / 020-04
N 4254	1216.3 1442 / 270.43 75.20 / 100.11 -4.54	5A / 1.91 / 11 -1	5.0 5.0 9 / 0.87 33.	10.29 / 1 0.05 / 0.08	2400. 2323.	272 426. 428.	16 10 300	1.56 24 10	16.8 31.13 / -3.3 18.7 -1.5	-20.84 24.5	10.53 10.01 11.11	0.30 0.079 3.83	M 99 / 7345 03-31-099 2
N 4256	1216.4 6611 / 128.21 50.89 / 50.70 10.65	3A / 0.21 / 42 -0+12	4.2 3.0 2 / 0.21 89.	11.7 / 3 0.67 / 0.02	2531. 2680.		77		39.1 32.96 / 24.3 29.7 7.2	-21.26 34.3	10.70		7351 11-15-045 2

Table of galaxy data (page 101). Each galaxy occupies three lines corresponding to the three-row labels in each column heading.

(1) Name	(2) α / l / SGL	(3) δ / b / SGB	(4) Type / ρxyz / Group	(5) D25 / D25b,i / source	(6) d/D / i	(7) $B_T^{b,i}$ / $H_{-0.5}$ / B-H	(8) source / A_B^{i-o} / A_B^b	(9) V_h / V_o / Tel	(10) W_{20} / W_R / W_D	(11) e_v / e_w / source	(12) F_c / source,e_f / f_h	(13) R / μ	(14) SGX / SGY / SGZ	(15) $M_B^{b,i}$ / Δ_{25}	(16) log L_B / log M_H / log M_T	(17) M_H/L_B / M_H/M_T / M_T/L_B	(18) Alternate Names / UGC·ESO MCG AGN? RCII
U 7352	1216.5	908	8	2.1	0.56	13.9	3	2464	205	40	0.46	16.8	-5.1	-17.23	9.08	0.68	IC 776
	278.69	70.19	1.31	1.9	60.		0.22	2363			28	31.13	18.3	9.3	8.91		
	105.47	-6.14	11 -1	2			0.00	6			1.06		-2.0				
N 4258	1216.5	4735	4X	17.1	0.42	8.56	1	449	440	7	2.08	6.8	2.4	-20.59	10.43	0.21	7352 02-31-088
	138.31	68.84	0.57	14.2	71.	6.07	0.39	521	426.		8	29.15	6.3	28.2	9.75	0.037	L/S
	68.77	5.53	14 -4	9		2.49	0.00	4	428.		1.36		0.7		11.17	5.54	
U 7354	1216.6	408	-5	0.7	0.81	14.3	3	1582		39		14.6	-5.0	-16.52	8.80		7353 08-22-104 2
	283.21	65.48	0.81	0.6	40.		0.00	1461				30.82	13.6	2.6			MARK 49, HARO 8
	110.30	-7.55	11 +4 +1	2			0.01						-1.9				ZW COMPACT
1216+14	1216.7	1409	13 P					-235		12	-0.19	16.8	-2.7		8.26		7354 01-31-050 2
	271.79	74.76	1.63				0.00	-314			218	31.13	14.3				RMB 56
	100.67	-4.62	11 -1				0.08				31		-1.2				
N 4260	1216.8	623	1B	2.2	0.59	12.44	2	1846		74		35.1	-10.7	-20.29	10.31		02-31-000 2
	281.55	67.64	0.79	2.0	59.		0.21	1734				32.73	32.5	20.5			
	108.14	-6.87	11 -24	2			0.00						-4.1				
N 4261	1216.8	606	-5	3.9	0.93	11.32	2	2202		75		35.1	-11.0	-21.41	10.76		7361 01-31-054 2
	281.79	67.37	0.84	3.9	26.	7.95	0.00	2089				32.73	33.0	40.0			
	108.41	-6.95	11 -24	9		3.37	0.00						-4.3				
N 4262	1217.0	1509	-2B	2.0	0.95	12.32	2	1364	402	9	0.12	16.8	-2.9	-18.81	9.72	0.07	7360 01-31-052 2
	270.13	75.67	3.00	2.1	22.		0.00	1290		20	210	31.13	17.0	10.3	8.57		
	99.73	-4.25	11 -1	2			0.08			223	25		-1.3				
N 4267	1217.2	1305	-2B	2.8	0.94	11.72	2	1260		75		16.8	-3.5	-19.41	9.96		7365 03-31-101 2
	274.00	73.87	3.58	2.8	23.		0.00	1177				31.13	16.7	13.7			
	101.73	-4.82	11 -1	9			0.08						-1.4				
N 4270	1217.3	545	-2	2.0	0.49	13.10	1	2347		50		35.1	-11.3	-19.63	10.04		7373 02-32-004 2
	282.38	67.07	0.83	1.7	66.	9.96	0.00	2233				32.73	33.2	17.4			
	108.79	-6.93	11 -24	9		3.14	0.00						-4.2				
N 4273	1217.4	537	5B	2.5	0.63	12.15	1	2349	302	24	0.62	35.1	-11.3	-20.58	10.42	0.19	7376 01-32-007 2
	282.55	66.96	0.83	2.3	55.		0.17	2234		10	217	32.73	33.1	23.6	9.71		
	108.92	-6.94	11 -24	9			0.00			217	20		-4.3				
N 4275	1217.4	2754	12	1.0	1.00	13.4	3	2324		108		36.7	1.5	-19.42	9.96		7380 01-32-008 2
	207.09	82.98	0.09	1.0			0.00	2307				32.82	36.7	10.7			
	87.61	-0.27	13 -0	2			0.02						-0.2				
N 4274	1217.4	2953	2B	6.5	0.41	10.71	1	922		10	1.04	9.7	0.7	-19.22	9.88	0.14	7382 05-29-058 2
	191.43	82.64	1.25	5.5	72.		0.41	914			51	29.93	9.7	15.6	9.01		7377 05-29-060 2
	85.72	0.33	14 -1	9			0.08				31		0.1				
N 4278	1217.6	2934	-5	3.8	1.00	11.11	1	651	427	18	-0.01	9.7	0.7	-18.82	9.72	0.02	7386 05-29-062 2
	193.72	82.77	1.25	3.8			0.00	642		40	216	29.93	9.4	10.8	7.96		L2
	86.03	0.28	14 -1	2			0.04			216	21		0.0				

Catalog data table (astronomical galaxy catalog). Columns are not labelled in the source; values are reproduced as read, with stacked sub-values separated by " / ".

Name	Position A	Position B	Type	Dimensions	b/a, PA	mag (± errors)	Velocity		N		μ₀	X/Y/Z	M, err	phot.	single	Other names
1217+12	1217.8 / 275.41 / 102.36	1228 / 73.38 / -4.86	13 P / 2.65 / 11 -1			0.00 / 0.07	272. / 187.		30		16.8 / 31.13	-3.3 / 15.0 / -1.3				RMB 169 / 02-32-000 2
N 4281	1217.8 / 282.75 / 108.90	540 / 67.03 / -6.83	-2 / 0.81 / 11 -24	3.1 / 2.7 / 9	0.55 / 61.	12.21 / 0.00 / 0.00	2602. / 2488.		50		35.1 / 32.73	-11.5 / 33.6 / -4.2	-20.52 / 27.7	10.40		7389 01-32-012 2
N 4283	1217.8 / 193.50 / 86.03	2935 / 82.81 / 0.33	-5 / 1.20 / 14 -1	1.2 / 1.2 / 2	1.00	12.91 / 0.00 / 0.04	1133. / 1124.		23		9.7 / 29.93	0.7 / 9.9 / 0.0	-17.02 / 3.4	9.00		7390 05-29-063 2
N 4291	1218.1 / 125.55 / 41.33	7539 / 41.60 / 12.99	-5 / 0.36 / 42 -13	2.2 / 2.1 / 2	0.85 / 35.	12.12 / 0.00 / 0.08	1785. / 1968.		43		29.4 / 32.34	21.5 / 18.9 / 6.6	-20.22 / 18.0	10.28		7397 13-09-024 2
N 4288	1218.2 / 138.53 / 69.82	4635 / 69.88 / 5.52	7B / 0.75 / 14 -4	2.9 / 2.6 / 2	0.64 / 55.	13.09 / 0.17 / 0.00	534. / 602. / 1	230	15	0.96 / 11 / 1.07	8.0 / 29.52	2.7 / 7.5 / 0.8	-16.43 / 6.1	8.76 / 8.77	1.01	DDO 119 / 7399 08-23-006 2
N 4290	1218.4 / 130.69 / 58.42	5822 / 58.58 / 8.85	3B / 0.14 / 42 -0	2.5 / 2.3 / 2	0.78 / 43.	12.4 / 0.08 / 0.00	2766. / 2885.		90		42.1 / 33.12	21.8 / 35.5 / 6.5	-20.72 / 28.3	10.48		7402 10-18-029 2
N 4293	1218.7 / 262.85 / 96.50	1840 / 78.83 / -2.80	1.98 / 11 -0 +1	6.3 / 5.4 / 9	0.49 / 66.	11.11 / 0.00 / 0.08	882. / 824.		42	0.51 / 24 46	17.0 / 31.15	-1.9 / 16.9 / -0.8	-20.04 / 26.8	10.21 / 8.97	0.06	7405 03-32-006 2
U 7408	1218.8 / 138.73 / 70.31	4606 / 70.37 / 5.47	10A / 0.60 / 14 -4	2.6 / 2.4 / 2	0.63 / 56.	13.27 / 0.17 / 0.00	464. / 530. / 1	59 50. 60.	10	0.40 / 13 / 1.05	6.7 / 29.13	2.2 / 6.3 / 0.6	-15.86 / 4.7	8.54 / 8.05 / 8.53	0.33 / 0.331 / 0.99	DDO 120 / 7408 08-23-013 2
N 4294	1218.8 / 277.14 / 103.09	1147 / 72.85 / -4.83	6B / 2.65 / 11 -1	2.9 / 2.4 / 2	0.41 / 72.	12.13 / 10.72 / 1.41 / 0.42 / 0.04	415. / 327.		79	0.81 / 217	16.8 / 31.13	-3.5 / 15.2 / -1.3	-19.00 / 11.8	9.79 / 9.26	0.29	TURN 61A / 7407 02-32-009 2
N 4298	1219.0 / 272.36 / 100.13	1453 / 75.67 / -3.87	5A / 3.60 / 11 -1	2.9 / 2.6 / 2	0.60 / 58.	11.79 / 0.19 / 0.09	1118. / 1044.		19		16.8 / 31.13	-3.0 / 16.6 / -1.1	-19.34 / 12.8	9.93		TURN 62A / 7412 03-32-007 2
N 4302	1219.2 / 272.53 / 100.15	1453 / 75.69 / -3.82	5 / 3.60 / 11 -1	4.7 / 3.4 / 2	0.21 / 89.	11.79 / 0.67 / 0.09	1118. / 1044.		19		16.8 / 31.13	-3.0 / 16.6 / -1.1	-19.34 / 16.7	9.93		TURN 62B / 7418 03-32-009 2
N 4299	1219.2 / 277.44 / 103.12	1147 / 72.89 / -4.73	7B / 2.65 / 11 -1	1.8 / 1.8 / 2	1.00	12.83 / 0.00 / 0.04	238. / 150. / 6		20	0.63 / 17 / 1.06	16.8 / 31.13	-3.5 / 14.9 / -1.3	-18.30 / 8.8	9.51 / 9.08	0.37	TURN 61B / 7414 02-32-010 2
N 4303	1219.4 / 284.40 / 109.90	445 / 66.28 / -6.71	4X / 1.06 / 11 +4 +1	5.9 / 5.9 / 2	0.97 / 17.	10.20 / 0.01 / 0.00	1581. / 1464.	177	4 12 217	1.30 / 201 12	15.2 / 30.91	-5.1 / 14.2 / -1.8	-20.71 / 26.2	10.48 / 9.66	0.15	M 61 / 7420 01-32-022 2
N 4308	1219.4 / 186.89 / 85.42	3020 / 82.89 / 0.88	-5 / 1.25 / 14 -1	1.0 / 1.0 / 2	0.88 / 31.	14.1 / 0.00 / 0.04	606. / 602.		150		9.7 / 29.93	0.7 / 9.4 / 0.1	-15.83 / 2.8	8.52		7426 05-29-069 2

(1) Name	(2) α / ℓ / SGL	(3) δ / b / SGB	(4) Type / ρ_{xyz} / Group	(5) $D_{25}^{b,i}$ / $D_{25}^{b,i}$ / source	(6) d/D / i	(7) $B_T^{b,i}$ / $H_{-0.5}$ / $B-H$	(8) source / A_B^{i-o} / A_B^b	(9) V_h / V_o / Tel	(10) W_{20} / W_R / W_D	(11) e_v / e_w / source	(12) F_c / source e_f / f_h	(13) R / μ	(14) SGX / SGY / SGZ	(15) $M_B^{b,i}$ / $\Delta_{25}^{b,i}$	(16) $\log L_B$ / $\log M_H$ / $\log M_T$	(17) M_H/L_B / M_H/M_T / M_T/L_B	(18) Alternate Names UGC/ESO MCG / AGN? RCII
N 4307	1219.6 / 280.63 / 105.52	919 / 70.63 / -5.36	1 / 3.45 / 11 -1	3.3 / 2.4 / 2	0.23 / 86.	12.5	3 / 0.67 / 0.00	1305. / 1207.		90		16.8 / 31.13	-4.6 / 16.4 / -1.6	-18.63 / 11.8	9.64		7431 02-32-12A 2
N 4304	1219.6 / 295.98 / 147.68	-3315 / 28.95 / -15.29	3B / 0.19 / 23 -5	3.5 / 3.7 / 6	1.00	12.46	2 / 0.00 / 0.29	2631. / 2395. / 2	260	15	0.91 / 15 / 1.02	36.0 / 32.78	-29.4 / 18.6 / -9.5	-20.32 / 38.9	10.32 / 10.02	0.50	380-20 -5-29-034 2
N 4319	1219.6 / 125.44 / 41.40	7536 / 41.66 / 13.07	2B / 0.33 / 42 -13	3.0 / 2.9 / 2	0.77 / 44.	12.0	3 / 0.09 / 0.08	1685. / 1868.		73		28.2 / 32.25	20.6 / 18.2 / 6.4	-20.25 / 23.9	10.29		7429 13-09-025 2
N 4310	1219.9 / 193.25 / 86.27	2929 / 83.27 / 0.73	-2XP / 1.25 / 14 -1	2.5 / 2.2 / 2	0.53 / 63.	12.8	3 / 0.00 / 0.09	901. / 893.		150		9.7 / 29.93	0.6 / 9.6 / 0.1	-17.13 / 6.2	9.04		NGC 4311 / 7440 05-29-074 2
N 4314	1220.1 / 187.69 / 85.63	3010 / 83.09 / 0.98	1B / 1.25 / 14 -1	4.2 / 4.3 / 2	0.98 / 15.	11.28	3 / 0.01 / 0.06	883. / 878.		85		9.7 / 29.93	0.7 / 9.7 / 0.2	-18.65 / 12.2	9.65		7443 05-29-075 2
1220+12	1220.1 / 277.26 / 102.56	1226 / 73.57 / -4.33	13 P / 2.28 / 11 -1				/ 0.01 / 0.06	44. / -40.		30		16.8 / 31.13	-3.3 / 14.7 / -1.1				RMB 175
N 4318	1220.2 / 281.86 / 106.36	829 / 69.89 / -5.46	-5 / 1.47 / 11 -1	1.0 / 0.9 / 2	0.76 / 45.	14.0	3 / 0.00 / 0.00	-199. / -300.		220		16.8 / 31.13	-4.1 / 14.0 / -1.4	-17.13 / 4.4	9.04		02-32-000 2 / ARAK 359
N 4331	1220.3 / 125.21 / 40.54	7628 / 40.81 / 13.29	10 P / 0.22 / 42 -13	2.2 / 1.8 / 2	0.31 / 79.	13.8	3 / 0.64 / 0.12	1570. / 1756. / 6	176	15	0.44 / 30 / 1.05	26.6 / 32.13	19.7 / 16.8 / 6.1	-18.33 / 14.0	9.52 / 9.29	0.58	7446 02-32-015
N 4321	1220.4 / 271.16 / 99.07	1606 / 76.90 / -3.18	4X / 2.95 / 11 -1	6.1 / 6.0 / 2	0.87 / 37.	10.00	1 / 0.06 / 0.04	1590. / 1522.	283 / 444. / 445.	5 / 12 / 217	1.25 / 201 10	16.8 / 31.13	-2.8 / 17.4 / -1.0	-21.13 / 29.4	10.64 / 9.70 / 11.23	0.11 / 0.030 / 3.82	7449 13-09-026 2 / M 100
N 4324	1220.6 / 284.56 / 109.23	532 / 67.11 / -6.20	-2A / 0.73 / 11 -24	3.2 / 2.6 / 9	0.42 / 71.	12.6	3 / 0.00 / 0.00	1695. / 1582.		35	0.06 / 223 43	35.1 / 32.73	-11.2 / 32.1 / -3.7	-20.13 / 26.6	10.24 / 9.15	0.08	7450 03-32-015 2 / 7451 01-32-032 2
1220-13	1220.6 / 292.58 / 127.97	-1339 / 48.38 / -11.21	10 / 0.18 / 22 -0	1.3 / 1.3 / 2	0.92 / 26.		/ 0.03 / 0.08	1158. / 974. / 6	86	15	0.08 / 33 / 1.03	19.7 / 31.47	-11.9 / 15.2 / -3.8	7.5	8.67		UGCA 278 / -2-32-000
N 4339	1221.0 / 284.19 / 108.46	622 / 67.94 / -5.87	-5 / 0.29 / 11 -5 +1	2.4 / 2.4 / 9	0.94 / 23.	12.35	2 / 0.00 / 0.00	1278. / 1169.		100		25.5 / 32.03	-8.0 / 24.0 / -2.6	-19.68 / 17.9	10.06		7461 01-32-036 2 / TURN 64B
1221-34	1221.0 / 296.46 / 148.85	-3421 / 27.90 / -15.17	5A / 0.21 / 23 -5	2.8 / 2.4 / 6	0.40 / 72.	11.7	5 / 0.43 / 0.26	2724. / 2487. / 2	395	30	0.86 / 21 / 1.02	37.0 / 32.84	-30.6 / 18.5 / -9.7	-21.14 / 25.9	10.65 / 10.00	0.22	IC 3253 / 380-24 -6-27-021 2

Name	Coordinates	(ref.)	Type	Dim.	B_T	V	eV	HI vel.	HI flux	Dist.	SG X Y Z	M	L	(extra)	Other designations
N 4346	1221.0 / 136.59 / 69.30	4716 / 69.39 / 6.17	-2B / 0.96 / 12 -1	3.5 2.9 2 / 0.42 71.	12.2 / 0.00 0.00 / 3	849. 922.	90			17.0 / 31.15	5.9 15.6 1.8	-18.95 14.4	9.77		7463 08-23-016 2
N 4342	1221.1 / 283.48 / 107.53	720 / 68.87 / -5.57	-2 / 2.64 / 11 -1	1.3 1.2 9 / 0.59 58.	13.54 / 0.00 0.00 / 1	714. 609.	50			16.8 / 31.13	-4.8 15.3 -1.6	-17.59 5.9	9.23		IC 3256 / 7466 01-32-039 2
N 4340	1221.1 / 269.77 / 98.26	1700 / 77.76 / -2.76	-2B / 2.89 / 11 -1	2.8 2.8 9 / 0.96 20.	11.96 / 0.00 0.05 / 2	854. 790.	90			16.8 / 31.13	-2.4 16.3 -0.8	-19.17 13.7	9.86		7467 03-32-021 2
U 7470	1221.2 / 277.70 / 102.33	1245 / 73.97 / -3.98	9BP / 1.41 / 11 -1	1.9 1.9 2 / 0.86 34.	13.6 / 0.05 0.08 / 3	-435. -517.	30 15 300	197 287. 289.	-0.12 218	16.8 / 31.13	-3.0 14.0 -1.0	-17.53 9.3	9.20 8.33 10.35	0.13 0.010 13.92	IC 3258 / 7470 02-32-021 2
N 4348	1221.3 / 289.60 / 117.72	-310 / 58.71 / -8.42	3 / 0.20 / 11+26+24	3.3 2.5 4 / 0.23 86.	11.9 / 0.67 0.09 / 5	2202. 2055.	90			33.8 / 32.64	-15.5 29.6 -4.9	-20.74 24.7	10.49		00-32-003 2
N 4350	1221.4 / 270.15 / 98.31	1658 / 77.77 / -2.70	-2A / 2.72 / 11 -1	3.0 2.8 9 / 0.62 56.	11.85 8.42 3.43 / 0.00 0.05 / 2	1184. 1120.	60			16.8 / 31.13	-2.5 16.8 -0.8	-19.28 13.7	9.90		TURN 65B
N 4351	1221.5 / 278.31 / 102.61	1229 / 73.75 / -3.99	2A / 1.97 / 11 -1	1.9 1.8 2 / 0.76 45.	12.67 / 0.09 0.04 / 2	2297. 2214.	10 30 201	422	0.17 201 31	16.8 / 31.13	-4.1 18.3 -1.3	-18.46 8.8	9.58 8.62	0.11	7473 03-32-023 2
U 7477	1221.6 / 284.17 / 108.00	653 / 68.48 / -5.58	13 P / 0.26 / 11 -0 +1	0.9 0.9 2 / 1.00	14.1 / 0.00 0.00 / 3	764. 657.	150			21.8 / 31.69	-6.7 20.6 -2.1	-17.59 5.7	9.23		7476 02-32-024 2
N 4359	1221.6 / 174.89 / 84.16	3148 / 82.60 / 1.77	5B / 1.13 / 14 -1	3.4 2.7 2 / 0.31 79.	12.5 11.22 1.28 / 0.64 0.01 / 3	1253. 1256. 6	10	220 187. 190.	0.72 19 1.11	9.7 / 29.93	1.0 10.0 0.3	-17.43 7.6	9.16 8.69 9.89	0.33 0.063 5.32	IC 3268 / 7477 01-32-045 2
1221+00	1221.7 / 288.02 / 113.85	50 / 62.64 / -7.25	10 / 0.32 / 11-25+24	1.3 1.2 4 / 0.79 41.	/ 0.08 0.01	2059. 1928. 1	20	88	0.24 26 1.02	32.6 / 32.56	-13.1 29.5 -4.1	11.4	9.27		7483 05-29-079 2
1221+04	1221.7 / 285.93 / 110.31	430 / 66.19 / -6.23	-2 / 0.27 / 11 +2 +1	1.0 0.8 2 / 0.39 73.	14.9 / 0.00 0.00 / 3	945. 829.	95			22.6 / 31.77	-7.8 21.1 -2.5	-16.87 5.3	8.94		DDO 121 / 00-32-004 2
N 4365	1221.9 / 283.79 / 107.33	736 / 69.18 / -5.31	-5 / 2.93 / 11 -1	6.5 6.2 9 / 0.76 45.	10.61 7.36 3.25 / 0.00 0.00 / 2	1177. 1073.	28			16.8 / 31.13	-5.0 16.1 -1.6	-20.52 30.4	10.40		UGCA 279, MARK 51 / 01-32-046 2 / 7488 01-32-048 2
1222+67	1222.0 / 126.93 / 49.32	6743 / 49.48 / 11.55	13 P / 0.01 / 12 -0 +1	0.8 0.7 2 / 0.70 50.	15.38 / 0.00 0.02 / 2	770. 926.	220			16.1 / 31.03	10.3 11.9 3.2	-15.65 3.3	8.45		UGCA 280, MARK 206 / 11-15-000 2
N 4369	1222.1 / 145.75 / 76.67	3940 / 76.52 / 4.19	1A / 0.29 / 12 -6 +1	2.4 2.4 2 / 1.00	11.8 / 0.00 0.00 / 3	1052. 1091.	10 50 211	140	0.35 211 50	21.6 / 31.67	5.0 21.0 1.6	-19.87 15.1	10.14 9.02	0.08	MARK 439 / 7489 07-26-004 2

(1) Name	(2) α / ℓ / SGL	(3) δ / b / SGB	(4) Type / Q_{xyz} / Group	(5) D_{25} / $D_{25}^{b,i}$ / source	(6) d/D / i	(7) $B_T^{b,i}$ / $H_{-0.5}$ / $B-H$	(8) source / A_B^{i-o} / A_B^b	(9) V_h / V_o / Tel	(10) W_{20} / W_R / W_D	(11) e_v / e_w / source	(12) F_c / source-e_f / f_h	(13) R / μ	(14) SGX / SGY / SGZ	(15) $M_B^{b,i}$ / $\Delta_{25}^{b,i}$	(16) $\log L_B$ / $\log M_H$ / $\log M_T$	(17) M_H/L_B / M_H/M_T / M_T/L_B	(18) Alternate Names (UGC/ESO, MCG) / AGN? RCII
U 7490	1222.1 / 126.24 / 46.43	7037 / 46.62 / 12.20	9A / 0.07 / 44 -0	3.7 / 3.7 / 2	1.00	13.09	7 / 0.00 / 0.01	470. / 636. / 1	83	8	0.64 / 13 / 1.15	11.1 / 30.22	7.5 / 7.9 / 2.3	-17.13 / 12.0	9.04 / 8.73	0.49	DDO 122 / 7490 12-12-008 / 2
N 4386	1222.4 / 125.17 / 41.23	7548 / 41.49 / 13.28	-5 / 0.33 / 42-13	3.0 / 2.8 / 2	0.62 / 56.	12.3	3 / 0.00 / 0.09	1682. / 1866.		61		28.1 / 32.24	20.6 / 18.0 / 6.5	-19.94 / 23.0	10.17		7491 13-09-027 / 2
N 4371	1222.4 / 279.69 / 103.16	1159 / 73.37 / -3.93	-2B / 4.09 / 11 -1	5.0 / 4.3 / 2	0.48 / 66.	11.83 / 8.43 / 3.40	1 / 0.00 / 0.06	982. / 897.		38		16.8 / 31.13	-3.8 / 16.1 / -1.1	-19.30 / 21.1	9.91		7493 02-32-033 / 2
N 4374	1222.5 / 278.18 / 102.03	1310 / 74.48 / -3.56	-5 / 3.99 / 11 -1	6.4 / 6.4 / 9	0.93 / 26.	10.18 / 6.99 / 3.19	1 / 0.00 / 0.13	933. / 854.		17		16.8 / 31.13	-3.4 / 16.2 / -1.0	-20.95 / 31.4	10.57 / 9.26	0.05	M 84 / 7494 02-32-034 / 2
N 4378	1222.7 / 286.08 / 109.70	512 / 66.93 / -5.80	1A / 0.73 / 11-24	3.1 / 3.1 / 2	0.94 / 23.	12.2	3 / 0.02 / 0.00	2560. / 2447.	367	14 / 12 / 223	0.58 / 223 29	35.1 / 32.73	-11.9 / 33.4 / -3.6	-20.53 / 31.8	10.40 / 9.67	0.18	7497 01-32-052 / 2
N 4377	1222.7 / 275.37 / 100.25	1502 / 76.20 / -2.97	-2A / 3.38 / 11 -1	1.7 / 1.6 / 9	0.79 / 42.	12.58 / 9.45 / 3.13	2 / 0.00 / 0.07	1336. / 1265.		68		16.8 / 31.13	-3.1 / 16.9 / -0.9	-18.55 / 7.8	9.61		III ZW 65
N 4379	1222.7 / 273.77 / 99.44	1553 / 76.97 / -2.72	-5 / 2.89 / 11 -1	1.9 / 1.8 / 9	0.84 / 37.	12.46	2 / 0.00 / 0.04	1038. / 971.		90		16.8 / 31.13	-2.7 / 16.5 / -0.8	-18.67 / 8.8	9.66		7501 03-32-025 / 2
N 4380	1222.8 / 281.94 / 104.82	1017 / 71.81 / -4.33	2A / 3.60 / 11 -1	3.3 / 3.0 / 2	0.66 / 53.	12.2 / 9.76 / 2.44	3 / 0.15 / 0.00	964. / 872.	329	10 / 30 / 223	0.04 / 223 29	16.8 / 31.13	-4.2 / 16.0 / -1.3	-18.93 / 14.7	9.76 / 8.49	0.05	7502 03-32-026 / 2
N 4384	1222.8 / 131.11 / 62.08	5447 / 62.20 / 8.48	1 P / 0.13 / 42+11	1.5 / 1.4 / 2	0.81 / 40.	13.2	3 / 0.07 / 0.01	2312. / 2417.		220		36.6 / 32.82	16.9 / 32.0 / 5.4	-19.62 / 15.0	10.04		MARK 207 / 7506 09-20-168 / 2
N 4382	1222.9 / 267.75 / 96.98	1828 / 79.24 / -1.91	-2AP / 2.04 / 11 -1	7.4 / 7.2 / 9	0.82 / 39.	10.02 / 7.06 / 2.96	1 / 0.00 / 0.08	773. / 717.		30		16.8 / 31.13	-2.0 / 16.2 / -0.5	-21.11 / 35.3	10.64		M 85, TURN 66A / 7508 03-32-029 / 2
N 4383	1222.9 / 272.12 / 98.63	1645 / 77.76 / -2.42	-2 P / 2.34 / 11 -1	2.0 / 1.8 / 2	0.51 / 64.	12.5	3 / 0.00 / 0.04	1609. / 1545.		90		16.8 / 31.13	-2.7 / 17.5 / -0.7	-18.63 / 8.8	9.64		7507 03-32-030 / 2
N 4389	1223.1 / 136.74 / 70.65	4558 / 70.74 / 6.15	6BP / 0.80 / 12 -1	2.5 / 2.4 / 2	0.72 / 48.	12.4	3 / 0.12 / 0.00	718. / 786. / 6	220	40	0.25 / 32 / 1.09	17.0 / 31.15	5.2 / 14.8 / 1.7	-18.75 / 11.9	9.69 / 8.71	0.10	7514 08-23-028 / 2
N 4385	1223.2 / 288.81 / 113.94	51 / 62.74 / -6.88	1B / 0.41 / 11-25+24	1.9 / 1.7 / 2	0.62 / 56.	12.84	2 / 0.18 / 0.02	2141. / 2011.	135. / 129. / 133.	10 / 12 / 223	0.18 / 223 16	33.5 / 32.62	-13.5 / 30.4 / -4.0	-19.78 / 16.6	10.10 / 9.23 / 9.90	0.13 / 0.212 / 0.63	MARK 52 / 7515 00-32-009 / 2

Name	Coord	V / l / b	Type	Size / PA	Mag	V / W	N	HI	D / μ / X Y Z	M / err	B mags	Flux	Other names
U 7513	1223.2 284.74 107.52	730 69.18 -5.03	5 2.93 11 -1	3.3 2.3 2 ; 0.14 90.	13.5 10.67 2.83 ; 3 0.67 0.00	1001. 898. 1 ; 292 254. 256.	10	0.80 20 1.06	16.8 31.13 ; -5.0 15.8 -1.5	-17.63 11.3	9.24 9.25 10.32	1.02 0.084 12.04	IC 3322A ; 7513 01-32-054 2
U 7512	1223.2 288.01 112.41	226 64.28 -6.45	10 1.13 11 -4 +1	1.3 1.0 2 ; 0.40 72.	14.8 ; 3 0.43 0.01	1503. 1379. 1 ; 87 71. 79.	20 12 226	0.39 226 11 1.01	15.1 30.89 ; -5.7 13.9 -1.7	-16.09 4.4	8.63 8.75 8.81	1.31 0.868 1.51	7512 00-32-008
N 4387	1223.2 278.89 102.16	1305 74.46 -3.42	-5 3.62 11 -1	1.6 1.5 9 ; 0.68 51.	12.82 9.89 2.93 ; 2 0.00 0.13	511. 432.	65		16.8 31.13 ; -3.3 15.5 -0.9	-18.31 7.4	9.52		7517 02-32-039 2
1223+15	1223.3 275.60 100.12	1513 76.43 -2.78	13 P 3.21 11 -1		; 0.00 0.05	462. 392.	30		16.8 31.13 ; -2.8 15.5 -0.8				RMB 46 ; 03-32-000 2
N 4393	1223.3 207.16 88.06	2750 84.29 0.95	7X 1.25 14 -1	3.3 3.3 2 ; 0.87 33.	12.4 ; 0.05 0.07	755. 741. 1 ; 137 199. 202.	8	0.94 9 1.11	9.7 29.93 ; 0.4 9.5 0.1	-17.53 9.3	9.20 8.91 10.03	0.51 0.076 6.71	IC 3323 ; 7521 05-29-083 2
N 4388	1223.3 279.18 102.31	1256 74.33 -3.44	3 P 1.56 11 -1	5.6 4.5 2 ; 0.31 79.	11.20 8.78 2.42 ; 1 0.64 0.13	2607. 2527. 1 ; 404 373. 375.	19 15 300	0.58 201 21	16.8 31.13 ; -4.1 18.8 -1.2	-19.93 22.1	10.16 9.03 10.95	0.07 0.012 6.11	7520 02-32-041 2 S2
N 4394	1223.4 268.23 97.00	1829 79.32 -1.79	3B 2.08 11 -1	3.4 3.4 2 ; 0.95 22.	11.64 ; 1 0.02 0.08	945. 889. ; 187		0.34 201 36	16.8 31.13 ; -2.0 16.5 -0.5	-19.49 16.7	9.99 8.79	0.06	TURN 66B
N 4395	1223.4 162.05 82.35	3349 81.55 2.73	9A 0.40 14 -7	13.8 13.3 2 ; 0.83 38.	10.64 ; 1 0.06 0.00	318. 332. 1 ; 131 170. 174.	10 15 201	2.08 211 3 1.81	3.6 27.78 ; 0.5 3.6 0.2	-17.14 14.0	9.05 9.19 10.07	1.40 0.133 10.49	7523 03-32-035 2
N 4406	1223.7 279.13 102.07	1313 74.63 -3.27	-5 1.41 11 -1	9.8 9.1 9 ; 0.64 55.	9.98 6.99 2.99 ; 1 0.00 0.13	-340. -419.	5 4 211		16.8 31.13 ; -3.0 14.1 -0.8	-21.15 44.6	10.65		7524 06-27-053 2
1223+48	1223.8 134.17 67.98	4846 68.08 7.03	13 P 0.36 14 -7	1.0 0.9 2 ; 0.89 31.	14.50 ; 2 0.00 0.00	287. 367.	35	0.88 8 1.08	4.7 28.36 ; 1.7 4.3 0.6	-13.86 1.2	7.74		M 86 ; 7532 02-32-046 2
U 7534	1223.8 129.29 58.39	5835 58.51 9.58	10B 0.18 12 -0 +1	3.3 3.0 2 ; 0.61 57.	13.8 ; 3 0.19 0.00	724. 845. 1 ; 86 80. 87.	43		15.0 30.88 ; 7.7 12.6 2.5	-17.08 13.1			UGCA 281, MARK 209 ; HARO 29, I ZW 36 ; 08-23-035 2
N 4411	1224.0 283.91 105.99	909 70.82 -4.37	7X 3.45 11 -1	2.3 2.3 2 ; 0.96 19.	13.54 ; 1 0.01 0.00	1282. 1186. 1	10	1.05	16.8 31.13 ; -4.7 16.4 -1.3	-17.59 11.3	9.02 9.23 9.39	1.62 0.700 2.31	DDO 123 ; 7534 10-18-045 2
N 4413	1224.0 279.84 102.41	1253 74.35 -3.29	3B 2.67 11 -1	2.4 2.3 2 ; 0.73 47.	12.8 ; 3 0.11 0.12	94. 14.	20	-0.05 223 51	16.8 31.13 ; -3.2 14.8 -0.9	-18.33 11.3	9.23	0.08	NGC 4411A ; 7537 02-32-048 2
N 4414	1224.0 174.50 84.60	3130 83.19 2.17	5A 1.19 14 -1	4.5 4.1 2 ; 0.70 50.	10.81 7.82 2.99 ; 2 0.13 0.01	720. 723. 1 ; 418 497. 498.	10 20	1.19 11 1.16	9.7 29.93 ; 0.9 9.5 0.4	-19.12 11.6	9.84 9.16 10.92 ; 9.52 8.40	0.21 0.018 12.02	7538 02-32-049 2 ; 7539 05-29-085 2

Astronomical data table.

(1) Name	(2) α / ℓ / SGL	(3) δ / b / SGB	(4) Type / Q_{xyz} / Group	(5) $D_{25}^{b,i}$ / $D_{25}^{b,i}$ / source	(6) d/D / i	(7) $B_T^{b,i}$ / $H_{-0.5}$ / B-H	(8) source / A_B^{i-v} / A_B^b	(9) V_h / V_o / Tel	(10) W_{20} / W_R / W_D	(11) e_v / e_w / source	(12) F_c / source/e_f / f_h	(13) R / μ	(14) SGX / SGY / SGZ	(15) $M_B^{b,i}$ / Δ_{25}	(16) log L_B / log M_H / log M_T	(17) M_H/L_B / M_T/L_B / M_T/L_B	(18) Alternate Names / UGC/ESO MCG / AGN? RCII
N 4412	1224.1 / 287.53 / 110.73	414 / 66.08 / -5.73	4BP / 0.69 / 11+24	1.7 / 1.6 / 2	0.89 / 31.	13.0	3 / 0.04 / 0.00	2301. / 2185.		90		35.6 / 32.75	-12.5 / 33.1 / -3.5	-19.75 / 16.6	10.09		7536 01-32-062 2
N 4417	1224.2 / 283.41 / 105.33	951 / 71.50 / -4.12	-2B / 3.44 / 11 -1	3.5 / 3.0 / 9	0.49 / 66.	12.00	1 / 0.00 / 0.00	826. / 733.		90		16.8 / 31.13	-4.3 / 15.7 / -1.2	-19.13 / 14.7	9.84		7542 02-32-053 2
U 7547	1224.3 / 281.37 / 103.42	1151 / 73.41 / -3.52	10 / 4.16 / 11 -1	1.7 / 1.6 / 2	0.72 / 48.	14.9	3 / 0.12 / 0.06	1100. / 1016. / 1	93 / 98. / 104.	10 / 7 / 226	0.62 / 226 8 / 1.02	16.8 / 31.13	-3.9 / 16.3 / -1.0	-16.23 / 7.8	8.68 / 9.07 / 9.34	2.44 / 0.538 / 4.53	IC 3356, KAR 130 / 7547 02-32-000
U 7546	1224.3 / 284.12 / 106.01	909 / 70.84 / -4.30	7A / 3.45 / 11 -1	3.0 / 3.0 / 2	1.00	13.02	1 / 0.00 / 0.00	1265. / 1170. / 1		20	/ / 1.06	16.8 / 31.13	-4.7 / 16.4 / -1.3	-18.11 / 14.7	9.44		NGC 4411B / 7546 02-32-055 2
N 4419	1224.4 / 276.45 / 100.10	1519 / 76.63 / -2.49	1B / 1.60 / 11 -1	3.0 / 2.4 / 2	0.36 / 75.	11.41	2 / 0.52 / 0.03	-273. / -342.	242	42 / 25 / 223	-0.19 / 223 32	16.8 / 31.13	-2.5 / 14.3 / -0.6	-19.72 / 11.8	10.08 / 8.26	0.02	7551 03-32-038 2
N 4420	1224.4 / 288.52 / 112.17	246 / 64.67 / -6.07	5 / 0.75 / 11 -4 +1	2.2 / 1.9 / 2	0.49 / 66.	12.5	3 / 0.30 / 0.00	1679. / 1557.		10	0.47 / 201 27	13.9 / 30.72	-5.2 / 12.8 / -1.4	-18.22 / 7.7	9.48 / 8.76	0.19	7549 01-32-064 2
N 4421	1224.5 / 275.79 / 99.71	1544 / 77.02 / -2.35	1B / 2.92 / 11 -1	2.5 / 2.3 / 2	0.75 / 45.	12.31	2 / 0.10 / 0.04	1692. / 1625.		250		16.8 / 31.13	-3.0 / 17.6 / -0.7	-18.82 / 11.3	9.72		7554 03-32-039 2
N 4423	1224.6 / 286.64 / 108.92	609 / 67.97 / -5.08	9 / 0.29 / 11 -5 +1	2.2 / 1.6 / 2	0.19 / 90.	13.6	3 / 0.67 / 0.00	1092. / 984. / 6	187	25	0.59 / 32 / 1.05	23.0 / 31.81	-7.5 / 21.7 / -2.0	-18.21 / 10.7	9.48 / 9.31	0.68	7556 01-32-065
U 7559	1224.6 / 148.62 / 78.97	3725 / 78.74 / 4.01	10B / 0.37 / 14 -7	4.2 / 3.9 / 2	0.70 / 50.	14.09	7 / 0.13 / 0.00	222. / 252. / 1	76 / 76. / 83.	8	0.77 / 9 / 1.14	2.8 / 27.24	0.5 / 2.7 / 0.2	-13.15 / 3.2	7.45 / 7.66 / 8.73	1.63 / 0.086 / 18.86	DDO 126 / 7559 06-27-055 2
U 7563	1224.6 / 275.01 / 99.29	1611 / 77.44 / -2.19	10 / 1.79 / 11 -1	2.1 / 1.9 / 2	0.61 / 57.	13.9	3 / 0.19 / 0.04	2336. / 2271. / 1	131	10	0.22 / 23 / 1.04	16.8 / 31.13	-3.1 / 18.7 / -0.7	-17.23 / 9.3	9.08 / 8.67	0.39	IC 3365 / 7563 03-32-041 2
U 7557	1224.7 / 285.72 / 107.59	732 / 69.32 / -4.66	9 / 2.64 / 11 -1	3.1 / 3.0 / 2	0.83 / 38.	13.2	3 / 0.06 / 0.00	935. / 833. / 1	165 / 217. / 220.	10 / 6 / 226	0.78 / 226 8 / 1.09	16.8 / 31.13	-5.0 / 15.7 / -1.3	-17.93 / 14.7	9.36 / 9.23 / 10.30	0.74 / 0.085 / 8.69	7557 01-32-066
N 4424	1224.7 / 283.92 / 105.51	942 / 71.40 / -4.04	2BP / 3.06 / 11 -1	3.4 / 3.0 / 9	0.53 / 63.	12.02	1 / 0.26 / 0.00	459. / 366.		77		16.8 / 31.13	-4.2 / 15.1 / -1.1	-19.11 / 14.7	9.84		7561 02-32-058 2
N 4425	1224.7 / 280.25 / 102.33	1301 / 74.53 / -3.03	-2 / 3.44 / 11 -1	2.8 / 2.3 / 9	0.40 / 73.	12.68 / 9.66 / 3.02	1 / 0.00 / 0.12	1883. / 1804.		50		16.8 / 31.13	-3.9 / 17.7 / -1.0	-18.45 / 11.3	9.57		7562 02-32-059 2

Name			Type														Other names
N 4428	1224.9 / 292.76 / 122.57	-754 / 54.22 / -8.78	5X / 0.00 / 11-34	1.9 / 1.5 / 4	0.37 / 74.	12.4	5 / 0.48 / 0.05	3036. / 2874.		90		43.3 / 33.18	-23.0 / 36.1 / -6.6	-20.78 / 19.0	10.50		-1-32-012 2
N 4429	1224.9 / 282.38 / 103.91	1123 / 73.01 / -3.51	-2A / 4.16 / 11 -1	5.4 / 4.7 / 9	0.47 / 67.	11.05 / 7.66 / 3.39	1 / 0.00 / 0.05	1114. / 1028.		65		16.8 / 31.13	-4.0 / 16.3 / -1.0	-20.08 / 23.1	10.22		7568 02-32-061 2
1225-39	1225.0 / 297.93 / 153.81	-3904 / 23.30 / -15.07	-5 / 0.20 / 23 -0 +5	2.5 / 2.6 / 6	0.83 / 37.	11.77	2 / 0.00 / 0.33	2973. / 2730.		90		39.9 / 33.00	-34.6 / 17.0 / -10.4	-21.23 / 30.3	10.68		IC 3370 / 322-14 -6-27-029 2
N 4433	1225.1 / 292.87 / 122.68	-800 / 54.13 / -8.76	4X / 0.16 / 11-34	2.0 / 1.7 / 4	0.41 / 71.	12.3	5 / 0.41 / 0.07	2917. / 2755.		23		41.8 / 33.11	-22.3 / 34.8 / -6.4	-20.81 / 20.7	10.52		-1-32-013 2
N 4435	1225.1 / 280.15 / 102.04	1321 / 74.88 / -2.90	-2B / 3.99 / 11 -1	3.3 / 3.2 / 9	0.77 / 44.	11.62 / 8.10 / 3.52	1 / 0.00 / 0.08	869. / 792.		100		16.8 / 31.13	-3.4 / 16.1 / -0.8	-19.51 / 15.7	10.00		7575 02-32-064 2 / VV 188
U 7577	1225.2 / 137.79 / 72.88	4346 / 72.95 / 5.90	10 / 0.46 / 14 -7	4.2 / 3.9 / 2	0.70 / 50.	12.77	7 / 0.13 / 0.00	195. / 254. / 3	49 / 38. / 51.	5	0.80 / 8 / 1.14	3.0 / 27.40	0.9 / 2.9 / 0.3	-14.63 / 3.4	8.04 / 7.75 / 8.16	0.51 / 0.397 / 1.29	DDO 125 / 7577 07-26-006 2
N 4438	1225.3 / 280.42 / 102.11	1317 / 74.84 / -2.87	0AP / 2.67 / 11 -1	9.1 / 7.2 / 9	0.32 / 78.	10.77 / 7.74 / 3.03	1 / 0.00 / 0.08	259. / 182.		28		16.8 / 31.13	-3.2 / 15.1 / -0.8	-20.36 / 35.3	10.34		ARP 120, VV 188 / 7574 02-32-065 2
N 4442	1225.5 / 284.16 / 105.20	1005 / 71.82 / -3.75	-2B / 3.69 / 11 -1	4.4 / 3.6 / 9	0.39 / 73.	11.29 / 7.75 / 3.54	1 / 0.00 / 0.00	580. / 489.		100		16.8 / 31.13	-4.2 / 15.3 / -1.0	-19.84 / 17.7	10.13		7583 02-32-068 2
N 4448	1225.8 / 195.30 / 87.20	2854 / 84.67 / 1.79	2B / 1.21 / 14 -1	3.8 / 3.2 / 2	0.41 / 71.	11.50	2 / 0.41 / 0.10	693. / 686.		65		9.7 / 29.93	0.5 / 9.5 / 0.3	-18.43 / 9.1	9.56		7591 05-29-089 2
N 4449	1225.8 / 136.83 / 72.33	4422 / 72.41 / 6.17	10B / 0.46 / 14 -7	5.4 / 5.1 / 2	0.77 / 43.	9.76	1 / 0.09 / 0.00	200. / 262.		5	2.35 / 96 / 19	3.0 / 27.42	0.9 / 2.9 / 0.3	-17.66 / 4.5	9.26 / 9.30	1.12	7592 07-26-009 2
N 4444	1225.9 / 298.52 / 157.85	-4259 / 19.42 / -15.42	3 / 0.13 / 23 -0	3.2 / 3.4 / 6	1.00		0.00 / 0.37	2915. / 2666. / 2	200	30	0.75 / 27 / 1.01	38.8 / 32.94	-34.6 / 14.1 / -10.3	38.5	9.93		268-10 -7-26-007
N 4450	1226.0 / 273.94 / 98.25	1722 / 78.65 / -1.53	2A / 1.88 / 11 -1	5.0 / 4.7 / 2	0.70 / 50.	10.77 / 8.03 / 2.74	1 / 0.13 / 0.04	1958. / 1899.	290 / 331. / 333.	12 / 30 / 223	0.47 / 223 / 64	16.8 / 31.13	-2.6 / 18.1 / -0.5	-20.36 / 23.1	10.34 / 8.92 / 10.86	0.04 / 0.011 / 3.38	7594 03-32-048 2
U 7599	1226.0 / 147.15 / 78.97	3730 / 78.80 / 4.30	9 / 0.41 / 14 -7	2.0 / 1.8 / 2	0.53 / 63.	14.71	7 / 0.25 / 0.00	280. / 312. / 6	86 / 75. / 83.	10	0.42 / 17 / 1.05	3.5 / 27.72	0.7 / 3.4 / 0.3	-13.01 / 1.8	7.40 / 7.51 / 8.48	1.29 / 0.107 / 12.06	DDO 127 / 7599 06-27-057 2
N 4451	1226.1 / 285.10 / 105.77	932 / 71.33 / -3.76	-2A / 3.44 / 11 -1	1.7 / 1.5 / 2	0.68 / 51.	13.30	2 / 0.00 / 0.00	693. / 600.		220		16.8 / 31.13	-4.4 / 15.5 / -1.1	-17.83 / 7.4	9.32		ARAK 368 / 7600 02-32-079 2

(1) Name	(2) α / ℓ / SGL	(3) δ / b / SGB	(4) Type / Q_{xyz} / Group	(5) $D_{25}^{b,i}$ / $D_{25}^{b,i}$ / source	(6) d/D / i	(7) $B_T^{b,i}$ / $H_{-0.5}$ / B-H	(8) source / $A_B^{i,o}$ / A_B^b	(9) V_h / V_o / Tel	(10) W_{20} / W_R / W_D	(11) e_v / e_w / source	(12) F_c / source/e_F / f_h	(13) R / μ	(14) SGX / SGY / SGZ	(15) $M_B^{b,i}$ / Δ_{25}	(16) $\log L_B$ / $\log M_H$ / $\log M_T$	(17) M_H/L_B / M_H/M_T / M_T/L_B	(18) Alternate Names / Alternate Names / UGC/ESO MCG AGN? RCII
N 4452	1226.2 / 282.72 / 103.38	1202 / 73.73 / -3.02	-2 / 3.00 / 11 -1	2.9 / 2.1 / 2	0.24 / 86.	13.10	2 / 0.00 / 0.05	212. / 130.		90		16.8 / 31.13	-3.5 / 14.9 / -0.8	-18.03 / 10.3	9.40		7601 02-32-080 2
U 7605	1226.2 / 150.96 / 80.42	3600 / 80.13 / 3.91	10 / 0.41 / 14 -7	1.7 / 1.6 / 2	0.72 / 48.	14.8	3 / 0.12 / 0.00	304. / 329. / 6	45 / 30. / 46.	10 / 8 / 226	0.17 / 226 13 / 1.04	3.7 / 27.83	0.6 / 3.6 / 0.3	-13.03 / 1.7	7.40 / 7.31 / 7.65	0.80 / 0.448 / 1.78	7605 06-27-058
N 4455	1226.2 / 251.59 / 92.78	2306 / 83.29 / 0.18	7 / 0.17 / 14 +1	2.6 / 2.2 / 2	0.41 / 71.	12.5 / 11.57 / 0.93	3 / 0.41 / 0.08	644. / 611. / 1	157 / 133. / 137.	10	0.89 / 11 / 1.05	6.2 / 28.95	-0.3 / 6.2 / 0.0	-16.45 / 4.0	8.77 / 8.47 / 9.31	0.50 / 0.146 / 3.45	7603 04-30-001 2
N 4454	1226.3 / 291.45 / 116.59	-140 / 60.42 / -6.82	1B / 0.25 / 11+26+24	2.5 / 2.4 / 2	0.89 / 31.	12.90	2 / 0.04 / 0.06	2358. / 2220.		90		35.7 / 32.77	-15.9 / 31.7 / -4.2	-19.87 / 25.0	10.14		7606 00-32-014 2
U 7608	1226.3 / 137.42 / 73.19	4330 / 73.26 / 6.02	10 / 0.74 / 14 -4	3.7 / 3.6 / 2	0.89 / 31.	13.68	7 / 0.04 / 0.00	543. / 601. / 1	74 / 111. / 116.	8	0.83 / 8 / 1.14	7.6 / 29.40	2.2 / 7.2 / 0.8	-15.72 / 8.0	8.48 / 8.59 / 9.46	1.29 / 0.137 / 9.45	DDO 129
N 4457	1226.4 / 289.12 / 111.26	351 / 65.84 / -5.29	1.28 / 11 +2 +1	3.2 / 3.1 / 2	0.84 / 37.	11.65	2 / 0.00 / 0.00	868. / 752.	177 / 241. / 243.	21 / 35 / 223	0.24 / 223 29	17.4 / 31.20	-6.3 / 16.1 / -1.6	-19.55 / 15.8	10.01 / 8.72 / 10.42	0.05 / 0.020 / 2.58	7609 01-32-075 2
N 4458	1226.4 / 281.07 / 101.97	1331 / 75.15 / -2.55	-5 / 3.21 / 11 -1	1.7 / 1.7 / 9	0.94 / 23.	12.79 / 10.04 / 2.75	2 / 0.00 / 0.11	383. / 308.		250		16.8 / 31.13	-3.2 / 15.3 / -0.7	-18.34 / 8.3	9.53		TURN 68A / 7610 02-32-082 2
N 4460	1226.4 / 135.76 / 71.60	4509 / 71.69 / 6.48	-2B / 0.75 / 14 -4	4.6 / 3.5 / 2	0.30 / 80.	12.3	3 / 0.00 / 0.00	558. / 624.		90		8.1 / 29.53	2.5 / 7.6 / 0.9	-17.23 / 8.3	9.08		7611 08-23-041 2
U 7612	1226.5 / 289.60 / 112.09	300 / 65.01 / -5.50	9B / 0.91 / 11 -4 +1	2.2 / 1.9 / 2	0.54 / 62.	14.0	3 / 0.24 / 0.02	1571. / 1452. / 1	193 / 179. / 182.	15	0.66 / 18 / 1.04	14.6 / 30.82	-5.5 / 13.4 / -1.4	-16.82 / 8.1	8.92 / 8.99 / 9.88	1.17 / 0.129 / 9.03	DDO 128 / 7612 01-32-077 2
N 4459	1226.5 / 280.15 / 101.27	1415 / 75.84 / -2.32	-2A / 4.06 / 11 -1	4.2 / 4.0 / 9	0.77 / 43.	11.27 / 7.93 / 3.34	2 / 0.00 / 0.08	1111. / 1039.		75		16.8 / 31.13	-3.3 / 16.5 / -0.7	-19.86 / 19.6	10.14		7614 02-32-083 2
N 4461	1226.5 / 281.23 / 102.02	1328 / 75.11 / -2.54	-2B / 3.44 / 11 -1	3.3 / 2.8 / 9	0.43 / 70.	11.92 / 8.67 / 3.25	2 / 0.00 / 0.11	1887. / 1811.		40		16.8 / 31.13	-3.8 / 17.7 / -0.8	-19.21 / 13.7	9.88		TURN 68B / 7613 02-32-084 2
N 4462	1226.7 / 296.35 / 137.51	-2254 / 39.41 / -11.85	2B / 0.12 / 22 -0 +1	3.5 / 3.0 / 6	0.39 / 73.	11.7	5 / 0.46 / 0.28	1866. / 1658.		90		27.7 / 32.21	-20.0 / 18.3 / -5.7	-20.51 / 24.3	10.40		506-13 -4-30-002 2
N 4464	1226.8 / 286.50 / 106.88	826 / 70.32 / -3.91	-5 / 3.31 / 11 -1	1.1 / 1.0 / 2	0.72 / 48.	13.60 / 9.92 / 3.68	2 / 0.00 / 0.00	1199. / 1102.		50		16.8 / 31.13	-4.9 / 16.2 / -1.2	-17.53 / 4.9	9.20		7619 01-32-078 2

Name	Coordinates		Type	Mag / axes	Velocity	V1 / V2	—	extra	—	Dist					Alt. designations
N 4469	1226.9 286.12 106.32	901 70.89 -3.72	2.84 11 -1	3.1 2.5 2 ; 0.38 74.	12.4 ; 3 0.00 0.00	498. 404.	90		16.8 31.13	-4.4 15.1 -1.0	-18.73 12.3	9.68				7622 02-32-089 2
N 4467	1227.0 286.76 107.05	816 70.17 -3.91	-5 3.40 11 -1	0.7 0.6 2 ; 0.79 42.	15.40 ; 2 0.00 0.00	1474. 1377.	300		16.8 31.13	-5.1 16.6 -1.2	-15.73 2.9	8.48				01-32-080 2
N 4472	1227.2 286.89 107.05	817 70.20 -3.86	-5 3.31 11 -1	11.4 11.0 9 ; 0.85 35.	9.31 6.18 3.13 ; 1 0.00 0.00	944. 847.	23	0.78 24 45	16.8 31.13	-4.8 15.8 -1.1	-21.82 54.0	10.92 9.23	0.02		M 49, ARP 134 / 7629 02-32-083 2
N 4473	1227.3 281.64 101.85	1342 75.39 -2.29	-5 2.17 11 -1	4.8 4.4 9 ; 0.63 56.	10.94 7.57 3.37 ; 2 0.00 0.09	2279. 2205.	15		16.8 31.13	-3.9 18.4 -0.8	-20.19 21.6	10.27			7631 02-32-093 2
N 4474	1227.4 280.84 101.24	1421 76.01 -2.08	-2 3.80 11 -1	2.4 2.2 9 ; 0.70 50.	12.52 9.52 3.00 ; 2 0.00 0.08	1526. 1455.	50		16.8 31.13	-3.4 17.2 -0.6	-18.61 10.8	9.64			7634 02-32-094 2
N 4476	1227.5 283.13 102.89	1238 74.40 -2.55	-2A 2.82 11 -1	1.8 1.7 9 ; 0.74 46.	13.07 10.16 2.91 ; 2 0.00 0.08	1978. 1899.	90		16.8 31.13	-4.1 17.8 -0.8	-18.06 8.3	9.42			7637 02-32-096 2
N 4477	1227.5 281.54 101.66	1355 75.61 -2.18	-2B 4.06 11 -1	3.7 3.7 9 ; 0.92 26.	11.26 ; 2 0.00 0.09	1263. 1190.	75		16.8 31.13	-3.5 16.7 -0.7	-19.87 18.2	10.14			7638 02-32-097 2
U 7639	1227.5 133.19 69.08	4748 69.17 7.37	10 0.35 14 -4	3.4 3.0 2 ; 0.51 64.	13.5 ; 3 0.28 0.00	384. 462. 1	10 / 50	-0.15 38 1.08	5.8 28.82	2.1 5.4 0.7	-15.32 5.1	8.32 7.38	0.11		7639 08-23-043 2
N 4478	1227.8 283.43 102.94	1236 74.39 -2.49	-5 3.92 11 -1	1.9 1.9 9 ; 0.85 35.	12.05 8.90 3.15 ; 2 0.00 0.08	1482. 1403.	75		16.8 31.13	-3.9 17.0 -0.8	-19.08 9.3	9.82			7645 02-32-099 2
N 4479	1227.8 281.90 101.75	1351 75.58 -2.13	-2B 4.00 11 -1	1.7 1.7 9 ; 0.94 24.	13.36 10.55 2.81 ; 2 0.00 0.09	822. 749.	100		16.8 31.13	-3.3 16.0 -0.6	-17.77 8.3	9.30			7646 02-32-100 2
1228+124	1228.0 283.41 102.80	1246 74.56 -2.39	-5 P 3.92 11 -1	0.7 0.7 ; 1.00 .	14.25 ; 2 0.00 0.07	1486. 1408.	43		16.8 31.13	-3.9 17.0 -0.7	-16.88 3.4	8.94			NGC 4486B, I ZW 38 / 02-32-101 2
N 4483	1228.1 286.78 106.13	918 71.24 -3.35	-2B 3.83 11 -1	1.7 1.5 2 ; 0.62 56.	13.2 ; 3 0.00 0.00	995. 903.	90		16.8 31.13	-4.6 15.9 -1.0	-17.93 7.4	9.36			7649 02-32-103 2
N 4485	1228.2 137.92 74.77	4158 74.83 5.94	10B 0.48 14 -4	2.2 2.1 9 ; 0.68 51.	12.19 ; 1 0.14 0.00	671. 724.	69		9.3 29.85	2.4 8.9 1.0	-17.65 5.7	9.25			ARP 269, VV 30 / TURN 69A / 7648 07-26-013 2
N 4486	1228.3 283.79 102.92	1240 74.49 -2.35	-5 P 4.17 11 -1	9.7 9.2 9 ; 0.74 47.	9.49 6.56 2.93 ; 1 0.00 0.07	1259. 1181.	16		16.8 31.13	-3.8 16.7 -0.7	-21.64 45.1	10.85			M 87, ARP 152 / 7654 02-32-105 L2

(1) Name	(2) α ℓ SGL	(3) δ b SGB	(4) Type ρ_{xyz} Group	(5) D_{25} $D_{25}^{b,i}$ source	(6) d/D i	(7) $B_T^{b,i}$ $H_{-0.5}$ $B-H$	(8) source A_B^{i-o} A_B^b	(9) V_h V_o Tel	(10) W_{20} W_R W_D	(11) e_v e_w source	(12) F_c source/e_f f_h	(13) R μ	(14) SGX SGY SGZ	(15) $M_B^{b,i}$ Δ_{25}	(16) $\log L_B$ $\log M_H$ $\log M_T$	(17) M_H/L_B M_H/M_T M_T/L_B	(18) Alternate Names Alternate Names UGC/ESO MCG · AGN? RCII
1228+121	1228.3 284.17 103.25	1219 74.16 -2.45	13 P 4.17 11 -1				0.00 0.07	1250. 1171.		30		16.8 31.13	-3.9 16.6 -0.7				RMB 132 02-32-000 2
N 4490	1228.3 137.91 74.83	4155 74.88 5.94	7BP 0.77 14 -4	5.6 4.8 9	0.50 65.	9.91	1 0.29 0.00	577. 629. 1	246	10 7 51	2.01 25 4 1.24	7.8 29.46	2.0 7.5 0.8	-19.55 10.9	10.01 9.79	0.61	ARP 269, VV 30 TURN 69B 7651 07-26-014 2
N 4487	1228.4 294.20 122.70	-748 54.45 -7.92	6X 0.33 11-14+10	3.9 3.6 4	0.72 49.	11.3	5 0.12 0.06	1037. 878. 1	225 251. 253.	30 13 79	0.91 217 8 1.13	19.9 31.49	-10.7 16.5 -2.8	-20.19 20.9	10.27 9.51 10.58	0.17 0.084 2.07	-1-32-021 2
N 4492	1228.5 287.77 107.07	821 70.34 -3.53	1A 3.24 11 -1	2.1 2.1	1.00	13.20	2 0.00 0.00	1735. 1639.		200		16.8 31.13	-5.2 17.0 -1.1	-17.93 10.3	9.36		7656 01-32-089 2
N 4494	1228.9 228.61 90.13	2603 85.31 1.62	-5 1.04 14 -1	5.0 5.0 2	0.96 20.	10.70	2 0.00 0.05	1307. 1289.		15		9.7 29.93	0.0 9.7 0.2	-19.23 14.2	9.88		7662 04-30-002 2
U 7666	1229.0 281.24 100.60	1508 76.88 -1.48	10 1.46 11 -1	1.4 1.0 2	0.14 90.	14.1	3 0.67 0.06	2556. 2489.		19		16.8 31.13	-3.5 19.0 -0.5	-17.03 4.9	9.00		IC 3453
N 4496	1229.1 290.57 111.11	412 66.32 -4.54	9B 0.36 11 -4 +1	3.7 3.5 2	0.77 43.	11.56	1 0.09 0.00	1738. 1625. 1	179 211. 214.	10	1.08 201 5 1.13	13.1 30.59	-4.7 12.2 -1.0	-19.03 13.4	9.80 9.31 10.24	0.32 0.119 2.71	VV 76 7668 01-32-090 2
1229+04	1229.1 290.57 111.11	412 66.32 -4.54	10 0.60 11 -4 +1	1.1 1.0 2	0.80 41.		0.08 0.00	1772. 1659.		50		13.4 30.64	-4.8 12.5 -1.1	3.9			NGC 4496B, VV 76 01-32-000 2
N 4498	1229.1 277.86 98.69	1708 78.75 -0.89	7B 2.34 11 -1	3.3 2.9 2	0.52 64.	12.2 10.73 1.47	3 0.27 0.03	1506. 1448. 1	190 173. 177.	20	0.46 17 1.08	16.8 31.13	-2.7 17.3 -0.3	-18.93 14.2	9.76 8.91 10.09	0.14 0.066 2.13	7669 03-32-056 2
1229-51	1229.1 299.85 166.67	-5128 11.01 -15.80	5 0.07 23 -0	2.4 1.9 6	0.07 90.		0.67 0.56	2632. 2376. 7		20		34.5 32.69	-32.3 7.7 -9.4	19.1			218-08
1229-02	1229.2 293.22 117.79	-242 59.52 -6.39	13 P 0.20 11+26+24	0.6 0.5 4	0.67 53.		0.00 0.06	2240. 2100.		220		34.4 32.68	-15.9 30.2 -3.8	5.0			ARAK 373 00-32-018
N 4501	1229.4 282.27 101.04	1442 76.51 -1.51	3A 2.04 11 -1	6.3 5.5 9	0.50 65.	9.90 7.10 2.80	1 0.28 0.08	2285. 2217. 1	534 548. 549.	15 11 201	1.04 201 9 1.28	16.8 31.13	-3.6 18.5 -0.5	-21.23 27.0	10.68 9.49 11.37	0.06 0.013 4.86	M 88 7675 03-32-059 2
U 7673	1229.5 180.54 86.38	3000 84.99 2.87	10 1.21 14 -1	2.0 1.9 2	0.82 39.		3 0.07 0.04	639. 639. 1	73 89. 95.	10	0.34 18 1.04	9.7 29.93	0.6 9.4 0.4	5.4	8.31 9.09	0.167	DDO 131 7673 05-30-014 2

N 4503	1229.6	1127	-2B	3.5	0.47	12.01	2	1417.		90		16.8	-4.2	-19.12	9.84		
	286.13	73.41	3.92	3.0	67.		0.00	1335.				31.13	16.8	14.7			
	104.17	-2.39	11 -1	9			0.04						-0.7				7680 02-32-118 2
N 4504	1229.7	-717	6A	3.3	0.67	11.5	5	1003.	250	8	1.35	19.5	-10.4	-19.95	10.17	0.58	
	294.64	55.00	0.34	3.1	52.		0.15	847.	269.		217 7	31.45	16.4	17.6	9.93	0.230	
	122.28	-7.47	11-14+10	4			0.04	1	271.		1.09		-2.5		10.57	2.51	-1-32-022 2
U 7685	1229.9	38	8B	4.3	0.67	12.47	1	1532.	177	10	1.05	14.5	-6.0	-18.34	9.53	0.70	NGC 4517A, REINMUTH 80
	292.49	62.84	0.65	4.0	52.	11.60	0.15	1405.	181.		10	30.81	13.1	16.9	9.37	0.147	
	114.61	-5.33	11 +4 +1	2		0.87	0.03	1	184.		1.13		-1.3		10.21	4.77	7685 00-32-019 2
1230+09	1230.0	927	-5	0.6	0.76	15.0	3	1290.		22		16.8	-4.7	-16.13	8.64		UGCA 284, ARAK 375
	288.09	71.50	3.91	0.5	45.		0.00	1200.				31.13	16.4	2.5			
	106.12	-2.86	11 -1	2			0.00						-0.9				02-32-124 2
U 7690	1230.0	4259	10 P	2.3	0.80	13.0	3	540.	103	10	0.68	7.5	2.1	-16.37	8.74	0.49	
	135.65	73.95	0.74	2.2	41.		0.08	598.	125.		12	29.37	7.2	4.8	8.43	0.123	
	73.88	6.53	14 -4	2			0.00	1	129.		1.05		0.9		9.34	3.97	7690 07-26-021
N 4517	1230.2	23	6A	9.5	0.17	10.38	1	1131.	317	7	1.49	9.8	-4.1	-19.58	10.02	0.28	
	292.74	62.61	0.12	6.7	90.		0.67	1004.	279.		217 7	29.96	8.9	19.2	9.47	0.069	
	114.87	-5.32	14-18	2			0.03	4	281.		1.10		-0.9		10.64	4.10	7694 00-32-020 2
U 7698	1230.4	3149	10	6.4	0.74	13.03	7	335.	83	7	1.09	3.8	0.3	-14.85	8.13	1.31	DDO 133
	164.34	84.01	0.40	6.0	46.		0.10	343.	89.		6	27.88	3.7	6.7	8.25	0.116	
	84.69	3.57	14 -7	2			0.02	1	95.		1.31		0.2		9.18	11.28	7698 05-30-016 2
U 7699	1230.4	3754	5B	3.7	0.31	12.2	3	503.	209	10	0.81	6.2	1.2	-16.76	8.90	0.32	
	142.22	78.80	0.38	2.9	79.	11.48	0.64	539.	176.		15	28.96	6.1	5.3	8.39	0.053	
	78.82	5.24	14 -4	2		0.72	0.00	1	180.		1.08		0.6		9.67	5.99	7699 06-28-008 2
1230+11	1230.6	1137	3XP	0.7	0.75	15.65	2	108.		40		16.8	-3.7	-15.48	8.38		IC 3483, VV 43
	286.81	73.64	2.23	0.7	46.		0.10	27.				31.13	14.7	3.4			
	104.08	-2.11	11 -1	4			0.05						-0.6				02-32-129 2
1230+13	1230.7	1308	13 P			14.8	3	374.		220		16.8	-3.4	-16.33	8.72		IC 3492, ARAK 376
	285.41	75.11	2.97				0.10	300.				31.13	15.2				
	102.63	-1.66	11 -1				0.07						-0.5				02-32-131
1230-04	1230.9	-436	8	2.5	0.19			1297.	148	20	0.53	24.0	-11.8				UGCA 286
	294.54	57.70	0.47	1.8	90.		0.67	1151.			21	31.90	20.7	12.6	9.29		
	119.75	-6.49	11+16+10	2			0.08	1			1.04		-2.7				-1-32-023
N 4519	1231.0	856	7B	3.6	0.69	12.22	1	1228.	227	10	1.08	16.8	-4.9	-18.91	9.76	0.60	
	289.19	71.05	3.40	3.3	51.	10.73	0.13	1136.	245.	15	201 6	31.13	16.3	16.2	9.53	0.120	
	106.69	-2.77	11 -1	2		1.49	0.00		248.	201			-0.8		10.45	4.94	7709 02-32-135 2
N 4522	1231.1	926	4	4.0	0.30	12.3	3	2331.	316	30	0.25	16.8	-5.3	-18.83	9.72	0.09	
	288.93	71.54	1.92	3.0	80.	10.53	0.67	2241.			201 24	31.13	18.1	14.7	8.70		
	106.21	-2.61	11 -1	2		1.77	0.00	6			1.14		-0.9				7711 02-32-137 2
N 4523	1231.3	1526	9X	2.5	0.89	14.33	7	262.	129	10	0.64	16.8	-2.8	-16.80	8.91	1.51	DDO 135
	283.16	77.35	2.82	2.4	31.		0.04	198.	201.		23	31.13	15.3	11.8	9.09	0.089	
	100.47	-0.87	11 -1	2			0.05	1	204.		1.06		-0.2		10.14	16.89	7713 03-32-068 2

(1) Name	(2) α / ℓ / SGL	(3) δ / b / SGB	(4) Type / ϱ_{xyz} / Group	(5) D_{25} / $D_{25}^{b,i}$ / source	(6) d/D / i	(7) $B_T^{b,i}$ / $H_{-0.5}$ / B-H	(8) source / A_B^{i-o} / A_B^b	(9) V_h / V_o / Tel	(10) W_{20} / W_R / W_D	(11) e_v / e_w / source	(12) F_c / source/e_f / f_h	(13) R / μ	(14) SGX / SGY / SGZ	(15) $M_B^{b,i}$ / Δ_{25}	(16) log L_B / log M_H / log M_T	(17) M_H/L_B / M_H/M_T / M_T/L_B	(18) Alternate Names UGC/ESO / MCG / AGN? RCII
N 4525	1231.3 / 172.65 / 85.94	3034 / 85.00 / 3.41	5 / 1.17 / 14 -1	2.9 / 2.6 / 2	0.57 / 60.	12.4	3 / 0.22 / 0.03	1131. / 1134.		150		9.7 / 29.93	0.7 / 9.9 / 0.6	-17.53 / 7.4	9.20		7714 05-30-020 2
N 4526	1231.5 / 290.15 / 107.64	759 / 70.15 / -2.92	-2X / 2.45 / 11 -1	7.4 / 5.9 / 9	0.37 / 74.	10.58 / 7.08 / 3.50	2 / 0.00 / 0.00	450. / 354.		26	0.86 / 24 22	16.8 / 31.13	-4.8 / 14.9 / -0.8	-20.55 / 28.9	10.41 / 9.31	0.08	7718 01-32-130 2
U 7719	1231.5 / 138.92 / 77.54	3917 / 77.57 / 5.82	8 / 0.67 / 14 -4	1.9 / 1.6 / 2	0.36 / 75.	14.5	3 / 0.52 / 0.00	686. / 728. / 1	89 / 72. / 80.	10	0.35 / 20 / 1.02	8.9 / 29.75	1.9 / 8.6 / 0.9	-15.25 / 4.2	8.29 / 8.25 / 8.80	0.91 / 0.284 / 3.19	7719 07-26-025
N 4527	1231.6 / 292.60 / 112.50	256 / 65.18 / -4.29	4X / 0.60 / 11 -4 +1	6.3 / 5.3 / 9	0.46 / 68.	10.93	1 / 0.34 / 0.02	1730. / 1614.		10	1.14 / 67 17	13.5 / 30.65	-5.2 / 12.4 / -1.0	-19.72 / 20.9	10.08 / 9.40	0.21	7721 01-32-101 2
N 4534	1231.7 / 145.57 / 80.92	3548 / 80.83 / 4.92	8 / 0.80 / 14 -6	4.2 / 3.9 / 2	0.76 / 45.	12.2	3 / 0.09 / 0.00	803. / 830. / 1	143 / 162. / 166.	8	1.20 / 7 / 1.15	9.8 / 29.95	1.5 / 9.6 / 0.8	-17.75 / 11.2	9.29 / 9.18 / 9.93	0.78 / 0.179 / 4.34	7723 06-28-010 2
N 4532	1231.8 / 291.04 / 108.84	645 / 68.95 / -3.19	10B / 1.89 / 11 -3 +1	3.3 / 2.7 / 2	0.38 / 73.	11.86 / 10.39 / 1.47	1 / 0.46 / 0.00	2010. / 1909. / 1	251 / 223. / 225.	20	1.06 / 201 5 / 1.05	15.5 / 30.95	-5.0 / 14.6 / -0.9	-19.09 / 12.2	9.83 / 9.44 / 10.25	0.41 / 0.157 / 2.62	7726 01-32-103 2
N 4535	1231.8 / 290.08 / 107.19	828 / 70.63 / -2.71	5X / 2.33 / 11 -1	6.7 / 6.2 / 9	0.70 / 50.	10.53 / 8.44 / 2.09	1 / 0.13 / 0.00	1966. / 1873. / 1	291 / 332. / 334.	8	1.40 / 201 5 / 1.43	16.8 / 31.13	-5.4 / 17.4 / -0.9	-20.60 / 30.4	10.43 / 9.85 / 10.99	0.26 / 0.073 / 3.60	7727 01-32-104 2
N 4536	1231.9 / 292.95 / 112.98	228 / 64.73 / -4.35	4X / 0.43 / 11 -4 +1	6.3 / 5.3 / 2	0.44 / 69.	10.60 / 8.26 / 2.34	1 / 0.37 / 0.02	1866. / 1748. / 1	337 / 320. / 321.	7 / 10 / 211	1.34 / 211 15	13.3 / 30.62	-5.2 / 12.2 / -1.0	-20.02 / 20.6	10.20 / 9.59 / 10.79	0.24 / 0.063 / 3.86	7732 00-32-023 2
U 7739	1232.2 / 291.40 / 109.05	634 / 68.79 / -3.15	10 / 1.89 / 11 -3 +1	1.5 / 1.5 / 2	0.94 / 24.	14.71	1 / 0.02 / 0.00	2032. / 1931. / 1		20	1.03	15.2 / 30.91	-5.0 / 14.4 / -0.8	-16.20 / 6.7	8.67		A 1232, HOLMBERG VII / DDO 13 / 7739 01-32-107 2
N 4540	1232.3 / 283.64 / 100.15	1550 / 77.80 / -0.52	6X / 2.97 / 11 -1	2.1 / 2.1 / 2	0.83 / 38.	12.2	3 / 0.06 / 0.05	1286. / 1224.		10	0.26 / 201 27 / 1.01	16.8 / 31.13	-3.0 / 16.9 / -0.2	-18.93 / 10.3	9.76 / 8.71	0.09	7742 03-32-074 2
U 7737	1232.3 / 284.11 / 100.47	1530 / 77.48 / -0.62	10 / 3.14 / 11 -1	1.3 / 1.2 / 2	0.54 / 62.	14.43	2 / 0.24 / 0.05	661. / 598. / 1	160	20	0.40 / 25 / 1.01	16.8 / 31.13	-2.9 / 15.8 / -0.2	5.9	8.85		IC 3522, DDO 136 / 7737 03-32-072 2
N 4545	1232.3 / 126.10 / 53.45	6348 / 53.50 / 11.75	6B / 0.16 / 42 -0 +12	2.8 / 2.5 / 2	0.59 / 59.	12.5 / 10.79 / 1.71	3 / 0.21 / 0.00	2716. / 2860. / 1	295 / 300. / 302.	10	0.71 / 23 / 1.06	41.5 / 33.09	24.2 / 32.6 / 8.5	-20.59 / 30.3	10.43 / 9.95 / 10.90	0.33 / 0.112 / 2.95	7747 11-15-064 2
N 4546	1232.9 / 295.22 / 118.83	-331 / 58.84 / -5.72	-2B / 0.27 / 11 +16 +10	3.0 / 2.7 / 2	0.55 / 61.	11.24	2 / 0.00 / 0.06	1022. / 882.		35		20.5 / 31.56	-9.8 / 17.9 / -2.0	-20.32 / 16.2	10.32		UGCA 288 / -1-32-027 2

Name	Pos (α/l/SGL)	Vel	Type	Size	Axis	B_T	Notes	Flux			HI	Dist	X Y Z	M	mag	corr	Identifications
N 4548	1232.9 285.68 101.22	1446 76.82 -0.68	3B 2.82 11 -1	5.0 4.9 2	0.84 37.	10.85	1 0.06 0.07	472. 406.	250. 352. 354.	10 8 300	0.60 201 22	16.8 31.13	-3.1 15.5 -0.2	-20.28 24.0	10.30 9.05 10.94	0.06 0.013 4.29	7753 03-32-075 2
1233-07	1233.0 296.12 122.82	-737 54.77 -6.77	9 0.34 11-14+10	4.0 3.5 2	0.48 67.		0.31 0.05	993. 838. 1	182 161. 165.	15	0.76 17 1.11	19.4 31.44	-10.4 16.3 -2.3	-1.60 19.8	0.64 9.34 10.17	0.146	UGCA 289, -1-32-028
N 4550	1233.0 288.11 103.40	1230 74.64 -1.30	-2 2.97 11 -1	3.4 2.6 9	0.29 81.	12.31 9.14 3.17	2 0.00 0.09	350. 274.		50		16.8 31.13	-3.6 15.1 -0.4	-18.82 12.8	9.72		TURN 71A, 7757 02-32-147 2
N 4551	1233.1 288.17 103.38	1232 74.67 -1.27	-5 4.09 11 -1	1.8 1.7 9	0.80 41.	12.76	2 0.00 0.09,	978. 903.		300		16.8 31.13	-3.8 16.2 -0.4	-18.37 8.3	9.54		TURN 71B, 7759 02-32-148 2
N 4552	1233.1 287.90 103.09	1250 74.97 -1.18	-5 2.97 11 -1	3.7 3.7 2	1.00	10.73 7.41 3.32	2 0.00 0.08	239. 165.		42		16.8 31.13	-3.5 15.0 -0.3	-20.40 18.2	10.35		7760 02-32-149 2, M 89
N 4562	1233.1 231.75 90.32	2608 86.23 2.55	6 1.00 14 -1	2.4 2.0 2	0.37 74.	13.4	3 0.49 0.04	1392. 1377.		100		9.7 29.93	0.0 10.2 0.5	-16.53 5.7	8.80		7758 04-30-004 2
N 4559	1233.5 198.44 88.32	2814 86.47 3.22	6X 1.01 14 -1	11.3 9.6 2	0.44 69.	9.86 8.30 1.56	1 0.37 0.07	816. 810. 4	251 228. 231.	10 8 79	1.85 217 4 1.10	9.7 29.93	0.2 9.6 0.6	-20.07 27.2	10.22 9.82 10.61	0.40 0.162 2.47	7766 05-30-030 2
N 4561	1233.6 277.86 96.63	1936 81.44 0.84	8B 0.16 11 -0 +1	1.5 1.6 9	0.93 25.	12.7	3 0.03 0.13	1474. 1430.		90		12.3 30.46	-1.4 12.3 0.2	-17.76 5.7	9.30		IC 3569, 7768 03-32-076 2
N 4565	1233.8 230.88 90.25	2615 86.42 2.74	3A 1.00 14 -1	16.2 11.5 9	0.14 90.	9.59 6.63 2.96	1 0.67 0.04	1228. 1213. 1	524 486. 487.	15	1.76	9.7 29.93	-0.1 10.0 0.5	-20.34 32.6	10.33 11.35	10.48	7772 04-30-006 2
N 4564	1233.9 289.54 104.21	1143 73.92 -1.31	-5 4.09 11 -1	3.0 2.7 9	0.54 62.	11.83	1 0.00 0.07	1020. 942.		34		16.8 31.13	-4.1 16.2 -0.4	-19.30 13.2	9.91		7773 02-32-150 2
U 7774	1233.9 135.65 76.69	4017 76.74 6.53	7 0.59 14 -4	3.5 2.5 2	0.16 90.	13.6 12.46 1.14	3 0.67 0.00	526. 574. 6	203 167. 171.	10	0.78 13 1.11	6.8 29.17	1.6 6.6 0.8	-15.57 5.0	8.42 8.45 9.60	1.05 0.068 15.26	7774 07-26-031
N 4567	1234.0 289.77 104.40	1132 73.75 -1.33	4A 2.67 11 -1	2.9 2.8 9	0.81 40.	11.93	1 0.07 0.08	2218. 2139.		18		16.8 31.13	-4.7 18.1 -0.4	-19.20 13.7	9.87 9.18	0.20	VV 219, TURN 72B, 7777 02-32-151 2
N 4568	1234.0 289.78 104.41	1131 73.73 -1.34	4A 2.67 11 -1	5.0 4.4 9	0.51 64.	11.32	1 0.28 0.08	2247. 2168.		24		16.8 31.13	-4.7 18.1 -0.4	-19.81 21.6	10.12		VV 219, TURN 72A, 7776 02-32-152 2
U 7781	1234.1 292.54 108.86	654 69.20 -2.60	9 2.67 11 -1	2.9 2.9 2	1.00	13.2	3 0.00 0.00	1076. 978. 1	67	10	0.65 5 10 1.09	16.8 31.13	-5.4 15.8 -0.8	-17.93 14.2	9.36 9.10	0.55	IC 3576, DDO 138, 7781 01-32-112 2

(1) Name	(2) α / ℓ / SGL	(3) δ / b / SGB	(4) Type / Q_{xyz} / Group	(5) D_{25} / $D_{25}^{b,i}$ / source	(6) d/D / i	(7) $B_T^{b,i}$ / $H_{-0.5}$ / B-H	(8) source / $A_B^{i,o}$ / A_B^b	(9) V_h / V_o / Tel	(10) W_{20} / W_R / W_D	(11) e_v / e_w / source	(12) F_c / source,e_F / f_h	(13) R / μ	(14) SGX / SGY / SGZ	(15) $M_B^{b,i}$ / $\Delta_{25}^{b,i}$	(16) log L_B / log M_H / log M_T	(17) M_H/L_B / M_H/M_T / M_T/L_B	(18) Alt. Names UGC/ESO / MCG / AGN?,RCII
U 7784	1234.2 / 288.28 / 102.49	1332 / 75.71 / -0.73	10B / 3.60 / 11 -1	2.5 / 2.3 / 2	0.64 / 55.	13.6	3 / 0.17 / 0.09	1125. / 1055. / 6	160	15	0.28 / 29 / 1.08	16.8 / 31.13	-3.7 / 16.5 / -0.2	-17.53 / 11.3	9.20 / 8.73	0.34	IC 3583 / 02-32-154 / 2
N 4569	1234.3 / 288.47 / 102.59	1326 / 75.62 / -0.73	2X / 1.15 / 11 -1	8.8 / 7.8 / 9	0.51 / 64.	9.86 / 7.41 / 2.45	1 / 0.28 / 0.09	-312. / -383.		23	0.83 / 24 33	16.8 / 31.13	-3.2 / 14.1 / -0.2	-21.27 / 38.3	10.70 / 9.28	0.04	M 90, ARP 76 / 7786 02-32-155 / 2, L/S
N 4570	1234.4 / 292.46 / 108.28	731 / 69.82 / -2.36	-2 / 2.66 / 11 -1	3.7 / 2.9 / 9	0.34 / 77.	11.70 / 8.32 / 3.38	2 / 0.00 / 0.00	1730. / 1634.		75		16.8 / 31.13	-5.6 / 16.9 / -0.7	-19.43 / 14.2	9.96		7785 01-32-114 / 2
N 4571	1234.4 / 287.53 / 101.59	1429 / 76.64 / -0.42	6A / 2.82 / 11 -1	3.6 / 3.6	0.93 / 25.	11.75	1 / 0.02 / 0.06	348. / 282. / 1	180	25 / 13 / 201	0.51 / 201 16 / 1.14	16.8 / 31.13	-3.1 / 15.3 / -0.1	-19.38 / 17.7	9.94 / 8.96	0.10	7788 02-32-156 / 2
N 4578	1235.0 / 291.70 / 106.10	950 / 72.13 / -1.57	-2A / 2.36 / 11 -1	3.1 / 2.9 / 9	0.73 / 47.	12.27 / 9.01 / 3.26	2 / 0.00 / 0.00	2282. / 2196.		50		16.8 / 31.13	-5.2 / 18.1 / -0.5	-18.86 / 14.2	9.74		7793 02-32-159 / 2
N 4579	1235.2 / 290.40 / 103.93	1206 / 74.36 / -0.89	3X / 3.26 / 11 -1	5.4 / 5.3 / 2	0.85 / 36.	10.46	1 / 0.05 / 0.10	1805. / 1729.	390 / 605. / 606.	46 / 12 / 300	0.91 / 24 30	16.8 / 31.13	-4.3 / 17.5 / -0.3	-20.67 / 26.0	10.46 / 9.36 / 11.44	0.08 / 0.008 / 9.57	M 58 / 7796 02-32-160 / 2
N 4580	1235.3 / 293.87 / 110.14	539 / 68.01 / -2.66	4X / 0.21 / 11 -0 +1	2.5 / 2.4 / 2	0.72 / 48.	12.6	3 / 0.12 / 0.00	1290. / 1187.		90		25.6 / 32.04	-8.8 / 24.0 / -1.2	-19.44 / 17.9	9.97		7794 01-32-117 / 2
N 4589	1235.5 / 124.23 / 42.73	7428 / 42.90 / 13.88	-5 / 0.31 / 42 -13	3.5 / 3.4 / 9	0.78 / 42.	11.75	2 / 0.00 / 0.05	1825. / 2007.		75		30.0 / 32.39	21.4 / 19.8 / 7.2	-20.64 / 29.8	10.45		SCI 118 / 7797 12-12-013 / 2, L2
1235-35	1235.7 / 299.95 / 150.37	-3520 / 27.19 / -12.36	5X / 0.17 / 23 -0 +5	1.3 / 1.2 / 6	0.60 / 58.	12.77	6 / 0.19 / 0.21	2931. / 2700.		20		39.7 / 32.99	-33.7 / 19.2 / -8.5	-20.22 / 13.9	10.28		380-50 / -6-28-C08 / 2
N 4586	1235.9 / 294.63 / 111.21	435 / 66.97 / -2.81	1A / 1.52 / 11 +2 +1	3.6 / 2.9 / 2	0.35 / 76.	12.05	1 / 0.54 / 0.01	829. / 722.		90		17.5 / 31.22	-6.3 / 16.3 / -0.9	-19.17 / 14.8	9.86		7804 01-32-122 / 2
1236+33	1236.3 / 148.24 / 83.87	3301 / 83.75 / 5.09	10 / 0.41 / 14 -7	1.1 / 1.1 / 2	0.73 / 47.		0.11 / 0.03	312. / 329. / 1	46 / 33. / 47.	7 / 4 / 225	0.51 / 225 9 / 1.01	3.6 / 27.77	0.4 / 3.5 / 0.3	1.2	7.62 / 7.56	1.148	CVNI DW A / 06-28-000
N 4592	1236.7 / 296.40 / 115.94	-16 / 62.18 / -3.94	8A / 0.12 / 14 -18	4.7 / 3.7 / 2	0.34 / 77.	11.4 / 10.49 / 0.91	3 / 0.57 / 0.03	1079. / 954. / 1	208 / 176. / 180.	8	1.55 / 7 / 1.13	9.6 / 29.90	-4.2 / 8.6 / -0.7	-18.50 / 10.4	9.59 / 9.51 / 9.97	0.84 / 0.351 / 2.38	7819 00-32-032 / 2
U 7822	1236.9 / 293.95 / 107.77	814 / 70.62 / -1.56	10 / 2.36 / 11 -1	1.3 / 1.2 / 2	0.62 / 56.	14.14	7 / 0.18 / 0.00	2086. / 1995. / 1	187	40	0.25 / 20 / 1.01	16.8 / 31.13	-5.6 / 17.6 / -0.5	-16.99 / 5.9	8.99 / 8.70	0.52	IC 3617, DDO 140, KAR 173? / 01-32-127 / 2

Name	Position	Vel/Pos	Type	Dim	Axis	m	(det)	V₁ / V₂	V₃	N	D	X Y Z	col	Abs m	L/S	Other names
N 4593	1237.0 / 297.45 / 120.60	-504 / 57.40 / -5.14	3B / 0.21 / 11-29	3.3 / 3.0 / 4	0.60 / 58.	11.4	5 / 0.19 / 0.04	2703. / 2560.		192	39.5 / 32.98	-20.0 / 33.9 / -3.5	-21.58 / 34.6	10.82		-1-32-032 2
N 4594	1237.3 / 298.43 / 126.72	-1121 / 51.15 / -6.69	1AP / 0.32 / 11-14+10	8.4 / 8.3 / 1	0.88 / 79.	8.52	1 / 0.64 / 0.11	1127. / 962.	762. / 738. / 739.	25	20.0 / 31.50	-11.9 / 15.9 / -2.3	-22.98 / 48.5	11.38 / 3.13	5.50	M 104, UGCA 293 L2 ; -2-32-020 2
N 4596	1237.4 / 293.30 / 105.67	1027 / 72.83 / -0.83	-2B / 2.87 / 11 -1	5.0 / 4.9 / 2	0.89 / 30.	11.46	1 / 0.00 / 0.01	2020. / 1939.		90	16.8 / 31.13	-5.0 / 17.7 / -0.3	-19.67 / 24.0	10.06		7828 02-32-170 2
N 4595	1237.4 / 289.60 / 100.75	1534 / 77.86 / 0.58	5 / 3.14 / 11 -1	1.8 / 1.6 / 2	0.68 / 52.	12.7	3 / 0.14 / 0.03	660. / 601.		90	16.8 / 31.13	-3.0 / 15.8 / 0.2	-18.43 / 7.8	9.56		7826 03-32-081 2
N 4597	1237.5 / 297.75 / 121.09	-532 / 56.95 / -5.15	9B / 0.19 / 11 -0+10	3.3 / 2.8 / 4	0.43 / 70.	12.0	5 / 0.38 / 0.03	1049. / 904. / 1	171. / 147. / 151.	8	17.0 / 31.15	-8.7 / 14.5 / -1.5	-19.15 / 13.9	9.85 / 9.53 / 9.94	0.48 / 0.390 / 1.22	-1-32-034 2
N 4605	1237.8 / 125.33 / 55.50	6153 / 55.47 / 12.00	5BP / 0.37 / 14 -0	5.7 / 4.8 / 9	0.44 / 69.	10.61	2 / 0.36 / 0.00	140. / 279. / 1	199	30	4.0 / 28.02	2.2 / 3.2 / 0.8	-17.41 / 5.6	9.16 / 8.31	0.14	7831 10-18-074 2
N 4602	1238.0 / 297.88 / 120.47	-452 / 57.62 / -4.85	4X / 0.21 / 11-29	2.9 / 2.3 / 4	0.30 / 79.	11.5	5 / 0.67 / 0.03	2559. / 2417. / 1	422	50	37.9 / 32.89	-19.2 / 32.6 / -3.2	-21.39 / 25.5	10.75 / 9.62	0.07	-1-32-036 2
N 4603	1238.3 / 300.82 / 155.88	-4042 / 21.86 / -12.80	5A / 0.00 / 23 -1	3.5 / 3.7 / 6	0.78 / 43.	11.4	5 / 0.09 / 0.52	2562. / 2323. / 2	495	60	46.4 / 33.33	-40.1 / 18.0 / -10.1	-21.93 / 50.1	10.96 / 10.26	0.20	322-52 -7-26-028 2
N 4608	1238.7 / 294.39 / 105.79	1025 / 72.84 / -0.54	-2B / 2.56 / 11 -1	3.3 / 3.3 / 2	1.00	12.09	1 / 0.00 / 0.00	1870. / 1789.		90	16.8 / 31.13	-4.9 / 17.4 / -0.2	-19.04 / 16.2	9.81		7842 02-32-177 2
N 4612	1239.0 / 295.73 / 108.53	735 / 70.04 / -1.24	-2X / 2.66 / 11 -1	3.1 / 2.9 / 9	0.73 / 47.	12.9	3 / 0.00 / 0.00	1832. / 1740.		90	16.8 / 31.13	-5.7 / 17.1 / -0.4	-18.23 / 14.2	9.48		7850 01-32-134 2
N 4618	1239.2 / 130.53 / 75.84	4125 / 75.84 / 7.78	9B / 0.74 / 14 -4	4.2 / 4.0 / 2	0.80 / 41.	11.19	1 / 0.08 / 0.00	546. / 602. / 1	157	8	7.3 / 29.31	1.8 / 7.0 / 1.0	-18.12 / 8.5	9.44 / 9.04	0.40	ARP 23, VV 73 ; 7853 07-26-037 2
1239-40	1239.4 / 301.03 / 155.76	-4033 / 22.01 / -12.57	5 / 0.00 / 23 -1	1.6 / 1.7 / 6	0.74 / 47.		0.11 / 0.50	2635. / 2397.		180	46.4 / 33.33	-40.2 / 18.1 / -9.8	23.0			NGC 4603D ; 322-55 -7-26-029 2
N 4621	1239.5 / 294.35 / 104.40	1155 / 74.36 / 0.07	-5 / 2.60 / 11 -1	5.8 / 5.6 / 9	0.82 / 38.	10.68 / 7.35 / 3.33	2 / 0.00 / 0.07	414. / 340.		125	16.8 / 31.13	-3.9 / 15.2 / 0.0	-20.45 / 27.5	10.37		M 59 ; 7858 02-32-183 2
N 4625	1239.5 / 130.22 / 75.72	4133 / 75.72 / 7.87	9XP / 0.72 / 14 -4	1.6 / 1.6 / 2	0.94 / 23.	12.88	1 / 0.02 / 0.00	610. / 667. / 6	86	10	8.2 / 29.57	2.0 / 7.9 / 1.1	-16.69 / 3.8	8.87 / 8.67	0.64	IC 3675 ; 7861 07-26-038 2

(1) Name	(2) α ℓ SGL	(3) δ b SGB	(4) Type ϱ_{xyz} Group	(5) D_{25} $D_{25}^{b,i}$ source	(6) d/D i	(7) $B_T^{b,i}$ $H_{-0.5}$ B-H	(8) source A_B^{i-o} A_B^b	(9) V_h V_o Tel	(10) W_{20} W_R W_D	(11) e_v e_w source	(12) F_c source/e_F f_h	(13) R μ	(14) SGX SGY SGZ	(15) $M_B^{b,i}$ $\Delta_{25}^{b,i}$	(16) $\log L_B$ $\log M_H$ $\log M_T$	(17) M_H/L_B M_H/M_T M_T/L_B	(18) Alternate Names Alternate Names AGN? UGC/ESO MCG RCII
N 4623	1239.6 296.04 108.22	757 70.42 -1.00	-2 2.36 11 -1	2.4 1.9 2	0.34 77.	13.3	3 0.00 0.00	1963. 1873.		90		16.8 31.13	-5.7 17.3 -0.3	-17.83 9.3	9.32		7862 01-32-135 2
U 7866	1239.7 132.10 78.45	3846 78.47 7.23	10 0.35 14 -7	3.7 3.4 2	0.66 53.	13.2	3 0.15 0.00	352. 396. 1	71 67. 75.	8	0.70 5 8 1.11	4.5 28.28	0.9 4.4 0.6	-15.08 4.5	8.22 8.01 8.76	0.61 0.175 3.47	IC 3687, DDO 141 7866 07-26-039 2
N 4631	1239.8 142.59 84.26	3249 84.23 5.75	7B 0.41 14 -6	14.7 10.8 2	0.24 85.	9.07	1 0.67 0.02	613. 631. 4	320 283. 285.	7 5 211	2.15 4 1.20	6.9 29.19	0.7 6.8 0.7	-20.12 21.8	10.24 9.83 10.70	0.39 0.133 2.91	ARP 281, TURN 76B 7865 06-28-020 2
N 4632	1240.0 298.09 115.73	11 62.70 -3.02	5A 0.66 11 -8 +1	3.1 2.6 2	0.41 71.	12.0	3 0.41 0.01	1693. 1572.		30		14.2 30.76	-6.2 12.8 -0.7	-18.76 10.8	9.70		7870 00-32-038 2
N 4630	1240.0 297.31 111.83	414 66.73 -1.92	9B 1.55 11 +2 +1	1.9 1.8 2	0.75 45.	13.1	3 0.10 0.00	692. 587.		90		17.9 31.26	-6.6 16.6 -0.6	-18.16 9.4	9.46		7871 01-32-136 2
U 7872	1240.0 123.75 41.63	7535 41.80 14.32	10 0.28 42-13	2.0 1.9 2	0.77 44.		 0.09 0.04	1887. 2073. 6	98 112. 117.	15	0.24 27 1.06	30.7 32.44	22.3 19.8 7.6	17.0	9.21 9.79	0.263	7872 13-09-030
N 4635	1240.2 286.90 96.46	2013 82.54 2.50	7A 0.12 11 -0 +1	1.9 1.8 2	0.80 41.	13.2	3 0.08 0.14	981. 944. 6	212	30	0.27 33 1.05	23.5 31.86	-2.6 23.4 1.0	-18.66 12.4	9.66 9.01	0.22	7876 03-32-087 2
N 4638	1240.3 295.17 104.65	1143 74.18 0.20	-2 3.71 11 -1	2.4 2.3 9	0.72 48.	11.94	2 0.00 0.06	1080. 1006.		150		16.8 31.13	-4.2 16.3 0.1	-19.19 11.3	9.87		7880 02-32-187 2
N 4636	1240.3 297.76 113.07	258 65.48 -2.19	-5 1.33 11 +2 +1	6.3 6.0 9	0.77 44.	10.47 7.49 2.98	2 0.00 0.03	979. 869.		26		17.0 31.15	-6.7 15.6 -0.6	-20.68 29.8	10.46		7878 01-32-137 2
N 4639	1240.4 294.34 102.91	1332 75.99 0.72	3X 3.54 11 -1	3.0 2.9 2	0.72 49.	12.03	1 0.12 0.05	983. 917. 6	302 353. 354.	20 9 300	0.69 201 12 1.11	16.8 31.13	-3.7 16.2 0.2	-19.10 14.2	9.83 9.14 10.71	0.20 0.027 7.57	7884 02-32-189 2
N 4643	1240.8 298.19 113.79	215 64.77 -2.27	 0.25 11 +6 +1	2.9 2.9 2	1.00	11.55	2 0.00 0.00	1346. 1233.		59	-0.05 223 56	25.7 32.05	-10.3 23.5 -1.0	-20.50 21.8	10.39 8.77	0.02	7895 00-33-005 2
N 4647	1241.0 295.74 104.57	1151 74.34 0.40	5X 3.49 11 -1	2.8 2.7 2	0.84 36.	11.79	1 0.06 0.06	1358. 1285.		64		16.8 31.13	-4.3 16.7 0.1	-19.34 13.2	9.93		ARP 116, VV 206 TURN 78A 7896 02-33-001 2
N 4649	1241.2 295.92 104.60	1150 74.32 0.44	-5 3.49 11 -1	8.5 8.2 9	0.80 40.	9.77 6.48 3.29	1 0.00 0.06	1200. 1127.		26		16.8 31.13	-4.3 16.5 0.1	-21.36 40.2	10.74		M 60, ARP 116, VV 206 TURN 78B 7898 02-33-002 2

Name	Coordinates	Type	Size	Mag	(6)	Vel	(8)	(9)	(10)	(11)	(12)	(13)	(14)	(15)	Other names
N 4651	1241.2 1639 293.08 79.10 99.96 1.76	5A 2.46 11 -1	3.6 0.69 59. 3.3	11.06 8.62 2.44	1 0.21 0.03 0.03 0.01	800. 748. 1	377 437. 439.	25 8 300	1.26 201 5 1.14	16.8 31.13	-2.8 16.1 0.5	-20.07 16.2	10.22 9.71 10.95	0.31 0.057 5.40	ARP 189, VV 56 7901 03-33-001 2
N 4653	1241.3 -16 298.86 62.27 116.25 -2.83	6X 0.17 11 -0	2.4 0.93 25. 2.4 2	12.77	1 0.03 0.01	2628. 2506. 1	204	10	0.76 16 1.08	39.1 32.96	-17.3 35.0 -1.9	-20.19 27.4	10.27 9.94	0.47	7900 00-33-006 2
N 4654	1241.4 1323 295.41 75.87 103.12 0.91	6X 2.93 11 -1	4.8 0.58 59. 4.4 2	10.82 8.76 2.06	1 0.21 0.07	1044. 978. 1	309 316. 318.	10 8 79	1.13 217 7 1.17	16.8 31.13	-3.8 16.3 0.3	-20.31 21.6	10.32 9.58 10.80	0.18 0.061 3.02	7902 02-33-004 2
N 4645	1241.4 -4129 301.47 21.09 156.77 -12.36	-5 0.00 23 -1	2.2 0.60 58. 2.2 6	12.3	5 0.00 0.48	2651. 2412.		90		46.4 33.33	-40.6 17.4 -9.6	-21.03 29.8	10.60		322-66 -7-26-037 2
1241-05	1241.5 -524 299.56 57.15 121.22 -4.15	9 0.68 11-11+10	3.3 0.67 52. 3.1 4	12.69	7 0.15 0.04	1429. 1287. 1	142 143. 147.	10	0.93 11 1.09	25.4 32.02	-13.1 21.6 -1.8	-19.33 23.0	9.92 9.74 10.13	0.65 0.403 1.62	DDO 142 -1-33-001 2
N 4656	1241.6 3226 140.21 84.72 84.73 6.02	9BP 0.38 14 -6	18.8 0.17 90. 13.2 2	10.09	1 0.67 0.01	649. 667. 4	183	7 6 211	1.88 211 5 1.13	7.2 29.30	0.7 7.2 0.8	-19.21 27.8	9.88 9.59	0.52	7907 05-30-066 2
N 4650	1241.6 -4028 301.48 22.11 155.75 -12.14	 0.00 23 -1	3.5 0.89 31. 3.8 6		0.00 0.48	2634. 2397.		180		46.4 33.33	-40.3 18.2 -9.5	51.5			322-67 -7-26-038 2
U 7911	1241.9 45 299.05 63.29 115.31 -2.41	9B 0.34 11 +6 +1	2.8 0.75 45. 2.6 2	13.7	3 0.10 0.00	1184. 1066. 1	119 132. 136.	10	0.49 22 1.07	23.5 31.86	-10.1 21.2 -1.1	-18.16 17.8	9.45 9.23 9.95	0.60 0.189 3.17	DDO 144 7911 00-33-007 2
N 4658	1242.0 -949 300.15 52.74 125.53 -5.19	5B 0.22 11-27	2.1 0.33 77. 1.7 4	11.9	5 0.58 0.09	2407. 2250.		90		35.7 32.76	-20.6 28.9 -3.2	-20.86 17.7	10.54		-2-33-001 2
1242-402	1242.0 -4026 301.56 22.15 155.73 -12.06	13 P 0.00 23 -1	1.8 0.41 72. 1.7 6		0.00 0.48	2475. 2238.		180		46.4 33.33	-40.0 18.1 -9.4	23.0			NGC 4650A
N 4660	1242.0 1128 296.78 73.98 105.00 0.53	-5 3.37 11 -1	2.7 0.88 31. 2.6 2	11.83 8.46 3.37	2 0.00 0.02	1017. 943.		30		16.8 31.13	-4.3 16.1 0.2	-19.30 12.8	9.91		322-69 -7-26-000 2
1242+28	1242.1 2845 171.01 87.96 88.33 5.18	12 0.93 14 -1	0.9 0.61 57. 0.8 2	14.55	2 0.19 0.06	956. 958.		37		9.7 29.93	0.3 9.8 0.9	-15.38 2.3	8.34		7914 02-33-006 2
U 7916	1242.1 3440 134.05 82.58 82.58 6.68	10 0.53 14 -6	2.5 0.79 42. 2.4 2		0.08 0.01	612. 640. 1	82 96. 101.	10	0.77 9 1.06	7.2 29.28	0.9 7.1 0.8	5.0	8.48 9.13	0.226	UGCA 294, HARO 33 05-30-070 2
1242-08	1242.4 -852 300.25 53.70 124.63 -4.84	10 0.75 11-10	1.1 0.90 29. 1.1 2		0.04 0.09	1378. 1225. 6	126	10	0.38 21 1.02	24.2 31.92	-13.7 19.8 -2.0	7.8	9.15		I ZW 42, VV 127 DDO 143 7916 06-28-024 2 UGCA 295 -1-33-000

(1) Name	(2) α / l / SGL	(3) δ / b / SGB	(4) Type / ρxyz / Group	(5) D_{25} / D^b_{25} / source	(6) d/D / i	(7) $B^{b,i}_T$ / $H_{-0.5}$ / B-H	(8) source / A^{i-0}_B / A^b_B	(9) V_h / V_o / Tel	(10) W_{20} / W_R / W_D	(11) e_v / e_w / source	(12) F_c / source/e_f / f_h	(13) R / μ	(14) SGX / SGY / SGZ	(15) $M^{b,i}_B$ / $\Delta^{b,i}_{25}$	(16) log L_B / log M_H / log M_T	(17) M_H/L_B / M_H/M_T / M_T/L_B	(18) Alternate Names / UGC/ESO MCG / AGN? RCII
1242-403	1242.5 / 301.67 / 155.87	-4033 / 22.03 / -11.99	2A / 0.00 / 23 -1	1.0 / 0.9 / 6	0.36 / 75.		0.51 / 0.48	2498. / 2261.		180		46.4 / 33.33	-40.0 / 18.0 / -9.3	12.2			NGC 4650B / 322-72 -7-26-040 2
N 4666	1242.6 / 299.55 / 116.26	-11 / 62.37 / -2.49	5X / 0.54 / 11 -8 +1	4.2 / 3.3 / 2	0.35 / 76.	10.98	1 / 0.53 / 0.04	1516. / 1395.		59		14.1 / 30.75	-6.2 / 12.6 / -0.6	-19.77 / 13.6	10.10		7926 00-33-008 2
N 4665	1242.6 / 299.09 / 112.87	320 / 65.88 / -1.54	1.08 / 11 +2 +1	4.2 / 4.2 / 2	1.00	11.9	3 / 0.00 / 0.03	785. / 678.		50		17.9 / 31.26	-6.9 / 16.5 / -0.5	-19.36 / 22.0	9.94		7924 01-33-305 2
N 4670	1242.8 / 212.66 / 89.68	2724 / 88.61 / 4.98	0BP / 0.52 / 14 +2 +1	1.7 / 1.7 / 2	0.89 / 31.	12.97	2 / 0.00 / 0.08	1072. / 1068.	269	9 / 50 / 203	0.50 / 203 20	11.0 / 30.21	0.1 / 11.0 / 1.0	-17.24 / 5.5	9.09 / 8.58	0.31	ARP 163
1243-05	1243.1 / 300.32 / 121.73	-549 / 56.75 / -3.88	10 / 0.76 / 11 -11 +10	3.3 / 3.3 / 4	0.86 / 34.		0.05 / 0.03	1479. / 1337.	148 / 211. / 214.	20	0.47 / 23 / 1.11	25.8 / 32.06	-13.5 / 21.9 / -1.7	24.9	9.29 / 10.51	0.061	DDO 146 / -1-33-003 2
1243-33	1243.3 / 301.64 / 148.91	-3334 / 29.02 / -10.48	10 / 0.13 / 14-15	3.5 / 3.1 / 6	0.40 / 72.		0.43 / 0.27	588. / 365.		20		5.1 / 28.53	-4.3 / 2.6 / -0.9	4.6			381-20 -6-28-017
1243+71	1243.7 / 123.60 / 45.74	7136 / 45.80 / 14.09	13 P / 0.07 / 12 -0	0.5 / 0.5 / 2	1.00	14.9	3 / 0.00 / 0.03	1054. / 1228.		220		20.1 / 31.52	13.6 / 14.0 / 4.9	-16.62 / 2.9	8.84		UGCA 296, MARK 223 / VII ZW 4 / 12-12-000 2
U 7941	1244.0 / 123.87 / 52.64	6450 / 52.56 / 13.18	7 / 0.17 / 42 -0+12	4.4 / 3.1 / 2	0.20 / 90.	13.2 / 12.47 / 0.73	3 / 0.67 / 0.02	2294. / 2445.	263 / 225. / 228.	15	0.62 / 23 / 1.10	36.2 / 32.79	21.4 / 28.0 / 8.3	-19.59 / 32.8	10.03 / 9.74 / 10.68	0.51 / 0.114 / 4.51	7941 11-16-006
1244+48	1244.1 / 125.25 / 69.03	4831 / 68.86 / 10.23	12 / 0.02 / 12 -0 +1	0.9 / 0.7 / 2	0.33 / 78.		3 / 0.60 / 0.00	1126. / 1214.		220		22.1 / 31.72	7.8 / 20.3 / 3.9	4.5			UGCA 297, MARK 224 / 08-23-092 2
U 7943	1244.2 / 299.70 / 110.19	614 / 68.80 / -0.37	5 / 1.83 / 11 +2 +1	2.5 / 2.4 / 2	0.82 / 39.	13.6	3 / 0.07 / 0.00	837. / 743. / 1	133	15	0.35 / 47 / 1.06	17.3 / 31.19	-6.0 / 16.2 / -0.1	-17.59 / 12.1	9.23 / 8.83	0.40	7943 01-33-007
N 4677	1244.2 / 302.03 / 156.70	-4119 / 21.27 / -11.81	-2 / 0.00 / 23 -1	1.5 / 1.4 / 6	0.47 / 68.	13.23	2 / 0.00 / 0.47	3215. / 2978.		90		46.4 / 33.33	-41.4 / 17.8 / -9.5	-20.10 / 19.0	10.23		322-78 -7-26-044 2
1244+26	1244.3 / 241.22 / 90.32	2650 / 88.80 / 5.15	-5 P / 0.67 / 14 +1	0.8 / 0.7 / 2	0.68 / 51.	14.6	3 / 0.00 / 0.05	876. / 870.		59		8.9 / 29.74	0.0 / 8.9 / 0.8	-15.14 / 1.8	8.25		UGCA 298, A 1244 / 05-30-079 2
U 7949	1244.6 / 128.39 / 80.65	3646 / 80.59 / 7.68	10 / 0.37 / 14 -7	2.1 / 1.9 / 2	0.61 / 57.		0.19 / 0.01	333. / 371. / 1	52 / 39. / 52.	8	0.66 / 9 / 1.04	4.2 / 28.11	0.7 / 4.1 / 0.6	2.3	7.91 / 8.01	0.784	DDO 147, KAR 200 / 7949 06-28-030 2

Name	Position	Type	Size	Orient	Mag		Vel									Alt. names
U 7950	1244.6 5155 / 124.63 65.47 / 65.66 11.00	10 / 0.36 / 14 -4	1.9 1.8 2	0.75 45.	15.00	2 0.10 0.00	510. 612. 1	15	104 115. 120.	0.29 25 1.03	8.9 29.76	3.6 8.0 1.7	-14.76 4.7	8.10 8.19 9.25	1.24 0.086 14.38	HARO 36; 7950 09-21-047 2
N 4684	1244.7 -227 / 300.84 60.13 / 118.58 -2.60	-2B / 0.37 / 11 -0+10	2.8 2.2 2	0.36 75.	12.6	3 0.00 0.02	1589. 1461.	90			27.4 32.19	-13.1 24.0 -1.2	-19.59 17.6	10.03		7951 00-33-011 2
1244-53	1244.7 -5340 / 302.36 8.93 / 169.21 -13.70	10 / 0.09 / 23 -0 +7	1.7 2.0 6	0.70 50.		0.13 1.05	1873. 1622. 7	20			24.7 31.97	-23.6 4.5 -5.9	14.4			172-09
N 4687	1245.0 3538 / 128.65 81.72 / 81.78 7.49	13 P / 0.61 / 14 -6	1.1 1.0 2	0.80 41.	14.1	3 0.00 0.01	690. 724.	100			8.4 29.61	1.2 8.2 1.1	-15.51 2.5	8.40		MARK 442; 7958 06-28-031 2
N 4688	1245.3 436 / 300.62 67.18 / 111.83 -0.55	6B / 1.35 / 11 +2 +1	4.1 4.1 2	1.00	12.9	3 0.00 0.00	991. 891. 1	7	70	0.95 217 7 1.18	17.1 31.16	-6.3 15.8 -0.2	-18.26 20.5	9.50 9.42	0.83	7961 01-33-013 2
N 4689	1245.3 1402 / 299.13 76.60 / 102.75 2.00	5A / 2.39 / 11 -1	3.7 3.7 2	0.89 31.	11.51	1 0.04 0.04	1776. 1715.	90			16.8 31.13	-4.0 17.5 0.6	-19.62 18.2	10.04		7965 02-33-022 2
N 4691	1245.7 -304 / 301.38 59.52 / 119.25 -2.52	3BP / 0.42 / 11 -0+10	3.5 3.4 2	0.88 32.	11.64	2 0.04 0.02	1119. 989.	26			22.5 31.76	-11.0 19.6 -1.0	-20.12 22.3	10.24		UGCA 299; 00-33-013 2
N 4694	1245.8 1115 / 300.18 73.83 / 105.46 1.37	-2BP / 2.76 / 11 -1	3.4 3.0 9	0.57 60.	12.2	3 0.00 0.03	1193. 1121.	29 12 223	109 100. 105.	0.08 223 25	16.8 31.13	-4.5 16.4 0.4	-18.93 14.7	9.76 8.53 9.63	0.06 0.080 0.73	7969 02-33-023 2
1245-45	1245.8 -4521 / 302.41 17.25 / 160.81 -12.24	10 / 0.32 / 23 -0 +2	1.2 1.3 6	0.92 26.		0.03 0.50	3177. 2935. 7	20			42.0 33.12	-38.8 13.5 -8.9	15.9			268-43
N 4698	1245.9 846 / 300.61 71.35 / 107.86 0.72	2A / 2.91 / 11 -1	3.3 2.9 2	0.55 68.	11.05 8.36 2.69	1 0.34 0.00	988. 905.	7 12 223	427 443. 445.	0.66 205 31	16.8 31.13	-5.1 15.8 0.2	-20.08 14.2	10.22 9.11 10.91	0.08 0.016 4.83	7970 02-33-024 2
N 4697	1246.0 -532 / 301.63 57.06 / 121.65 -3.11	-5 / 0.60 / 11-11+10	7.1 6.7 9	0.76 44.	10.17	2 0.00 0.03	1308. 1169.	65			23.3 31.84	-12.2 19.8 -1.3	-21.67 45.7	10.85 0.53 0.27		UGCA 300; -1-33-010 2
N 4696	1246.1 -4102 / 302.41 21.56 / 156.48 -11.41	-5 / 0.00 / 23 -1	3.3 3.4 2	0.75 45.	11.28	2 0.00 0.47	2811. 2575.	26			46.4 33.33	-40.8 17.8 -9.0	-22.05 46.1	11.01		322-91 -7-26-051 2
N 4707	1246.1 5126 / 124.11 65.65 / 66.19 11.13	9 / 0.55 / 14 -4	2.5 2.4 2	0.89 31.	13.40	7 0.04 0.00	468. 569. 1	10	83 127. 131.	0.59 13 1.06	8.0 29.52	3.2 7.2 1.5	-16.12 5.6	8.64 8.40 9.42	0.57 0.095 6.01	I ZW 43, DDO 150; 7971 09-21-050 2
1246-04	1246.1 -459 / 301.65 57.61 / 121.12 -2.94	10 / 0.70 / 11-11+10	1.8 1.8 4	1.00		0.00 0.05	1343. 1206. 1	10	65	0.01 28 1.04	24.5 31.94	-12.6 20.9 -1.3	12.9	8.79		DDO 148, KAR 205; -1-33-011 2

(1) Name	(2) α ℓ SGL	(3) δ b SGB	(4) Type ϱ_{xyz} Group	(5) D_{25} $D_{25}^{b,i}$ source	(6) d/D i	(7) $B_T^{b,i}$ $H_{-0.5}$ B-H	(8) source A_B^{i-o} A_B^b	(9) V_h V_o Tel	(10) W_{20} W_R W_D	(11) e_v e_w source	(12) F_c source/e_F f_h	(13) R μ	(14) SGX SGY SGZ	(15) $M_B^{b,i}$ $\Delta_{25}^{b,i}$	(16) $\log L_B$ $\log M_H$ $\log M_T$	(17) M_H/L_B M_H/M_T M_T/L_B	(18) Alternate Names Alternate Names AGN? UGC/ESO MCG RCII
N 4699	1246.4 301.90 124.44	-824 54.19 -3.77	3X 0.78 11-10	3.1 3.0 2	0.80 41.	10.31	2 0.08 0.07	1507. 1358.		10	0.80 67 32	25.7 32.05	-14.5 21.1 -1.7	-21.74 22.5	10.89 9.62	0.05	UGCA 301 -1-33-013 2
N 4700	1246.5 302.02 127.09	-1108 51.46 -4.46	7B 0.64 11-10	3.1 2.2 4	0.15 90.	11.6	5 0.67 0.09	1406. 1248. 1	167 134. 139.	30	0.84 19 1.06	24.0 31.90	-14.4 19.1 -1.9	-20.30 15.4	10.31 9.60 9.90	0.19 0.496 0.39	 -2-33-013 2
N 4701	1246.7 301.58 112.83	340 66.26 -0.47	6A 0.49 11 +2 +1	3.4 3.3 2	0.85 35.	12.72	2 0.05 0.00	727. 624. 1	185 260. 262.	10	1.12 9 1.11	20.5 31.55	-7.9 18.9 -0.2	-18.83 19.8	9.72 9.74 10.59	1.05 0.143 7.31	 7975 01-33-015 2
1246-09	1246.8 302.11 125.87	-951 52.75 -4.05	9B 0.78 11-10	4.8 4.5 4	0.72 49.	11.9	5 0.12 0.09	1318. 1164. 1	161	15	0.94 17 1.19	23.3 31.83	-13.6 18.8 -1.6	-19.93 30.6	10.16 9.67	0.32	A 1247 -2-33-015 2
N 4710	1247.2 300.91 101.52	1526 78.02 2.82	-2A 2.00 11 -1	4.7 3.7 2	0.33 78.	11.77 8.41 3.36	2 0.00 0.03	1129. 1076.		24		16.8 31.13	-3.4 16.6 0.8	-19.36 18.2	9.94		 7980 03-33-009 2
N 4713	1247.5 302.00 111.03	535 68.18 0.24	7X 1.41 11 +2 +1	3.1 2.8 2	0.66 53.	12.05 10.43 1.62	2 0.15 0.00	655. 560. 1	193 197. 200.	10 8 217	1.14 217 6 1.08	17.9 31.26	-6.4 16.7 0.1	-19.21 14.6	9.88 9.65 10.22	0.59 0.269 2.19	 7985 01-33-018 2
1247-41	1247.5 302.70 156.75	-4115 21.35 -11.19	-2 0.00 23 -1	1.5 1.3 6	0.43 70.		 0.00 0.45	2684. 2449.		90		46.4 33.33	-40.8 17.5 -8.8	 17.6			 323-05 -7-26-058 2
1247-10	1247.9 302.56 126.67	-1036 52.00 -3.99	8 0.25 11-27	3.8 3.2 4	0.43 70.		 0.38 0.08	2407. 2251. 1	285 263. 265.	15	0.95 15 1.10	35.7 32.76	-21.3 28.5 -2.5	 33.4	 10.06 10.83	 0.170	DDO 151 -2-33-020 2
N 4725	1248.1 295.96 91.57	2546 88.34 5.71	2XP 0.21 14 -2 +1	10.5 10.0 2	0.77 43.	9.82 7.05 2.77	1 0.09 0.03	1207. 1199. 2	410 542. 543.	10 6 211	1.41 211 10 1.14	12.4 30.47	-0.3 12.4 1.2	-20.65 36.2	10.45 9.60 11.49	0.14 0.013 10.89	 7989 04-30-022 2
N 4750	1248.3 123.07 44.19	7309 44.25 14.61	2A 0.14 42-13	2.3 2.3 2	0.96 19.	11.8	3 0.01 0.03	1518. 1698.		60		26.1 32.09	18.1 17.6 6.6	-20.29 17.5	10.31		 7994 12-12-019 2
N 4731	1248.4 302.73 122.38	-608 56.47 -2.69	6B 0.80 11-11+10	6.2 5.3 2	0.48 67.	10.9	5 0.31 0.05	1497. 1358. 6	246 227. 230.	10 7 79	1.44 217 4 1.36	25.9 32.07	-13.8 21.8 -1.2	-21.17 40.1	10.66 10.27 10.78	0.41 0.311 1.31	UGCA 302 -1-33-026 2
N 4736	1248.6 123.31 76.28	4123 76.02 9.49	2A 0.42 14 -7	12.2 11.8 2	0.87 33.	8.80	1 0.05 0.00	307. 368. 2	241 373. 375.	15 8 207	1.21 207 10 1.23	4.3 28.17	1.0 4.1 0.7	-19.37 14.8	9.94 8.48 10.78	0.03 0.005 6.86	M 94 L2 7996 07-26-058 2
N 4730	1249.0 303.00 156.42	-4052 21.73 -10.84	-5 0.00 23 -1	1.2 1.3 6	1.00		 0.00 0.45	2268. 2034.		90		46.4 33.33	-40.2 17.6 -8.4	 17.6			 323-17 -7-27-002 2

Name	Coordinates (1)	Coordinates (2)	Type	Size		Bmag	Flux idx	Vel				Dist			Mag		Alternate names
N 4742	1249.2 303.08 126.35	-1011 52.42 -3.57	-5 0.73 11-10	2.2 2.0 2	0.72 48.	12.02	2 0.00 0.08	1321. 1168.		50		23.2 31.83	-13.8 18.7 -1.4	-19.81 13.5	10.12		UGCA 303 / -2-33-032 2
N 4747	1249.3 305.86 91.38	2602 88.62 6.03	5BP 0.18 14 -2 +1	3.3 2.8 2	0.47 67.	12.66	1 0.33 0.02	1189. 1183. 1	197	10	0.80 56 8 1.06	12.3 30.46	-0.3 12.3 1.3	-17.80 10.1	9.31 8.98	0.47	ARP 159
N 4743	1249.5 303.10 156.71	-4108 21.47 -10.80	-2 0.00 23 -1	1.6 1.4 6	0.37 74.		0.00 0.44	3054. 2820.		90		46.4 33.33	-41.3 17.7 -8.5	19.0			8005 04-30-023 2
1249-09	1249.8 303.33 125.69	-928 53.13 -3.24	6 0.19 11-27	2.5 1.8 2	0.13 90.		0.67 0.09	2258. 2108. 1		20	0.16 66 1.04	34.0 32.66	-19.8 27.6 -1.9	17.9	9.22		323-21 -7-27-005 2
N 4754	1249.8 303.72 105.41	1135 74.18 2.40	-2B 2.62 11 -1	4.3 3.8 9	0.53 63.	11.39 7.94 3.45	2 0.00 0.05	1461. 1393.		75		16.8 31.13	-4.6 16.8 0.7	-19.74 18.6	10.09		UGCA 304 / -1-33-032
N 4753	1249.8 303.42 117.46	-56 61.67 -0.96	13 0.49 11 +9 +1	4.3 3.9 2	0.60 58.	10.84	2 0.00 0.01	1221. 1102.		61		15.1 30.89	-7.0 13.4 -0.3	-20.05 17.2	10.21		8010 02-33-030 2
1250-06	1250.0 303.45 122.39	-602 56.57 -2.28	10 0.80 11-11+10	2.0 2.0 4	1.00		0.00 0.05	1536. 1398. 1	105	10	0.44 17 1.05	26.3 32.10	-14.1 22.2 -1.0	15.4	9.28		8009 00-33-016 2
U 8011	1250.0 305.44 95.42	2155 84.51 5.15	10 0.12 11 -0 +1	1.6 1.6 2	0.82 39.		0.07 0.13	770. 747. 1	143 182. 185.	10	0.28 23 1.03	21.4 31.66	-2.0 21.3 1.9	10.0	8.94 9.98	0.091	DDO 152, KAR 211 / -1-33-033 2
N 4758	1250.3 304.60 101.06	1607 78.71 3.72	3BP 1.71 11 -0 +1	2.9 2.2 2	0.30 80.	12.9 11.15 1.75	3 0.67 0.04	1244. 1196. 1	205	10	0.43 19 1.05	16.2 31.05	-3.1 15.9 1.1	-18.15 10.4	9.45 8.85	0.25	8011 04-30-000
N 4762	1250.4 304.26 105.52	1130 74.10 2.52	-2B 2.65 11 -1	8.3 6.0 9	0.22 87.	11.09	2 0.00 0.03	942. 874.		18		16.8 31.13	-4.4 16.0 0.7	-20.04 29.4	10.21		8014 03-33-015 2
N 4765	1250.7 304.10 112.07	444 67.33 0.78	1 0.46 11 +2 +1	1.6 1.5 2	0.76 45.	12.8	3 0.09 0.00	783. 687.		83		21.2 31.63	-8.0 19.6 0.3	-18.83 9.3	9.72		8016 02-33-033 2
1250-04	1250.8 303.85 121.16	-442 57.90 -1.73	10 0.78 11-11+10	1.1 1.1 2	1.00		0.00 0.03	1410. 1278. 6	80	20	-0.02 33 1.02	25.3 32.01	-13.1 21.6 -0.8	8.1	8.79		ARAK 391 / 8018 01-33-020 2
N 4772	1250.9 304.13 114.30	226 65.03 0.21	1A 0.99 11 +2 +1	2.8 2.5 2	0.52 64.	12.3	3 0.27 0.00	1087. 982.		90		16.3 31.07	-6.7 14.9 0.1	-18.77 11.9	9.70		UGCA 305, KAR 213 / -1-33-000
N 4767	1251.1 303.44 155.09	-3927 23.15 -10.16	-5 0.00 23 -1	2.5 2.3 6	0.43 70.	12.2	5 0.00 0.38	3131. 2901.		37		46.4 33.33	-41.0 19.0 -8.1	-21.13 31.2	10.64		8021 00-33-018 2 / 323-36 -6-28-023 2

122

(1) Name	(2) α / ℓ / SGL	(3) δ / b / SGB	(4) Type / ρ_xyz / Group	(5) D_25 / D_25^{b,i} / source	(6) d/D / i	(7) B_T^{b,i} / H_{-0.5} / B-H	(8) source / A_B^{i-o} / A_B^b	(9) V_h / V_o / Tel	(10) W_20 / W_R / W_D	(11) e_v / e_w / source	(12) F_c / source,e_F / f_h	(13) R / μ	(14) SGX / SGY / SGZ	(15) M_B^{b,i} / Δ_25^{b,i}	(16) log L_B / log M_H / log M_T	(17) M_H/L_B / M_H/M_T / M_T/L_B	(18) Alternate Names; UGC/ESO MCG AGN? RCII
N 4775	1251.2 / 303.99 / 122.78	−621 / 56.25 / −2.08	7A / 0.70 / 11-11+10	2.3 / 2.3 / 2	0.92 / 26.	11.8	5 / 0.03 / 0.05	1564. / 1426.	146	10 / 12 / 217	0.87 / 217 7	26.6 / 32.12	−14.4 / 22.3 / −1.0	−20.32 / 17.9	10.32 / 9.72	0.25	UGCA 306; −1-33-043 2
1251-11	1251.4 / 303.93 / 128.07	−1149 / 50.78 / −3.48	10 / 0.09 / 11-14+10	2.3 / 1.9 / 2	0.44 / 70.		/ 0.37 / 0.07	824. / 667. / 1	103 / 87. / 93.	8	0.81 / 12 / 1.04	17.3 / 31.19	−10.6 / 13.5 / −1.1	−1.73 / 9.5	0.69 / 9.28 / 9.33	0.918	UGCA 307, DDO 153; −2-33-047 2
U 8024	1251.6 / 34.76 / 90.16	2725 / 89.40 / 6.87	10 / 0.29 / 14 +3	2.5 / 2.4 / 2	0.89 / 31.	13.90	7 / 0.04 / 0.03	378. / 380. / 1	103 / 160. / 163.	7	1.24 / 5 6 / 1.06	4.0 / 28.01	0.0 / 4.0 / 0.5	−14.11 / 2.8	7.84 / 8.44 / 9.32	4.06 / 0.134 / 30.36	NGC 4789A, DDO 154; 8024 05-30-120 2
N 4781	1251.8 / 304.13 / 126.60	−1016 / 52.33 / −2.97	7B / 0.72 / 11-10	3.8 / 3.1 / 4	0.37 / 74.	11.2	5 / 0.49 / 0.08	1265. / 1113. / 1	272 / 243. / 246.	20	1.02 / 14 / 1.10	22.5 / 31.76	−13.4 / 18.0 / −1.2	−20.56 / 20.4	10.42 / 9.72 / 10.54	0.20 / 0.152 / 1.34	−2-33-049 2
N 4793	1252.2 / 101.79 / 88.44	2912 / 88.06 / 7.43	5X / 0.16 / 45 +2	3.2 / 2.9 / 2	0.56 / 60.	12.05	2 / 0.22 / 0.03	2488. / 2498.		45		38.9 / 32.95	1.0 / 38.5 / 5.0	−20.90 / 32.9	10.55		8033 05-31-003 2
N 4810	1252.3 / 304.99 / 113.92	255 / 65.50 / 0.68	10 P / 0.46 / 11 +2 +1	0.9 / 0.8 / 2	0.67 / 52.	14.3	3 / 0.15 / 0.05	881. / 779.		95		21.6 / 31.67	−8.7 / 19.7 / 0.3	−17.37 / 5.0	9.14		ARP 27, VV 313; TURN 81B; 01-33-023 2
N 4790	1252.3 / 304.34 / 126.34	−958 / 52.62 / −2.78	5B / 0.79 / 11-10	1.8 / 1.7 / 4	0.67 / 52.	12.3	5 / 0.15 / 0.10	1354. / 1204. / 1	255	20	0.67 / 30 / 1.03	23.6 / 31.87	−14.0 / 19.0 / −1.1	−19.57 / 11.7	10.02 / 9.42	0.25	−2-33-056 2
N 4809	1252.3 / 304.99 / 113.91	256 / 65.52 / 0.68	10 P / 0.37 / 11 +2 +1	1.9 / 1.5 / 2	0.37 / 74.	14.1	3 / 0.49 / 0.05	945. / 843. / 1	173 / 145. / 149.	15	0.72 / 16 / 1.02	22.2 / 31.73	−9.0 / 20.3 / 0.3	−17.63 / 9.7	9.24 / 9.41 / 9.77	1.47 / 0.436 / 3.38	WITH NGC 4810 = ARP 27; VV 313, TURN 81A; 8034 01-33-223 2
U 8036	1252.3 / 308.61 / 97.95	1927 / 82.01 / 5.04	5B / 0.15 / 11 -0 +1	3.7 / 2.8 / 2	0.26 / 84.	12.6 / 11.72 / 0.88	3 / 0.67 / 0.05	926. / 893. / 1	216 / 181. / 184.	10	0.86 / 14 / 1.08	23.0 / 31.81	−3.2 / 22.7 / 2.0	−19.21 / 18.8	9.88 / 9.58 / 10.25	0.51 / 0.215 / 2.38	IC 3881; 8036 03-33-016
N 4800	1252.3 / 121.30 / 71.03	4648 / 70.59 / 11.24	3A / 0.17 / 12 -0 +1	1.8 / 1.6 / 2	0.73 / 47.	12.19	2 / 0.11 / 0.00	746. / 831.		50		15.2 / 30.91	4.8 / 14.1 / 3.0	−18.72 / 7.1	9.68		8035 08-24-024 2
U 8041	1252.6 / 304.98 / 116.38	23 / 62.97 / 0.07	7B / 0.39 / 11 +9 +1	3.5 / 3.2 / 2	0.63 / 56.	12.3	3 / 0.17 / 0.00	1330. / 1218. / 1	202 / 201. / 204.	20	0.71 / 27 / 1.09	14.2 / 30.76	−6.3 / 12.7 / 0.0	−18.46 / 13.3	9.58 / 9.01 / 10.19	0.27 / 0.067 / 4.13	A 1253; 8041 00-33-021 2
1252-49	1252.6 / 303.60 / 165.49	−4948 / 12.80 / −11.89	7B / 0.17 / 23 -8 +7	2.9 / 3.4 / 6	0.83 / 37.		/ 0.06 / 0.95	2132. / 1888. / 7		20		28.4 / 32.26	−26.9 / 7.0 / −5.8	28.2			IC 3896A; 219-08 2
1252-44	1252.7 / 303.69 / 160.65	−4458 / 17.63 / −10.97	10 / 0.18 / 23 -7	1.7 / 1.8 / 6	0.85 / 36.		/ 0.05 / 0.41	2223. / 1985. / 2	133 / 182. / 185.	20	0.56 / 16	30.0 / 32.38	−27.7 / 9.7 / −5.7	15.8	9.51 / 10.18	0.216	269-09 −7-27-000

Name	Designation															ID
UGCA 308	-1-33-000		9.07	8.8	-13.8 / 18.6 / -1.1	23.2 / 31.83	0.34 / 37 / 1.02	15	86	1321. / 1171. / 1	0.22 / 0.10		1.4 0.57 / 1.3 60. / 2	10 / 0.76 / 11-10	1252.9 -1006 / 304.58 52.49 / 126.51 -2.67	1252-10
8051 10-19-003 2			10.30	-20.27 / 35.6	19.6 / 32.8 / 9.1	39.3 / 32.97		65		2531. / 2663.	2 / 0.10 / 0.00	12.70	3.3 0.74 / 3.1 46. / 2	3A / 0.15 / 42 -0	1253.2 5837 / 121.94 58.78 / 59.12 13.34	N 4814
8053 01-33-027		1.02 / 0.209 / 4.85	9.16 / 9.17 / 9.85	-17.43 / 10.0	-7.8 / 18.7 / 0.5	20.2 / 31.53	0.56 / 19 / 1.03	10	131 / 156. / 160.	717. / 621. / 1	3 / 0.08 / 0.02	14.1	1.8 0.79 / 1.7 42. / 2	8X / 0.45 / 11 +2 +1	1253.3 417 / 305.73 66.86 / 112.67 1.28	U 8053
8054 01-33-028 2		0.93 / 0.329 / 2.83	9.95 / 9.92 / 10.40	-19.40 / 12.8	-8.0 / 19.3 / 0.5	20.9 / 31.60	1.28 / 217 8 / 1.04	15	284 / 261. / 263.	762. / 668. / 1	3 / 0.39 / 0.01	12.2 / 9.92 / 2.28	2.5 0.43 / 2.1 71. / 2	6A / 0.46 / 11 +2 +1	1253.3 435 / 305.76 67.16 / 112.39 1.36	N 4808
UGCA 309, CVNII DW B	06-28-000		8.59	5.0	1.9 / 15.2 / 2.5	15.6 / 30.97	0.20 / 225 15 / 1.01	10 / 6 / 225	49	730. / 766. / 1	0.04 / 0.01		1.1 0.90 / 1.1 29. / 2	10 / 0.04 / 43 -1	1253.9 3455 / 115.36 82.41 / 82.90 9.09	1253+34
IC 3896	219-12 2		10.47	-20.70 / 22.0	-28.7 / 7.2 / -6.1	30.2 / 32.40		90		2274. / 2031.	5 / 0.00 / 0.94	11.7	2.2 0.72 / 2.5 48. / 6	-5 / 0.16 / 23 -8 +7	1253.9 -5005 / 303.81 12.51 / 165.82 -11.73	1253-50
-1-33-057 2			10.46	-20.66 / 17.0	-12.3 / 17.6 / -0.7	21.5 / 31.66		90		1155. / 1012.	5 / 0.60 / 0.08	11.0	3.4 0.33 / 2.7 78. / 4	2X / 0.57 / 11 -0+10	1254.3 -815 / 305.25 54.33 / 124.82 -1.84	N 4818
M 64	8062 04-31-001 2	0.03	9.85 / 8.32	-19.15 / 8.4	-0.4 / 4.1 / 0.4	4.1 / 28.06	1.09 / 51 17 / 1.13	20	311	414. / 394. / 4	1 / 0.31 / 0.14	8.91 / 6.06 / 2.85	8.0 0.49 / 7.0 66. / 9	2A / 0.20 / 14 +3	1254.3 2157 / 315.75 84.42 / 95.64 6.12	N 4826
UGCA 310	-1-33-000	0.387	9.46 / 9.88	12.5	-13.6 / 23.0 / -0.3	26.7 / 32.13	0.61 / 17 / 1.03	10	133 / 144. / 148.	1538. / 1411. / 1	0.11 / 0.03		1.8 0.73 / 1.6 47. / 2	10 / 0.61 / 11 -0+10	1254.6 -354 / 305.69 58.67 / 120.64 -0.60	1254-03
DDO 156	8074 01-33-031 2	1.13	9.11 / 9.16	-17.29 / 10.8	-8.9 / 19.8 / 0.5	21.7 / 31.69	0.49 / 22 / 1.05	10	134	911. / 811. / 6	3 / 0.02 / 0.01	14.4	1.7 0.94 / 1.7 23. / 2	9 / 0.41 / 11 +2 +1	1255.2 258 / 306.74 65.52 / 114.07 1.39	U 8074
UGCA 311	-1-33-060	0.121	9.53 / 10.44	17.7	-14.8 / 20.4 / -0.9	25.3 / 32.01	0.72 / 16 / 1.11	15	269 / 231. / 234.	1487. / 1341. / 6	0.67 / 0.06		3.4 0.13 / 2.4 90. / 2	5 / 0.73 / 11-10	1255.2 -922 / 305.55 53.20 / 125.95 -1.93	1255-09
269-19 2		0.24	10.61 / 10.00	-21.05 / 31.0	-27.5 / 9.1 / -5.5	29.5 / 32.35	1.06 / 26 / 1.02	20	380	2185. / 1947. / 2	0.67 / 0.52	11.3	4.6 0.17 / 3.6 90. / 6	5X / 0.19 / 23 -7	1255.3 -4559 / 304.74 16.60 / 161.76 -10.73	N 4835
DDO 157	8078 00-33-025 2		10.02	-19.56 / 16.4	-6.6 / 14.1 / 0.3	15.6 / 30.96		90		1228. / 1124.	3 / 0.67 / 0.00	11.4	4.8 0.29 / 3.6 81. / 2	1A / 0.49 / 11 -0 +1	1255.4 150 / 306.70 64.39 / 115.18 1.13	N 4845
DDO 158	8084 01-33-032 2	0.79	9.67 / 9.57	-18.70 / 20.4	-16.7 / 37.5 / 1.1	41.1 / 33.07	0.34 / 22 / 1.03	30	160	2769. / 2670. / 1	7 / 0.04 / 0.01	14.37	1.8 0.89 / 1.7 31. / 2	8B / 0.16 / 11+36	1255.8 304 / 307.12 65.61 / 114.01 1.56	U 8084

(1) Name	(2) α / ℓ / SGL	(3) δ / b / SGB	(4) Type / ϱ_{xyz} / Group	(5) D_{25} / $D^{b,i}_{25}$ / source	(6) d/D / i	(7) $B^{b,i}_T$ / $H_{-0.5}$ / $B{-}H$	(8) source / A^{i-o}_B / A^b_B	(9) V_h / V_o / Tel	(10) W_{20} / W_R / W_D	(11) e_v / e_w / source	(12) F_c / source/e_f / f_h	(13) R / μ	(14) SGX / SGY / SGZ	(15) $M^{b,i}_B$ / $\Delta^{b,i}_{25}$	(16) $\log L_B$ / $\log M_H$ / $\log M_T$	(17) M_H/L_B / M_H/M_T / M_T/L_B	(18) Alternate Names / UGC/ESO MCG AGN? RCII
U 8091	1256.2 / 310.76 / 103.01	1429 / 76.97 / 4.67	10 / 0.19 / 14+12	1.3 / 1.2 / 2	0.83 / 38.	14.56	7 / 0.06 / 0.04	216. / 165. / 3	46 / 40. / 52.	7	0.31 / 13 / 1.01	1.7 / 26.17	-0.4 / 1.7 / 0.1	-11.61 / 0.6	6.84 / 6.77 / 7.44	0.86 / 0.213 / 4.03	DDO 155, GR 8 / 8091 02-33-041 2
1256-11	1256.5 / 305.89 / 128.53	-1157 / 50.61 / -2.31	10 / 0.63 / 11-13+10	1.1 / 1.0 / 2	0.60 / 58.		/ 0.20 / 0.06	1307. / 1153. / 6	97 / 90. / 96.	10	0.42 / 18 / 1.01	22.6 / 31.77	-14.1 / 17.7 / -0.9	/ 6.6	/ 9.13 / 9.19	0.867	UGCA 312
1256-49	1256.5 / 304.25 / 165.16	-4921 / 13.23 / -11.18	10 / 0.10 / 23 -0 +7	1.7 / 1.8 / 6	0.50 / 65.		/ 0.29 / 1.01	1895. / 1653. / 7		20		25.5 / 32.03	-24.1 / 6.4 / -4.9	/ 13.4			-2-33-C00 / 219-16
N 4856	1256.7 / 305.77 / 131.26	-1446 / 47.79 / -3.01	/ 0.35 / 11-15+10	3.8 / 3.1 / 2	0.37 / 74.	11.36	2 / 0.00 / 0.04	1251. / 1087.	(blank)	75		21.1 / 31.63	-13.9 / 15.9 / -1.1	-20.27 / 19.1	10.30		UGCA 313 / -2-33-078 2
N 4861	1256.7 / 311.47 / 82.82	3508 / 82.09 / 9.70	9B / 0.28 / 43 -1	4.2 / 3.5 / 2	0.41 / 71.	12.38	2 / 0.41 / 0.02	847. / 885. / 1	116 / 97. / 103.	7	0.98 / 20 7 / 1.11	17.8 / 31.25	2.1 / 17.5 / 3.0	-18.87 / 18.2	9.74 / 9.48 / 9.70	0.55 / 0.610 / 0.90	IC 3961, ARP 266 / MARK 59, I ZW 49 / 8098 06-29-003 2
N 4866	1257.0 / 311.59 / 103.09	1427 / 76.91 / 4.85	-2A / 1.08 / 11 -0 +1	6.6 / 4.9 / 2	0.24 / 86.	11.96	2 / 0.00 / 0.04	1963. / 1912.		17	0.50 / 205 32	16.0 / 31.02	-3.6 / 15.6 / 1.4	-19.06 / 22.9	9.82 / 8.91	0.12	8102 02-33-045 2
1257-12	1257.7 / 306.34 / 128.72	-1204 / 50.48 / -2.06	10B / 0.49 / 11-13+10	2.0 / 1.8 / 2	0.53 / 63.		/ 0.25ᵢ / 0.06	1582. / 1428. / 1	200	12	0.69 / 15 / 1.03	26.0 / 32.07	-16.2 / 20.3 / -0.9	/ 13.7	9.52		UGCA 314 / -2-33-085
N 4880	1257.7 / 311.34 / 104.79	1245 / 75.21 / 4.57	-2A / 0.85 / 11 -0 +1	3.7 / 3.5 / 2	0.80 / 41.	12.3	3 / 0.00 / 0.05	1557. / 1499.		90		15.7 / 30.98	-4.0 / 15.1 / 1.3	-18.68 / 16.0	9.66		8109 02-33-047 2
1258-15	1258.1 / 306.22 / 132.01	-1527 / 47.09 / -2.87	10 / 0.39 / 11-15+10	1.8 / 1.8 / 4	1.00	1.00	/ 0.00 / 0.04	1385. / 1220. / 6	103	15	0.05 / 36 / 1.07	22.9 / 31.80	-15.3 / 17.0 / -1.1	/ 12.0	8.77		DDO 159, KAR 217 / -2-33-088 2
N 4900	1258.2 / 308.50 / 114.46	246 / 65.27 / 2.06	5B / 0.70 / 11 +2 +1	2.5 / 2.5 / 2	1.00	12.10	2 / 0.00 / 0.00	968. / 870. / 1	130	20	0.64 / 217 11 / 1.07	17.3 / 31.19	-7.2 / 15.7 / 0.6	-19.09 / 12.6	9.83 / 9.12	0.19	8116 01-33-035 2
N 4899	1258.3 / 306.44 / 130.30	-1340 / 48.87 / -2.35	5X / 0.37 / 11-30	2.6 / 2.3 / 4	0.57 / 60.	12.1	5 / 0.22 / 0.05	2656. / 2497. / 1	277	15	0.75 / 16 / 1.06	38.4 / 32.92	-24.8 / 29.2 / -1.6	-20.82 / 25.8	10.52 / 9.92	0.25	-2-33-090 2
N 4891	1258.3 / 306.48 / 129.82	-1310 / 49.37 / -2.21	3 P / 0.33 / 11-30	2.7 / 2.7 / 2	0.94 / 23.		/ 0.02 / 0.05	2546. / 2389. / 1	236	10	0.90 / 16 / 1.08	37.1 / 32.85	-23.7 / 28.5 / -1.4	/ 29.2	10.04		UGCA 316 / -2-33-089 2
N 4902	1258.4 / 306.43 / 130.87	-1415 / 48.29 / -2.48	3B / 0.36 / 11-30	2.6 / 2.6 / 2	0.97 / 17.	11.83	2 / 0.01 / 0.06	2724. / 2563.		61		39.2 / 32.97	-25.6 / 29.6 / -1.7	-21.14 / 29.8	10.65		UGCA 315 / -2-33-092 2

Name / Designation	ID	a-ratios	mag	col	vel (3)	dist	params (3)	N	col	V	col2	extra	size / PA	type	coords
UGCA 317, MARK 234 / 11-16-000 2	1258+64			6.6	18.7 / 24.8 / 8.2	32.2 / 32.54		220		1969. / 2125.	0.00 / 0.02		0.8 / 0.7 ; 0.53 / 63.	0 / 0.16 / 42 -0	1258.6 121.31 52.96 / 6443 52.65 14.71
DDO 160 / -1-33-070 2	1258-04		9.33	13.8	-22.5 / 36.7 / 0.2	43.1 / 33.17	0.06 / 43 / 1.02	30	160	2981. / 2855. / 1	0.67 / 0.03	10.09	1.5 / 1.1 / 4 ; 0.25 / 84.	10 / 0.13 / 11 -0	1258.9 307.66 121.50 / -429 58.03 0.27
219-21	1259-50		10.55	25.3	-17.6 / 4.4 / -3.5	18.4 / 31.33		20	256 / 219. / 221.	1372. / 1131. / 7	0.67 / 0.97		5.4 / 4.7 / 6 ; 0.23 / 87.	5 / 0.13 / 23 -9	1259.4 304.71 165.97 / -5004 12.50 -10.86
UGCA 319 / -3-33-027	1259-16		7.62	2.9	-4.9 / 5.1 / -0.4	7.1 / 29.25	-0.08 / 37 / 1.02	10	49	738. / 569. / 1	0.10 / 0.11		1.5 / 1.4 / 2 ; 0.74 / 46.	10 / 0.13 / 14-17	1259.5 306.59 133.55 / -1657 45.58 -2.94
8146 10-19-015	U 8146	0.91 / 0.247 / 3.69	9.22 / 9.17 / 9.78	-17.56 / 10.4	7.1 / 11.8 / 3.5	14.2 / 30.76	0.87 / 13 / 1.07	10	176 / 142. / 146.	669. / 805. / 1	3 / 0.67 / 0.00	13.2 / 12.34 / 0.86	3.6 / 2.5 / 2 ; 0.16 / 90.	6 / 0.10 / 12 -0 +1	1300.0 120.29 58.88 / 5858 58.38 14.26
8153 01-33-039	U 8153	0.49	9.88 / 9.57	-19.23 / 23.5	-16.6 / 38.9 / 2.2	42.3 / 33.13	0.32 / 22 / 1.04	15	143	2869. / 2779. / 1	3 / 0.00 / 0.02	13.9	1.9 / 1.9 / 2 ; 1.00	6 / 0.16 / 11+36	1300.3 310.33 113.18 / 416 66.70 3.03
UGCA 320, DDO 161 / -3-33-030 2	1300-17	2.20 / 0.483 / 4.56	8.80 / 9.14 / 9.46	-16.52 / 9.5	-4.9 / 5.1 / -0.3	7.1 / 29.26	1.44 / 8 / 1.20	7	129 / 102. / 108.	746. / 578. / 1	7 / 0.67 / 0.11	12.74	6.4 / 4.6 / 2 ; 0.19 / 90.	10 P / 0.13 / 14-17	1300.7 306.98 133.80 / -1708 45.38 -2.72
323-74 -7-27-029	N 4930	0.040	10.07 / 11.47	57.2	-31.9 / 13.4 / -5.3	35.0 / 32.72	0.98 / 14 / 1.03	15	277 / 420. / 421.	2591. / 2364. / 2	0.05 / 0.41		5.4 / 5.6 / 6 ; 0.86 / 35.	5B / 0.09 / 23 -0	1301.3 305.49 157.20 / -4109 21.39 -8.63
UGCA 321 / -1-33-077 2	N 4941	0.07	9.12 / 7.94	-17.31 / 7.1	-3.4 / 5.4 / 0.1	6.4 / 29.03	0.33 / 6 74	24		720. / 593.	2 / 0.15 / 0.03	11.72	4.2 / 3.8 / 2 ; 0.66 / 53.	2X / 0.10 / 14 -0	1301.6 308.79 122.45 / -517 57.17 0.70
-2-33-104 2	N 4939	0.19 / 0.086 / 2.20	11.16 / 10.44 / 11.51	-22.43 / 59.5	-26.7 / 35.4 / -0.4	44.3 / 33.23	1.15 / 217 11 / 1.19	10 / 7 / 211	453 / 431. / 433.	3117. / 2973. / 1	5 / 0.49 / 0.11	10.8 / 8.99 / 1.81	5.6 / 4.6 / 4 ; 0.37 / 74.	4A / 0.00 / 11 -0	1301.7 308.13 127.06 / -1004 52.41 -0.57
-1-33-078 2	N 4942		9.30	12.5	-16.1 / 23.5 / 0.1	28.5 / 32.27	0.39 / 22 / 1.02	20	166	1751. / 1617. / 1	0.09 / 0.09	13.83	1.6 / 1.5 / 4 ; 0.77 / 43.	6A / 0.39 / 11-17+10	1301.7 308.50 124.49 / -724 55.06 0.15
UGCA 322, A 1302 / 00-33-028 2	1301-03	0.300	9.82 / 10.34	24.8	-12.7 / 21.5 / 0.6	25.0 / 31.99	1.02 / 10 / 1.11	8	134 / 174. / 178.	1368. / 1249. / 1	0.06 / 0.01		3.5 / 3.4 / 2 ; 0.83 / 38.	9 / 0.48 / 11 -0+10	1301.9 309.29 120.56 / -318 59.14 1.31
NGC 4948A, DDO 162 / -1-33-080 2	1302-075	0.43	9.50 / 9.13	-18.26 / 10.7	-15.0 / 21.5 / 0.1	26.2 / 32.09	0.29 / 23 / 1.02	15	103	1553. / 1417. / 1	7 / 0.03 / 0.09		1.4 / 1.4 / 4 ; 0.91 / 28.	7B / 0.68 / 11-17+10	1302.4 308.73 125.00 / -753 54.57 0.19
DDO 163 / -1-33-082 2	1302-073		9.17	8.0	-12.0 / 17.2 / 0.1	21.0 / 31.61	0.53 / 26 / 1.02	15	218	1123. / 988. / 1	0.46 / 0.10		1.6 / 1.3 / 4 ; 0.38 / 73.	9B / 0.40 / 11-12+10	1302.5 308.82 124.75 / -737 54.83 0.28

(1) Name	(2) α / ℓ / SGL	(3) δ / b / SGB	(4) Type / ϱ_xyz / Group	(5) D25 / D25^{b,i} / source	(6) d/D / i	(7) B_T^{b,i} / H_{-0.5} / B-H	(8) source / A_B^{i-o} / A_B^b	(9) V_h / V_o / Tel	(10) W_20 / W_R / W_D	(11) e_v / e_w / source	(12) F_c / source e_f / f_h	(13) R / μ	(14) SGX / SGY / SGZ	(15) M_B^{b,i} / Δ_25	(16) log L_B / log M_H / log M_T	(17) M_H/L_B / M_H/M_T / M_T/L_B	(18) Alternate Names UGC/ESO MCG / AGN? RCII
N 4945	1302.5 / 305.27 / 165.21	-4912 / 13.34 / -10.20	6B / 0.15 / 14-15	18.6 / 15.5 / 9	0.19 / 90.	7.95	2 / 0.67 / 0.84	563. / 325. / 7		20		5.2 / 28.60	-5.0 / 1.3 / -0.9	-20.65 / 23.5	10.45		219-24 / 2
N 4951	1302.5 / 309.04 / 123.42	-614 / 56.21 / 0.66	6X / 0.43 / 11-12+10	1.3 / 1.0 / 4	0.30 / 80.	11.6 / 9.59 / 2.01	5 / 0.67 / 0.06	1178. / 1048. /	261 / 227. / 229.	10 / 15 / 79	0.93 / 217 10 /	22.2 / 31.73	-12.2 / 18.5 / 0.3	-20.13 / 6.5	10.24 / 9.62 / 9.99	0.24 / 0.433 / 0.55	-1-33-C81 / 2
N 4947	1302.6 / 306.14 / 151.30	-3504 / 27.45 / -6.94	3B / 0.18 / 11 -0	2.8 / 2.5 / 6	0.51 / 64.	11.9	5 / 0.27 / 0.16	2409. / 2194. / 2	380	70	0.52 / 24 / 1.01	33.3 / 32.61	-29.0 / 15.9 / -4.0	-20.71 / 24.3	10.48 / 9.56	0.12	IC 3974 / 382-05 -6-29-006 / 2
N 4958	1303.2 / 309.10 / 124.91	-744 / 54.70 / 0.42	-2B / 0.40 / 11-12+10	2.2 / 2.3 / 4	1.00	11.30	2 / 0.00 / 0.10	1119. / 984. /		10	0.16 / 205 21 /	20.9 / 31.60	-12.0 / 17.2 / 0.2	-20.30 / 14.0	10.31 / 8.80	0.03	-1-33-084 / 2
N 4961	1303.4 / 44.48 / 90.19	2800 / 86.75 / 9.55	6B / 0.24 / 45 +2	1.7 / 1.6 / 2	0.72 / 48.	13.73	2 / 0.12 / 0.04	2531. / 2543. /	229 / 258. / 260.	10 / 15 / 79	0.54 / 217 10 /	39.4 / 32.98	-0.1 / 38.9 / 6.5	-19.25 / 18.4	9.89 / 9.73 / 10.55	0.69 / 0.151 / 4.56	8185 05-31-126 / 2
U 8188	1303.5 / 107.69 / 80.37	3752 / 79.10 / 11.59	9 / 0.28 / 14 -7	6.4 / 6.2 / 2	0.87 / 33.	12.6	3 / 0.05 / 0.00	326. / 380. / 1	56 / 69. / 77.	7 / 5 / 211	1.16 / 226 2 / 1.35	4.4 / 28.22	0.7 / 4.3 / 0.9	-15.62 / 8.0	8.44 / 8.45 / 9.04	1.02 / 0.254 / 3.99	IC 4182 / 8188 06-29-031 / 2
1303-17	1303.7 / 307.99 / 134.09	-1714 / 45.22 / -2.05	10 / 0.28 / 11 -0+10	1.8 / 1.9 / 4	1.00		/ 0.00 / 0.13	1463. / 1296. / 1	115	10	0.56 / 18 / 1.04	23.6 / 31.87	-16.4 / 17.0 / -0.8	13.1	9.31		DDO 164
1304-28	1304.1 / 307.02 / 144.78	-2817 / 34.20 / -4.91	7B / 0.21 / 11-23+22	2.1 / 2.3 / 6	0.92 / 26.		/ 0.03 / 0.36	2217. / 2019. / 2	115	15	0.55 / 20 / 1.01	31.7 / 32.50	-25.8 / 18.2 / -2.7	21.3	9.55		-3-33-032 / UGCA 324 / 2
N 4965	1304.5 / 307.16 / 144.49	-2757 / 34.53 / -4.74	9A / 0.21 / 11-23+22	2.4 / 2.5 / 6	0.93 / 25.		/ 0.02 / 0.33	2265. / 2068. / 2	190	15	0.66 / 20 / 1.02	32.2 / 32.54	-26.1 / 18.7 / -2.7	23.5	9.68		UGCA 326 / 443-69 -5-31-035 / 443-70 -5-31-036 / 2
U 8201	1304.5 / 120.77 / 49.65	6758 / 49.37 / 15.55	10 / 0.46 / 14+10	3.5 / 3.3 / 2	0.70 / 50.	12.64	7 / 0.13 / 0.03	33. / 201. / 1	68 / 66. / 74.	8	0.82 / 10 / 1.10	2.8 / 27.22	1.7 / 2.0 / 0.7	-14.58 / 2.7	8.02 / 7.71 / 8.53	0.49 / 0.152 / 3.23	VII ZW 499, DDO 165 / 8201 11-16-010 / 2
1305-22	1305.1 / 307.84 / 139.33	-2235 / 39.87 / -3.18	9 / 0.17 / 11-33	3.3 / 2.5 / 6	0.17 / 90.		/ 0.67 / 0.28	2616. / 2434. / 6	253 / 215. / 218.	15	0.64 / 26 / 1.08	37.1 / 32.85	-28.1 / 24.1 / -2.1	27.1	9.78 / 10.56	0.165	UGCA 327 / 508-11 -4-31-035
N 4976	1305.7 / 305.80 / 165.37	-4915 / 13.26 / -9.70	-5 / 0.12 / 23 -9	4.5 / 4.8 / 6	0.58 / 59.	10.29	2 / 0.00 / 0.88	1369. / 1132. /		26		18.6 / 31.34	-17.7 / 4.6 / -3.1	-21.05 / 26.1	10.61		219-29 / 2
U 8215	1305.8 / 114.61 / 71.10	4707 / 70.01 / 13.56	10 / 0.39 / 14 -7	1.1 / 1.1 / 2	0.82 / 39.		/ 0.07 / 0.00	221. / 314. / 1	52 / 53. / 63.	10 / 8 / 226	226 13 / 1.01	3.9 / 27.96	1.2 / 3.6 / 0.9	1.3	7.18 / 8.01	0.149	8215 08-24-048

Name	Alt. names																	
N 4981	-1-34-003 2	0.27 / 0.090 / 2.97	10.36 / 9.79 / 10.83	-20.42 / 21.9	-15.5 / 23.0 / 0.7	27.8 / 32.22	0.90 / 17 / 1.07	15	283 / 327. / 329.	1677. / 1549. / 1	5 / 0.12 / 0.08	11.8	2.8 / 2.7 / 4	0.72 / 49.	4X / 0.47 / 11-17+10	1306.2 / 310.62 / 123.95	-631 / 55.83 / 1.47
N 4984	-2-34-004 2		10.21	-20.04 / 13.7	-14.3 / 15.7 / -0.3	21.3 / 31.64		90		1259. / 1101.	5 / 0.08 / 0.08	11.6	2.3 / 2.2 / 4	0.80 / 41.	1X / 0.26 / 11 -0+10	1306.3 / 309.14 / 132.35	-1515 / 47.14 / -0.91
1306-24	UGCA 328 / 508-15 -4-31-037		9.47	18.7	-31.1 / 25.3 / -2.3	40.1 / 33.02	0.26 / 34 / 1.02	20	119	2872. / 2687. / 6	0.09 / 0.41		1.5 / 1.6 / 6	0.78 / 43.	10 / 0.26 / 11-33	1306.7 / 308.15 / 140.91	-2407 / 38.31 / -3.24
N 4995	UGCA 329 / -1-34-007 2		10.40	-20.51 / 18.0	-16.1 / 22.9 / 0.7	28.0 / 32.23		58		1708. / 1577.	2 / 0.09 / 0.10	11.72	2.3 / 2.2 / 2	0.76 / 45.	3X / 0.41 / 11-17+10	1307.1 / 310.79 / 125.02	-734 / 54.76 / 1.39
1307-10	UGCA 330 / -2-34-006		9.22	19.6	-13.2 / 17.2 / 0.3	21.7 / 31.68	0.55 / 25 / 1.09	15	190	1213. / 1073. / 1	0.13 / 0.10		3.3 / 3.1 / 2	0.69 / 51.	5 / 0.42 / 11 -0+10	1307.2 / 310.34 / 127.41	-1003 / 52.29 / 0.73
1307-46	269-57			40.4	-38.4 / 12.2 / -6.2	40.7 / 33.05		20		3075. / 2843. / 7	0.12 / 0.48		3.3 / 3.4 / 6	0.71 / 49.	1 / 0.32 / 23 +2	1307.2 / 306.28 / 162.39	-4610 / 16.32 / -8.75
1307-23	508-19 -4-31-040		9.68	21.7	-32.0 / 26.1 / -2.2	41.3 / 33.08	0.45 / 25 / 1.01	20		2967. / 2783.	0.32 / 0.41		1.9 / 1.8 / 6	0.48 / 67.	10 P / 0.23 / 11-33	1307.3 / 308.34 / 140.81	-2358 / 38.44 / -3.07
U 8246	8246 06-29-048 2	0.76 / 0.206 / 3.66	9.27 / 9.15 / 9.83	-17.69 / 12.1	1.8 / 17.0 / 3.5	17.3 / 31.19	0.67 / 16 / 1.06	20	170 / 139. / 143.	813. / 855. / 1	3 / 0.67 / 0.01	13.5 / 12.79 / 0.71	3.2 / 2.4 / 2	0.28 / 81.	6B / 0.30 / 43 -1	1307.7 / 94.53 / 83.96	3427 / 81.89 / 11.77
1308-23	UGCA 331 / 508-24 -4-31-041		9.77	32.7	-30.8 / 25.4 / -2.0	40.0 / 33.01	0.57 / 18 / 1.01	15	188	2855. / 2672. / 2	0.03 / 0.31		2.6 / 2.8 / 6	0.91 / 28.	5B / 0.26 / 11-33	1308.0 / 308.59 / 140.50	-2336 / 38.79 / -2.81
N 5005	8256 06-29-052 2	0.28	10.70 / 10.15	-21.27 / 30.5	3.2 / 20.6 / 4.6	21.3 / 31.64	1.49 / 51 / 43	26	127 / 113. / 118.	1015. / 1070.	1 / 0.27 / 0.00	10.37	5.6 / 4.9 / 9	0.52 / 64.	4X / 0.28 / 43 -1	1308.5 / 101.72 / 81.11	3719 / 79.27 / 12.46
U 8261	8261 06-29-053	0.402	8.90 / 9.29	5.3	2.3 / 17.5 / 3.8	18.1 / 31.29	0.38 / 25 / 1.01	10		855. / 903. / 1	0.25 / 0.00		1.1 / 1.0 / 2	0.53 / 63.	10 / 0.32 / 43 -1	1308.7 / 97.67 / 82.66	3547 / 80.63 / 12.22
U 8263	A 1309 / 8263 01-34-005 2	0.77	9.77 / 9.65	-18.94 / 11.7	-18.2 / 40.3 / 3.8	44.4 / 33.24	0.36 / 217 / 20	9 / 20 / 217	195	3036. / 2949.	3 / 0.67 / 0.03	14.3	1.3 / 0.9 / 2	0.15 / 90.	5B / 0.00 / 11+36	1308.8 / 315.14 / 114.29	341 / 65.82 / 4.85
N 5012	8270 04-31-012 2	0.42	10.28 / 9.90	-20.23 / 30.7	-3.6 / 39.7 / 6.8	40.4 / 33.03	0.69 / 35 / 1.06	25	407	2625. / 2620. / 1	3 / 0.18 / 0.03	12.8	2.8 / 2.6 / 2	0.62 / 56.	5X / 0.29 / 45 -1	1309.1 / 351.35 / 95.23	2312 / 83.81 / 9.74
N 5014	MARK 449 / 8271 06-29-055 2		9.71	-18.79 / 9.3	3.1 / 22.0 / 4.9	22.8 / 31.79		71		1105. / 1157.	3 / 0.45 / 0.00	13.0	1.7 / 1.4 / 2	0.39 / 73.	1 / 0.22 / 43 -1	1309.2 / 99.21 / 81.91	3633 / 79.90 / 12.46

(1) Name	(2) α / ℓ / SGL	(3) δ / b / SGB	(4) Type / ϱ_{xyz} / Group	(5) D_{25} / $D_{25}^{b,i}$ / source	(6) d/D / i	(7) $B_T^{b,i}$ / $H_{-0.5}$ / $B{-}H$	(8) source / A_B^{i-o} / A_B^b	(9) V_h / V_o / Tel	(10) W_{20} / W_R / W_D	(11) e_v / e_w / source	(12) F_c / source,e_F / f_h	(13) R / μ	(14) SGX / SGY / SGZ	(15) $M_B^{b,i}$ / $\Delta_{25}^{b,i}$	(16) $\log L_B$ / $\log M_H$ / $\log M_T$	(17) M_H/L_B / M_H/M_T / M_T/L_B	(18) Alternate Names — UGC/ESO MCG AGN? RCII
1309-11	1309.4	-1149	6B	1.7	0.68			2107.	125	15	0.27	32.0	-20.3		9.28		UGCA 332
	310.86	50.47	0.11	1.6	52.		0.14	1963.			26	32.53	24.8	15.0			-2-34-008
	129.26	0.76	11 -0	2			0.06	1			1.03		0.4				
1309-06	1309.4	-644	8B	3.6	0.59			1485.	173	8	1.04	25.6	-14.4		9.86	0.390	UGCA 333
	311.97	55.51	0.50	3.2	59.		0.21	1359.	163.		11	32.04	21.1	23.9	10.27		-1-34-011
	124.37	2.17	11-17+10	2			0.08	1	167.		1.09		1.0				
N 5016	1309.7	2422	5X	1.9	0.76	13.6	3	2613.	247	15	0.43	40.3	-2.8	-19.43	9.96	0.47	8279 04-31-013 2
	1.07	84.44	0.30	1.8	45.		0.09	2614.	299.		34	33.03	39.6	21.2	9.64	0.080	
	94.11	10.13	45 -1	2			0.02	1	301.		1.03		7.1		10.74	5.97	
U 8280	1309.8	3556	5	2.6	0.22	13.3	3	815.	194	15	0.42	17.4	2.2	-17.90	9.35	0.35	IC 4213
	96.91	80.39	0.30	1.9	87.		0.67	865.	159.		20	31.20	16.9	9.7	8.90	0.113	8280 06-29-057
	82.55	12.47	43 -1	2			0.00	1	163.		1.04		3.8		9.85	3.15	
1309+26	1309.9	2657	12	1.1	0.61	14.2	3	886.	142	70		18.1	-0.6	-17.09	9.03	0.51	05-31-153 2
	28.66	85.33	0.18	1.0	57.		0.19	898.				31.29	17.9	5.3			
	91.55	10.73	43 +1	2			0.02						3.4				
U 8285	1310.0	727	9	2.0	0.29	14.2	3	896.	142	15	0.22	17.5	-6.2	-17.01	9.00		8285 01-34-008
	317.97	69.44	0.42	1.5	81.		0.67	825.			28	31.21	16.3	7.7	8.71		
	110.71	6.13	11 -0 +1	2			0.02	1			1.03		1.9				
N 5011	1310.0	-4250	-5	2.4	0.89	11.96	6	3063.		64		40.9	-37.9	-21.10	10.63		269-65 -7-27-042 2
	307.08	19.60	0.27	2.6	30.		0.00	2838.				33.06	14.4	31.1			
	159.24	-7.46	23 -0 +2	6			0.44						-5.3				
N 5023	1310.2	4418	5	6.8	0.14	11.6	3	400.	193	10	1.16	6.0	1.6	-17.29	9.11	0.41	8286 07-27-043 2
	110.36	72.58	0.36	4.7	90.	10.69	0.67	484.	158.		8	28.89	5.6	8.2	8.72	0.087	
	74.08	13.88	14 -5	2		0.91	0.00	1	161.		1.21		1.4		9.78	4.65	
U 8290	1310.2	2306	9 P	1.6	0.76	14.4	3	2604.	190	40	0.33	40.1	-3.7	-18.62	9.64	0.79	8290 04-31-014
	352.27	83.56	0.29	1.5	45.		0.09	2599.			20	33.02	39.3	17.6	9.54		
	95.39	9.96	45 -1	2			0.03	1			1.02		6.9				
1310-32	1310.2	-3226	8B	3.1	0.21			2382.	261	15	0.68	33.3	-28.5		9.72	0.165	UGCA 334
	308.16	29.95	0.23	2.3	90.		0.67	2178.	223.		23	32.61	17.0	22.4	10.51		443-83 -5-31-039 2
	149.15	-4.73	11+32	6			0.21	2	226.		1.01		-2.7				
N 5017	1310.3	-1630	-5	1.4	0.80	13.0	5	2543.		90		36.8	-25.5	-19.83	10.12		-3-34-016 2
	310.33	45.80	0.32	1.4	41.		0.00	2384.				32.83	26.5	15.0			
	133.82	-0.33	11-31	4			0.11						-0.2				
N 5018	1310.3	-1915	-5	4.5	0.75	11.60	2	2897.		75		40.9	-29.7	-21.46	10.78		UGCA 335
	309.89	43.07	0.29	4.4	45.		0.00	2729.				33.06	28.2	52.5			576-10 -3-34-017 2
	136.46	-1.10	11 -0	6			0.20						-0.8				
1310-15	1310.9	-1511	8A	2.7	0.94			2503.	85	8	0.50	36.4	-24.7		9.62		UGCA 338
	310.77	47.09	0.33	2.7	23.		0.02	2348.			18	32.81	26.8	28.7			-2-34-010
	132.59	0.17	11-31	2			0.07	1			1.08		0.1				

ID	Names	(1)	(2)	(3)	(4)	(5)	(6)	(7)	(8)	(9)	(10)	(11)	(12)	(13)	(14)	(15)
U 83C3	A 1311, HO VIII / DDO 166 / 8303 06-29-061 2	0.58	9.40 / 9.16	-18.02 / 16.0	2.7 / 19.6 / 4.5	20.3 / 31.54	0.55 / 15 / 1.05	10	106	949. / 1002. / 1	1 / 0.00 / 0.00	13.52	2.7 / 2.7 / 2	1.00	10 / 0.30 / 43 -1	1311.0 97.17 82.06 / 3628 79.81 12.80
N 5033	8307 06-29-062 2 / L1.9	0.37 / 0.053 / 7.07	10.60 / 10.17 / 11.45	-21.03 / 47.5	2.6 / 18.1 / 4.2	18.7 / 31.36	1.63 / 211 8 / 1.11	10 / 7 / 211	446 / 454 / 456	877. / 931. / 4	1 / 0.27 / 0.00	10.33 / 7.47 / 2.86	10.1 / 8.7 / 2	0.52 / 64.	5A / 0.33 / 43 -1	1311.2 97.99 81.68 / 3651 79.45 12.91
U 8308	DDO 167 / 8308 08-24-090 2	0.233	7.05 / 7.68	1.2	1.0 / 2.9 / 0.8	3.2 / 27.53	0.04 / 23 / 1.02	7	50. / 37.	165. / 259. / 1	0.16 / 0.00		1.4 / 1.3 / 2	0.65 / 54.	10 / 0.37 / 14 -7	1311.2 111.61 71.78 / 4635 70.32 14.40
U 8313	8313 07-27-048 2	0.33	8.52 / 8.04	-15.83 / 4.0	2.2 / 8.7 / 2.2	9.2 / 29.83	0.11 / 29 / 1.02	15	127	621. / 699. / 1	3 / 0.64 / 0.00	14.0	1.9 / 1.5 / 2	0.31 / 79.	5B / 0.36 / 14 -5	1311.6 107.48 75.99 / 4228 74.25 13.89
1312-22	UGCA 339 / 508-30 -4-31-000	0.334	9.19 / 9.67	12.4	-18.0 / 15.1 / -0.7	23.5 / 31.85	0.45 / 25 / 1.01	20	137 / 114 / 119.	1508. / 1330. / 2	0.46 / 0.30		2.1 / 1.8 / 6	0.38 / 73.	10 / 0.21 / 11 -19	1312.2 309.92 140.06 / -2252 39.42 -1.68
U 8320	DDO 168 / 8320 08-24-093 2	0.98 / 0.417 / 2.35	8.37 / 8.36 / 8.74	-15.45 / 3.6	1.1 / 3.3 / 0.9	3.6 / 27.78	1.25 / 5 6 / 1.10	7	88 / 73. / 81.	198. / 291. / 1	7 / 0.37 / 0.00	12.33	4.0 / 3.4 / 2	0.44 / 70.	10B / 0.39 / 14 -7	1312.3 110.80 72.21 / 4611 70.66 14.52
N 5037	-3-34-029 2		10.28	-20.21 / 12.7	-20.1 / 21.0 / 0.1	29.0 / 32.31		90		1887. / 1730.	5 / 0.67 / 0.11	12.1	2.1 / 1.5 / 4	0.22 / 87.	1A / 0.20 / 11 -20	1312.3 311.05 133.79 / -1620 45.90 0.17
U 8323	MARK 450 / 8323 06-29-065 2		8.90	-16.76 / 6.3	2.0 / 17.3 / 3.9	17.9 / 31.26		100		840. / 888.	3 / 0.10 / 0.00	14.5	1.3 / 1.2 / 2	0.75 / 45.	10 P / 0.33 / 43 -1	1312.5 91.54 83.44 / 3509 80.77 12.87
N 5044	-3-34-034 2		10.60	-21.01 / 27.3	-26.8 / 28.1 / 0.3	38.9 / 32.95		90		2704. / 2548.	2 / 0.00 / 0.11	11.94	2.4 / 2.4 / 2	0.96 / 20.	-5 / 0.38 / 11 -31	1312.7 311.22 133.61 / -1607 46.11 0.33
N 5042	UGCA 340 / 508-31 -4-31-043 2	0.155	9.74 / 10.55	26.0	-16.9 / 13.7 / -0.7	21.7 / 31.69	1.07 / 19 / 1.21	15	233 / 218. / 221.	1390. / 1210. / 6	0.27 / 0.26	9.96	4.5 / 4.1 / 6	0.52 / 64.	5X / 0.14 / 11 -19	1312.8 309.97 140.94 / -2345 38.53 -1.80
N 5049	UGCA 343 / -3-34-037 2		9.99	-19.49 / 16.1	-27.1 / 28.4 / 0.3	39.3 / 32.97		65		2744. / 2588.	2 / 0.00 / 0.12	13.48	1.8 / 1.4 / 2	0.31 / 79.	-2 / 0.37 / 11 -31	1313.3 311.43 133.67 / -1608 46.07 0.46
U 8331	DDO 169 / 8331 08-24-097 2	0.93 / 0.249 / 3.72	7.91 / 7.88 / 8.48	-14.29 / 3.1	1.5 / 4.2 / 1.2	4.6 / 28.29	0.55 / 13 / 1.05	8	74 / 58. / 68.	258. / 358. / 1	3 / 0.49 / 0.00	14.0	2.9 / 2.3 / 2	0.37 / 74.	10A / 0.40 / 14 -5	1313.3 111.50 70.61 / 4746 69.09 14.91
U 8333	DDO 170 / 8333 04-31-016 2	0.374	8.89 / 9.31	6.1	-0.9 / 18.6 / 3.8	19.0 / 31.39	0.33 / 27 / 1.01	15	132 / 108. / 113.	941. / 950. / 1	0.54 / 0.04		1.4 / 1.1 / 2	0.35 / 76.	10 / 0.15 / 43 -0 +1	1313.5 17.15 92.97 / 2542 84.26 11.26
N 5055	M 63 / 8334 07-27-054 2	0.26 / 0.032 / 8.05	10.25 / 9.65 / 11.15	-20.14 / 24.6	1.7 / 6.7 / 1.8	7.2 / 29.27	1.94 / 211 5 / 1.20	5 / 4 / 211	405 / 447. / 448.	497. / 575. / 4	1 / 0.17 / 0.00	9.13 / 6.28 / 2.85	13.0 / 11.7 / 2	0.63 / 55.	4A / 0.40 / 14 -5	1313.5 106.03 76.23 / 4217 74.30 14.21

(1) Name	(2) α / l / SGL	(3) δ / b / SGB	(4) Type / ϱ_{xyz} / Group	(5) D_{25} / $D^{b:i}_{25}$ / source	(6) d/D / i	(7) $B^{b:i}_T$ / $H_{-0.5}$ / $B-H$	(8) source / A^{i-o}_B / A^b_B	(9) V_h / V_o / Tel	(10) W_{20} / W_R / W_D	(11) e_v / e_w / source	(12) F_c / source/e_f / f_h	(13) R / μ	(14) SGX / SGY / SGZ	(15) $M^{b:i}_B$ / $\Delta^{b:i}_{25}$	(16) $\log L_B$ / $\log M_H$ / $\log M_T$	(17) M_H/L_B / M_H/M_T / M_T/L_B	(18) Alternate Names UGC/ESO — MCG — AGN? RCII
N 5054	1314.4 / 311.76 / 133.97	-1622 / 45.80 / 0.65	3A / 0.23 / 11-20	4.6 / 4.3 / 2	0.62 / 56.	11.0	5 / 0.18 / 0.13	1743. / 1587. / 1	347 / 371. / 372.	25	0.83 / 217 13 / 1.16	27.3 / 32.18	-19.0 / 19.7 / 0.3	-21.18 / 34.3	10.66 / 9.70 / 11.14	0.11 / 0.037 / 2.97	UGCA 344 / -3-34-039 / 2
U 8355	1315. / 116.20 / 60.27	5748 / 59.26 / 16.13	-2 / 0.12 / 42 -0	1.8 / 1.7 / 2	0.76 / 45.	13.5	3 / 0.00 / 0.00	2696. / 2834.		105		41.5 / 33.09	19.8 / 34.6 / 11.5	-19.59 / 20.6	10.03		IC 875, MARK 249 / 8355 10-19-059 / 2
N 5061	1315.3 / 310.24 / 143.81	-2634 / 35.66 / -2.05	-5 / 0.31 / 11-22	2.5 / 2.7 / 6	1.00	11.10	6 / 0.00 / 0.25	1961. / 1775.		61		28.7 / 32.29	-23.1 / 16.9 / -1.0	-21.19 / 22.6	10.67		508-38 / -4-31-048 / 2
1316-08	1316.0 / 314.37 / 126.22	-811 / 53.82 / 3.34	10 / 0.35 / 11 -0+10	1.5 / 1.5 / 2	0.71 / 49.	13.83	7 / 0.12 / 0.09	1310. / 1183. / 1	67	15	0.16 / 22 / 1.02	23.3 / 31.84	-13.8 / 18.8 / 1.4	-18.01 / 10.2	9.40 / 8.89	0.32	DDO 171 / -1-34-014 / 2
N 5064	1316.0 / 307.70 / 164.21	-4739 / 14.70 / -7.66	3A / 0.35 / 23 -2	2.4 / 2.4 / 6	0.50 / 65.	11.8	5 / 0.29 / 0.66	2982. / 2752.		90		39.5 / 32.98	-37.7 / 10.7 / -5.3	-21.18 / 27.7	10.66		220-02 / / 2
N 5068	1316.3 / 311.51 / 138.33	-2047 / 41.37 / -0.18	6X / 0.14 / 14+17	7.6 / 7.9 / 6	0.90 / 29.	10.2	5 / 0.04 / 0.32	679. / 510.	113	8	1.58 / 5 / 1.12	6.7 / 29.13	-5.0 / 4.4 / 0.0	-18.93 / 15.5	9.76 / 9.23	0.29	UGCA 345 / 576-29 -3-34-046 / 2
U 8365	1316.5 / 104.02 / 76.4C	4212 / 74.17 / 14.75	7B / 0.27 / 43 -2 +1	2.5 / 2.2 / 2	0.64 / 55.	14.23	7 / 0.17 / 0.00	1215. / 1294. / 1	195	60	0.35 / 25 / 1.05	23.7 / 31.88	5.4 / 22.3 / 6.0	-17.65 / 15.2	9.25 / 9.10	0.70	DDO 172 / 8365 07-27-058 / 2
N 5077	1316.9 / 313.55 / 130.34	-1224 / 49.63 / 2.36	-5 / 0.23 / 11 -0	1.9 / 1.9 / 2	0.79 / 42.	12.52	2 / 0.00 / 0.08	2823. / 2682.		27		40.6 / 33.04	-26.3 / 30.9 / 1.7	-20.52 / 22.5	10.40		UGCA 347 / -2-34-027 / L2 / 2
1317-47	1317.0 / 307.94 / 163.64	-4701 / 15.32 / -7.33	5 / 0.33 / 23 -2	2.6 / 2.6 / 6	0.66 / 53.		/ 0.15 / 0.56	2850. / 2622. / 7		20		37.9 / 32.89	-36.1 / 10.6 / -4.8	/ 28.8			269-85
N 5089	1317.3 / 61.40 / 88.31	3031 / 83.07 / 13.03	5 P / 0.09 / 45 -0	2.1 / 1.9 / 2	0.51 / 64.	13.5	3 / 0.28 / 0.03	2154. / 2186.		100		35.1 / 32.72	1.0 / 34.1 / 7.9	-19.22 / 19.5	9.88		8371 05-31-175
1317-35	1317.4 / 309.43 / 152.81	-3548 / 26.45 / -4.22	8 / 0.18 / 16 -4	3.9 / 3.5 / 6	0.50 / 65.		/ 0.29 / 0.16	1455. / 1248. / 2	162 / 143. / 147.	15	0.77 / 15 / 1.01	21.2 / 31.63	-18.8 / 9.7 / -1.6	/ 21.7	9.42 / 10.11	0.206	382-45 -6-29-028
1317-77	1317.4 / 304.61 / 193.72	-7716 / -14.75 / -13.87	7 / 0.15 / 55 -0	3.2 / 2.5 / 6	0.15 / 90.		/ 0.67 / 0.62	2655. / 2416. / 7		20		33.1 / 32.60	-31.2 / -7.6 / -7.9	/ 24.2			040-07
N 5084	1317.6 / 311.76 / 139.17	-2134 / 40.55 / -0.12	-2 / 0.29 / 11-21	12.5 / 9.3 / 6	0.15 / 90.	11.7	5 / 0.00 / 0.31	1728. / 1558.		23		26.4 / 32.11	-20.0 / 17.3 / -0.1	-20.41 / 71.7	10.36		576-33 -4-32-004 / 2

N 5085	1317.6	-2409	5A	3.9	0.90	11.6	5	1958.	254	15	1.09	28.9	-22.7	-20.71	10.48	0.35	UGCA 349
	311.29	37.99	0.33	4.0	29.		0.04	1780.			12	32.31	18.0	33.8	10.01		
	141.64	-0.86	11-22	6			0.27	6			1.24		-0.4				508-50 -4-32-005 2
N 5087	1317.7	-2021	-2A	2.5	0.83	11.71	2	1832.		150		27.8	-20.7	-20.51	10.40		UGCA 350
	312.03	41.75	0.31	2.6	37.		0.00	1666.				32.22	18.6	21.1			
	138.01	0.26	11-21	6			0.29						0.1				576-35 -3-34-050 2
N 5088	1317.7	-1219	5A	2.5	0.23	12.2	5	1464.		90		24.5	-15.8	-19.74	10.09		
	313.87	49.68	0.29	1.8	87.		0.67	1324.				31.94	18.6	12.9			
	130.31	2.57	11 -0+10	4			0.07						1.1				-2-34-034 2
U 8385	1318.2	1003	9X	2.5	0.47	13.75	7	1133.	217	20	0.60	22.4	-7.1	-18.00	9.39	0.81	DDO 173
	326.02	71.34	0.19	2.2	67.		0.33	1079.	195.		21	31.75	21.0	14.4	9.30	0.126	
	108.70	8.75	11 -0 +1	2			0.00	1	198.		1.04		3.4		10.20	6.43	8385 02-34-008 2
N 5101	1319.0	-2711		5.4	1.00	11.24	6	1864.	200	10	0.71	27.4	-22.3	-20.95	10.57	0.10	UGCA 351
	311.14	34.94	0.31	5.7			0.00	1679.			12	32.19	15.9	45.6	9.59		
	144.64	-1.44	11-22	6			0.28	2			1.06		-0.7				508-58 -4-32-008 2
N 5102	1319.1	-3622	-2A	8.3	0.42	10.16	2	459.	222	9	1.23	3.5	-3.2	-17.59	9.23	0.12	
	309.73	25.84	0.17	7.1	71.		0.00	251.	196.	10	204 9	27.75	1.6	7.3	8.32	0.026	
	153.45	-4.05	14-15	6			0.19		199.	204			-0.3		9.91	4.78	382-50 -6-29-031 2
N 5107	1319.2	3848	6B	1.9	0.31	13.0	3	943.	177	10	0.57	19.9	3.4	-18.49	9.59	0.38	
	95.97	76.97	0.29	1.5	79.	12.19	0.64	1010.	146.	15	217 15	31.49	19.0	8.7	9.17	0.273	
	79.96	14.78	43 -1	2		0.81	0.00		150.	217			5.1		9.73	1.39	8396 07-28-001 2
N 5112	1319.6	3900	6B	3.7	0.73	11.8	3	978.	217	10	1.07	20.5	3.5	-19.76	10.10	0.40	
	96.13	76.76	0.28	3.5	47.		0.11	1046.	246.	7	217 7	31.56	19.5	21.0	9.69	0.134	
	79.76	14.89	43 -1	2			0.00	1	248.	79	1.12		5.3		10.57	2.95	8403 07-28-003 2
U 8409	1320.6	2334	8A	2.5	0.47	14.1	3	2812.	197	20	0.26	42.7	-4.0	-19.05	9.81	0.51	
	6.49	81.90	0.00	2.2	67.		0.33	2817.			38	33.15	41.5	27.4	9.52		
	95.44	12.39	45 +1	2			0.00	1			1.04		9.2				8409 04-32-009
N 5116	1320.6	2715	5B	2.3	0.35	13.0	3	2834.		100		43.0	-1.3	-20.17	10.26		
	33.59	82.98	0.00	1.8	76.		0.54	2854.				33.17	41.9	22.6			
	91.74	13.12	45 -0	2			0.06						9.8				8410 05-32-009 2
N 5117	1320.6	2835	5B	2.2	0.49	13.5	3	2435.		100		38.3	-0.3	-19.41	9.96		
	44.46	82.92	0.18	1.9	66.		0.30	2461.				32.91	37.2	21.2			
	90.39	13.37	45 -0	2			0.01						8.8				8411 05-32-010
1321-24	1321.3	-2422	10	2.4	0.67			2056.	113	25	0.38	30.2	-23.8				UGCA 352, DDO 241
	312.30	37.65	0.25	2.3	53.		0.15	1880.			25	32.40	18.6	20.3	9.34		
	142.09	-0.12	11-22	6			0.26	2			1.01		-0.1				508-66 -4-32-000 2
N 5121	1321.9	-3725	-2A	2.2	0.80	12.0	5	1532.		90		22.1	-19.9	-19.72	10.08		
	310.19	24.73	0.16	2.2	41.		0.00	1324.				31.72	9.4	14.2			
	154.62	-3.81	16 -4	6			0.22						-1.5				382-57 -6-29-035 2
1322-19	1322.0	-1927	6X	4.2	0.56			1975.	209	30	0.73	29.8	-21.9				UGCA 353
	313.57	42.47	0.22	3.9	61.		0.23	1815.	197.		18	32.37	20.1	33.9	9.68	0.126	
	137.44	1.49	11-21	6			0.26	6	200.		1.13		0.8		10.58		576-50 -3-34-069

(1) Name	(2) α / ℓ / SGL	(3) δ / b / SGB	(4) Type / ϱ_{xyz} / Group	(5) D_{25} / $D^{b,i}_{25}$ / source	(6) d/D / i	(7) $B^{b,i}_T$ / $H_{-0.5}$ / B-H	(8) source / A^{i-o}_B / A_B	(9) V_h / V_o / Tel	(10) W_{20} / W_R / W_D	(11) e_v / e_w / source	(12) F_c / source/e_f / f_h	(13) R / μ	(14) SGX / SGY / SGZ	(15) $M^{b,i}_B$ / $\Delta^{b,i}_{25}$	(16) log L_B / log M_H / log M_T	(17) M_H/L_B / M_H/M_T / M_T/L_B	(18) Alternate Names UGC/ESO / MCG / AGN? RCII
N 5128	1322.4 / 309.49 / 159.77	-4245 / 19.43 / -5.25	-2 P / 0.20 / 14-15	16.7 / 17.5 / 9	0.78 / 43.	7.48	1 / 0.00 / 0.48	541. / 323.		8		4.9 / 28.45	-4.6 / 1.7 / -0.4	-20.97 / 25.0	10.58		ARP 153, CENTAURUS A / 270-09 -7-28-001 2
N 5134	1322.6 / 313.43 / 138.85	-2053 / 41.04 / 1.20	1A / 0.26 / 11-21	3.9 / 3.8 / 6	0.68 / 51.	12.0	5 / 0.14 / 0.27	1696. / 1532.		90		26.1 / 32.09	-19.7 / 17.2 / 0.5	-20.09 / 29.0	10.23		576-52 -3-34-073 2
N 5145	1323.0 / 102.13 / 75.20	4331 / 72.48 / 16.10	12 / 0.27 / 43 -2 +1	2.3 / 2.2 / 2	0.80 / 41.	12.9	3 / 0.08 / 0.00	1225. / 1313. / 1	233 / 300. / 302.	15	0.54 / 26 / 1.05	23.7 / 31.88	5.8 / 22.1 / 6.6	-18.98 / 15.2	9.78 / 9.29 / 10.60	0.32 / 0.049 / 6.53	8439 07-28-009
1323-47	1323.1 / 308.88 / 164.84	-4758 / 14.25 / -6.59	5 / 0.17 / 23 +2	1.9 / 1.7 / 6	0.30 / 79.		0.67 / 0.58	2629. / 2402. / 7		20		35.0 / 32.72	-33.6 / 9.1 / -4.0	17.4			220-08
U 8441	1323.5 / 114.21 / 60.04	5805 / 58.73 / 17.25	10 / 0.09 / 42 -0 +3	3.3 / 3.1 / 2	0.74 / 46.	13.3	3 / 0.10 / 0.00	1519. / 1661. / 1	131 / 144. / 148.	15	0.58 / 15 / 1.09	26.8 / 32.14	12.8 / 22.1 / 7.9	-18.84 / 24.3	9.73 / 9.44 / 10.16	0.51 / 0.187 / 2.73	DDO 175 / 8441 10-19-070 2
N 5147	1323.8 / 322.89 / 116.56	222 / 63.62 / 8.11	8B / 0.16 / 11 -0	1.9 / 1.9 / 2	0.90 / 29.	12.22	2 / 0.04 / 0.03	1100. / 1020.		7 / 12 / 211	0.69 / 217 15 / 1.01	21.6 / 31.67	-9.6 / 19.1 / 3.1	-19.45 / 12.0	9.97 / 9.36	0.24	8443 00-34-033 2
1323-21	1323.9 / 313.57 / 139.96	-2157 / 39.93 / 1.17	10 / 0.20 / 11-21	2.1 / 2.0 / 6	0.58 / 59.		0.21 / 0.24	1440. / 1273. / 2	104	30	0.19 / 31 / 1.01	22.7 / 31.78	-17.4 / 14.6 / 0.5	13.3	8.90		UGCA 356, DDO 174 / 576-59 -4-32-017 2
1324+38	1324.0 / 90.59 / 80.88	3802 / 77.08 / 15.60	10 / 0.23 / 43 -1	1.0 / 0.9 / 2	0.77 / 44.		0.09 / 0.00	1170. / 1237. / 1	170 / 199. / 202.	20	0.52 / 14 / 1.01	23.3 / 31.83	3.6 / 22.1 / 6.3	6.1	9.25 / 9.85	0.256	UGCA 357
1324-29	1324.2 / 312.02 / 147.49	-2949 / 32.17 / -1.12	10 P / 0.23 / 11 -0+22	2.1 / 1.9 / 6	0.60 / 58.		0.19 / 0.21	1894. / 1705. / 6	140 / 132. / 136.	15	0.54 / 21 / 1.06	27.5 / 32.20	-23.2 / 14.8 / -0.5	15.3	9.42 / 9.89	0.341	UGCA 358 / 444-37 -5-32-000
U 8449	1324.4 / 100.64 / 75.73	4302 / 72.80 / 16.30	9 / 0.28 / 43 -2 +1	1.5 / 1.2 / 2	0.40 / 72.		0.43 / 0.00	1237. / 1324. / 1	193	25	0.33 / 30 / 1.01	23.9 / 31.89	5.6 / 22.2 / 6.7	8.4	9.09		8449 07-28-010
1324-37	1324.6 / 310.69 / 155.25	-3755 / 24.15 / -3.45	7 / 0.17 / 16 -4	3.8 / 2.9 / 6	0.22 / 87.		0.67 / 0.23	1443. / 1235. / 2	215 / 179. / 182.	15	1.23 / 12 / 1.01	20.7 / 31.58	-18.8 / 8.7 / -1.2	17.5	9.86 / 10.21	0.447	324-23 -6-30-003
1324-41	1324.7 / 310.18 / 158.42	-4113 / 20.89 / -4.40	10 / 0.21 / 14-15	3.5 / 3.6 / 6	0.76 / 45.		0.10 / 0.34	525. / 311. / 2	112 / 125. / 130.	10	1.06 / 8	4.6 / 28.30	-4.2 / 1.7 / -0.4	4.8	8.39 / 9.34	0.111	324-24 -7-28-000
1325-27	1325.6 / 312.88 / 145.20	-2719 / 34.58 / -0.07	10 / 0.26 / 11-22	2.1 / 2.1 / 6	0.77 / 44.		0.09 / 0.18	1837. / 1656. / 7		20		27.1 / 32.16	-22.2 / 15.4 / 0.0	16.6			509-26 -4-32-000

ID	(1)	(2)	(3)	(4)	(5)	(6)	(7)	(8)	(9)	(10)	(11)	(12)	(13)	(14)	(15)	(16)	Names
N 5156	1325.7	-4839	1B	2.4	0.93	11.9	5	2983.		20		39.5	-38.0	-21.08	10.62		220-13 2
	309.22	13.51	0.38	2.7	24.		0.02	2757.				32.98	9.7	31.1			
	165.62	-6.37	23 +2	6			0.63	7					-4.4				
N 5169	1326.0	4656	3B	2.1	0.37	14.0	3	2482.	335	40	0.60	39.0	11.7	-18.95	9.77	1.02	8465 08-25-004 2
	105.03	69.16	0.48	1.7	74.		0.49	2585.			12	32.95	35.4	19.4	9.78		
	71.71	16.96	42 -2 +1	2			0.00	1			1.03		11.4				
N 5161	1326.3	-3256	6X	6.1	0.50	11.5	5	2391.	356	15	1.26	33.5	-29.1	-21.12	10.64	0.47	UGCA 359
	311.95	29.02	0.21	5.4	65.	9.16	0.29	2196.	351.		10	32.62	16.4	52.8	10.31	0.108	383-04 -5-32-031 2
	150.60	-1.63	11-32	6		2.34	0.17	2	352.		1.04		-1.0		11.28	4.32	
N 5173	1326.3	4651	-5	1.1	1.00	13.5	3	2404.		50		38.0	11.4	-19.40	9.95		8468 08-25-005 2
	104.80	69.21	0.43	1.1			0.00	2507.				32.90	34.5	12.2			
	71.80	17.01	42 -2 +1	2			0.00						11.1				
N 5170	1327.1	-1742	3A	8.3	0.15	10.9	5	1498.	629	10	1.31	24.0	-17.3	-21.00	10.59	0.30	UGCA 360
	315.65	43.96	0.20	6.1	90.	8.30	0.67	1347.	591.		15	31.90	16.6	42.7	10.07	0.027	576-65 -3-34-084 2
	136.13	3.16	11-18	6		2.60	0.23	1	592.		1.25		1.3		11.64	11.08	
U 8489	1327.6	4539	8X	2.3	0.35	13.91	7	1302.	154	10	0.51	24.6	6.9	-18.05	9.41	0.76	DDO 176
	102.75	70.21	0.21	1.8	76.		0.54	1401.	127.		5 17	31.96	22.5	12.9	9.29	0.324	8489 08-25-011 2
	73.07	17.13	43 -2 +1	2			0.00	1	131.		1.03		7.3		9.78	2.34	
N 5204	1327.7	5840	9A	4.8	0.65	11.61	1	200.	127	5	1.44	4.8	2.3	-16.79	8.91	0.78	8490 10-19-078 2
	113.50	58.02	0.29	4.4	53.	10.29	0.15	346.	125.	4	212 4	28.40	3.9	6.2	8.80	0.227	
	59.44	17.82	14 -9	2		1.32	0.00	3	130.	212	1.19		1.5		9.45	3.45	
N 5194	1327.8	4727	4AP	13.6	0.52	8.68	1	467.	195	10	1.69	7.7	2.4	-20.75	10.49	0.09	M 51, ARP 85, VV 1
	104.83	68.56	0.33	11.8	64.		0.27	573.		6	106 6	29.43	7.0	26.5	9.46		TURN 85A L/S
	71.19	17.31	14 -5	9			0.00	4			1.17		2.3				8493 08-25-012 2
N 5195	1327.9	4732	13 P	6.3	0.74	10.53	1	552.		16		9.3	2.9	-19.32	9.92		ARP 85, VV 1, TURN 85B L2
	104.88	68.48	0.13	5.9	46.		0.00	658.				29.85	8.4	16.0			8494 08-25-014 2
	71.11	17.33	14 -5	2			0.00						2.8				
N 5198	1328.1	4656	-5	2.2	0.85	12.7	3	2488.		44		39.0	11.7	-20.26	10.30		I ZW 59
	104.11	69.01	0.48	2.1	35.		0.00	2592.				32.96	35.4	23.9			8499 08-25-015 2
	71.74	17.32	42 -2 +1	2			0.00						11.6				
N 5205	1328.2	6246	12	3.3	0.61	12.7	3	1781.	253	15	0.56	29.9	16.3	-19.68	10.06	0.28	8501 11-17-003
	115.37	54.05	0.23	3.0	57.	10.38	0.19	1940.	256.		16	32.38	23.4	26.2	9.51	0.065	
	55.13	17.89	42 +8 +3	2		2.32	0.03	1	258.		1.08		9.2		10.70	4.30	
U 8507	1328.6	1942	13 P	1.7	0.60	13.9	3	1020.	163	40	0.08	22.8	-3.7	-17.89	9.35	0.28	VV 88
	354.77	78.11	0.09	1.5	58.		0.00	1014.			32	31.79	21.8	10.0	8.80		8507 03-34-044
	99.72	13.41	41 -0	2			0.02	1			1.02		5.3				
N 5188	1328.6	-3432	1BP	3.8	0.38	12.1	5	2366.		90		32.9	-29.1	-20.49	10.39		383-09 -6-30-007 2
	312.18	27.36	0.19	3.1	74.		0.48	2169.				32.59	15.3	29.8			
	152.27	-1.66	11-32	6			0.16						-1.0				
U 8508	1328.8	5510	10A	1.8	0.63	14.4	3	68.	73	10	0.58	2.7	1.1	-12.73	7.28	1.44	I ZW 60
	111.13	61.31	0.34	1.6	56.		0.17	202.	67.		12	27.13	2.3	1.3	7.44	0.169	8508 09-22-075
	63.12	17.89	14 -9	2			0.00	1	75.		1.02		0.8		8.22	8.54	

(1) Name	(2) α / ℓ / SGL	(3) δ / b / SGB	(4) Type / ϱ_{xyz} / Group	(5) D_{25} / $D_{25}^{b,i}$ / source	(6) d/D / i	(7) $B_T^{b,i}$ / $H_{-0.5}$ / B–H	(8) source / A_B^{i-0} / A_B^b	(9) V_h / V_o / Tel	(10) W_{20} / W_R / W_D	(11) e_v / e_w / source	(12) F_c / source/e_f / f_h	(13) R / μ	(14) SGX / SGY / SGZ	(15) $M_B^{b,i}$ / Δ_{25}	(16) $\log L_B$ / $\log M_H$ / $\log M_T$	(17) M_H/L_B / M_H/M_T / M_T/L_B	(18) Alternate Names / UGC/ESO MCG / AGN? RCII
1331−45	1331.7 / 310.82 / 162.70	−4517 / 16.67 / −4.40	10 / 0.06 / 14+15	12.7 / 9.7 / 6	0.12 / 90.		0.67 / 0.43	829. / 612. / 2	158. / 127. / 131.	8	1.65 / 6 / 1.03	10.8 / 30.17	−10.3 / 3.2 / −0.8	/ 30.6	9.72 / 10.16	0.364	FOURCADE-FIGUERO / 270-17 -7-28-004
1331+69	1331.7 / 117.36 / 48.45	6907 / 47.81 / 18.03	13 P / 0.02 / 42 −0	0.6 / 0.6 / 2	0.77 / 44.		0.00 / 0.02	1350. / 1529.		220		24.2 / 31.92	15.3 / 17.2 / 7.5	/ 4.2			UGCA 362, MARK 263 / 12-13-000 2
1331+60	1331.8 / 113.64 / 57.35	6039 / 55.97 / 18.34	10 / 0.35 / 42 −8 +3	1.1 / 1.0 / 2	0.80 / 41.		0.08 / 0.01	2072. / 2226. / 1	56. / 58. / 67.	8	−0.02 / 25 / 1.01	33.7 / 32.64	17.3 / 27.0 / 10.6	/ 9.8	9.04 / 8.98	1.130	UGCA 363 / 10-19-089
N 5229	1332.0 / 103.93 / 70.50	4810 / 67.61 / 18.06	5B / 0.33 / 14 −5	3.5 / 2.5 / 2	0.20 / 90.	13.2	3 / 0.67 / 0.00	365. / 476. / 1	157. / 126. / 130.	10	0.81 / 12 / 1.07	6.4 / 29.02	2.0 / 5.7 / 2.0	−15.82 / 4.7	8.52 / 8.42 / 9.33	0.86 / 0.123 / 6.50	8550 08-25-019 2
N 5238	1332.7 / 107.40 / 66.61	5152 / 64.19 / 18.35	8X / 0.33 / 14 −9	2.0 / 1.9 / 2	0.82 / 39.	13.3	3 / 0.07 / 0.00	243. / 368. / 1	59. / 66. / 74.	8	−0.26 / 19 / 1.04	4.9 / 28.46	1.9 / 4.3 / 1.5	−15.16 / 2.7	8.26 / 7.64 / 8.54	0.24 / 0.127 / 1.90	I ZW 64 / 8565 09-22-082
U 8575	1333.3 / 335.01 / 110.43	914 / 69.04 / 12.15	10 / 0.26 / 41 −8 +7	2.5 / 1.8 / 2	0.22 / 87.	14.5	3 / 0.67 / 0.00	1167. / 1122. / 1	147. / 117. / 122.	20 / 12 / 226	0.50 / 21 / 1.04	22.9 / 31.80	−7.7 / 21.1 / 4.8	−17.30 / 12.0	9.11 / 9.22 / 9.68	1.28 / 0.347 / 3.69	8575 02-35-010
U 8578	1333.3 / 47.68 / 89.97	2929 / 80.04 / 16.26	13 P / 0.14 / 43 +1	1.3 / 0.9 / 2	0.27 / 82.	15.19	2 / 0.00 / 0.01	840. / 878.		16	−0.11 / 10 78	17.7 / 31.24	0.0 / 17.0 / 4.9	−16.05 / 4.7	8.61 / 8.39	0.59	HARO 38 / 8578 05-32-041 2
U 8588	1333.7 / 100.75 / 72.60	4611 / 69.24 / 18.22	9 / 0.17 / 43 −0	1.5 / 1.5 / 2	1.00	14.3	3 / 0.00 / 0.05	1447. / 1551. / 1	53	8	0.04 / 24 / 1.03	26.4 / 32.10	7.5 / 23.9 / 8.2	−17.80 / 11.6	9.31 / 8.88	0.37	DDO 178 / 8588 08-25-023 2
N 5236	1334.2 / 314.58 / 147.97	−2937 / 31.97 / 1.01	5X / 0.18 / 14-15	11.5 / 11.7 / 9	0.94 / 24.	8.04	2 / 0.02 / 0.14	518. / 337.	287	5 / 10 / 69	2.67 / 25 / 6	4.7 / 28.35	−4.0 / 2.5 / 0.1	−20.31 / 16.1	10.32 / 10.01	0.50	M 83, UGCA 366 / 444-81 -5-32-050 2
U 8597	1334.2 / 100.92 / 72.31	4628 / 68.95 / 18.33	7B / 0.57 / 42 −2 +1	1.9 / 1.8 / 2	0.85 / 35.	14.08	7 / 0.05 / 0.00	2434. / 2540. / 1	119. / 163. / 166.	15	0.53 / 16 / 1.03	38.4 / 32.92	11.1 / 34.7 / 12.1	−18.84 / 20.2	9.73 / 9.70 / 10.19	0.93 / 0.321 / 2.91	DDO 177 / 8597 08-25-025 2
1334−11	1334.4 / 320.27 / 130.81	−1135 / 49.49 / 6.69	7X / 0.15 / 11-28	1.9 / 1.9 / 2	0.86 / 34.		0.05 / 0.11	2474. / 2349. / 1	117	15	0.37 / 25 / 1.04	36.5 / 32.81	−23.7 / 27.4 / 4.2	/ 20.3	9.49		UGCA 367 / -2-35-006
1334−27	1334.5 / 315.11 / 146.27	−2748 / 33.73 / 1.65	10 / 0.13 / 14-15	1.4 / 1.4 / 6	0.81 / 40.		0.07 / 0.21	585. / 409. / 7		20		5.6 / 28.75	−4.7 / 3.1 / 0.2	/ 2.3			444-84 -5-32-000
U 8611	1334.8 / 98.78 / 73.70	4509 / 70.04 / 18.35	7X / 0.41 / 42 +1	1.7 / 1.7 / 2	0.94 / 23.	14.4	3 / 0.02 / 0.00	2640. / 2741. / 1	185	20	0.34 / 21 / 1.03	40.9 / 33.06	10.9 / 37.3 / 12.9	−18.66 / 20.3	9.66 / 9.56	0.81	8611 08-25-029

Name	(coordinates)																Other designations
U 8614	1334.9 / 334.25 / 111.86	754 / 67.68 / 12.21	10 / 0.24 / 41 -8 +7	3.7 / 3.3 / 2	0.53 / 63.	12.94	7 / 0.25 / 0.01	1053. / 1004. / 1	227 / 214. / 216.	15	0.81 / 14 / 1.10	21.6 / 31.67	-7.8 / 19.6 / 4.5	-18.73 / 20.8	9.68 / 9.48 / 10.44	0.62 / 0.109 / 5.72	DDO 179 / 8614 01-35-020 2
N 5248	1335.1 / 335.96 / 110.65	908 / 68.74 / 12.56	4X / 0.30 / 41 -8 +7	6.1 / 5.7 / 2	0.76 / 45.	10.71	1 / 0.09 / 0.00	1156. / 1112. / 4	291 / 361. / 362.	10	1.34 / 217 6 / 1.06	22.7 / 31.78	-7.8 / 20.7 / 5.0	-21.07 / 37.8	10.62 / 10.05 / 11.15	0.27 / 0.079 / 3.43	8616 02-35-015 2
N 5247	1335.4 / 318.34 / 136.67	-1738 / 43.58 / 5.06	5A / 0.14 / 11 -18	4.6 / 4.9 / 6	1.00	10.79	2 / 0.00 / 0.31	1360. / 1216. / 1	160	10 / 8 / 217	1.20 / 9 / 1.21	22.2 / 31.74	-16.1 / 15.2 / 2.0	-20.95 / 31.8	10.57 / 9.89	0.21	UGCA 368 / 577-14 -3-35-011 2
1335-09	1335.5 / 321.56 / 128.94	-933 / 51.37 / 7.55	9 / 0.13 / 11 -0+10	2.3 / 2.4 / 2	1.00	12.65	7 / 0.00 / 0.10	1300. / 1184. / 1	110	8	0.65 / 14 / 1.06	22.7 / 31.78	-14.2 / 17.5 / 3.0	-19.13 / 15.9	9.84 / 9.36	0.33	DDO 180 / -2-35-010 2
N 5254	1336.9 / 321.32 / 130.66	-1114 / 49.66 / 7.38	5A / 0.12 / 11 -28	2.9 / 2.6 / 4	0.53 / 63.		/ 0.25 / 0.14	2315. / 2194. / 1	346 / 346. / 348.	20	0.77 / 26 / 1.07	34.6 / 32.70	-22.4 / 26.0 / 4.4	26.3	9.85 / 10.96	0.077	-2-35-012 2
U 8639	1337.0 / 105.84 / 66.82	5141 / 64.09 / 19.01	10 / 0.16 / 42 -0 +3	1.7 / 1.6 / 2	0.77 / 44.	14.4	3 / 0.09 / 0.00	1708. / 1834. / 1	134	15	-0.03 / 58 / 1.03	29.4 / 32.34	10.9 / 25.5 / 9.6	-17.94 / 13.7	9.37 / 8.91	0.35	8639 09-22-089 2
N 5253	1337.1 / 314.86 / 149.86	-3124 / 30.09 / 1.02	-3 P / 0.18 / 14 -15	4.5 / 3.6 / 6	0.33 / 77.	10.87	2 / 0.00 / 0.12	417. / 233. / 2	100 / 81. / 88.	10	1.10 / 51 7 / 1.03	3.2 / 27.53	-2.8 / 1.6 / 0.1	-16.66 / 3.4	8.86 / 8.11 / 8.81	0.18 / 0.201 / 0.89	UGCA 369 / 445-04 -5-32-069 2
1337-48	1337.5 / 311.31 / 165.67	-4805 / 13.74 / -4.33	6A / 0.42 / 23 -3	3.2 / 3.5 / 6	1.00		/ 0.00 / 0.48	2847. / 2628.		13		37.8 / 32.89	-36.5 / 9.3 / -2.9	38.6			NGC 5266A / 220-30 2
U 8651	1337.8 / 89.72 / 78.10	4100 / 73.10 / 18.58	10 / 0.33 / 14 +8 +7	2.9 / 2.5 / 2	0.53 / 63.	13.8	3 / 0.25 / 0.00	200. / 287. / 3	56 / 42. / 54.	5	0.51 / 10 / 1.06	3.4 / 27.66	0.4 / 2.1 / 0.7	-13.86 / 2.5	7.74 / 7.57 / 8.10	0.69 / 0.295 / 2.33	DDO 181 / 8651 07-28-046 2
U 8658	1338.7 / 107.93 / 63.76	5435 / 61.34 / 19.30	4X / 0.33 / 42 -0 +3	2.6 / 2.4 / 2	0.63 / 56.	13.18	1 / 0.17 / 0.00	2035. / 2172. / 1	246 / 252. / 255.	10	0.72 / 18 / 1.05	33.4 / 32.62	13.9 / 28.2 / 11.0	-19.44 / 23.4	9.97 / 9.77 / 10.63	0.63 / 0.136 / 4.64	A 1339, HOLMBERG V / ScI 132 / 8658 09-22-091 2
N 5264	1338.8 / 315.72 / 148.34	-2940 / 31.70 / 1.94	10 / 0.20 / 14 -15	2.8 / 2.8 / 6	0.86 / 35.	12.42	7 / 0.05 / 0.17	478. / 300. / 2	55 / 64. / 73.	8	0.44 / 14 / 1.01	4.2 / 28.10	-3.5 / 2.2 / 0.1	-15.68 / 3.4	8.46 / 7.69 / 8.61	0.17 / 0.119 / 1.40	UGCA 370, DDO 242 / 445-12 -5-32-066 2
1339+30	1339.7 / 52.93 / 88.86	3046 / 78.43 / 17.82	13 P / 0.04 / 43 -0 +1	0.5 / 0.5 / 2	0.92 / 26.		/ 0.00 / 0.00	1065. / 1113.		95		22.4 / 31.75	0.4 / 21.3 / 6.8	3.3			UGCA 372, MARK 67 / 05-32-000 2
N 5283	1339.7 / 115.82 / 49.64	6756 / 48.76 / 18.85	-2 / 0.08 / 42 -0	1.2 / 1.2 / 2	0.92 / 27.	14.1	3 / 0.00 / 0.04	2697. / 2875.		43		41.4 / 33.08	25.3 / 29.8 / 13.4	-18.98 / 14.5	9.78		MARK 270 S2 / 8672 11-17-007 2
N 5266	1339.9 / 311.74 / 165.66	-4756 / 13.80 / -3.90	-2A / 0.35 / 23 -3	2.9 / 3.0 / 6	0.71 / 49.	11.76	6 / 0.00 / 0.51	3082. / 2865.		50		40.8 / 33.05	-39.4 / 10.1 / -2.8	-21.29 / 35.7	10.71		220-33 2

136

(1) Name	(2) α / ℓ / SGL	(3) δ / b / SGB	(4) Type / ρ_{xyz} / Group	(5) D_{25} / $D_{25}^{b,i}$ / source	(6) d/D / i	(7) $B_T^{b,i}$ / $H_{-0.5}$ / B-H	(8) source / A_B^{i-o} / A_B^b	(9) V_h / V_o / Tel	(10) W_{20} / W_R / W_D	(11) e_v / e_w / source	(12) F_c / source/e_F / f_h	(13) R / μ	(14) SGX / SGY / SGZ	(15) $M_B^{b,i}$ / $\Delta_{25}^{b,i}$	(16) log L_B / log M_H / log M_T	(17) M_H/L_B / M_H/M_T / M_T/L_B	(18) Alternate Names / LGC/ESO MCG / AGN? RCII
N 5273	1339.9 / 74.39 / 83.48	3555 / 76.24 / 18.50	-2A / 0.15 / 43 +1	3.0 / 2.9 / 2	0.83 / 37.	12.43	2 / 0.00 / 0.00	1021. / 1090.		17		21.3 / 31.65	2.3 / 20.1 / 6.8	-19.22 / 18.0	9.88		TURN 89A / 8675 06-30-072 / 2
U 8683	1340.4 / 85.90 / 79.29	3955 / 73.57 / 18.98	10 / 0.27 / 14 -5	2.3 / 2.3 / 2	1.00	14.3	3 / 0.00 / 0.00	663. / 748. / 1	44	7	0.21 / 19 / 1.06	10.2 / 30.04	1.3 / 6.9 / 2.4	-15.74 / 6.9	8.49 / 8.23	0.55	DDO 182 / 8683 07-28-055 / 2
1342-41	1342.0 / 313.50 / 159.82	-4137 / 19.90 / -1.45	10 / 0.15 / 14-15	3.2 / 2.9 / 6	0.50 / 65.		0.29 / 0.29	543. / 338. / 7		20		5.3 / 28.62	-5.0 / 1.8 / -0.1	4.5			325-11
N 5290	1343.1 / 89.32 / 77.16	4158 / 71.72 / 19.64	4BP / 0.78 / 42 -1	3.4 / 2.6 / 2	0.29 / 81.	11.9 / 9.17 / 2.73	3 / 0.67 / 0.00	2579. / 2673. / 1	479	30	0.46 / 55 / 1.07	37.8 / 32.89	8.1 / 35.8 / 13.1	-20.99 / 28.7	10.59 / 9.61	0.11	8700 07-28-061 / 2
N 5296	1344.2 / 93.01 / 74.91	4406 / 69.95 / 19.96	-2 / 0.77 / 42 -1	1.1 / 1.0 / 2	0.58 / 60.	14.7	3 / 0.00 / 0.00	2281. / 2384.		87		37.8 / 32.89	9.1 / 33.8 / 12.7	-18.19 / 11.0	9.47		TURN 90A / 07-28-062 / 2
N 5297	1344.3 / 93.00 / 74.89	4407 / 69.93 / 19.97	3X / 0.79 / 42 -1	5.3 / 3.8 / 2	0.21 / 89.	11.5	3 / 0.67 / 0.00	2404. / 2507. / 1	422 / 384. / 386.	20	1.09 / 13 / 1.14	37.8 / 32.89	9.3 / 34.5 / 13.0	-21.39 / 41.9	10.75 / 10.24 / 11.25	0.31 / 0.098 / 3.20	TURN 90B / 8709 07-28-063 / 2
N 5301	1344.4 / 96.67 / 72.52	4621 / 68.12 / 20.08	3A / 0.14 / 43 -0	4.0 / 2.9 / 2	0.22 / 87.	11.9	3 / 0.67 / 0.02	1562. / 1673. / 1		71	0.82 / 11 / 1.00	27.7 / 32.21	7.8 / 24.8 / 9.5	-20.31 / 23.5	10.32 / 9.70	0.24	8711 08-25-041 / 2
U 8716	1344.8 / 110.97 / 57.31	6038 / 55.45 / 19.93	13 P / 0.45 / 42 -8 +3	1.3 / 1.1 / 2	0.37 / 74.		0.00 / 0.00	2063. / 2222. / 1	128	15	0.09 / 32 / 1.01	33.5 / 32.63	17.0 / 26.5 / 11.4	10.8	9.14		8716 10-20-000 / 2
N 5308	1345.4 / 111.23 / 56.69	6113 / 54.88 / 19.97	-2 / 0.45 / 42 -8 +3	3.0 / 2.1 / 2	0.18 / 90.	12.19	2 / 0.00 / 0.01	1972. / 2133.		46		32.4 / 32.55	16.7 / 25.5 / 11.1	-20.36 / 19.9	10.34		8722 10-20-029 / 2
N 5300	1345.7 / 335.73 / 116.23	412 / 63.15 / 13.87	5A / 0.27 / 41 -7	3.6 / 3.3 / 2	0.64 / 55.	12.3 / 10.36 / 1.94	3 / 0.17 / 0.02	1179. / 1123. / 1	232 / 238. / 241.	10	0.62 / 24 / 1.10	23.1 / 31.82	-9.9 / 20.1 / 5.5	-19.52 / 22.3	10.00 / 9.35 / 10.56	0.22 / 0.061 / 3.66	8727 01-35-038 / 2
1346-48	1346.0 / 312.66 / 166.43	-4823 / 13.15 / -3.09	5 / 0.43 / 23 -3	2.2 / 1.9 / 6	0.30 / 80.		0.67 / 0.59	2930. / 2715. / 7		20		38.9 / 32.95	-37.8 / 9.1 / -2.1	21.6			221-02
1346-35	1346.2 / 315.81 / 154.66	-3549 / 25.36 / 1.32	10 / 0.12 / 14-15	4.6 / 4.7 / 6	0.92 / 27.		0.03 / 0.20	330. / 141. / 2	59	10	0.85 / 8 / 1.05	1.9 / 26.39	-1.7 / 0.8 / 0.0	2.6	7.41		383-87 -6-30-025
U 8733	1346.5 / 91.24 / 75.41	4339 / 70.04 / 20.35	5B / 0.81 / 42 -1	2.5 / 2.3 / 2	0.61 / 57.	13.4	3 / 0.19 / 0.00	2345. / 2447. / 1	215	30	0.66 / 17 / 1.05	37.8 / 32.89	8.9 / 34.2 / 13.1	-19.49 / 25.4	9.99 / 9.81	0.67	8733 07-28-070

Name	Position data	Type	Axis ratios	b/a, PA	mag	—	V (km/s)	—	N	—	Dist	—	—	—	—	Cross IDs
U 8742	1347.4 3909 / 80.74 73.10 / 80.21 20.27	10 / 0.65 / 42 -1	1.7 / 1.6 / 2	0.89 / 31.		3 / 0.04 / 0.00	2266. / 2352. / 1	89 / 137. / 141.	15 / 12 / 226	-0.01 / 226 32 / 1.03	37.8 / 32.89	5.9 / 34.2 / 12.8	17.7	9.14 / 9.98	0.145	8742 07-28-000
N 5322	1347.6 6026 / 110.27 55.50 / 57.50 20.29	-5 / 0.43 / 42 -8 +3	6.6 / 6.1 / 2	0.68 / 51.	10.84	2 / 0.00 / 0.01	1902. / 2061.		75		31.6 / 32.50	15.9 / 25.0 / 10.9	-21.66 / 56.3	10.86		8745 10-20-035 2
N 5313	1347.6 4014 / 83.41 72.36 / 79.06 20.37	3 / 0.75 / 42 -1	1.7 / 1.5 / 2	0.60 / 58.	12.4	3 / 0.20 / 0.00	2606. / 2696.		90		37.8 / 32.89	6.9 / 35.9 / 13.6	-20.49 / 16.6	10.39		8744 07-28-074 2
N 5320	1348.2 4137 / 86.38 71.33 / 77.59 20.57	5 / 0.77 / 42 -1	3.4 / 3.0 / 2	0.53 / 63.	12.4 / 10.26 / 2.14	3 / 0.25 / 0.00	2613. / 2709. / 1	308. / 304. / 306.	15	0.77 / 20 / 1.08	37.8 / 32.89	7.9 / 35.8 / 13.8	-20.49 / 33.1	10.39 / 9.92 / 10.95	0.34 / 0.095 / 3.63	8749 07-28-076 2
1348-33	1348.4 -3334 / 316.92 27.43 / 152.71 2.53	7B / 0.12 / 16 -5	6.1 / 5.7 / 6	0.63 / 56.		0.18 / 0.18	1388. / 1206. / 7		20		20.5 / 31.55	-18.2 / 9.4 / 0.9	34.1			384-02 -5-33-000
1348+62	1348.5 6259 / 111.80 53.11 / 54.78 20.19	10 / 0.32 / 42 -8 +3	1.1 / 1.1 / 2	1.00		0.00 / 0.03	2132. / 2299. / 1	48	10	-0.17 / 35 / 1.01	34.4 / 32.68	18.6 / 26.3 / 11.9	11.0	8.90		UGCA 374 / 11-17-000
U 8760	1348.7 3816 / 77.78 73.45 / 81.17 20.46	10 / 0.31 / 14 +8 +7	2.3 / 1.8 / 2	0.30 / 80.	13.71	7 / 0.67 / 0.00	189. / 273. / 3	45. / 23. / 41.	7	0.41 / 5 11 / 1.03	3.3 / 27.58	0.5 / 3.0 / 1.1	-13.87 / 1.7	7.74 / 7.45 / 7.43	0.51 / 1.052 / 0.48	DDO 183 / 8760 06-30-105 2
N 5326	1348.7 3949 / 81.92 72.48 / 79.52 20.56	1 / 0.70 / 42 -1	2.2 / 1.9 / 2	0.49 / 66.	12.6	3 / 0.30 / 0.00	2564. / 2653.		90		37.8 / 32.89	6.6 / 35.8 / 13.7	-20.29 / 21.0	10.31		8764 07-28-082 2
1349+40	1349.3 4028 / 83.27 71.96 / 78.83 20.71	13 P / 0.83 / 42 -1	1.1 / 1.1 / 2	1.00	15.0	3 / 0.00 / 0.00	2370. / 2462.		100		37.8 / 32.89	6.8 / 34.6 / 13.3	-17.89 / 12.1	9.35		MARK 462 / 07-29-002 2
N 5324	1349.4 -548 / 328.98 53.73 / 126.35 11.96	4A / 0.00 / 41 -0	2.5 / 2.5 / 4	1.00	12.2	5 / 0.00 / 0.06	3045. / 2954. / 1	232	15	0.78 / 13 / 1.07	44.0 / 33.22	-25.5 / 34.7 / 9.1	-21.02 / 32.1	10.60 / 10.07	0.29	-1-35-016 2
N 5334	1350.4 -53 / 333.06 58.10 / 121.59 13.62	5B / 0.16 / 41 -0	4.2 / 4.1 / 2	0.80 / 41.	12.3	3 / 0.08 / 0.12	1383. / 1311. / 1	223 / 286. / 288.	15	0.84 / 15 / 1.15	24.7 / 31.96	-12.6 / 20.5 / 5.8	-19.66 / 29.6	10.06 / 9.63 / 10.85	0.37 / 0.060 / 6.17	8790 00-35-024 2
1350+64	1350.4 6437 / 112.49 51.52 / 53.03 20.24	13 P / 0.17 / 42 +8 +3	0.8 / 0.6 / 2	0.37 / 74.	15.2	3 / 0.00 / 0.03	1656. / 1828.		220		28.4 / 32.26	16.0 / 21.3 / 9.8	-17.06 / 5.0	9.02		UGCA 375, MARK 277 / VII ZW 528 / 11-17-009 2
U 8798	1350.7 4404 / 90.39 69.23 / 74.99 21.12	8 / 0.76 / 42 -1	1.7 / 1.5 / 2	0.53 / 63.		0.25 / 0.00	2282. / 2388. / 1	203 / 188. / 191.	20	0.40 / 20 / 1.02	37.8 / 32.89	9.0 / 33.5 / 13.4	16.6	9.55 / 10.23	0.211	8798 07-29-006
1350-82	1350.9 -8250 / 305.04 -20.48 / 199.62 -13.70	6 / 0.17 / 55 -0	2.1 / 2.3 / 6	1.00		0.00 / 0.43	2451. / 2221. / 7		20		30.3 / 32.40	-27.7 / -9.9 / -7.2	20.3			008-04

(1) Name	(2) α / ℓ / SGL	(3) δ / b / SGB	(4) Type / Q_{xyz} / Group	(5) D_{25} / $D_{25}^{b,i}$ / source	(6) d/D / i	(7) $B_T^{b,i}$ / $H_{-0.5}$ / B–H	(8) source / $A_B^{i\to o}$ / A_B^b	(9) V_h / V_o / Tel	(10) W_{20} / W_R / W_D	(11) e_v / e_w / source	(12) F_c / source/e_f / f_h	(13) R / μ	(14) SGX / SGY / SGZ	(15) $M_B^{b,i}$ / Δ_{25}	(16) log L_B / log M_H / log M_T	(17) M_H/L_B / M_H/M_T / M_T/L_B	(18) Alt. Names UGC/ESO / MCG / AGN? RCII
N 5347	1351.1 / 62.17 / 86.03	3344 / 75.23 / 20.56	2B / 0.19 / 42 -0 +1	1.7 / 1.7 / 2	0.84 / 37.	13.34	2 / 0.06 / 0.00	2296. / 2364.		90		36.7 / 32.82	2.4 / 34.3 / 12.9	-19.48 / 18.2	9.98		8805 06-31-007 / 2
N 5350	1351.3 / 82.79 / 78.71	4036 / 71.59 / 21.10	4B / 0.80 / 42 -1	3.2 / 3.1 / 2	0.81 / 40.	12.13	2 / 0.07 / 0.00	2316. / 2410. / 1	293 / 400. / 401.	15	0.85 / 16 / 1.10	37.8 / 32.89	6.9 / 34.2 / 13.4	-20.76 / 34.2	10.50 / 10.00 / 11.20	0.32 / 0.064 / 5.07	8810 07-29-009 / 2
N 5333	1351.3 / 313.57 / 166.63	-4816 / 13.05 / -2.23	-2 / 0.34 / 23 -4 +3	1.9 / 1.9 / 6	0.50 / 65.		/ 0.00 / 0.77	2750. / 2538.		90		36.6 / 32.82	-35.6 / 8.5 / -1.4	20.3			221-17
N 5353	1351.4 / 82.59 / 78.78	4032 / 71.62 / 21.11	-2 / 0.50 / 42 -1	3.0 / 2.8 / 2	0.65 / 54.	12.05	2 / 0.00 / 0.00	2022. / 2116.		23		37.8 / 32.89	6.5 / 32.5 / 12.9	-20.84 / 30.9	10.53		8813 07-29-010 / 2
N 5348	1351.7 / 340.05 / 115.34	529 / 63.49 / 15.64	3 / 0.20 / 41 -7	3.3 / 2.4 / 2	0.21 / 89.	13.2 / 11.46 / 1.74	3 / 0.67 / 0.01	1457. / 1411. / 1	193 / 158. / 161.	15	0.49 / 19 / 1.06	26.2 / 32.09	-10.8 / 22.8 / 7.1	-18.89 / 18.4	9.75 / 9.33 / 10.12	0.38 / 0.160 / 2.37	8821 01-35-051 / 2
U 8833	1352.7 / 69.69 / 83.56	3605 / 73.94 / 21.09	10 P / 0.22 / 14 +8 +7	1.3 / 1.2 / 2	0.83 / 38.		3 / 0.06 / 0.00	225. / 303. / 3	38 / / 34.	5	0.10 / 18 / 1.01	3.6 / 27.78	0.4 / 3.3 / 1.3	1.3	7.21		8833 06-31-023
N 5362	1352.8 / 84.45 / 77.69	4134 / 70.74 / 21.42	3 P / 0.78 / 42 -1	2.4 / 2.0 / 2	0.46 / 68.	12.7	3 / 0.34 / 0.00	2263. / 2361.		90		37.8 / 32.89	7.3 / 33.7 / 13.6	-20.19 / 22.1	10.27		8835 07-29-016 / 2
U 8837	1352.9 / 103.71 / 64.17	5409 / 60.80 / 21.38	10B / 0.31 / 14 -9	4.6 / 3.5 / 2	0.30 / 80.	13.13 / 12.43 / 0.70	7 / 0.67 / 0.00	141. / 283. / 1	101 / 81. / 88.	8 / 4 / 212	0.72 / 212 5 / 1.12	3.9 / 27.95	1.6 / 3.3 / 1.4	-14.82 / 4.0	8.12 / 7.90 / 8.88	0.61 / 0.105 / 5.75	A 1353, HOLMBERG IV / DDO 185 / 8837 09-23-017 / 2
U 8839	1353.0 / 3.91 / 102.61	1802 / 72.55 / 18.74	10 / 0.09 / 41 -0	3.2 / 3.0 / 2	0.78 / 43.	13.3	3 / 0.08 / 0.01	965. / 971. / 1	115 / 134. / 138.	10	0.77 / 12 / 1.09	22.2 / 31.73	-4.6 / 20.7 / 7.2	-18.43 / 19.4	9.56 / 9.46 / 10.01	0.79 / 0.287 / 2.76	8839 03-36-001 / 2
N 5360	1353.1 / 340.41 / 115.69	514 / 63.09 / 15.91	13BP / 0.30 / 41 -7	2.1 / 1.9 / 2	0.60 / 58.	14.1	3 / 0.00 / 0.04	1177. / 1131.		31		22.9 / 31.80	-9.5 / 19.8 / 6.3	-17.70 / 12.7	9.27		8838 01-36-001 / 2
N 5371	1353.6 / 82.14 / 78.62	4042 / 71.19 / 21.54	4X / 0.74 / 42 -1	4.2 / 4.0 / 2	0.80 / 41.	11.32	2 / 0.08 / 0.00	2558. / 2654.	419 / 585. / 585.	10 / 15 / 217	0.97 / 217 13	37.8 / 32.89	7.1 / 35.4 / 14.3	-21.57 / 44.2	10.82 / 10.12 / 11.64	0.20 / 0.030 / 6.63	8846 07-29-020 / L2 / 2
N 5363	1353.6 / 340.96 / 115.45	530 / 63.25 / 16.10	13 / 0.28 / 41 -7	4.7 / 4.3 / 9	0.68 / 52.	11.18	1 / 0.00 / 0.02	1125. / 1081.		37		22.4 / 31.75	-9.2 / 19.4 / 6.2	-20.57 / 28.1	10.42		8847 01-36-002 / 2
N 5376	1353.6 / 108.56 / 58.17	5945 / 55.81 / 21.09	3X / 0.47 / 42 -8 +3	1.8 / 1.6 / 2	0.68 / 52.	12.7	3 / 0.14 / 0.01	2064. / 2224.		90		33.6 / 32.63	16.5 / 26.7 / 12.1	-19.93 / 15.7	10.16		8852 10-20-047 / 2

Name	(1)	(2)	(3)	(4)	(5)	(6)	(7)	(8)	(9)	(10)	(11)	(12)	(13)	(14)	(15)	ID
N 5364	1353.7 340.72 115.69	516 63.04 16.07	4AP 0.26 41 -7	5.3 4.9 9	0.66 53.	10.86	1 0.15 0.04	1393. 1348.	150			25.5 32.03	-10.6 22.1 7.1	-21.17 36.5	10.66		8853 01-36-003 2
N 5377	1354.3 94.91 71.33	4729 66.21 21.78	1B 0.19 42 -0 +3	4.2 3.7 2	0.60 58.	11.77	2 0.20 0.02	1830. 1951.	100			31.0 32.46	9.2 27.3 11.5	-20.69 33.5	10.47		8863 08-25-052 2
N 5452	1354.5 118.87 38.48	7828 38.41 18.36	7X 0.09 42+13	2.3 2.2 2	0.80 41.	13.2	3 0.08 0.04	2066. 2272. 1	15	190 238. 240.	0.40 22 1.05	33.0 32.59	24.5 19.5 10.4	-19.39 21.2	9.95 9.44 10.54	0.31 0.079 3.93	8867 13-10-014
N 5375	1354.6 44.76 90.72	2924 75.45 20.85	2B 0.10 42 -0	3.3 3.1 2	0.79 42.	12.2	3 0.08 0.00	2391. 2444. 1	30	302	0.38 25 1.11	37.8 32.89	-0.4 35.3 13.5	-20.69 34.2	10.47 9.53	0.12	8865 05-33-027 2
N 5365	1354.8 315.39 162.59	-4341 17.33 0.02	-2B 0.10 23 -0	4.3 4.1 6	0.62 56.	12.2	5 0.00 0.29	2433. 2232.	64			33.0 32.59	-31.5 9.9 0.0	-20.39 39.5	10.35		271-08 -7-29-002 2
N 5383	1355.0 84.80 77.13	4206 70.07 21.85	3BP 0.75 42 -1	3.3 3.1 2	0.71 49.	11.93	2 0.12 0.00	2253. 2355.	14		0.56 3	37.8 32.89	7.7 33.6 13.8	-20.96 34.2	10.58 9.71	0.14	MARK 281 / 8875 07-29-023 2
1355-29	1355.1 319.93 149.01	-2904 31.36 5.48	3A 0.19 11-35	6.1 4.5 6	0.15 90.	11.4 8.58 2.82	5 0.67 0.19	2661. 2496. 6	30	523 485. 486.	1.10 20 1.33	-37.3 32.86	-31.8 19.1 3.6	-21.46 49.0	10.78 10.24 11.52	0.29 0.053 5.60	IC 4351, UGCA 376 / 445-84 -5-33-034 2
U 8892	1356.0 105.95 60.81	5715 57.88 21.61	10 0.25 42 -6 +3	2.4 2.3 2	0.77 44.	14.1	3 0.09 0.00	1748. 1901. 1	15	116 133. 137.	0.45 18 1.05	29.7 32.37	13.5 24.1 11.0	-18.27 19.9	9.50 9.40 10.01	0.79 0.243 3.24	8892 10-20-052.
U 8894	1356.4 110.90 53.97	6339 52.14 21.00	9 0.23 42 +8 +3	2.2 2.0 2	0.67 52.		0.15 0.03	1771. 1943. 1	20	144	0.24 36 1.04	29.8 32.37	16.4 22.5 10.7	17.4	9.19		8894 11-17-011
1357-48	1357.0 314.58 166.77	-4802 13.04 -1.26	5 0.41 23 -4 +3	2.4 2.0 6	0.17 90.		0.67 0.80	2869. 2661. 7	20			38.1 32.90	-37.1 8.7 -0.8	22.3			221-22
1357-45	1357.7 315.49 164.16	-4510 15.76 -0.06	5B 0.11 16 -1	2.3 2.3 6	0.86 35.	12.5	5 0.05 0.28	1502. 1300. 2	20	155	0.49 22 1.02	21.0 31.61	-20.2 5.7 0.0	-19.11 14.1	9.84 9.13	0.20	A 1358 / 271-10 -7-29-005 2
1358-30	1358.3 320.34 150.22	-3005 30.18 5.75	7X 0.19 11-35	2.6 2.6 6	0.88 33.		0.04 0.17	2594. 2428. 2	15	187 283. 285.	0.75 12 1.02	36.3 32.80	-31.4 18.0 3.6	27.6	9.87 10.81	0.116	UGCA 378 / 445-89 -5-33-036
N 5398	1358.5 319.44 152.78	-3249 27.56 4.77	8BP 0.08 16 -5	3.4 3.2 6	0.58 59.	12.3	5 0.21 0.24	1226. 1053. 2	15	137 127. 131.	0.78 12 1.02	18.3 31.31	-16.2 8.4 1.5	-19.01 17.1	9.80 9.30 9.90	0.32 0.252 1.28	UGCA 379 / 384-32 -5-33-037 2
N 5422	1358.9 103.50 62.76	5524 59.28 22.16	0 0.36 42 -6 +3	3.1 2.2 2	0.19 90.	13.0	3 0.00 0.00	1837. 1986.	90			30.9 32.45	13.1 25.5 11.7	-19.45 19.9	9.97		8935 09-23-024 2

Columns: (1) Name · (2) α / ℓ / SGL · (3) δ / b / SGB · (4) Type / Q_xyz / Group · (5) D_{25} / D_{25}^i / source · (6) d/D / i · (7) B_T^{b,i} / H_{-0.5} / B-H · (8) source / A_B^{i-o} / A_B^b · (9) V_h / V_o / Tel · (10) W_{20} / W_R / W_D · (11) e_v / e_w / source · (12) F_c / source,e_F / f_h · (13) R / μ · (14) SGX / SGY / SGZ · (15) M_B^{b,i} / Δ_{25}^{b,i} · (16) log L_B / log M_H / log M_T · (17) M_H/L_B / M_H/M_T / M_T/L_B · (18) Alternate Names UGC/ESO · MCG · AGN? · RCII

Each galaxy occupies three stacked sub-lines; values below are given top / middle / bottom.

(1) Name	(2)	(3)	(4)	(5)	(6)	(7)	(8)	(9)	(10)	(11)	(12)	(13)	(14)	(15)	(16)	(17)	(18) UGC/ESO · MCG · AGN? · RCII
1359+37	1359.3 / 70.48 / 82.61	3703 / 72.32 / 22.47	13 / 0.28 / 42 +1	0.6 / 0.6 / 5	1.00	14.9	3 / 0.00 / 0.00	2703. / 2789.		71		41.7 / 33.10	5.0 / 38.2 / 15.9	-18.20 / 7.3	9.47		MARK 465 · · ·
N 5408	1400.3 / 317.15 / 160.63	-4109 / 19.49 / 1.92	10 / 0.19 / 14-15	2.1 / 2.0 / 6	0.62 / 57.		/ 0.18 / 0.24	510. / 318. / 2	114 / 108. / 113.	8	1.18 / 7 / 1.02	4.9 / 28.43	-4.6 / 1.6 / 0.2	/ 2.9	8.56 / / 8.99	0.375	325-47 · -7-29-006 · · 2
N 5426	1400.8 / 333.25 / 127.26	-550 / 52.51 / 14.65	5AP / 0.15 / 41-12	2.4 / 2.1 / 1	0.55 / 61.	12.44	1 / 0.24 / 0.07	2378. / 2296.		95	1.01 / 217 12	35.9 / 32.77	-21.0 / 27.6 / 9.1	-20.33 / 22.0	10.32 / 10.12	0.63	UGCA 381, ARP 271; VV 21 · -1-36-004 · · 2
N 5427	1400.8 / 333.27 / 127.23	-548 / 52.54 / 14.66	5AP / 0.19 / 41-12	2.3 / 2.3 / 1	0.97 / 17.	11.97	1 / 0.01 / 0.07	2565. / 2483.		16		38.1 / 32.91	-22.3 / 29.4 / 9.6	-20.94 / 25.6	10.57		UGCA 380, ARP 271; VV 21 · -1-36-003 · · 2
N 5448	1400.9 / 95.73 / 69.21	4925 / 64.01 / 22.83	1X / 0.32 / 42 -4 +3	4.0 / 3.5 / 2	0.50 / 65.	11.88	2 / 0.29 / 0.03	1970. / 2101.		50		32.6 / 32.57	10.7 / 28.1 / 12.7	-20.69 / 33.3	10.47		8969 · 08-26-003 · · 2
N 5457	1401.5 / 102.03 / 63.59	5436 / 59.76 / 22.60	6X / 0.29 / 14 -9	23.8 / 23.8 / 2	1.00	8.20	1 / 0.00 / 0.00	231. / 379. / 2	194	7 / 6 / 211	2.59 / 25 2 / 1.91	5.4 / 28.65	2.2 / 4.4 / 2.1	-20.45 / 37.5	10.37 / 10.05	0.48	M 101, ARP 26, VV 344; 8981 · 09-23-028 · · 2
1402-24	1402.1 / 323.39 / 145.40	-2436 / 35.10 / 8.56	6B / 0.08 / 11 -0	2.2 / 2.3 / 6	0.96 / 19.	13.8	3 / 0.01 / 0.26	2337. / 2191. / 6	195	15	0.95 / 12 / 1.08	33.6 / 32.63	-27.3 / 18.9 / 5.0	/ 22.6	10.00		UGCA 383; 510-59 · -4-33-040 · ·
U 8995	1402.4 / 349.82 / 112.40	903 / 64.76 / 19.08	8X / 0.26 / 41 +7	2.3 / 2.0 / 2	0.55 / 61.		3 / 0.23 / 0.01	1233. / 1210. / 1	164 / 150. / 154.	15	0.36 / 22 / 1.04	23.7 / 31.87	-8.5 / 20.7 / 7.7	-18.07 / 13.8	9.42 / 9.11 / 9.96	0.49 / 0.142 / 3.43	8995 · 02-36-031 · · 2
N 5473	1403.0 / 102.25 / 63.00	5508 / 59.19 / 22.77	-2X / 0.48 / 42 -6 +3	2.4 / 2.3 / 2	0.78 / 43.	12.30	2 / 0.00 / 0.00	2006. / 2156.		38		33.0 / 32.59	13.8 / 27.1 / 12.8	-20.29 / 22.2	10.31		9011 · 09-23-031 · · 2
N 5474	1403.3 / 100.82 / 64.33	5354 / 60.19 / 22.92	6AP / 0.24 / 14 -9	5.9 / 5.6 / 2	0.83 / 37.	11.29	1 / 0.06 / 0.00	277. / 423. / 1	61 / 71. / 79.	7 / 4 / 211	1.45 / 212 4 / 1.24	6.0 / 28.87	2.4 / 4.9 / 2.3	-17.58 / 9.8	9.22 / 9.01 / 9.16	0.61 / 0.707 / 0.86	VV 344; 9013 · 09-23-032 · L2 · 2
N 5477	1403.8 / 101.60 / 63.45	5442 / 59.49 / 22.92	9 / 0.24 / 14 -9	1.9 / 1.8 / 2	0.76 / 45.	14.25	7 / 0.09 / 0.00	294. / 443.		3	0.50 / 212 6	6.4 / 29.01	2.6 / 5.2 / 2.5	-14.76 / 3.4	8.10 / 8.11	1.04	DDO 186; 9018 · 09-23-034 · · 2
N 5468	1404.0 / 334.88 / 126.90	-513 / 52.69 / 15.60	6X / 0.16 / 41 -13+12	2.3 / 2.3 / 2	1.00	12.0	5 / 0.00 / 0.07	2840. / 2763. / 1	152	10	0.89 / 217 8 / 1.06	41.5 / 33.09	-24.0 / 32.0 / 11.2	-21.09 / 27.9	10.63 / 10.13	0.31	UGCA 384 · -1-36-007 · · 2
N 5464	1404.2 / 321.85 / 150.42	-2947 / 30.06 / 7.05	10B / 0.18 / 11 -35	1.3 / 1.2 / 6	0.57 / 60.	12.80	6 / 0.22 / 0.22	2688. / 2528.		29		37.5 / 32.87	-32.4 / 18.4 / 4.6	-20.07 / 13.1	10.22		446-11 · -5-33-045 · · 2

ID	Other names														Coordinates
N 5480	TURN 96A / 9026 09-23-035 2	0.14	10.16 / 9.30	-19.92 / 14.9	11.2 / 27.1 / 12.6	31.9 / 32.52	0.29 / 217 25	23		1907. / 2045.	3 / 0.17 / 0.00	12.6	1.8 0.63 / 1.6 56. / 2	5AP / 0.36 / 42 -4 +3	1404.5 5058 / 96.87 62.44 / 67.49 23.32
N 5481	TURN 96B / 9029 09-23-036 2		10.14	-19.86 / 19.8	12.5 / 30.2 / 14.1	35.6 / 32.76		70		2212. / 2350.	3 / 0.00 / 0.00	12.9	2.0 0.74 / 1.9 47. / 2	-2 / 0.36 / 42 -4 +3	1404.8 5058 / 96.79 62.41 / 67.49 23.37
N 5485	9033 09-23-037 2		10.26	-20.18 / 25.9	13.8 / 26.8 / 12.9	32.8 / 32.58		50		1985. / 2137.	2 / 0.00 / 0.00	12.40	2.9 0.75 / 2.7 46. / 2	-2AP / 0.48 / 42 -6 +3	1405.5 5514 / 101.79 58.91 / 62.85 23.11
N 5486	9036 09-23-038 2	0.71	9.53 / 9.38	-18.34 / 11.2	10.8 / 20.9 / 10.0	25.6 / 32.04	0.56 / 19 / 1.02	20	203	1407. / 1559. / 1	3 / 0.17 / 0.00	13.7 / 12.40 / 1.30	1.6 0.64 / 1.5 55. / 2	5 P / 0.08 / 42 -0 +3	1405.7 5520 / 101.85 58.81 / 62.74 23.13
1406-45	271-18			17.5	-34.1 / 8.8 / 0.7	35.2 / 32.73		20		2626. / 2427. / 7	0.19 / 0.43		1.7 0.60 / 1.7 58. / 6	9B / 0.20 / 23 -0 +3	1406.9 -4559 / 316.85 14.52 / 165.54 1.10
N 5483	271-19 -7-29-008 2	0.91	10.30 / 10.26	-20.26 / 34.6	-23.6 / 7.2 / 1.0	24.7 / 31.96	1.47 / 6 / 1.04	10	185	1770. / 1578. / 2	5 / 0.03 / 0.27	11.7	4.6 0.92 / 4.8 27. / 6	5A / 0.16 / 16 -2	1407.4 -4305 / 317.86 17.25 / 162.93 2.36
U 9057	9057 00-36-023	0.69	9.80 / 9.64	-19.03 / 18.6	-14.3 / 21.0 / 7.9	26.6 / 32.13	0.79 / 11 / 1.05	15	244	1573. / 1510. / 1	3 / 0.49 / 0.17	13.1	2.9 0.37 / 2.4 74. / 2	8B / 0.25 / 41 -4 +1	1407.6 -220 / 338.61 54.75 / 124.32 17.32
N 5489	271-21			16.2	-38.4 / 9.9 / 1.0	39.7 / 32.99		60		2970. / 2773.	0.10 / 0.39		1.4 0.75 / 1.4 45. / 6	1 / 0.28 / 23 -0 +3	1408.8 -4551 / 317.21 14.54 / 165.56 1.46
N 5493	UGCA 386 / -1-36-013 2		10.46	-20.67 / 19.3	-22.4 / 29.8 / 11.3	38.9 / 32.95		75		2627. / 2555.	2 / 0.00 / 0.07	12.28	1.8 0.66 / 1.7 53. / 6	-2 P / 0.18 / 41 -13+12	1408.9 -449 / 336.95 52.45 / 126.90 16.88
1408-49	221-32			27.0	-37.7 / 7.6 / 0.1	38.5 / 32.93		20		2901. / 2697. / 7	0.03 / 1.01		2.0 0.92 / 2.4 27. / 6	5 / 0.32 / 23 -0 +3	1408.9 -4909 / 316.18 11.40 / 168.57 0.11
N 5496	9079 00-36-026 2	0.55 / 0.275 / 1.98	10.24 / 9.97 / 10.53	-20.11 / 23.9	-13.7 / 21.0 / 8.2	26.4 / 32.11	1.13 / 7 10 / 1.10	10	260 / 222. / 225.	1544. / 1488. / 1	3 / 0.67 / 0.16	12.0 / 11.17 / 0.83	4.2 0.21 / 3.1 89. / 2	7 / 0.30 / 41 -4 +1	1409.0 -55 / 340.44 55.77 / 123.00 18.07
1409-65	097-13 2		10.68	-21.23 / 13.2	-4.2 / -0.2 / -0.5	4.2 / 28.11		10		439. / 214.	2 / 0.29 / 4.08	6.88	5.4 0.50 / 10.8 65. / 6	3A / 0.13 / 14+20	1409.3 -6506 / 311.33 -3.80 / 183.17 -6.39
1409-87	001-06			20.1	-24.2 / -11.0 / -7.0	27.5 / 32.19		20		2253. / 2027. / 7	0.67 / 0.37		3.1 0.28 / 2.5 81. / 6	3 / 0.15 / 55 -4 +3	1409.3 -8733 / 303.93 -25.09 / 204.37 -14.80
U 9083	VV 125 / 9083 08-26-009	0.64	9.32 / 9.13	-17.82 / 10.2	10.9 / 27.0 / 13.0	31.9 / 32.52	0.12 / 31 / 1.01	30	145	1905. / 2044. / 1	3 / 0.07 / 0.03	14.7	1.1 0.82 / 1.1 39. / 2	8 / 0.36 / 42 -4 +3	1409.6 5027 / 94.82 62.32 / 67.99 24.16

(1) Name	(2) α ℓ SGL	(3) δ b SGB	(4) Type ϱ_{xyz} Group	(5) D_{25} $D_{25}^{b,i}$ source	(6) d/D i	(7) $B_T^{b,i}$ $H_{-0.5}$ $B-H$	(8) source A_B^{i-o} A_B^b	(9) V_h V_o Tel	(10) W_{20} W_R W_D	(11) e_v e_w source	(12) F_c source/e_F f_h	(13) R μ	(14) SGX SGY SGZ	(15) $M_B^{b,i}$ $\Delta_{25}^{b,i}$	(16) $\log L_B$ $\log M_H$ $\log M_T$	(17) M_H/L_B M_H/M_T M_T/L_B	(18) Alternate Names UGC/ESO MCG AGN? RCII
N 5506	1410.7 339.16 125.21	−259 53.80 17.87	9 0.25 41 −4 +1	2.5 2.1 2	0.35 76.		0.54 0.13	1753. 1690.		95		28.7 32.29	−15.7 22.3 8.8	17.6			UGCA 387 S2 00-36-028 2
N 5507	1410.7 339.22 125.14	−255 53.85 17.89	−5 0.11 41 −0 +1	1.7 1.5 2	0.55 61.		0.00 0.13	2232. 2169.		95		34.3 32.68	−18.8 26.7 10.5	15.0			UGCA 388 00-36-029 2
1412-43	1412.0 318.48 163.85	−4343 16.38 2.86	5B 0.21 16 −2	3.5 3.2 6	0.44 69.		0.36 0.32	1882. 1691. 2	239 216. 219.	10	1.05 9 1.02	26.0 32.08	−25.0 7.2 1.3	24.3	9.88 10.52	0.231	IC 4386 2 271-25 −7-29-011
N 5523	1412.6 32.12 95.36	2533 71.22 24.38	6A 0.07 43 −0	4.4 3.5 2	0.34 77.	12.3 10.35 1.95	3 0.57 0.00	1048. 1099. 1	285 254. 256.	10	1.05 217 8 1.11	21.5 31.66	−1.8 19.5 8.9	−19.36 22.0	9.94 9.71 10.61	0.60 0.126 4.76	9119 04-34-008 2
U 9126	1413.3 8.79 104.87	1647 67.82 23.28	10 0.11 41 −0	1.9 1.8 2	0.90 29.	15.08	7 0.04 0.00	2277. 2294. 1	92	30	0.11 56 1.04	36.0 32.78	−8.5 32.0 14.2	−17.70 18.9	9.27 9.22	0.89	DDO 188 9126 03-36-098 2
N 5529	1413.4 65.16 83.38	3627 70.00 25.27	5 0.00 42 −0 +1	5.7 4.0 2	0.14 90.	11.9 9.25 2.65	3 0.67 0.00	2878. 2971. 1	597 559. 560.	15 9 217	0.96 217 7 1.16	43.9 33.21	4.6 39.5 18.7	−21.31 51.3	10.72 10.24 11.67	0.34 0.038 8.94	9127 06-31-085 2
U 9128	1413.6 25.55 97.86	2317 70.47 24.33	10 0.10 14+12	1.9 1.8 2	0.90 29.	14.30	7 0.04 0.01	153. 196. 3	54	7	0.50 10 1.04	2.2 26.68	−0.3 2.0 0.9	−12.38 1.2	7.14 7.18	1.10	DDO 187 9128 04-34-009 2
N 5530	1415.4 319.28 163.60	−4309 16.71 3.66	4A 0.13 16 −1	4.3 3.9 6	0.44 70.	11.1	5 0.37 0.39	1200. 1013. 2	292 271. 273.	25	1.10 29 1.04	16.9 31.14	−16.2 4.8 1.1	−20.04 19.2	10.21 9.56 10.61	0.22 0.088 2.54	272-03 −7-29-013 2
1415-47	1415.6 317.80 167.55	−4730 12.60 1.81	7 0.13 16 −1	1.7 1.6 6	0.40 72.		0.43 0.73	1284. 1087. 7		20		17.7 31.24	−17.3 3.8 0.6	8.3			222-01
U 9164	1416.4 22.71 99.29	2203 69.48 24.81	13 P 0.08 41 −0	1.3 1.2 2	0.75 45.	15.0	3 0.00 0.03	2510. 2551.		70		39.0 32.96	−5.7 35.0 16.4	−17.96 13.7	9.38		MARK 674 9164 04-34-014
U 9169	1417.3 356.68 112.73	936 62.62 22.79	10 0.30 41 −0 +1	4.2 3.0 2	0.21 89.	13.0	3 0.67 0.03	1281. 1273. 1	161 129. 134.	10	0.78 10 1.10	24.0 31.90	−8.6 20.4 9.3	−18.90 21.0	9.75 9.54 10.01	0.61 0.342 1.80	9169 02-37-001 2
N 5556	1417.6 325.31 150.88	−2901 29.73 10.03	7X 0.08 16 −0	4.6 4.8 6	0.83 37.	11.8	3 0.06 0.32	1384. 1236. 6	181 242. 244.	10	1.06 11 1.17	20.9 31.61	−18.0 10.0 3.6	−19.81 29.3	10.12 9.70 10.70	0.38 0.101 3.81	UGCA 389, DDO 243 446-50 −5-34-009 2
N 5566	1417.8 349.27 118.45	410 58.57 21.59	2B 0.52 41 −3 +1	5.6 4.7 2	0.41 71.	10.88	2 0.41 0.05	1518. 1489.		68		26.4 32.11	−11.7 21.6 9.7	−21.23 36.2	10.68		ARP 286 9175 01-37-002 2

Name																	Alt names
1418-46	1418.0 318.70 166.44	-4604 13.81 2.81	4 0.13 16 -3 +2	5.0 3.9 6	0.18 90.		0.67 0.45	1653. 1460. 7	20			22.9 31.80	-22.3 5.4 1.1	26.1			IC 4402 272-05
N 5585	1418.2 100.99 60.74	5657 56.48 24.64	7X 0.17 14 -9	5.5 5.1 2	0.69 51.	11.26 10.14 1.12	1 0.13 0.00	303. 466. 3	5 4 212	162 167. 171.	1.47 212 4 1.26	7.0 29.22	3.1 5.5 2.9	-17.96 10.4	9.38 9.16 9.93	0.61 0.172 3.55	9179 10-20-094 2
N 5574	1418.4 348.64 119.22	328 57.93 21.55	-2B 0.50 41 -3 +1	1.2 1.2 2	0.74 47.	13.16	2 0.00 0.04	1716. 1685.	50			28.7 32.29	-13.0 23.3 10.5	-19.13 10.1	9.84		9181 01-37-006 2
N 5576	1418.5 348.72 119.19	330 57.94 21.58	-5 0.54 41 -3 +1	3.3 3.1 2	0.78 43.	11.71	2 0.00 0.04	1528. 1497.	100			26.4 32.11	-12.0 21.4 9.7	-20.40 23.9	10.35		9183 01-37-007 2
N 5577	1418.7 348.99 119.03	340 58.04 21.67	4A 0.50 41 -3 +1	3.1 2.4 2	0.33 78.	12.7 10.74 1.96	3 0.60 0.04	1485. 1455.	10 25 217	256 223. 226.	0.49 217 25	25.9 32.07	-11.7 21.1 9.6	-19.37 18.2	9.94 9.32 10.42	0.24 0.079 3.01	9187 01-37-009 2
1418-75	1418.8 308.80 192.72	-7506 -13.52 -9.75	5 0.16 55 -0 +1	2.4 2.5 6	0.53 63.		0.25 0.73	2552. 2324. 7	20			32.0 32.53	-30.8 -6.9 -5.4	23.4			041-06
N 5584	1419.8 345.11 123.05	-9 54.87 20.88	6X 0.51 41 +2 +1	3.3 3.2 2	0.82 39.	11.9	3 0.07 0.12	1641. 1597. 1	15 11 79	219 292. 294.	0.89 217 8 1.10	27.4 32.19	-14.0 21.5 9.8	-20.29 25.6	10.31 9.77 10.80	0.29 0.092 3.11	9201 00-37-001 2
1420-49	1420.4 317.88 169.63	-4926 10.51 1.68	10 0.12 16 +1	1.5 1.7 6	0.76 44.	14.2	3 0.09 0.89	1545. 1347. 7	20			21.1 31.62	-20.8 3.8 0.6	10.5			222-04
U 9211	1420.6 84.12 73.18	4536 64.29 26.38	10 0.08 44 -0	2.5 2.4 2	0.82 39.	14.2	3 0.07 0.01	690. 819. 1	10	115 146. 150.	0.76 11 1.06	14.7 30.84	3.8 12.6 6.5	-16.64 10.3	8.85 9.09 9.80	1.76 0.195 9.04	DDO 189 9211 08-26-029 2
N 5608	1421.3 76.81 77.18	4200 66.19 26.73	10 P 0.07 44 -0	2.9 2.6 2	0.61 57.	13.2	3 0.19 0.00	662. 780. 1	15	159 151. 155.	0.50 19 1.06	12.7 30.52	2.5 11.1 5.7	-17.32 9.6	9.12 8.71 9.80	0.39 0.080 4.84	9219 07-30-009
N 5600	1421.4 7.44 107.32	1452 65.14 24.85	4 P 0.11 41 -0	1.5 1.5 2	1.00	13.20	2 0.00 0.00	2409. 2425.	83			37.5 32.87	-10.1 32.5 15.8	-19.67 16.4	10.06		ARAK 449 9220 03-37-013 2
N 5595	1421.5 332.78 139.44	-1630 40.70 15.80	5X 0.16 41 -14	2.1 1.9 4	0.50 65.	11.9	5 0.29 0.34	2672. 2568.	90			38.5 32.93	-28.1 24.1 10.5	-21.03 21.4	10.60		-3-37-001 2
N 5597	1421.7 332.82 139.49	-1632 40.65 15.83	6X 0.16 41 -14	1.9 2.0 4	1.00	12.27	2 0.00 0.33	2677. 2573.	90			38.6 32.93	-28.2 24.1 10.5	-20.66 22.5	10.46		-3-37-002 2
U 9240	1422.8 81.97 74.11	4444 64.49 26.83	10A 0.25 14 -0	2.0 2.0 2	1.00	13.30	7 0.00 0.00	153. 281. 3	7	62	0.75 8 1.05	3.7 27.82	0.9 3.1 1.7	-14.52 2.2	8.00 7.89	0.77	I ZW 87, DDO 190 9240 08-26-030 2

Table of galaxy data (columns 1–18):

(1) Name	(2) α / ℓ / SGL	(3) δ / b / SGB	(4) Type / Q_{xyz} / Group	(5) $D_{25}^{b,i}$ / D_{25}^i / source	(6) d/D / i	(7) $B_T^{b,i}$ / $H_{-0.5}$ / B-H	(8) source / A_B^{i-o} / A_B^b	(9) V_h / V_o / Tel	(10) W_{20} / W_R / W_D	(11) e_v / e_w / source	(12) F_c source/e_f / f_h	(13) R / μ	(14) SGX / SGY / SGZ	(15) $M_B^{b,i}$ / $\Delta_{25}^{b,i}$	(16) $\log L_B$ / $\log M_H$ / $\log M_T$	(17) M_H/L_B / M_H/M_T / M_T/L_B	(18) Alternate Names UGC/ESO MCG / AGN? / RCII
U 9242	1423.3 / 71.36 / 79.69	3945 / 66.91 / 27.20	6 / 0.09 / 43 -0	5.0 / 3.5 / 2	0.07 / 90.	13.4	3 / 0.67 / 0.00	1440. / 1551. /	205 / 169. / 172.	15	0.66 / 18 / 1.12	26.3 / 32.10	4.2 / 23.0 / 12.0	-18.70 / 26.9	9.67 / 9.50 / 10.35	0.67 / 0.142 / 4.74	I ZW 88 / 9242 07-30-011
1423-45	1423.7 / 319.80 / 166.51	-4539 / 13.85 / 3.88	10 / 0.17 / 16 -3 +2	1.3 / 1.5 / 6	0.93 / 24.		/ 0.02 / 0.52	1946. / 1758. / 7		20		26.7 / 32.13	-25.9 / 6.2 / 1.8	11.7			272-09
U 9249	1424.5 / 358.15 / 113.91	854 / 60.84 / 24.36	7 / 0.35 / 41 -0 +1	2.3 / 1.6 / 2	0.16 / 90.	14.1	3 / 0.67 / 0.02	1366. / 1361. / 1	175	25	0.35 / 40 / 1.03	24.8 / 31.97	-9.2 / 20.6 / 10.2	-17.87 / 11.6	9.34 / 9.14	0.63	9249 02-37-000
1424-46	1424.8 / 319.82 / 166.98	-4605 / 13.37 / 3.86	10 / 0.13 / 14-16+15				/ 0.02 / 0.58	393. / 204. / 7		20		3.1 / 27.48	-3.0 / 0.7 / 0.2				
N 5631	1425.0 / 99.52 / 60.72	5648 / 56.02 / 25.58	-2A / 0.38 / 42 -7 +3	2.2 / 2.2 / 2	1.00	12.50	2 / 0.00 / 0.00	1979. / 2144.		60		32.7 / 32.57	14.4 / 25.7 / 14.1	-20.07 / 21.0	10.22	0.28	9261 10-21-002 2
N 5630	1425.6 / 74.74 / 77.72	4129 / 65.75 / 27.56	9 / 0.24 / 42 -0 +1	2.5 / 1.9 / 2	0.31 / 79.	12.9 / 11.82 / 1.08	3 / 0.64 / 0.00	2658. / 2777. / 1	230	15	0.48 / 29 / 1.04	41.3 / 33.08	7.8 / 35.8 / 19.1	-20.18 / 22.9	10.26 / 9.71		9270 07-30-014
N 5633	1425.6 / 84.33 / 72.24	4622 / 63.14 / 27.18	3A / 0.41 / 42 +3	2.5 / 2.1 / 2	0.48 / 67.	12.51	2 / 0.31 / 0.03	2323. / 2458.		29	0.43 / 209 40	37.1 / 32.85	10.1 / 31.4 / 17.0	-20.34 / 22.7	10.33 / 9.57	0.17	I ZW 89 / 9271 08-26-034 2
U 9275	1425.9 / 7.20 / 108.48	1400 / 63.75 / 25.76	8X / 0.17 / 41 -0 +1	2.8 / 2.5 / 2	0.67 / 52.	13.2	3 / 0.15 / 0.00	1294. / 1311. / 1	220 / 231. / 234.	10	0.88 / 14 / 1.06	24.2 / 31.92	-6.9 / 20.7 / 10.5	-18.72 / 17.7	9.68 / 9.65 / 10.44	0.93 / 0.163 / 5.71	IC 1014 / 9275 02-37-012
1427-34	1427.1 / 325.19 / 156.35	-3401 / 24.33 / 9.72	7B / 0.07 / 11 -0	2.6 / 2.6 / 6	0.81 / 40.	12.93	6 / 0.07 / 0.26	3014. / 2858.		50		41.3 / 33.08	-37.3 / 16.3 / 7.0	-20.15 / 31.4	10.25		SCI 141 / 385-33 -6-32-009 2
N 5638	1427.2 / 351.70 / 119.87	327 / 56.49 / 23.66	-5 / 0.79 / 41 -3 +1	2.5 / 2.5 / 2	0.92 / 27.	12.15	2 / 0.00 / 0.05	1700. / 1676. / 1	160	39	0.20 / 29 / 1.03	28.4 / 32.27	-13.0 / 22.6 / 11.4	-20.12 / 20.7	10.24	0.53	TURN 105B / 9308 01-37-018 2
U 9310	1427.5 / 351.78 / 119.91	326 / 56.43 / 23.72	8B / 0.52 / 41 -3 +1	2.2 / 1.7 / 2	0.31 / 79.	14.3	3 / 0.64 / 0.05	1853. / 1829. / 1	163	15	0.51 / 15 / 1.05	30.2 / 32.40	-13.8 / 24.0 / 12.2	-18.10 / 15.0	9.43 / 9.16		9310 01-37-000
U 9324	1428.0 / 80.67 / 74.10	4440 / 63.76 / 27.76	9B / 0.18 / 42 -0 +1	2.5 / 2.3 / 2	0.61 / 57.	14.1	3 / 0.19 / 0.00	2746. / 2877. / 1	163	15		42.3 / 33.13	10.3 / 36.0 / 19.7	-19.03 / 28.4	9.80 / 9.76	0.91	DDO 192 / 9324 08-26-000 2
N 5660	1428.0 / 89.56 / 68.31	4950 / 60.66 / 27.16	5X / 0.48 / 42 -3	3.1 / 3.0 / 2	0.89 / 31.	12.25	2 / 0.04 / 0.01	2336. / 2483. / 1	145	20 / 12 / 211	0.75 / 211 9 / 1.10	37.2 / 32.85	12.2 / 30.7 / 17.0	-20.60 / 32.6	10.43 / 9.89	0.29	9325 08-26-039 2

145

N 5645 — 9328 01-37-019 2
```
1428.2   730      7B           3.1  0.56   12.55  2     1378.  193  15   0.63      24.8   -9.7  20.3  10.4   -19.42   9.96   0.29
357.35   59.22    0.39         2.7  60.    0.22   1371.       30            30                                19.6    9.42
115.64   24.92    41 +3 +1                 0.03   1                1.07                                       
```

N 5612 — 022-01 2
```
1428.2   -7811    -2A          1.7  0.61   12.5   5     2764.       90                34.6   -32.7  -9.3  -6.4  -20.20  10.27
308.14   -16.58   0.24         1.6  57.    0.00   2538.                              32.70                     16.2
195.78   -10.58   55 -2 +1     6           0.48                                      
```

1428-55 — 175-09
```
1428.4   -5515    9            1.3  0.86         9     3108.       20                40.6   -40.5  3.3  0.1   23.7
316.91   4.65     0.00         2.0  35.    0.05  2903.                               33.04
175.36   0.08     23 -0        6           2.57  7
```

1428-43 — IC 4444
```
1428.5   -4312    5A           1.4  0.94   11.8  5     1949.  190  20   0.70      26.9   -25.8  7.1  2.7   -20.35  10.33  0.17
321.58   15.79    0.18         1.5  24.    0.02  1770.            24   24                                    11.8    9.56
164.72   5.77     16 -0 +2     6           0.45  2                1.01
```

N 5643 — 272-14 -7-30-002 2 / 272-16 -7-30-003 2 S2
```
1429.4   -4359    5X           5.4  0.89   10.17  2    1200.  218  20   1.07      16.9   -16.3  4.2  1.6   -20.97  10.58  0.09
321.42   15.01    0.16         5.8  31.    0.04   1019. 357.       15   15                                   28.6    9.53   0.032
165.49   5.55     16 -1        9           0.49   2    358.        1.07                                             11.02  2.78
```

N 5665 — ARP 49 / 9352 01-37-024 2
```
1430.0   818      9BP          2.3  0.64   12.2  3     2266.       15                35.4   -13.4  29.0  15.3  -20.54  10.41
359.09   59.43    0.07         2.1  55.    0.17  2264.                               32.74                     21.7
114.90   25.54    41 -5 +1     2           0.05
```

1430+08 — NGC 5665A, ARP 49 / 01-37-000 2
```
1430.0   818      -5 P                           5     2188.       22                34.5   -13.1  28.2  14.9
359.09   59.43    0.07                     0.17  2186.                               32.69
114.90   25.54    41 -5 +1                 0.05
```

N 5669 — 9353 02-37-021 2
```
1430.3   1006     6X           4.2  0.76   12.1  3     1371.  212  10   1.03      24.9   -8.7   20.6  10.9  -19.88  10.14  0.48
1.92     60.55    0.29         4.0  45.    0.09  1376. 251.  8    217  7                                      29.1    9.82   0.125
112.97   26.02    41 -0 +1     2           0.03  1     253.  79   1.15                                               10.73  3.81
```

N 5678 — 9358 10-21-005 2
```
1430.6   5808     3X           3.3  0.52   12.05  3    2219.       61                35.6   16.4   27.4  15.6  -20.96  10.58
100.04   54.51    0.33         2.9  64.    0.27   2391.                              32.76                     30.1
59.10    26.05    42 -7 +3     2           0.00
```

N 5668 — 9363 01-37-028 2
```
1430.9   440      7A           3.3  0.87   12.05  2    1579.  114  6    1.00      26.9   -11.8  21.4  11.3  -20.10  10.23  0.42
354.59   56.75    0.80         3.2  33.    0.05   1563. 165.  7    217  10                                    25.1    9.86   0.365
118.85   24.87    41 -3 +1     2           0.05         169.  217                                                    10.30  1.16
```

N 5676 — 9366 08-26-043 2
```
1431.0   4941     4A           4.0  0.47   11.24  2    2125.  469  7    0.95      34.5   11.3   28.4  16.0  -21.45  10.77  0.18
88.69    60.38    0.43         3.4  67.    8.82   2273. 467.  15   217  20                                    34.3    10.03  0.049
68.40    27.66    42 -3        9           2.42        468.  217                                                    11.34  3.66
                                           0.33
                                           0.03
```

U 9380 — 9380 01-37-000
```
1432.1   429      10           2.1  0.56   14.1  3     1708.  127  15   0.17      28.5   -12.6  22.5  12.1  -18.17   9.46  0.42
354.59   56.41    0.79         1.9  60.    0.22  1692. 116.  14   226  24                                     15.8    9.08  0.195
119.13   25.11    41 -3 +1     2           0.05  1     121.  226  1.03                                                9.79  2.14
```

N 5670 — 272-19
```
1432.4   -4544    -2           2.0  0.43         3     2920.       90                39.0   -37.9  8.6  3.5   20.5
321.21   13.19    0.20         1.8  70.    0.00  2737.                               32.96
167.29   5.19     23 -6        6           0.60
```

U 9385 — 9385 01-37-037
```
1432.8   529      10           1.7  0.94         2     1644.  105  15   0.27      27.8   -11.8  22.1  12.0  13.8     9.16
356.10   56.99    0.80         1.7  23.    0.02  1633.       12   226  20                                  
118.11   25.53    41 -2 +1     2           0.07  1                1.03
```

146

(1) Name	(2) α / ℓ / SGL	(3) δ / b / SGB	(4) Type / Q_{xyz} / Group	(5) D_{25} / $D_{25}^{b,i}$ / source	(6) d/D / i	(7) $B_T^{b,i}$ / $H_{-0.5}$ / $B-H$	(8) source / A_B^{i-0} / A_B^b	(9) V_h / V_o / Tel	(10) W_{20} / W_R / W_D	(11) e_v / e_w / source	(12) F_c / source/e_F / f_h	(13) R / μ	(14) SGX / SGY / SGZ	(15) $M_B^{b,i}$ / $\Delta_{25}^{b,i}$	(16) $\log L_B$ / $\log M_H$ / $\log M_T$	(17) M_H/L_B / M_H/M_T / M_T/L_B	(18) Alternate Names UGC/ESO / MCG / AGN? RCII
N 5682	1433.0 / 87.00 / 69.24	4853 / 60.63 / 28.10	3B / 0.49 / 42 -3	2.0 / 1.5 / 2	0.23 / 86.	13.68	2 / 0.67 / 0.11	2267. / 2414.		32		36.3 / 32.80	11.4 / 30.0 / 17.1	-19.12 / 15.9	9.84		9388 08-27-002 2
U 9391	1433.2 / 101.19 / 57.46	5934 / 53.21 / 26.05	8B / 0.24 / 42 -7 +3	1.9 / 1.7 / 2	0.65 / 54.	14.66	7 / 0.16 / 0.00	1921. / 2097. / 1	140	15	0.33 / 20 / 1.03	31.8 / 32.51	15.4 / 24.1 / 14.0	-17.85 / 15.8	9.33 / 9.33	1.01	DDO 193 / 9391 10-21-011 2
N 5687	1433.3 / 95.39 / 62.77	5442 / 56.76 / 27.14	-2 / 0.37 / 42 +7 +3	2.5 / 2.4 / 2	0.75 / 46.	12.60	2 / 0.00 / 0.00	2119. / 2283.		75		34.4 / 32.68	14.0 / 27.2 / 15.7	-20.08 / 24.1	10.22		9395 09-24-020 2
N 5689	1433.7 / 87.00 / 69.13	4858 / 60.48 / 28.21	0.47 / 42 -3	3.8 / 3.0 / 2	0.31 / 79.	12.71	2 / 0.00 / 0.09	2205. / 2353.		50		35.6 / 32.76	11.2 / 29.3 / 16.8	-20.05 / 31.2	10.21		9399 08-27-004 2
U 9405	1433.9 / 98.72 / 59.70	5729 / 54.70 / 26.62	10 / 0.22 / 14 -9	2.5 / 2.2 / 2	0.64 / 55.		3 / 0.17 / 0.00	222. / 393. / 1	102 / 98. / 104.	15 / 8 / 212	0.18 / 212 7 / 1.05	5.7 / 28.79	2.6 / 4.4 / 2.6	3.7	7.69 / 9.01	0.048	DDO 194 / 9405 10-21-013 2
N 5690	1435.2 / 353.12 / 121.51	228 / 54.41 / 25.31	5 / 0.87 / 41 -2 +1	3.4 / 2.8 / 2	0.35 / 76.	11.85 / 9.74 / 2.11	2 / 0.54 / 0.10	1750. / 1729. / 1	310 / 280. / 282.	15	0.79 / 21 / 1.07	28.8 / 32.30	-13.6 / 22.2 / 12.3	-20.45 / 23.5	10.37 / 9.71 / 10.73	0.22 / 0.096 / 2.27	9416 00-37-019 2
N 5691	1435.3 / 350.19 / 124.32	-11 / 52.42 / 24.57	1XP / 0.59 / 41 -2 +1	1.9 / 1.9 / 2	0.86 / 34.	12.4	3 / 0.05 / 0.11	1881. / 1850.		90		30.2 / 32.40	-15.5 / 22.7 / 12.6	-20.00 / 16.8	10.19		9420 00-37-020 2
N 5692	1435.8 / 354.69 / 120.33	337 / 55.14 / 25.77	13 P / 0.73 / 41 -2 +1	1.1 / 1.0 / 2	0.64 / 55.	13.2	3 / 0.00 / 0.08	1816. / 1800.		220		29.7 / 32.36	-13.5 / 23.1 / 12.9	-19.16 / 8.7	9.86		ARAK 454
U 9426	1435.8 / 86.39 / 69.20	4851 / 60.28 / 28.56	10 / 0.47 / 42 -3	2.0 / 1.8 / 2	0.53 / 63.		0.25 / 0.09	2315. / 2463. / 1	62 / 50. / 60.	10	226 13 / 1.03	36.9 / 32.83	11.5 / 30.3 / 17.6	19.4	9.13 / 9.15	0.968	9426 08-27-008 2
N 5688	1436.2 / 322.23 / 166.82	-4449 / 13.76 / 6.22	5 / 0.17 / 23 -6	3.9 / 4.0 / 6	0.70 / 50.		0.13 / 0.51	2806. / 2628. / 2	445 / 533. / 534.	15	1.10 / 28 / 1.07	37.7 / 32.88	-36.5 / 8.5 / 4.1	44.0	10.25 / 11.56	0.049	272-22 -7-30-004
1436-08	1436.3 / 342.78 / 132.89	-825 / 45.76 / 22.12	10 / 0.15 / 41 -0 +1	2.8 / 2.1 / 4	0.16 / 90.		0.67 / 0.20	1823. / 1761. / 1	113	15	0.47 / 17 / 1.05	28.8 / 32.30	-18.1 / 19.5 / 10.8	17.7	9.39		DDO 195 / -1-37-010 2
N 5714	1436.4 / 82.90 / 71.44	4651 / 61.34 / 28.96	5 / 0.40 / 42 -3	2.9 / 2.0 / 2	0.15 / 90.	13.4	3 / 0.67 / 0.00	2239. / 2382. / 1	350 / 312. / 314.	20	0.74 / 23 / 1.05	36.1 / 32.78	10.0 / 29.9 / 17.5	-19.38 / 21.1	9.94 / 9.86 / 10.77	0.81 / 0.120 / 6.77	9431 08-27-011
U 9432	1436.6 / 354.39 / 120.87	310 / 54.68 / 25.84	10 / 0.71 / 41 -2 +1	1.7 / 1.7 / 2	0.94 / 23.	14.3	3 / 0.02 / 0.08	1573. / 1556. / 1	112	10	0.26 / 226 20 / 1.03	26.7 / 32.13	-12.3 / 20.6 / 11.6	-17.83 / 13.3	9.32 / 9.11	0.62	9432 01-37-040

Name	Coordinates	Type	Dimensions	Mag	Velocity		HI	D	Pos	M/D	Lum	Col	Alt. names
N 5701	1436.7 / 357.46 56.37 535 / 118.28 26.50	0.58 / 41 -2 +1	4.2 4.2 2 / 1.00	11.73 / 2 / 0.00 0.07	1505. 1498. 1	140 / 10	1.11 8 1.20	26.1 32.08	-11.0 20.5 11.6	-20.35 32.0	10.33 9.94	0.41	9436 01-37-042 2
1437+37	1437.1 / 62.73 65.30 3701 / 82.68 30.01	13 P / 0.00 / 44 -0		/ 0.00 0.00	550. 661.	100 / 100		9.6 29.92	1.1 8.3 4.8				MARK 475 / 06-32-000 2
N 5705	1437.2 / 350.40 51.85 -32 / 124.84 24.92	7B / 0.62 / 41 -2 +1	2.8 2.6 2 / 0.65 54.	13.3 11.42 1.88 / 3 / 0.16 0.12	1768. 1737. 1	235 244. 247. / 15	0.78 18 1.06	28.8 32.29	-14.9 21.4 12.1	-18.99 21.9	9.79 9.70 10.58	0.81 0.132 6.15	9447 00-37-021 2
N 5713	1437.6 / 351.01 52.12 -5 / 124.40 25.15	4XP / 0.61 / 41 -2 +1	3.1 3.0 2 / 0.78 43.	11.82 / 2 / 0.08 0.10	1900. 1871. 1	209 255. 257. / 30 9 217	1.05 217 7 1.09	30.4 32.42	-15.6 22.7 12.9	-20.60 26.6	10.43 10.02 10.70	0.38 0.206 1.86	9451 00-37-022 2
N 5727	1438.4 / 55.95 65.63 3412 / 85.92 30.37	8X / 0.09 / 43 -0	2.3 2.0 2 / 0.55 61.	13.7 / 3 / 0.23 0.00	1491. 1594. 1	191 178. 181. / 15	0.51 24 1.04	26.9 32.15	1.7 23.1 13.6	-18.45 15.7	9.57 9.37 10.16	0.63 0.162 3.87	9465 06-32-083
N 5728	1439.6 / 337.33 38.11 -1702 / 141.74 19.58	1X / C.18 / 41+15	2.3 2.1 4 / 0.50 65.	11.46 / 2 / 0.29 0.35	2970. 2879.	/ 90		42.2 33.13	-31.2 24.6 14.1	-21.67 25.9	10.86		-3-37-005 S2 2
U 9477	1439.9 / 100.07 52.69 5930 / 57.29 26.89	10 / 0.25 / 42 +7 +3	1.3 1.3 2 / 1.00	/ 0.00 0.00	2331. 2510. 1	89 / 20 14 226	-0.04 226 27 1.02	37.0 32.84	17.8 27.8 16.7	14.0	9.10		9477 10-21-023
1439-77	1439.9 / 308.92 -16.41 -7743 / 195.65 -9.83	5 / .25 / 55 -2 +1	1.8 1.9 6 / 0.73 47.	/ 0.11 0.44	2737. 2513. 7	/ 20		34.3 32.68	-32.5 -9.1 -5.9	19.0			022-03
U 9483	1440.4 / 357.95 55.36 506 / 119.07 27.27	12 / 0.81 / 41 -2 +1	2.5 2.0 2 / 0.36 75.	13.0 10.41 2.59 / 3 / 0.52 0.07	1633. 1627. 1	317 289. 291. / 15	0.59 23 1.04	27.6 32.20	-11.9 21.4 12.6	-19.20 16.1	9.87 9.47 10.59	0.40 0.076 5.24	IC 1048 / 9483 01-37-051 2
1441-49	1441.1 / 321.14 9.44 -4911 / 171.03 4.76	9 / 0.07 / 23 -0	2.4 2.4 6 / 0.33 77.	/ 0.58 1.10	2286. 2100. 7	/ 20		30.7 32.44	-30.2 4.8 2.5	21.5			222-15
N 5740	1441.9 / 354.51 52.85 154 / 122.64 26.76	3X / 0.64 / 41 -2 +1	3.0 2.8 2 / 0.60 58.	12.28 9.54 2.74 / 1 / 0.20 0.12	1575. 1558. 1	325 / 20	0.90 32 1.07	26.6 32.13	-12.8 20.0 12.0	-19.85 21.7	10.13 9.75	0.41	9493 00-38-003 2
N 5746	1442.5 / 355.00 52.94 210 / 122.40 26.98	3X / 0.83 / 41 -2 +1	8.1 5.8 9 / 0.17 90.	10.61 7.67 2.94 / 1 / 0.67 0.11	1801. 1785.	37 / 33		29.4 32.34	-14.0 22.1 13.3	-21.73 49.8	10.88		9499 00-38-005 2
U 9500	1442.9 / 2.72 56.86 805 / 115.97 28.60	9B / 0.44 / 41 -0 +1	2.9 2.9 2 / 1.00	/ 0.00 0.05	1690. 1698. 3	37 / 5	0.36 12 1.09	28.4 32.27	-10.9 22.4 13.6	24.0	9.27		DDO 196 / 9500 01-38-000 2
1442-20	1442.9 / 335.89 34.69 -2033 / 145.47 18.81	10 / 0.15 / 41 -0	1.5 1.5 6 / 0.88 32.	/ 0.04 0.35	2360. 2259. 2	218 345. 347. / 20	0.68 20 1.02	34.3 32.68	-26.7 18.4 11.1	15.0	9.75 10.71	0.109	UGCA 393 / 580-20 -3-38-005

(1) Name	(2) α / ℓ / SGL	(3) δ / b / SGB	(4) Type / Q_{xyz} / Group	(5) D_{25} / $D_{25}^{b,i}$ / source	(6) d/D / i	(7) $B_T^{b,i}$ / $H_{-0.5}$ / B-H	(8) source / A_B^{i-o} / A_B^b	(9) V_h / V_o / Tel	(10) W_{20} / W_R / W_D	(11) e_v / e_w / source	(12) F_c / source/e_f / f_h	(13) R / μ	(14) SGX / SGY / SGZ	(15) $M_B^{b,i}$ / $\Delta_{25}^{b,i}$	(16) log L_B / log M_H / log M_T	(17) M_H/L_B / M_H/M_T / M_T/L_B	(18) Alternate Names / UGC/ESO MCG / AGN? RCII
N 5750	1443.7 / 352.90 / 124.84	-1 / 51.15 / 26.63	0.33 / 41 -2 +1	2.9 / 2.6 / 2	0.61 / 57.	12.38	2 / 0.00 / 0.12	2023. / 2000.		90		31.8 / 32.51	-16.3 / 23.4 / 14.3	-20.13 / 24.1	10.24		9512 00-38-006 2
1444-17	1444.6 / 338.46 / 142.47	-1715 / 37.29 / 20.58	5 / 0.19 / 41 +9	3.1 / 2.4 / 2	0.15 / 90.		0.67 / 0.35	2210. / 2122. / 1	257	20	1.05 / 14 / 1.06	32.7 / 32.57	-24.3 / 18.7 / 11.5	22.9	10.08		UGCA 394 / -3-38-008
N 5756	1444.8 / 340.33 / 139.95	-1439 / 39.44 / 21.69	5 P / 0.17 / 41 +9	1.9 / 1.6 / 4	0.37 / 74.	11.8	5 / 0.48 / 0.36	2184. / 2106.		90		32.6 / 32.56	-23.2 / 19.5 / 12.0	-20.76 / 15.2	10.50		-2-38-012 2
N 5757	1445.0 / 337.48 / 144.08	-1852 / 35.87 / 19.98	3B / 0.17 / 41+15	2.1 / 2.2 / 6	0.88 / 32.	12.3	5 / 0.04 / 0.35	2771. / 2678.		90		39.5 / 32.98	-30.1 / 21.8 / 13.5	-20.68 / 25.4	10.46		580-33 -3-38-014 2
1446-09	1446.8 / 344.46 / 135.47	-957 / 43.00 / 23.97	10BP / 0.15 / 41 -0 +1	3.4 / 3.2 / 4	0.60 / 58.	12.78	7 / 0.19 / 0.31	1849. / 1790. / 1	200. / 195. / 197.	30	0.73 / 14 / 1.09	28.8 / 32.30	-18.8 / 18.5 / 11.7	-19.52 / 26.9	10.00 / 9.65 / 10.47	0.45 / 0.150 / 2.97	ARP 261, VV 140 / DDO 197 / -2-38-016 2
U 9560	1448.9 / 58.77 / 83.96	3547 / 63.25 / 32.45	13 P / 0.13 / 43 -3	1.0 / 0.8 / 2	0.35 / 76.	14.75	2 / 0.00 / 0.00	1202. / 1317.		9	-0.10 / 214	23.1 / 31.82	2.0 / 19.4 / 12.4	-17.07 / 5.4	9.02 / 8.63	0.40	II ZW 70, VV 324 / TURN 113A / 9560 06-33-002 2
U 9562	1449.2 / 58.68 / 84.00	3545 / 63.19 / 32.52	13 P / 0.13 / 43 -3	1.3 / 1.3 / 2	1.00 /	14.30	3 / 0.00 / 0.00	1250. / 1366.		9	0.31 / 214	23.8 / 31.88	2.1 / 19.9 / 12.8	-17.58 / 9.0	9.22 / 9.06	0.69	II ZW 71, VV 324 / TURN 113B / 9562 06-33-004 2
1450-03	1450.0 / 351.26 / 128.94	-321 / 47.62 / 27.06	8X / 0.40 / 41 +2 +1	2.7 / 2.7 / 2	0.80 / 41.		1 / 0.08 / 0.21	1955. / 1924. / 1	175 / 217. / 220.	8	0.84 / 11 / 1.07	30.7 / 32.44	-17.2 / 21.3 / 14.0	24.2	9.81 / 10.52	0.197	UGCA 396 / 00-38-010
N 5774	1451.4 / 359.41 / 121.32	347 / 52.49 / 29.51	7X / 0.67 / 41 -2 +1	3.2 / 3.2 / 2	0.84 / 37.	12.63	1 / 0.06 / 0.13	1589. / 1587.		20		26.8 / 32.14	-12.1 / 19.9 / 13.2	-19.51 / 25.0	10.00		9576 01-38-013 2
N 5775	1451.5 / 359.45 / 121.38	345 / 52.41 / 29.58	5B / 0.67 / 41 -2 +1	3.9 / 3.0 / 2	0.25 / 84.	11.44	1 / 0.67 / 0.13	1582. / 1580.		95		26.7 / 32.13	-12.1 / 19.8 / 13.2	-20.69 / 23.4	10.47		9579 01-38-014 2
N 5783	1451.9 / 88.82 / 64.74	5217 / 56.19 / 30.37	5X / 0.32 / 42 -5 +3	2.8 / 2.5 / 2	0.61 / 57.	13.0	3 / 0.19 / 0.02	2330. / 2497. / 1	260 / 264. / 266.	10	0.68 / 8 / 1.06	37.1 / 32.85	13.7 / 29.0 / 18.8	-19.85 / 27.1	10.13 / 9.82 / 10.74	0.49 / 0.120 / 4.04	9586 09-24-050 2
1452+42	1452.8 / 71.62 / 76.28	4213 / 60.82 / 32.52	13 P / 0.21 / 72 -0			14.3	3 / 0.19 / 0.02	2530. / 2669.		105		39.6 / 32.99	7.9 / 32.4 / 21.3	-18.69	9.67		I ZW 97 / 07-31-000 2
1454-82	1454.9 / 307.13 / 200.26	-8236 / -21.03 / -11.77	6 / 0.18 / 55 -0	2.0 / 1.9 / 6	0.54 / 62.		0.25 / 0.39	2517. / 2294. / 7		20		31.2 / 32.47	-28.6 / -10.6 / -6.4	17.3			008-07

Name	(2)	(3)	(4)	(5)	(5b)	(6)	i	(8a)	Vel	(9)	(10)	Size	Axis	Type	Coord 1	Coord 2	Alt Name
1455-47			27.0	14.7 / 30.83	-14.4 / 2.3 / 1.9		20		1052. / 878. / 7	0.05 / 0.96		5.4 / 6.3 / 6	0.86 / 35.	10 / 0.07 / 16 -0	1455.0 / 324.00 / 170.83	-4730 / 9.91 / 7.61	273-14
N 5792	0.46 / 0.056 / 8.36	10.52 / 10.19 / 11.45	-20.83 / 50.9	30.6 / 32.43	-16.0 / 21.4 / 15.0	1.22 / 12 / 1.26	20	467 / 435. / 436.	1930. / 1914. / 1	3 / 0.67 / 0.17	11.6 / 8.14 / 3.46	7.2 / 5.7 / 2	0.29 / 81.	3B / 0.52 / 41 +2 +1	1455.8 / 355.36 / 126.83	-53 / 48.43 / 29.23	9631 00-38-012 2
N 5796		10.56	-20.91 / 25.7	41.9 / 33.11	-30.7 / 23.1 / 16.7		90		2946. / 2871.	5 / 0.00 / 0.44	12.2	1.9 / 2.1 / 5	0.85 / 36.	-5 / 0.17 / 41+15	1456.6 / 342.00 / 143.00	-1626 / 36.35 / 23.52	-3-38-039 2
U 9638	0.81 / 0.297 / 2.72	9.59 / 9.50 / 10.03	-18.50 / 20.1	36.3 / 32.80	17.3 / 26.6 / 17.6	0.38 / 226 14 / 1.03	15 / 11 / 226	133 / 135. / 139.	2275. / 2461. / 1	3 / 0.14 / 0.01	14.3	2.0 / 1.9 / 2	0.68 / 52.	10 / 0.25 / 42 +9	1456.9 / 97.11 / 57.03	5904 / 51.43 / 29.09	9638 10-21-040
N 5806		10.34	-20.36 / 21.6	28.5 / 32.27	-13.3 / 19.8 / 14.1		65		1301. / 1298.	2 / 0.25 / 0.15	11.91 / 9.32 / 2.59	2.9 / 2.6 / 2	0.53 / 63.	3X / 0.85 / 41 -1	1457.5 / 359.09 / 123.72	205 / 50.18 / 30.54	9645 00-38-014 2
N 5832	0.42 / 0.116 / 3.59	9.47 / 9.09 / 10.03	-18.20 / 11.1	11.5 / 30.30	7.6 / 7.2 / 4.7	0.97 / 11 / 1.10	10	172 / 182. / 185.	451. / 660. / 1	3 / 0.13 / 0.04	12.1	3.5 / 3.3 / 2	0.70 / 50.	3BP / 0.09 / 44 -0	1457.6 / 110.10 / 43.73	7153 / 42.15 / 24.27	9649 12-14-015 2
1457-13		9.62	25.8	40.2 / 33.02	-27.9 / 23.4 / 17.0	0.41 / 21 / 1.05	10	89	2788. / 2726. / 1	0.00 / 0.38		2.0 / 2.2 / 2	1.00	10 / 0.13 / 41+15	1457.7 / 344.58 / 140.06	-1321 / 38.67 / 25.07	UGCA 397 / -2-38-024
1457-48			8.3	7.7 / 29.42	-7.5 / 1.1 / 1.0		20		586. / 412. / 7	0.10 / 1.09		3.2 / 3.7 / 6	0.75 / 45.	10 / 0.13 / 14-16+15	1457.7 / 324.11 / 171.58	-4806 / 9.17 / 7.65	223-09
N 5812		10.42	-20.58 / 23.1	31.7 / 32.50	-19.5 / 20.2 / 14.7		50		2066. / 2027.	2 / 0.00 / 0.28	11.92	2.4 / 2.5 / 2	0.96 / 20.	-5 / 0.19 / 41 -0 +1	1458.3 / 349.80 / 133.89	-716 / 43.31 / 27.61	UGCA 398 / -1-38-016 2
N 5813		10.50	-20.77 / 30.0	28.5 / 32.27	-13.8 / 20.4 / 14.7		65		1882. / 1879.	0.00 / 0.15	11.50	3.7 / 3.6 / 2	0.80 / 41.	-5 / 0.88 / 41 -1	1458.7 / 359.20 / 124.03	154 / 49.84 / 30.78	9655 00-38-016 2
U 9663	0.59 / 0.134 / 4.37	9.68 / 9.44 / 10.32	-18.71 / 17.8	38.2 / 32.91	14.4 / 29.3 / 19.9	0.28 / 27 / 1.03	15	133 / 200. / 203.	2422. / 2594. / 1	3 / 0.04 / 0.02	14.2	1.6 / 1.6 / 2	0.88 / 32.	10 / 0.23 / 42 -5 +3	1459.7 / 88.39 / 63.80	5248 / 54.92 / 31.38	DDO 198 / 9663 09-25-005 2
N 5831		10.17	-19.95 / 18.3	28.5 / 32.27	-13.8 / 19.9 / 14.8		50		1684. / 1682.	2 / 0.00 / 0.13	12.32	2.2 / 2.2 / 2	0.91 / 29.	-5 / 0.83 / 41 -1	1501.6 / 359.40 / 124.82	125 / 48.98 / 31.32	9678 00-38-020 2
N 5838		10.43	-20.60 / 27.5	28.5 / 32.27	-13.2 / 19.7 / 14.8		50		1427. / 1430.	2 / 0.00 / 0.13	11.67	3.8 / 3.3 / 2	0.45 / 69.	-2A / 0.75 / 41 -1	1502.9 / 0.74 / 123.94	218 / 49.32 / 31.90	9692 00-38-022 2
N 5845		9.91	-19.30 / 5.8	28.5 / 32.27	-13.7 / 20.0 / 15.1		53		1785. / 1786.	2 / 0.00 / 0.13	12.97	0.8 / 0.7 / 2	0.60 / 58.	-5 / 0.84 / 41 -1	1503.5 / 0.35 / 124.51	150 / 48.90 / 31.90	9700 00-38-024 2

(1) Name	(2) α / ℓ / SGL	(3) δ / b / SGB	(4) Type / Q_{xyz} / Group	(5) $D_{25}^{b,i}$ / $D_{25}^{b,i}$ / source	(6) d/D / i	(7) $B_T^{b,i}$ / $H_{-0.5}$ / $B-H$	(8) source / A_B^{i-o} / A_B^b	(9) V_h / V_o / Tel	(10) W_{20} / W_R / W_D	(11) e_v / e_w / source	(12) F_c / source,e_f / f_h	(13) R / μ	(14) SGX / SGY / SGZ	(15) $M_B^{b,i}$ / $\Delta_{25}^{b,i}$	(16) $\log L_B$ / $\log M_H$ / $\log M_T$	(17) M_H/L_B / M_H/M_T / M_T/L_B	(18) Alternate Names / UGC/ESO MCG / AGN? RCII
1503+01	1503.9 0.40 124.61	147 48.80 31.98	-5 0.78 41 -1	0.6 0.6 2	1.00	14.06	2 0.00 0.14	2192. 2193.		29		28.5 32.27	-14.0 20.3 15.4	-18.21 5.0	9.48		NGC 5846A, TURN 121A 00-38-026 2
N 5846	1503.9 0.42 124.59	148 48.81 31.99	-5 0.84 41 -1	4.9 4.9 9	0.89 31.	11.11	1 0.00 0.14	1784. 1785.	412	21 30 202	-0.31 202 41	28.5 32.27	-13.7 20.0 15.1	-21.16 40.8	10.66 8.60	0.01	TURN 121B 9706 00-38-025 2
N 5850	1504.6 0.52 124.72	144 48.63 32.13	3B 0.65 41 -1	4.6 4.6 2	0.91 28.	11.58	1 0.03 0.14	2530. 2532.	205	10 20 211	0.71 211 41	28.5 32.27	-14.4 20.7 15.9	-20.69 38.3	10.47 9.62	0.14	9715 00-39-002 2
N 5866	1505.1 92.03 60.01	5557 52.49 31.16	-2A 0.24 44 -1	7.3 6.2 2	0.48 67.	10.84	2 0.00 0.01	692. 875.		14		15.3 30.92	6.5 11.3 7.9	-20.08 27.7	10.22		9723 09-25-017 2
N 5854	1505.3 1.88 123.62	246 49.18 32.60	-2B 0.74 41 -1	2.4 1.9 2	0.29 80.	12.54	2 0.00 0.11	1626. 1632.		65		28.5 32.27	-13.2 19.8 15.3	-19.73 15.8	10.08		9726 01-39-001 2
1505-75	1505.7 311.29 194.71	-7540 -15.31 -7.48	2 0.28 55 -0 +1	2.8 2.3 6	0.29 81.		 0.67 0.45	2850. 2632. 7		20		35.9 32.77	-34.4 -9.0 -4.7	 24.1			IC 4522 042-02
1505-52	1505.9 323.16 175.87	-5222 4.81 6.29	7 0.07 16 -0	2.4 2.9 6	0.17 90.		 0.67 2.60	1409. 1230. 7		20		19.1 31.40	-18.9 1.4 2.1	 16.2			223-12
N 5861	1506.5 348.51 138.81	-1108 39.06 27.95	5X 0.15 41 -6 +1	3.1 3.0 4	0.53 63.	11.5 9.63 1.87	5 0.25 0.40	1867. 1821. 1	358. 360. 362.	20	0.93 15 1.08	28.9 32.30	-19.2 16.8 13.5	-20.80 25.3	10.51 9.85 10.98	0.22 0.075 2.93	-2-39-003 2
N 5833	1506.7 312.95 192.30	-7240 -12.77 -5.66	5 0.23 55 -1	3.2 3.3 6	0.75 45.		 0.10 0.48	2996. 2782. 7		20		38.0 32.90	-36.9 -8.0 -3.7	 36.6			042-03
N 5864	1507.0 2.86 123.21	315 49.16 33.15	-2B 0.74 41 -1	2.7 2.2 2	0.35 76.	12.8	3 0.00 0.10	1623. 1633.		33		28.5 32.27	-13.0 19.8 15.5	-19.47 18.3	9.98		9740 01-39-002 2
U 9749	1508.2 104.97 47.78	6723 44.84 27.06	-5 0.50 14 -12	52.0 47.5 2	0.64 55.	12.4	3 0.00 0.04	-189. 17.		70		0.1 20.00	0.1 0.1 0.0	-7.60 1.4	5.23		U MI, DDO 199 9749 11-18-030 2
N 5879	1508.4 93.25 58.45	5712 51.40 31.15	4A 0.21 44 -1	4.5 3.7 2	0.39 73.	11.64 9.25 2.39	2 0.45 0.02	775. 962. 4	285. 259. 261.	15 6 211	0.90 217 9 1.03	16.8 31.12	7.5 12.2 8.7	-19.48 18.2	9.98 9.35 10.55	0.23 0.063 3.66	9753 10-22-001 2
U 9760	1509.5 1.87 125.01	152 47.79 33.34	7 0.68 41 -1	2.7 2.0 2	0.09 90.	13.8	2 0.67 0.11	2016. 2023. 1	163	15	0.42 26 1.04	28.5 32.27	-13.9 19.8 15.9	-18.47 16.6	9.58 9.33	0.56	9760 00-39-011

Catalog data table (no column headers printed). Each galaxy occupies up to three stacked data lines.

ID															Designation
U 9762	1509.7	3250	9	1.3 0.92	0.03		2275.	148	20	0.12	36.3	1.4	12.7	9.24		9762 06-33-000
	51.85	59.24	0.07	1.2 26.	0.04		2396.		14	226 29	32.80	29.0				
	87.14	36.96	70 -0	2			1		226	1.02		21.8				
U 9764	1509.8	6505	8B	2.5 0.65	3	14.1	2246.	238	10	0.72	35.8	20.3	-18.67	9.66	1.47	9764 11-18-029
	102.42	46.30	0.10	2.3 54.	0.16	12.62	2448.	248.		19	32.77	24.2	24.0	9.83	0.157	
	50.04	28.20	42 -0	2	0.03	1.48	1	250.		1.05		16.9		10.63	9.38	
1510-87	1510.3	-8715	10	2.3 0.21	0.67		2269.		20		27.8	-24.5	14.6			008-08
	304.76	-25.14	0.17	1.8 89.	0.44		2046.				32.22	-11.1				
	204.42	-14.05	55 -4 +3	6			7					-6.7				
1510-17	1510.4	-1757	8B	2.1 0.96	0.01		2279.	205	15	0.56	33.4	-25.0	22.4	9.61		IC 4536, UGCA 401
	344.11	33.17	0.29	2.3 19.	0.45		2210.			25	32.62	16.8				581-24 -3-39-002
	146.13	25.77	41+10	6			1			1.05		14.5				
N 5894	1510.6	6000	9B	3.2 0.19	3	12.4	2485.	463	25	0.59	38.9	19.1	-20.55	10.41	0.23	9768 10-22-004
	96.51	49.50	0.23	2.2 90.	0.67	9.95	2679.	425.		41	32.95	27.6	25.0	9.77	0.045	
	55.31	30.37	42 +9	2	0.02	2.45	1	426.		1.06		19.7		11.12	5.07	
1510-20	1510.7	-2030	9	3.5 0.23	0.67		2277.	352	20	0.88	33.1	-25.7	27.1	9.92	0.107	UGCA 402
	342.35	31.10	0.29	2.8 87.	0.47		2199.	315.		34	32.60	15.7		10.89		581-25 -3-39-003
	148.64	24.61	41+11	6			2	316.		1.02		13.8				
U 9769	1510.7	5559	8A	2.9 0.79	3	14.2	844.	208	30	0.73	17.7	7.6	-17.04	9.01	1.65	NGC 5866B
	91.35	51.83	0.15	2.8 42.	0.08		1029.			14	31.24	13.0	14.5	9.23		
	59.66	31.88	44 -1	2	0.02		1			1.08		9.3				
1510-46	1510.8	-4638	5	11.2 0.17	0.67		518.		20		7.2	-7.0	20.0			9769 09-25-034 2
	326.80	9.33	0.12	9.5 90.	0.93		356.				29.28	1.0				274-01
	171.66	10.31	14-16+15	6			7					1.3				
N 5878	1511.0	-1405	3A	3.1 0.49	2	11.54	2111.		65		31.6	-22.2	-20.96	10.58		UGCA 403
	347.19	36.09	0.22	3.0 66.	0.30		2057.				32.50	17.1	27.7			-2-39-006 2
	142.35	27.67	41-10	2	0.47		1					14.7				
1511-15	1511.0	-1518	7A	3.4 0.79	5	11.7	2287.	169	20	0.81	33.7	-24.1	-20.94	10.57	0.20	A 1511, FATH 703
	346.24	35.15	0.25	3.5 41.	0.08		2229.	206.		14	32.64	17.8	34.4	9.87	0.173	-2-39-007 2
	143.57	27.13	41-10	4	0.45		1	209.		1.11		15.4		10.63	1.15	
U 9776	1511.7	5710	10	1.1 0.53	0.25		830.	75	10	-0.02	17.4	7.8	5.1	8.46		9776 10-22-005
	92.80	51.05	0.18	1.0 63.	0.02		1018.			226 25	31.21	12.6				
	58.30	31.58	44 -1	2			1			1.01		9.1				
1512-72	1512.2	-7229	10	1.3 0.50	0.29		3210.		20		40.7	-39.5	14.3			068-01
	313.40	-12.83	0.15	1.2 65.	0.52		2997.				33.05	-8.7				
	192.40	-5.22	55 -1	6			7					-3.7				
N 5885	1512.4	-953	5X	4.4 0.79	2	11.69	2005.	209	10	1.03	30.7	-19.9	-20.75	10.49	0.33	-2-39-013 2
	350.98	39.05	0.18	4.5 41.	0.08		1968.	262.	8	217 7	32.44	17.8	40.3	10.00	0.126	
	138.19	29.79	41 -6 +1	4	0.43		1	264.	217	1.18		15.3		10.90	2.59	
N 5899	1513.2	4214	5X	2.8 0.47	2	12.21	2549.	50	50	0.67	39.8	8.1	-20.79	10.51	0.23	9789 07-31-045 2
	69.40	57.22	0.30	2.4 67.	0.33		2701.			54 53	33.00	31.1	27.9	9.87		
	75.43	36.23	72 +1	2	0.06							23.5				

(1) Name	(2) α / ℓ / SGL	(3) δ / b / SGB	(4) Type / Q_xyz / Group	(5) D_25 / D_25^{b,i} / source	(6) d/D / i	(7) B_T^{b,i} / H_{-0.5} / B-H	(8) source / A_B^{i-o} / A_B^b	(9) V_h / V_o / Tel	(10) W_20 / W_R / W_D	(11) e_v / e_w / source	(12) F_c / source e_f / f_h	(13) R / μ	(14) SGX / SGY / SGZ	(15) M_B^{b,i} / Δ_25^{b,i}	(16) log L_B / log M_H / log M_T	(17) M_H/L_B / M_H/M_T / M_T/L_B	(18) Alternate Names UGC/ESO MCG / Alternate Names / AGN? RCII
N 5907	1514.6 / 91.59 / 58.84	5631 / 51.08 / 32.19	5A / 0.26 / 44 -1	11.2 / 7.9 / 2	0.17 / 90.	10.31 / 7.52 / 2.79	1 / 0.67 / 0.02	666. / 854. / 4	489 / 451. / 452.	7 / 6 / 211	1.79 / 211 7 / 1.14	14.9 / 30.86	6.5 / 10.8 / 7.9	-20.55 / 34.4	10.41 / 10.14 / 11.31	0.53 / 0.068 / 7.85	9801 09-25-040 2
N 5898	1515.3 / 340.98 / 152.46	-2355 / 27.72 / 23.82	-5 / 0.23 / 41-11	2.5 / 2.8 / 6	1.00	12.08	2 / 0.00 / 0.52	2260. / 2173.		82		32.6 / 32.57	-26.4 / 13.8 / 13.2	-20.49 / 26.7	10.39		UGCA 404
1515-23	1515.7 / 341.05 / 152.53	-2356 / 27.65 / 23.89	-5 / 0.27 / 41-11	1.3 / 1.2 / 6	0.42 / 71.	14.00	2 / 0.00 / 0.60	2380. / 2294.		90		34.2 / 32.67	-27.7 / 14.4 / 13.8	-18.67 / 12.0	9.66		514-02 -4-36-006 2
N 5903	1515.7 / 341.08 / 152.48	-2353 / 27.69 / 23.91	-5 / 0.20 / 41-11	2.9 / 3.0 / 6	0.71 / 49.	11.90	2 / 0.00 / 0.60	2524. / 2438.		80		35.9 / 32.78	-29.1 / 15.2 / 14.5	-20.88 / 31.4	10.54		UGCA 405 / 514-03 -4-36-007 2
1516-41	1516.0 / 330.65 / 167.62	-4103 / 13.55 / 14.37	3B / 0.12 / 16 -0	2.4 / 2.5 / 6	0.77 / 44.		0.09 / 0.44	1334. / 1191. / 2	178	25	0.65 / 19 / 1.02	19.1 / 31.40	-18.1 / 4.0 / 4.7	13.9	9.21		514-04 -4-36-008 2
1517-36	1517.5 / 333.37 / 164.18	-3645 / 16.97 / 17.13	4B / 0.07 / 40 -0	2.8 / 2.7 / 6	0.57 / 60.	12.40	6 / 0.22 / 0.44	3069. / 2940.		50		41.8 / 33.10	-38.4 / 10.9 / 12.3	-20.70 / 33.0	10.47		328-43 -7-31-011
N 5915	1518.8 / 349.90 / 142.13	-1255 / 35.74 / 29.89	2BP / 0.18 / 41-10	1.4 / 1.5 / 2	0.79 / 42.	11.8	5 / 0.08 / 0.53	2272. / 2229.		18		33.7 / 32.64	-23.1 / 17.9 / 16.8	-20.84 / 14.8	10.53		SCI 147 / 387-26 -6-34-002 2
N 5921	1519.5 / 8.14 / 121.92	515 / 47.93 / 36.70	4B / 0.13 / 41 -0 +1	4.9 / 4.9 / 2	0.88 / 32.	11.34	2 / 0.04 / 0.07	1480. / 1508.	190. / 294. / 296.	10 / 7 / 217	0.90 / 217 12	25.2 / 32.01	-10.7 / 17.1 / 15.0	-20.67 / 36.1	10.46 / 9.70 / 10.96	0.17 / 0.056 / 3.13	UGCA 407 / -2-39-019 2
1522-73	1522.1 / 313.29 / 193.86	-7346 / -14.30 / -5.46	6 / 0.26 / 55 -1	2.3 / 2.5 / 6	0.89 / 30.		0.04 / 0.47	2971. / 2759. / 7		20		37.5 / 32.87	-36.3 / -8.9 / -3.6	27.4			9824 01-39-021 2
U 9837	1522.6 / 92.93 / 56.46	5814 / 49.25 / 32.50	5X / 0.17 / 42+10	2.0 / 2.0 / 2	0.91 / 28.	13.6	3 / 0.03 / 0.02	2661. / 2857.	178	15	0.34 / 23 / 1.04	41.2 / 33.07	19.2 / 28.9 / 22.1	-19.47 / 24.1	9.98 / 9.57	0.39	042-07
1523-22	1523.2 / 343.86 / 151.80	-2206 / 28.05 / 26.32	4X / 0.24 / 41+11	2.9 / 2.9 / 6	0.67 / 53.		0.15 / 0.42	2325. / 2251. / 2	277	30	0.67 / 25 / 1.02	33.6 / 32.63	-26.6 / 14.2 / 14.9	28.5	9.72		SCI 149 / 9837 10-22-013 2
N 5929	1524.3 / 67.86 / 75.30	4151 / 55.31 / 38.33	-5 / 0.26 / 72 -1	1.1 / 1.1 / 2	0.91 / 29.		0.00 / 0.06	2448. / 2606.		33		38.5 / 32.93	7.7 / 29.2 / 23.9	12.4			UGCA 408 / ARP 90, I ZW 112 / TURN 126A / 9851 07-32-006 2
N 5930	1524.3 / 67.86 / 75.30	4151 / 55.31 / 38.33	3XP / 0.26 / 72 -1	2.0 / 1.7 / 2	0.44 / 70.	12.7	3 / 0.37 / 0.06	2715. / 2873.		33		41.9 / 33.11	8.3 / 31.8 / 26.0	-20.41 / 20.8	10.36		582-12 -4-36-014 / ARP 90, I ZW 112 / TURN 126B / 9852 07-32-007 2

Name	Cross-ID	Position 1	Position 2	Type	Dim 1	Dim 2	m_b	(p)	Mag	(g)	(v)	(t1)	(t2)	(t3)	(s1)	(s2)	(vel)
U 9857	9857 07-32-009	1524.7 / 67.94 / 75.20	4155 / 55.22 / 38.38	10B / 0.19 / 72 -1	1.9 / 1.6 / 2	0.36 / 75.		0.52 / 0.07	9.20		17.3	7.4 / 28.1 / 23.0	37.1 / 32.85	0.06 / 33 / 1.02	20	140	2332. / 2491. / 1
U 9858	9858 07-32-010	1524.8 / 65.90 / 76.66	4044 / 55.44 / 38.69	4X / 0.29 / 72 -1	4.2 / 3.0 / 2	0.21 / 89.	12.5 / 10.30 / 2.20	3 / 0.67 / 0.03	10.41 / 10.29 / 11.07	0.76 / 0.164 / 4.61	-20.54 / 35.6	7.3 / 30.8 / 25.4	40.6 / 33.04	1.07 / 13 / 1.10	10	376. / 338. / 340.	2619. / 2774. / 1
1525-42	274-16 -7-32-000	1525.6 / 331.28 / 170.02	-4237 / 11.25 / 14.84	10 / 0.11 / 16 -0	2.8 / 2.6 / 6	0.43 / 70.		0.39 / 0.59	9.33 / 9.89	0.276	14.5	-18.2 / 3.2 / 4.9	19.1 / 31.40	0.77 / 19	15	160 / 136. / 140.	1336. / 1195. / 2
N 5949	9866 11-19-008 2	1527.2 / 100.56 / 49.23	6455 / 44.99 / 29.91	5A / 0.10 / 44 -0	2.3 / 2.0 / 2	0.52 / 64.	12.4 / 10.32 / 2.08	3 / 0.27 / 0.06	9.33 / 8.26 / 9.89	0.09 / 0.023 / 3.64	-17.84 / 6.5	6.3 / 7.3 / 5.6	11.2 / 30.24	0.16 / 217 / 20	10 / 15 / 217	217 / 202. / 204.	436. / 645. / 1
U 9893	I ZW 115 / 9893 08-28-038 2	1531.3 / 75.20 / 69.03	4637 / 52.91 / 38.14	13 P / 0.09 / 44 -0 +1	1.3 / 1.1 / 2	0.37 / 74.	14.8	3 / 0.00 / 0.02	8.62 / 8.39	0.59	-16.07 / 4.8	4.2 / 11.0 / 9.2	15.0 / 30.87	0.04 / 209 / 20	10		665. / 840.
N 5951	9895 03-40-003 2	1531.4 / 23.51 / 110.15	1510 / 50.44 / 41.58	5 / 0.18 / 71 -1	3.3 / 2.4 / 2	0.23 / 86.	12.7 / 11.05 / 1.65	3 / 0.67 / 0.09	10.05 / 9.62 / 10.52	0.37 / 0.126 / 2.94	-19.65 / 20.7	-7.6 / 20.7 / 19.6	29.5 / 32.35	0.68 / 22 / 1.06	15	272 / 235. / 237.	1784. / 1860. / 1
N 5953	ARP 91, VV 244 / TURN 127A / 9903 03-40-005 2	1532.2 / 23.92 / 109.92	1522 / 50.35 / 41.80	-5 / 0.30 / 71 -1	1.8 / 1.7 / 2	0.78 / 43.	13.0	3 / 0.00 / 0.11	10.03	0.12	-19.59 / 16.4	-8.4 / 23.1 / 22.0	33.0 / 32.59		39		2069. / 2147.
N 5954	ARP 91, VV 244 / TURN 127B / 9904 03-40-006 2	1532.3 / 23.94 / 109.92	1522 / 50.33 / 41.82	5 / 0.33 / 71 -1	1.3 / 1.1 / 2	0.49 / 66.	13.2	3 / 0.30 / 0.11	9.92 / 9.00		-19.33 / 10.3	-8.1 / 22.5 / 21.4	32.1 / 32.53	-0.01 / 217 / 40	29 / 20 / 217	251	1988. / 2066.
N 5963	9906 09-25-058 2	1532.3 / 89.98 / 57.39	5645 / 48.87 / 34.32	12 P / 0.26 / 44 -1	3.7 / 3.5 / 2	0.77 / 44.	12.1	3 / 0.09 / 0.01	9.69 / 9.41 / 10.68	0.52 / 0.054 / 9.71	-18.75 / 15.1	6.6 / 10.3 / 8.3	14.8 / 30.85	1.07 / 11 / 1.12	15	265 / 330. / 331.	657. / 854. / 1
U 9912	VV 132 / 9912 03-40-007	1532.8 / 25.96 / 108.14	1643 / 50.77 / 42.10	8B / 0.11 / 41+16	1.8 / 1.7 / 2	0.89 / 31.	14.0	3 / 0.04 / 0.10	9.18 / 9.00 / 9.35	0.66 / 0.444 / 1.48	-17.47 / 9.8	-4.5 / 13.9 / 13.2	19.7 / 31.47	0.41 / 12 / 1.03	7	63 / 89. / 96.	1034. / 1117.
N 5961	ARAK 478 / 9918 05-37-005	1533.2 / 48.87 / 88.82	3101 / 54.19 / 42.10	13 P / 0.18 / 71 -6	1.1 / 0.9 / 2	0.43 / 71.	13.8	3 / 0.00 / 0.07	9.58		-18.48 / 7.5	0.4 / 21.2 / 19.1	28.5 / 32.28		220		1668. / 1800.
N 5962	9926 03-40-011 2	1534.2 / 26.24 / 108.12	1646 / 50.48 / 42.44	5A / 0.32 / 71 -1	2.8 / 2.6 / 2	0.74 / 46.	11.85	2 / 0.10 / 0.10	10.46 / 9.62 / 11.12	0.15 / 0.032 / 4.64	-20.66 / 24.1	-7.3 / 22.3 / 21.5	31.8 / 32.51	0.62 / 217 / 15	10 / 15 / 217	353 / 435. / 437.	1964. / 2048.
1534+30	MARK 689 / 05-37-008	1534.3 / 48.63 / 89.01	3051 / 53.94 / 42.36	13 P / 0.18 / 71 -6			14.7	3 / 0.10 / 0.07	9.22		-17.57	0.4 / 21.0 / 19.2	28.5 / 32.27		70		1655. / 1788.
U 9936	DDO 200 / 9936 07-32-039 2	1535.1 / 71.37 / 71.44	4424 / 52.83 / 39.50	9 / 0.25 / 72 +1	1.8 / 1.6 / 2	0.56 / 60.	14.9	3 / 0.22 / 0.02	9.45 / 9.56	1.28	-18.15 / 19.0	10.0 / 29.8 / 25.9	40.7 / 33.05	0.34 / 32 / 1.02	15	188	2628. / 2800. / 1

(1) Name	(2) α ℓ SGL	(3) δ b SGB	(4) Type ϱ_{xyz} Group	(5) D_{25} $D_{25}^{b,i}$ source	(6) d/D i	(7) $B_T^{b,i}$ $H_{-0.5}$ B-H	(8) source A_B^{i-o} A_B^b	(9) V_h V_o Tel	(10) W_{20} W_R W_D	(11) e_v e_w source	(12) F_c source/e_f f_h	(13) R μ	(14) SGX SGY SGZ	(15) $M_B^{b,i}$ Δ_{25}	(16) $\log L_B$ $\log M_H$ $\log M_T$	(17) M_H/L_B M_H/M_T M_T/L_B	(18) Alternate Names UGC/ESO MCG AGN? RCII
N 5964	1535.1 12.42 122.13	609 45.25 40.68	7B 0.10 41 -0 +1	4.1 4.0 2	0.79 42.	12.5	3 0.08 0.13	1447. 1492. 1	209 260. 262.	10 8 217	1.01 217 9 1.15	24.7 31.96	-10.0 15.9 16.1	-19.46 28.8	9.98 9.80 10.75	0.66 0.110 5.98	9935 01-40-008 2
1535+55	1535.8 87.85 58.56	5526 49.03 35.34	13 P 0.26 44 -1	0.5 0.5 2	0.89 31.	14.8	3 0.00 0.03	663. 860.		39		15.0 30.88	6.4 10.4 8.7	-16.08 2.2	8.62		UGCA 410, MARK 487 I ZW 123 09-26-000 2
U 9941	1536.0 21.42 113.10	1307 48.55 42.39	10 0.24 71 -2 +1	1.6 1.6 2	0.88 32.	14.4	3 0.04 0.12	1856. 1929. 1	172 263. 265.	20 12 226	0.36 226 17 1.03	30.3 32.41	-8.8 20.6 20.4	-18.01 14.2	9.40 9.32 10.45	0.85 0.074 11.41	9941 02-40-005
N 5970	1536.1 20.41 114.13	1221 48.18 42.28	5B 0.27 71 -2 +1	2.8 2.6 2	0.72 48.	11.95	2 0.12 0.08	1968. 2038.	318	9 50 211	0.59 211 51	31.6 32.50	-9.5 21.3 21.2	-20.55 24.0	10.41 9.59	0.15	9943 02-40-006 2
N 5981	1536.8 93.21 54.07	5933 47.00 33.53	5 0.07 72 -0	2.8 1.9 2	0.15 90.	13.21	2 0.67 0.02	1717. 1921.		145		29.2 32.33	14.3 19.7 16.1	-19.12 16.2	9.84		9948 10-22-027 2
N 5974	1537.0 50.50 87.47	3155 53.45 42.79	13 P 0.17 71 -6	0.8 0.8 2	0.57 60.	14.1	3 0.00 0.08	1737. 1875.		156		29.4 32.34	1.0 21.6 20.0	-18.24 6.9	9.49		ARAK 482
N 5985	1538.6 92.99 53.99	5930 46.83 33.75	3X 0.17 42+10	5.3 4.7 2	0.57 60.	11.55	2 0.22 0.22	2513. 2718.	519	10 25 211	0.84 211 60	39.2 32.97	19.2 26.4 21.8	-21.42 53.8	10.76 10.03	0.18	9969 10-22-030 2
U 9977	1539.5 7.35 129.18	52 41.31 40.13	5 0.19 71 -5	3.7 2.7 2	0.14 90.	13.0 11.33 1.67	3 0.67 0.28	1915. 1944. 1	291 253. 255.	20	0.68 16 1.08	30.2 32.40	-14.6 17.9 19.4	-19.40 23.8	9.95 9.64 10.65	0.49 0.099 4.94	9977 00-40-007
U 9979	1539.8 7.14 129.52	37 41.10 40.11	10 0.18 71 -5	1.6 1.4 2	0.37 74.	13.87	7 0.49 0.26	1978. 2006. 1	140	30	0.37 26 1.02	31.0 32.45	-15.1 18.3 19.9	-18.58 12.7	9.62 9.35	0.54	DDO 201
N 5984	1540.6 23.87 111.62	1423 48.10 43.68	7B 0.15 41+16	2.8 2.1 2	0.25 85.	12.5 10.88 1.62	3 0.67 0.09	1118. 1199. 1	237 201. 204.	10 15 217	0.50 217 18	20.7 31.58	-5.5 13.9 14.3	-19.08 12.7	9.82 9.13 10.17	0.20 0.091 2.23	9987 02-40-011 2
U 9992	1541.4 102.20 45.91	6725 42.37 29.79	10 0.10 44 -0	1.9 1.8 2	0.70 50.	14.4	3 0.13 0.07	429. 646. 3	52 43. 55.	7	0.40 8 1.03	11.2 30.24	6.7 7.0 5.5	-15.84 5.9	8.53 8.50 8.50	0.93 0.998 0.94	9992 11-19-011
N 5967	1541.9 313.29 196.05	-7531 -16.49 -5.66	6X 0.26 55 -0 +1	2.4 2.4 6	0.67 53.	12.0	5 0.15 0.31	2904. 2695.		90		36.6 32.81	-35.0 -10.1 -3.6	-20.81 25.6	10.52		042-10 2
U10014	1543.4 22.00 114.14	1239 46.73 44.09	10 0.14 41+16	1.4 1.4 2	0.86 34.	15.1	3 0.05 0.11	1126. 1203. 1	135 190. 193.	10 8 226	0.75 226 8 1.02	20.7 31.58	-6.1 13.6 14.4	-16.48 8.5	8.78 9.38 9.95	3.96 0.272 14.57	10014 02-40-013

Name	Coordinates	Type	Dim.	(i)	B	flags	V	W	N	HI	μ/log	dist.	M_B	log	color
U10020	1543.6 33.26 102.85 / 2044 49.79 44.93	7A 0.21 71 -0	2.3 2.3 2	0.96 19.	13.4	3 0.01 0.14	2089. 2195. 1	85	15	0.32 18 1.06	33.4 32.62	-5.3 23.0 23.6	-19.22 22.4	9.88 9.37	0.31	10020 04-37-031
1544+46	1544.9 73.50 68.52 / 4609 50.76 40.50	13 P 0.14 72 -0			14.7	3 0.01 0.03	2788. 2970.		109		42.7 33.15	11.9 30.2 27.8	-18.45	9.57		MARK 490 08-29-000 2
U10031	1545.0 95.22 51.23 / 6143 45.06 33.27	9B 0.17 44 -2 +1	1.8 1.7 2	0.89 31.		0.04 0.03	888. 1099. 1	71 105. 110.	10	0.26 19 226 1.03	18.2 31.30	9.5 11.9 10.0	9.0	8.78 9.46	0.209	10031 10-22-000
U10041	1546.5 13.62 124.23 / 520 42.42 43.18	8X 0.10 71 -0	2.9 2.7 2	0.61 57.	13.5	3 0.19 0.17	2172. 2224. 1	224 222. 225.	20	0.78 16 1.07	33.6 32.63	-13.8 20.3 23.0	-19.13 26.5	9.84 9.83 10.58	0.97 0.180 5.42	10041 01-40-015
N 6000	1546.8 343.30 161.74 / -2915 19.15 26.55	6 0.07 40 -0	1.9 2.1 6	0.87 33.		0.05 0.67	2110. 2029.		60		30.1 32.39	-25.6 8.4 13.5	18.5			450-20 -5-37-003
U10054	1548.0 116.28 32.96 / 8159 32.93 21.07	8B 0.09 42 -0	3.1 2.8 2	0.48 67.	13.28	7 0.31 0.16	1499. 1725. 1	193 172. 175.	20	0.57 24 1.07	25.6 32.04	20.1 13.0 9.2	-18.76 20.9	9.70 9.39 10.25	0.49 0.136 3.62	DDO 203 10054 14-07-029 2
U10058	1548.3 41.73 95.20 / 2605 50.17 45.89	9B 0.09 71 -0	1.3 1.3 2	0.77 44.		0.09 0.18	2154. 2282. 1	129 148. 152.	10	0.19 17 226 1.02	34.4 32.68	-2.2 23.8 24.7	13.1	9.26 9.92	0.221	10058 04-37-000
U10061	1548.8 27.88 109.03 / 1628 47.14 45.89	10 0.26 71 -3	2.3 2.2 2	0.88 32.		0.04 0.06	2087. 2182. 1	98 146. 150.	10	0.33 17 1.05	33.2 32.61	-7.5 21.8 23.8	21.3	9.37 10.12	0.179	DDO 202 10061 03-40-0C0 2
N 6015	1550.7 95.70 50.04 / 6228 44.12 33.41	6A 0.16 44 -2 +1	6.3 5.1 2	0.38 73.	11.17 9.11 2.06	1 0.46 0.02	834. 1048. 4	312 286. 288.	10 6 211	1.35 217 2 1.03	17.5 31.22	9.4 11.2 9.7	-20.05 26.1	10.21 9.84 10.79	0.42 0.111 3.79	10075 10-23-003 2 ARAK 489
U10086	1552.3 28.73 108.74 / 1645 46.47 46.75	13 P 0.18 71 -3	0.9 0.7 2	0.33 78.	14.2	0.00 0.07	2301. 2400.	123 105. 111.	220		35.9 32.77	-7.9 23.3 26.1	-18.57 7.3	9.62		10086 03-40-063
U10095	1553.8 72.12 68.42 / 4534 49.36 42.16	10 0.11 70 +1	1.5 1.3 2	0.47 67.		0.33 0.05	1869. 2055. 1		25 13 226	0.26 16 226 1.02	31.1 32.46	8.5 21.4 20.9	11.8	9.25 9.58	0.466	10095 08-29-022
U10133	1559.0 12.05 130.21 / 151 37.90 45.04	5X 0.23 71 +4	2.8 2.7 2	0.64 55.	13.0	3 0.17 0.32	1929. 1978.		9		30.3 32.41	-13.8 16.4 21.4	-19.41 23.9	9.96		IC 1158 10133 00-41-002 2
1559+18	1559.8 32.62 105.65 / 1857 45.61 48.68	13 P 0.12 71 -0	1.0 0.8 2	0.53 63.	14.8	3 0.00 0.06	2525. 2638.		39	-0.30 17100	38.7 32.94	-6.9 24.6 29.0	-18.14 9.0	9.45 8.88	0.27	UGCA 411, MARK 294 03-41-000 2
N 6035	1601.2 35.60 102.49 / 2102 45.98 49.05	5X 0.17 71 -0	1.2 1.2 2	0.83 38.	14.3	3 0.06 0.10	2236. 2357.		75		35.1 32.73	-5.0 22.5 26.5	-18.43 12.3	9.56		10154 04-38-018 2

(1) Name	(2) α / l / SGL	(3) δ / b / SGB	(4) Type / ρ_{xyz} / Group	(5) D_{25} / $D_{25}^{b,i}$ / source	(6) d/D / i	(7) $B_T^{b,i}$ / $H_{-0.5}$ / B-H	(8) source / A_B^{i-o} / A_B^b	(9) V_h / V_o / Tel	(10) W_{20} / W_R / W_D	(11) e_v / e_w / source	(12) F_c / source/e_f / f_h	(13) R / μ	(14) SGX / SGY / SGZ	(15) $M_B^{b,i}$ / $\Delta_{25}^{b,i}$	(16) log L_B / log M_H / log M_T	(17) M_H/L_B / M_H/M_T / M_T/L_B	(18) Alternate Names UGC/ESO MCG / AGN? RCII
1604+41	1604.0 / 65.59 / 72.65	4127 / 48.02 / 45.48	13 P / 0.15 / 70 -1	1.0 / 0.8 / 2	0.53 / 63.	14.7	3 / 0.00 / 0.00	2038. / 2221.		35		33.2 / 32.60	6.9 / 22.2 / 23.6	-7.90 / 7.8	9.35		TURN 133A
U10200	1604.0 / 65.64 / 72.61	4129 / 48.02 / 45.47	13 P / 0.15 / 70 -1	1.0 / 0.9 / 2	0.77 / 44.	13.6	3 / 0.00 / 0.00	1954. / 2137.		36		32.1 / 32.53	6.7 / 21.5 / 22.9	-18.93 / 8.4	9.76		07-33-040 ARAK 497, TURN 133B ZW COMPACT 10200 07-33-039
N 6070	1607.4 / 12.46 / 132.66	50 / 35.59 / 46.63	6A / 0.25 / 71 -4	3.7 / 3.5 / 2	0.55 / 61.	11.75 / 9.41 / 2.34	2 / 0.23 / 0.37	2005. / 2058.	407 / 421. / 422.	7 / 15 / 217	0.86 / 217 11	31.2 / 32.47	-14.5 / 15.7 / 22.7	-20.72 / 31.9	10.48 / 9.85 / 11.21	0.23 / 0.043 / 5.43	10230 00-41-004 2
U10288	1511.8 / 12.28 / 134.52	-5 / 34.17 / 47.29	5 / 0.23 / 71 -4	4.9 / 3.7 / 2	0.14 / 90.	12.6 / 10.29 / 2.31	3 / 0.67 / 0.38	2046. / 2099. / 1	370 / 332. / 334.	10	0.98 / 16 / 1.13	31.5 / 32.49	-15.0 / 15.3 / 23.2	-19.89 / 34.0	10.15 / 9.98 / 11.04	0.67 / 0.087 / 7.74	10288 00-41-009
U10290	1612.0 / -3.34 / 133.17	56 / 34.69 / 47.73	9 / 0.22 / 71 -4	2.0 / 2.1 / 2	0.95 / 21.	13.5	3 / 0.02 / 0.34	1982. / 2039. / 1	138	20 / 12 / 226	0.35 / 226 16 / 1.05	30.8 / 32.44	-14.2 / 15.1 / 22.8	-18.94 / 18.9	9.77 / 9.33	0.36	10290 00-41-010
U10310	1614.8 / 73.62 / 64.19	4710 / 45.53 / 44.65	10B / 0.08 / 44 -0	3.1 / 2.9 / 2	0.77 / 44.	13.49	7 / 0.09 / 0.00	715. / 917. / 1	130 / 150. / 153.	8	0.69 / 12 / 1.09	15.8 / 31.00	4.9 / 10.1 / 11.1	-17.51 / 13.4	9.20 / 9.09 / 9.94	0.78 / 0.140 / 5.56	ARP 2, DDO 204 10310 08-30-002 2
N 6106	1616.4 / 21.02 / 124.14	732 / 37.16 / 50.93	5A / 0.07 / 71 -0	2.5 / 2.3 / 2	0.55 / 61.	12.36 / 10.29 / 2.07	2 / 0.23 / 0.21	1456. / 1542.	260 / 254. / 256.	10 / 7 / 217	0.79 / 217 11	24.5 / 31.94	-8.7 / 12.8 / 19.0	-19.58 / 16.5	10.02 / 9.57 / 10.49	0.35 / 0.120 / 2.91	10328 01-41-016 2
1616-60	1616.6 / 326.10 / 188.44	-6022 / -7.42 / 7.81	10 / 0.08 / 14+16+15	3.2 / 3.4 / 6	0.38 / 74.		0.48 / 1.34	606. / 440. / 7		20		7.2 / 29.29	-7.1 / -1.0 / 1.0	7.1			137-18
N 6118	1619.3 / 11.47 / 138.51	-210 / 31.45 / 48.11	6A / 0.07 / 71 -0	4.6 / 4.3 / 2	0.44 / 70.	11.4 / 9.37 / 2.03	3 / 0.37 / 0.49	1578. / 1629. / 1	355 / 338. / 340.	15 / 12 / 79	0.87 / 217 11 / 1.14	25.4 / 32.02	-12.7 / 11.2 / 18.9	-20.62 / 31.9	10.44 / 9.68 / 11.02	0.17 / 0.045 / 3.84	10350 00-42-002 2
N 6140	1620.5 / 97.48 / 44.73	6530 / 39.86 / 34.01	5A / 0.12 / 44 -0 +1	7.2 / 6.8 / 2	0.75 / 45.	11.8	3 / 0.10 / 0.04	919. / 1147. / 1	230 / 271. / 273.	10 / 6 / 217	1.45 / 217 5 / 1.39	18.6 / 31.34	10.9 / 10.8 / 10.4	-19.54 / 36.9	10.01 / 9.99 / 10.90	0.96 / 0.124 / 7.72	10359 11-20-012 2
U10419	1628.1 / 46.98 / 90.50	2750 / 41.89 / 54.42	10 / 0.07 / 71 -0	1.6 / 1.6 / 2	0.88 / 32.		0.04 / 0.13	2623. / 2786. / 1	105 / 157. / 161.	15 / 11 / 226	0.07 / 226 18 / 1.03	40.1 / 33.02	-0.2 / 23.3 / 32.6	18.7	9.28 / 10.13	0.141	10419 05-39-000
N 6181	1630.2 / 37.17 / 104.14	1956 / 39.20 / 55.85	5X / 0.08 / 71 -0	2.5 / 2.2 / 2	0.44 / 70.	11.90 / 9.26 / 2.64	2 / 0.37 / 0.24	2375. / 2515. / 1	390 / 375. / 377.	8 / 15 / 217	0.74 / 217 20	36.7 / 32.82	-5.0 / 20.0 / 30.4	-20.92 / 23.6	10.56 / 9.87 / 10.98	C.20 / 0.077 / 2.65	10439 03-42-020 2
1634+52	1634.0 / 80.09 / 55.78	5220 / 41.74 / 44.19	13 P / 0.07 / 70 -0	0.5 / 0.5 / 2	0.82 / 39.	15.2	3 / 0.00 / 0.02	2662. / 2883.		105		41.0 / 33.07	16.5 / 24.3 / 28.6	-17.87 / 6.0	9.34		UGCA 412, I ZW 159 09-27-000 2

Name			Type														Designation
N 6217	1635.1 / 111.31 / 34.04	7818 / 33.36 / 25.13	4B / 0.15 / 44 -0 +5	3.5 / 3.6 / 2	1.00	11.70	2 / 0.00 / 0.15	1359. / 1592.	219	5 10 / 211	1.16 / 217 10	23.9 / 31.89	17.9 / 12.1 / 10.1	-20.19 / 25.1	10.27 / 9.92	0.45	ARP 185
1638+37	1638.1 / 59.89 / 74.22	3716 / 41.35 / 53.22	13 P / 0.07 / 70 -0			14.8	3 / 0.00 / 0.03	1905. / 2100.		220		31.3 / 32.48	5.1 / 18.1 / 25.1	-17.68	9.26		10470 13-12-008 2
N 6207	1641.3 / 59.55 / 74.23	3656 / 40.68 / 53.94	5A / 0.12 / 44 -4	3.1 / 2.6 / 2	0.40 / 72.	11.67	2 / 0.43 / 0.05	852. / 1048. / 4	225 / 198. / 201.	10	0.88 / 217 10 / 1.02	17.4 / 31.21	2.8 / 9.9 / 14.1	-19.54 / 13.2	10.01 / 9.36 / 10.18	0.23 / 0.153 / 1.47	ARAK 507 / 06-37-001
1643-57	1643.1 / 330.69 / 189.54	-5720 / -7.89 / 12.25	13 P / 0.06 / 19 -2	3.3 / 3.8 / 6	0.57 / 60.		0.00 / 1.35	840. / 694. / 7		20		10.9 / 30.20	-10.5 / -1.8 / 2.3	12.1			10521 06-37-007 2
N 6236	1645.0 / 102.74 / 38.50	7052 / 35.78 / 31.48	6X / 0.19 / 44 -6 +5	2.9 / 2.7 / 2	0.64 / 55.	12.1	0.17 / 0.18	1288. / 1525.	196 / 196. / 199.	15	0.76 / 12 / 1.07	23.3 / 31.83	15.5 / 12.4 / 12.2	-19.73 / 18.4	10.08 / 9.49 / 10.31	0.26 / 0.153 / 1.69	179-13
N 6215	1646.8 / 329.79 / 190.80	-5854 / -9.26 / 11.17	5A / 0.16 / 19 -3 +2	1.9 / 2.3 / 6	0.87 / 33.	10.7	5 / 0.05 / 0.96	1555. / 1406.		10		20.5 / 31.55	-19.7 / -3.8 / 4.0	-20.85 / 13.8	10.53		10546 12-16-008 2
N 6248	1646.8 / 102.19 / 38.65	7027 / 35.78 / 31.90	6B / 0.18 / 44 -6 +5	3.1 / 2.7 / 2	0.40 / 72.	12.7	3 / 0.43 / 0.18	1126. / 1364. / 1	206 / 179. / 182.	20	0.96 / 11 / 1.07	21.2 / 31.63	14.1 / 11.2 / 11.2	-18.93 / 16.7	9.76 / 9.61 / 10.19	0.71 / 0.264 / 2.67	137-46 2
N 6221	1648.4 / 329.74 / 191.10	-5908 / -9.57 / 11.07	5B / 0.14 / 19 -3 +2	3.9 / 4.2 / 6	0.64 / 55.	10.4	5 / 0.16 / 0.91	1481. / 1332.		11		19.4 / 31.44	-18.7 / -3.7 / 3.7	-21.04 / 23.8	10.61		10564 12-16-009
N 6239	1648.5 / 67.37 / 64.71	4249 / 39.76 / 52.01	3BP / 0.10 / 44 -0	2.5 / 2.1 / 2	0.48 / 67.	12.48	2 / 0.31 / 0.01	947. / 1160.		13		18.9 / 31.39	5.0 / 10.5 / 14.9	-18.91 / 11.6	9.76		138-03 2
N 6255	1652.9 / 59.48 / 72.81	3635 / 38.33 / 56.14	5B / 0.13 / 44 -4	3.3 / 2.8 / 2	0.44 / 70.	12.7 / 11.88 / 0.82	3 / 0.37 / 0.06	921. / 1124. / 1	220 / 195. / 198.	15	0.85 / 14 / 1.07	18.3 / 31.32	3.0 / 9.8 / 15.2	-18.62 / 15.0	9.64 / 9.37 / 10.22	0.54 / 0.144 / 3.78	10577 07-35-001 2
U10608	1653.2 / 80.75 / 52.08	5311 / 38.75 / 45.72	8B / 0.07 / 44 -0	1.4 / 1.3 / 2	0.72 / 48.	15.06	7 / 0.12 / 0.07	1094. / 1325. / 1	116 / 123. / 128.	8	0.47 / 15 / 1.02	21.0 / 31.61	9.0 / 11.6 / 15.0	-16.55 / 8.0	8.81 / 9.11 / 9.54	2.01 / 0.372 / 5.39	ZW COMPACT / 10606 06-37-014 2
1654-60	1654.6 / 329.43 / 192.30	-6008 / -10.80 / 10.59	5 / 0.11 / 19 -2	5.4 / 5.9 / 6	0.71 / 49.		0.12 / 0.84	1130. / 980. / 7		20		14.7 / 30.84	-14.1 / -3.1 / 2.7	25.3			DDO 206 / 10608 09-28-007 2 / 138-10
U10669	1701.0 / 101.65 / 37.54	7022 / 34.67 / 32.64	10 / 0.10 / 44 -3	1.5 / 1.5 / 2	1.00		0.00 / 0.15	440. / 682. / 3	52	8	226 18 / 1.03	11.6 / 30.33	7.8 / 6.0 / 6.3	5.1	8.13		KAR 235 / 10669 12-16-000
1702-62	1702.4 / 328.46 / 194.04	-6201 / -12.68 / 9.37	5 / 0.14 / 19 -3 +2	4.3 / 3.5 / 6	0.13 / 90.	10.83	0.67 / 0.61	1506. / 1353. / 7	238 / 201. / 204.	20		19.5 / 31.45	-18.7 / -4.7 / 3.2	19.9	10.37		138-14

(1) Name	(2) α / l / SGL	(3) δ / b / SGB	(4) Type / ρ_xyz / Group	(5) D_25 / D^{b,i}_25 / source	(6) d/D / i	(7) B^{b,i}_T / H_{-0.5} / B−H	(8) source / A^{i−o}_B / A^b_B	(9) V_h / V_o / Tel	(10) W_20 / W_R / W_D	(11) e_v / e_w / source	(12) F_c / source,e_f / f_h	(13) R / μ	(14) SGX / SGY / SGZ	(15) M^{b,i}_B / Δ_25	(16) log L_B / log M_H / log M_T	(17) M_H/L_B / M_H/M_T / M_T/L_B	(18) Alternate Names UGC/ESO MCG · AGN? · RCII
1706-77	1706.3 / 315.13 / 201.17	−7728 / −21.45 / −4.35	5 / 0.12 / 55 −0	4.3 / 4.3 / 6	0.73 / 47.		0.11 / 0.43	2927. / 2730. / 7		20		36.6 / 32.82	−34.1 / −13.2 / −2.8	— / 46.0			IC 4633; 044-03
U10736	1708.4 / 100.48 / 37.41	6932 / 34.27 / 33.69	8X / 0.10 / 44 −3	3.3 / 2.8 / 2	0.38 / 74.	13.2	3 / 0.47 / 0.17	491. / 735. / 1	159 / 133. / 137.	10	0.72 / 16 / 1.07	12.2 / 30.44	8.1 / 6.2 / 6.8	−17.24 / 10.0	9.09 / 8.89 / 9.71	0.64 / 0.153 / 4.18	10736 12-16-020
N 6340	1711.3 / 103.72 / 35.50	7222 / 33.35 / 31.34	— / 0.28 / 44 −5	3.2 / 3.2 / 2	0.92 / 26.	11.73	2 / 0.00 / 0.17	1198. / 1441. / 1	224	14 / 30 / 203	0.56 / 203 19	22.0 / 31.71	15.3 / 10.9 / 11.4	−19.98 / 20.6	10.18 / 9.24	0.12	10762 12-16-023 · 2
N 6300	1712.3 / 328.49 / 195.42	−6246 / −14.05 / 9.19	5BP / 0.11 / 19 −2	5.2 / 5.1 / 9	0.58 / 60.	10.45	2 / 0.21 / 0.47	1110. / 958.		14		14.3 / 30.78	−13.6 / −3.8 / 2.3	−20.33 / 21.3	10.32		101-25 · s2 · 2
N 6305	1713.7 / 331.72 / 194.03	−5906 / −12.16 / 12.60	−2 / 0.07 / 61 −0	1.5 / 1.5 / 6	0.60 / 58.		0.00 / 0.53	2770. / 2631.		90		35.8 / 32.77	−33.9 / −8.5 / 7.8	— / 15.7			138-19
1713-57	1713.8 / 332.83 / 193.49	−5748 / −11.45 / 13.79	9 / 0.07 / 19 −2				0.11 / 0.63	1087. / 952. / 7		20		14.3 / 30.78	−13.5 / −3.2 / 3.4				
N 6339	1715.5 / 65.57 / 61.72	4054 / 34.58 / 57.10	6B / 0.07 / 70 −0	3.2 / 2.9 / 2	0.59 / 59.	12.8 / 11.05 / 1.75	3 / 0.21 / 0.07	2111. / 2336. / 1	245 / 243. / 245.	20	0.51 / 18 / 1.08	33.7 / 32.64	8.7 / 16.1 / 28.3	−19.84 / 28.5	10.13 / 9.57 / 10.69	0.27 / 0.075 / 3.64	10790 07-35-059
U10792	1715.6 / 106.98 / 33.62	7517 / 32.31 / 28.91	10 / 0.26 / 44 −7 +5	2.0 / 2.0 / 2	1.00		0.00 / 0.17	1239. / 1481. / 1	60	8	0.25 / 226 14 / 1.05	22.4 / 31.75	16.3 / 10.9 / 10.8	13.1	8.95		10792 13-12-000
U10805	1717.6 / 36.11 / 116.50	1427 / 26.66 / 67.01	9B / 0.07 / 73 −0	1.9 / 2.1 / 2	0.90 / 29.	14.77	7 / 0.04 / 0.45	1557. / 1715. / 3	59	8	0.29 / 14 / 1.04	25.7 / 32.05	−4.5 / 9.0 / 23.6	−17.28 / 15.8	9.10 / 9.11	1.01	10805 02-44-002 · 2
U10806	1717.6 / 76.56 / 50.95	4955 / 35.04 / 50.66	8B / 0.07 / 44 −0	2.5 / 2.1 / 2	0.46 / 68.	13.1	3 / 0.34 / 0.06	936. / 1175. / 1	187 / 164. / 168.	15	0.76 / 14 / 1.04	18.8 / 31.37	7.5 / 9.2 / 14.5	−18.27 / 11.5	9.50 / 9.31 / 9.95	0.64 / 0.226 / 2.84	10806 08-31-040
U10822	1719.4 / 86.37 / 43.81	5758 / 34.72 / 44.21	−5 / 0.50 / 14 −12	66.9 / 61.1 / 2	0.62 / 56.	11.8	3 / 0.00 / 0.07	−277. / −31.		70		0.1 / 20.00	0.1 / 0.0 / 0.1	−8.20 / 1.8	5.47		DRACO, DDO 208; 10822 10-25-008 · 2
N 6368	1724.8 / 34.02 / 124.70	1135 / 23.88 / 68.26	3 / 0.07 / 70 −0	3.7 / 3.1 / 2	0.26 / 84.	11.8 / 9.54 / 2.26	3 / 0.67 / 0.60	2767. / 2920. / 1	403 / 367. / 369.	10	0.70 / 29 / 1.09	40.9 / 33.06	−8.6 / 12.5 / 38.0	−21.26 / 37.0	10.70 / 9.92 / 11.16	0.17 / 0.058 / 2.91	10856 02-44-004 · 2
U10862	1725.8 / 30.12 / 135.25	728 / 21.86 / 67.22	5B / 0.13 / 73 −3	3.2 / 3.3 / 2	0.92 / 26.	12.8	3 / 0.03 / 0.32	1694. / 1834. / 1	157	10	0.64 / 16 / 1.12	26.9 / 32.15	−7.4 / 7.3 / 24.8	−19.35 / 25.9	9.93 / 9.50	0.37	10862 01-44-007

N 6395	1727.1	7108	5 P	2.6	0.34	12.1	3	1167.	218	15	0.64	21.6	14.9	-19.58	10.02	0.19	
	101.97	32.41	0.23	2.1	77.	11.44	0.57	1415.	186.		15	31.68	10.4	13.2	9.31	0.153	
	34.86	33.00	44 -5	2		0.66	0.14	1	189.		1.04		11.8		10.12	1.26	10876 12-16-039 2
N 6384	1730.0	706	4X	6.3	0.69	10.81	1	1664.	377	9	1.29	26.6	-7.3	-21.31	10.72	0.27	
	30.27	20.77	0.14	6.2	51.	8.40	0.13	1805.	437.	12	217 10	32.12	6.8	48.2	10.14	0.052	
	137.23	68.02	73 -3	2		2.41	0.36		438.	217			24.6		11.43	5.13	10891 01-45-001 2
N 6412	1731.4	7544	5A	2.3	1.00	12.17	1	1328.	147	7	0.77	23.5	17.4	-19.69	10.07	0.28	ARP 38
	107.22	31.24	0.23	2.4			0.00	1573.		10	217 15	31.86	11.0	16.5	9.51		
	32.38	28.91	44 -7 +5	2			0.18			217			11.4				10897 13-12-026 2
1742-64	1742.2	-6437	10B	2.4	0.53	11.32	2	311.	128	7		3.0	-2.8	-16.06	8.62		IC 4662
	328.55	-17.84	0.08	2.2	63.		0.25	162.	114.	10		27.38	-1.0	1.9			
	199.21	8.64	14+16+15	6			0.28		119.	301			0.5		8.86	1.76	102-14 2
1742+40	1742.5	4053	13 P			14.3	3	1560.		220		26.6	7.6	-17.82	9.32		ARAK 527
	66.61	29.55	0.08				0.25	1800.				32.12	10.6				
	54.37	60.50	70 -0				0.11						23.1				07-36-033
U10991	1746.5	6721	10	1.7	0.77			1457.	56	8	0.03	25.3	16.6				
	97.33	31.12	0.10	1.7	44.		0.09	1712.	55.		226 18	32.02	11.5	12.6	8.84	0.623	
	34.68	37.15	44 -0	2			0.17	1	64.		1.03		15.3		9.04		10991 11-22-001
U11000	1747.7	3609	13 P	1.0	0.48	13.8	3	666.		220		14.5	3.2	-17.00	8.99		ARAK 532
	61.65	27.45	0.06	0.9	67.		0.00	900.				30.80	5.3	3.8			
	58.94	64.87	44 -0	2			0.12						13.1				11000 06-39-022
N 6503	1749.9	7010	6A	5.9	0.38	10.28	1	51.	173	20	1.54	6.1	4.2	-18.64	9.65	0.29	
	100.58	30.65	0.08	4.9	74.		0.47	305.		25	217 7	28.92	2.7	8.7	9.11		
	33.17	34.60	14 -0	9			0.14	4		79	1.08		3.5				11012 12-17-009 2
U11074	1756.7	709	7	2.6	0.45	13.2	3	1899.	215	15	0.59	29.3	-6.8	-19.13	9.84	0.48	
	33.43	14.87	0.18	2.6	69.		0.36	2059.	191.		16	32.33	4.4	22.2	9.52	0.142	
	146.86	73.87	73 -2	2			0.76	1	194.		1.08		28.1		10.37	3.37	11074 01-46-001
N 6509	1757.0	617	6	1.7	0.83	12.4	3	1817.	308	20	0.95	28.2	-6.9	-19.85	10.13	0.52	
	32.66	14.41	0.16	1.9	38.		0.06	1974.	443.		10	32.25	4.1	15.6	9.85	0.080	
	149.56	73.45	73 -2	2			0.75	1	444.		1.05		27.1		10.95	6.57	11075 01-46-002
U11093	1759.5	658	5	3.3	0.17	12.4	3	1961.	329	10	1.16	30.1	-6.9	-19.99	10.19	0.85	
	33.58	14.16	0.16	2.7	90.	10.13	0.67	2122.	291.		8	32.39	4.2	23.7	10.12	0.225	
	148.83	74.35	73 -2	2		2.27	0.80	1	293.		1.10		29.0		10.77	3.78	11093 01-46-003 2
U11105	1802.4	2138	8	2.5	0.75	13.8	3	2226.	195	15	0.48	34.2	0.3	-18.87	9.74	0.64	
	47.82	19.65	0.15	2.5	45.		0.10	2435.	224.		26	32.67	7.9	25.0	9.55	0.097	
	88.08	76.64	73 -0	2			0.41	1	227.		1.07		33.3		10.56	6.61	11105 04-42-024
U11113	1803.4	2316	6X	1.7	1.00	14.1	3	2333.	122	10	0.38	35.7	1.3	-18.66	9.66	0.68	
	49.52	20.06	0.13	1.9			0.00	2547.			17	32.76	8.5	19.8	9.49		
	81.50	76.08	73 -0	2			0.50	1			1.04		34.6				11113 04-43-003
N 6555	1805.6	1736	5X	2.1	0.87	11.8	3	2224.	220	10	0.33	33.9	-1.8	-20.85	10.53	0.07	
	44.23	17.36	0.21	2.3	33.		0.05	2424.	336.	15	217 25	32.65	6.5	22.8	9.39	0.033	
	105.15	78.53	73 +1	2			0.57		338.	217			33.3		10.87	2.19	11121 03-46-015 2

(1) Name	(2) α / l / SGL	(3) δ / b / SGB	(4) Type / Q_{vyz} / Group	(5) $D_{25}^{b,i}$ / D_{25} / source	(6) d/D / i	(7) $B_T^{b,i}$ / $H_{-0.5}$ / B-H	(8) source / A_B^{i-o} / A_B^b	(9) V_h / V_o / Tel	(10) W_{20} / W_R / W_D	(11) e_v / e_w / source	(12) F_c / source,e_f / f_h	(13) R / μ	(14) SGX / SGY / SGZ	(15) $M_B^{b,i}$ / Δ_{25}	(16) log L_B / log M_H / log M_T	(17) M_H/L_B / M_H/M_T / M_T/L_B	(18) Alternate Names UGC/ESO · MCG · AGN?RCII
U11124	1805.7 / 52.16 / 52.41	3533 / 23.78 / 67.46	6B / 0.08 / 70 -0	2.6 / 2.6 / 2	0.89 / 31.	13.0	3 / 0.04 / 0.17	1609. / 1852. / 1	175 / 278. / 280.	30	0.71 / 14 / 1.07	26.9 / 32.15	6.3 / 8.2 / 24.8	-19.15 / 20.4	9.85 / 9.57 / 10.66	0.52 / 0.081 / 6.44	11124 06-40-005 2
N 6570	1808.8 / 41.23 / 123.54	1405 / 15.20 / 79.27	9B / 0.23 / 73 -1	1.9 / 1.9 / 2	0.60 / 58.	12.4 / 10.71 / 1.69	3 / 0.20 / 0.69	2287. / 2478.	237 / 235. / 238.	10 / 15 / 217	0.43 / 217 20	34.5 / 32.69	-3.6 / 5.4 / 33.9	-20.29 / 19.1	10.31 / 9.51 / 10.49	0.16 / 0.105 / 1.51	11137 02-46-008 2
N 6438	1809.4 / 308.03 / 205.89	-8526 / -26.52 / -11.13	-5 P / 0.22 / 55 -3	2.6 / 2.6 / 9	0.77 / 44.	12.14	2 / 0.00 / 0.36	2431. / 2220.		29		29.8 / 32.37	-26.3 / -12.8 / -5.8	-20.23 / 22.6	10.28		010-01 2
U11141	1809.5 / 39.42 / 134.09	1204 / 14.19 / 78.92	8 / 0.17 / 73 +1	2.0 / 2.3 / 2	1.00	13.5	3 / 0.00 / 0.62	2241. / 2426. / 1	112	15	0.49 / 16 / 1.08	33.8 / 32.64	-4.5 / 4.7 / 33.2	-19.14 / 22.7	9.85 / 9.55	0.50	11141 02-46-009
N 6574	1809.6 / 42.14 / 118.90	1458 / 15.40 / 79.57	4X / 0.22 / 73 -1	1.4 / 1.5 / 2	0.72 / 48.	11.96	0.12 / 0.77	2315. / 2509.		21		35.0 / 32.72	-3.1 / 5.5 / 34.4	-20.76 / 15.3	10.50		11144 02-46-010 2
1809-85	1809.9 / 308.03 / 205.90	-8526 / -26.53 / -11.13	10 P / 0.19 / 55 -3	2.6 / 2.3 / 6	0.44 / 70.		0.37 / 0.36	2514. / 2303.		16		30.9 / 32.45	-27.3 / -13.3 / -6.0	20.8			NGC 6438A / 010-02 2
U11152	1810.3 / 45.64 / 98.94	1835 / 16.74 / 79.43	8B / 0.08 / 73 -0	2.1 / 2.1 / 2	0.56 / 60.	13.3	3 / 0.22 / 0.65	2730. / 2935. / 1	175	25	0.46 / 16 / 1.05	40.4 / 33.03	-1.2 / 7.3 / 39.7	-19.73 / 24.8	10.08 / 9.67	0.39	11152 03-46-018
1811-58	1811.1 / 336.18 / 201.09	-5813 / -18.31 / 15.67	5 / 0.17 / 61 -0	2.4 / 2.1 / 6	0.40 / 72.	14.4	3 / 0.43 / 0.35	2779. / 2664. / 7		20		35.8 / 32.77	-32.1 / -12.4 / 9.7	22.0			IC 4694 / 140-14
U11193	1815.7 / 101.50 / 30.40	7100 / 28.49 / 34.33	10 / 0.19 / 44 -8	1.7 / 1.6 / 2	0.66 / 53.		3 / 0.15 / 0.18	1528. / 1786. / 1	140 / 139. / 143.	25 / 13 / 226	0.16 / 226 20 / 1.02	26.1 / 32.09	18.6 / 10.9 / 14.7	-17.69 / 12.2	9.27 / 8.99 / 9.83	0.53 / 0.144 / 3.69	11193 12-17-020
N 6643	1821.2 / 105.54 / 29.21	7433 / 28.18 / 30.90	5A / 0.19 / 44 -8	3.7 / 3.4 / 2	0.51 / 64.	11.25 / 8.97 / 2.28	1 / 0.27 / 0.23	1489. / 1744.	359 / 357. / 358.	6 / 12 / 217	0.92 / 217 10	25.5 / 32.04	19.1 / 10.7 / 13.1	-20.79 / 25.3	10.51 / 9.73 / 10.97	0.17 / 0.058 / 2.90	11218 12-17-021 2
U11220	1821.8 / 68.76 / 40.10	4055 / 22.39 / 63.82	10 / 0.07 / 70 -0	1.5 / 1.6 / 2	1.00		0.00 / 0.17	1447. / 1706. / 1	94	10	0.45 / 226 8 / 1.03	24.9 / 31.98	8.4 / 7.1 / 22.3	11.6	9.24		11220 07-38-000
1823-67	1823.5 / 327.87 / 203.89	-6701 / -22.65 / 7.19	9B / 0.09 / 19 -1	3.0 / 3.1 / 6	0.82 / 39.	11.9	3 / 0.07 / 0.35	741. / 593.		10		8.9 / 29.75	-8.1 / -3.6 / 1.1	-17.85 / 8.1	9.33		IC 4710 / 103-22 2
N 6654	1825.2 / 103.97 / 29.12	7309 / 27.85 / 32.33	0.09 / 44 -8	2.7 / 2.7 / 2	0.84 / 37.	12.23	2 / 0.00 / 0.22	1806. / 2063.		72		29.5 / 32.35	21.8 / 12.2 / 15.8	-20.12 / 23.3	10.24		VII ZW 793 / 11238 12-17-023 2

Name	Coordinates (RA/Dec 1950; l,b; SGL,SGB)	Type	Dimensions	m	q	Index	Velocity	N	Pop	Ratios	Dist	X / Y / Z	M	Mod	Extra	Other names
1828-57	1828.4 -5742 / 337.56 -20.27 / 203.37 16.51	7B / 0.25 / 61 +4	1.8 0.68 51. / 1.8 / 6			0.14 / 0.38	2730. / 2622. / 7	20			35.1 / 32.72	-30.9 / -13.3 / 10.0	18.4			140-23
1830-58	1830.1 -5832 / 336.80 -20.75 / 203.70 15.71	6B / 0.06 / 61 -0	5.8 0.25 84. / 4.6 / 6	11.5	3	0.67 / 0.37	2245. / 2134.	13			28.8 / 32.30	-25.4 / -11.1 / 7.8	-20.80 38.7	10.51		IC 4721; 140-27 (2)
N 6667	1830.8 6756 / 98.16 26.93 / 29.36 37.56	2XP / 0.06 / 70 -0	3.1 0.74 46. / 3.0 / 2	12.1	3	0.10 / 0.28	2582. / 2847. / 1	30	400	0.79 / 24 / 1.10	39.5 / 32.98	27.3 / 15.3 / 24.1	-20.88 34.6	10.54 9.98	0.27	11269 11-22-053 (2)
U11283	1832.6 4914 / 77.98 22.95 / 32.67 56.13	8BP / 0.07 / 70 -0	2.2 0.88 32. / 2.3 / 2	13.3	3	0.04 / 0.26	1960. / 2230. / 1	15	209	0.67 / 17 / 1.06	31.5 / 32.49	14.8 / 9.5 / 26.1	-19.19 21.2	9.87 9.67	0.63	IC 1291; 11283 08-34-004 (2)
N 6689	1835.8 7029 / 101.05 26.83 / 28.53 35.07	7A / 0.09 / 44 -0	3.9 0.39 73. / 3.3 / 2	11.7 10.51 1.19	3	0.45 / 0.21	487. / 750. / 1	15	238 210. 213.	0.98 / 13 / 1.10	12.2 / 30.43	8.8 / 4.8 / 7.0	-18.73 11.8	9.68 9.15 10.18	0.29 0.095 3.11	NGC 6690; 11300 12-17-026 (2)
U11332	1840.3 7333 / 104.51 26.80 / 27.80 32.04	7BP / 0.17 / 44 -8	2.4 0.40 72. / 2.1 / 2			0.43 / 0.28	1565. / 1824. / 1	20		0.56 / 217 40 / 1.04	26.4 / 32.11	19.8 / 10.5 / 14.0	16.2	9.40		NGC 6654A; 11332 12-17-029 (2)
N 6684	1844.1 -6514 / 330.33 -24.20 / 205.85 9.13	-2B / 0.21 / 19 -1	3.9 0.60 58. / 3.7 / 6	11.12	2	0.00 / 0.23	876. / 741.	59			10.9 / 30.18	-9.7 / -4.7 / 1.7	-19.06 11.8	9.82		104-16 (2)
N 6703	1845.9 4530 / 74.90 19.67 / 29.17 60.08	-2 / 0.08 / 70 -0	2.8 0.90 30. / 2.9 / 9	12.21	2	0.00 / 0.24	2318. / 2591.	35			35.9 / 32.77	15.6 / 8.7 / 31.1	-20.56 30.4	10.42		11356 08-34-020 (2)
1847-64	1847.5 -6453 / 330.80 -24.47 / 206.20 9.50	10 / 0.17 / 19 -1	2.8 0.57 60. / 2.6 / 6			0.22 / 0.23	1003. / 870.	10			12.6 / 30.49	-11.1 / -5.5 / 2.1	9.6			NGC 6684A; 104-19 (2)
1850-64	1850.8 -6452 / 330.90 -24.81 / 206.55 9.52	10 / 0.18 / 19 -1	1.7 0.60 58. / 1.6 / 6			0.19 / 0.23	790. / 657. / 7	20			9.7 / 29.93	-8.5 / -4.3 / 1.6	4.5			104-22
N 6707	1851.3 -5353 / 342.52 -22.15 / 206.53 20.50	3 / 0.39 / 61 -4	1.9 0.39 73. / 1.7 / 6			0.45 / 0.29	2759. / 2676.	90			35.7 / 32.76	-30.0 / -14.9 / 12.5	17.7			183-25 (2)
N 6708	1851.6 -5347 / 342.64 -22.16 / 206.58 20.60	-2 / 0.39 / 61 -4	1.2 0.73 47. / 1.1 / 6			0.00 / 0.25	2622. / 2539.	90			35.7 / 32.76	-29.7 / -14.9 / 12.5	11.5			183-27 (2)
1852-541	1852.4 -5417 / 342.16 -22.42 / 206.71 20.10	-2A / 0.38 / 61 -4	1.3 0.58 59. / 1.2 / 6			0.00 / 0.29	3076. / 2991.	90			35.7 / 32.76	-30.3 / -15.2 / 12.4	12.5			IC 4796; 183-28 (2)
1852-542	1852.5 -5422 / 342.07 -22.45 / 206.72 20.02	-5 / 0.40 / 61 -4	2.2 0.44 69. / 2.0 / 6	12.01	2	0.00 / 0.29	2620. / 2535.	90			35.7 / 32.76	-29.8 / -15.0 / 12.2	-20.75 20.8	10.49		IC 4797; 183-29

(1) Name	(2) α / ℓ / SGL	(3) δ / b / SGB	(4) Type / ρ_{xyz} / Group	(5) D_{25} / D^b_{25} / source	(6) d/D / i	(7) $B^{b,i}_T$ / $H_{-0.5}$ / $B-H$	(8) source / A^{i-o}_B / A^b_B	(9) V_h / V_o / Tel	(10) W_{20} / W_R / W_D	(11) e_v / e_w / source	(12) F_c / source,e_f / f_h	(13) R / μ	(14) SGX / SGY / SGZ	(15) $M^{b,i}_B$ / $\Delta^{b,i}_{25}$	(16) $\log L_B$ / $\log M_H$ / $\log M_T$	(17) M_H/L_B / M_T/M_T / M_T/L_B	(18) Alternate Names (UGC/ESO, MCG, RCII) / AGN?
1852-543	1852.9 / 341.83 / 206.79	-5437 / -22.58 / 19.77	-2 / 0.40 / 61 -4	1.3 / 1.3 / 6	0.83 / 37.	12.51	2 / 0.00 / 0.29	2761. / 2675.		90		35.7 / 32.76	-30.1 / -15.2 / 12.1	-20.25 / 13.6	10.29		A 1853
1902-59	1902.7 / 336.95 / 208.07	-5933 / -25.06 / 14.81	5 / 0.15 / 61 +5	2.7 / 2.0 / 6	0.12 / 90.		/ 0.67 / 0.21	1847. / 1740. / 7		20		23.5 / 31.86	-20.1 / -10.7 / 6.0	/ 13.7			183-30 / IC 4819 / AGN 2
N 6744	1905.0 / 332.23 / 208.13	-6356 / -26.14 / 10.42	4X / 0.25 / 19 -1	20.8 / 19.5 / 9	0.64 / 55.	8.70	2 / 0.17 / 0.16	842. / 717.	340 / 369. / 370.	10 / 10 / 301		10.4 / 30.09	-9.0 / -4.8 / 1.9	-21.39 / 59.2	10.75 / / 11.37	4.18	141-27 / 104-42 / AGN 2
N 6764	1907.0 / 81.49 / 22.88	5051 / 18.23 / 54.65	4B / 0.08 / 70 -0	2.3 / 2.2 / 2	0.64 / 55.	12.3 / 10.40 / 1.90	3 / 0.17 / 0.25	2409. / 2691. / 1	292 / 311. / 313.	10	0.55 / 217 12 / 1.05	37.0 / 32.84	19.7 / 8.3 / 30.2	-20.54 / 23.8	10.41 / 9.69 / 10.82	0.19 / 0.073 / 2.61	11407 08-35-003 / L/S / AGN 2
1908-62	1908.7 / 334.25 / 208.67	-6210 / -26.27 / 12.15	10 / 0.23 / 19 -1	3.2 / 3.0 / 6	0.63 / 56.	11.9	/ 0.18 / 0.19	940. / 823. / 7		20		11.8 / 30.35	-10.1 / -5.5 / 2.5	/ 10.3			IC 4824 / 141-33
1911-54	1911.2 / 342.40 / 209.59	-5445 / -25.17 / 19.53	5B / 0.31 / 61 +4	2.6 / 2.3 / 6	0.47 / 67.		5 / 0.33 / 0.22	2697. / 2615. / 7		20		34.7 / 32.70	-28.4 / -16.1 / 11.6	-20.80 / 23.3	10.51		IC 4837 / 184-46 / AGN 2
1911-62	1911.6 / 334.00 / 208.99	-6227 / -26.64 / 11.85	10 / 0.24 / 19 -1	3.2 / 2.3 / 6	0.18 / 90.		5 / 0.67 / 0.19	890. / 772. / 7		20		11.1 / 30.23	-9.5 / -5.3 / 2.3	/ 7.5			141-42
1927-17	1927.1 / 21.06 / 221.32	-1747 / -16.29 / 55.54	10 / 0.35 / 14 -12	2.8 / 3.0 / 6	0.86 / 35.		/ 0.05 / 0.50	-73. / 30.		5	0.91 / 222 11	0.7 / 24.23	-0.3 / -0.3 / 0.6	/ 0.6	/ 6.60		SAGDIG, UKS 1927-177 / 594-04 -3-49-000
1931-57	1931.5 / 339.76 / 212.15	-5738 / -28.45 / 16.31	5 / 0.23 / 61 -5	3.3 / 2.4 / 6	0.12 / 90.		/ 0.67 / 0.20	1911. / 1819. / 7		20		24.3 / 31.93	-19.8 / -12.4 / 6.8	/ 17.0			IC 4872 / 142-24
1931-61	1931.6 / 335.79 / 211.56	-6108 / -28.83 / 12.86	10 / 0.15 / 61 +5	1.4 / 1.5 / 6	0.94 / 24.		/ 0.02 / 0.24	1795. / 1687. / 7		20		22.6 / 31.77	-18.8 / -11.5 / 5.0	/ 9.9			IC 4869 ? / 142-25
1938-58	1938.4 / 338.41 / 212.83	-5855 / -29.48 / 14.88	10 / 0.24 / 61 -5	1.3 / 1.3 / 6	0.80 / 41.		/ 0.08 / 0.20	1927. / 1831. / 7		20		24.4 / 31.94	-19.8 / -12.8 / 6.3	/ 9.3			142-32
N 6810	1939.4 / 338.57 / 212.99	-5847 / -29.60 / 14.99	1A / 0.22 / 61 -5	3.2 / 2.5 / 6	0.28 / 82.	11.4	5 / 0.67 / 0.18	1995. / 1900. / 7		56		25.3 / 32.01	-20.5 / -13.3 / 6.5	-20.61 / 18.5	10.44		142-35 / AGN 2
1939-60	1939.5 / 336.30 / 212.59	-6046 / -29.76 / 13.04	4 / 0.08 / 61 -0	2.0 / 1.5 / 6	0.25 / 84.		/ 0.67 / 0.23	2514. / 2409. / 7		20		31.8 / 32.51	-26.1 / -16.7 / 7.2	/ 13.9			IC 4885 ? / 142-36

Name	Position / coords	Type	Dimensions	Mag	Velocity				Abs.		Other names
N 6814	1939.9 -1027 -16.01 61.44 / 29.34 / 231.61	4X 0.18 66 +1	4.4 5.0 4 / 1.00	11.38 / 2 0.00 0.64	1557. 1698. / 97 / 10 7 217	0.88 217 10	22.8 31.79 / -6.8 -8.5 20.0	-20.41 33.3	10.36 9.60	0.17	IC 4889 / -2-50-001 2 / S1.2 2
1941-54	1941.3 -5428 -29.42 19.16 / 343.54 / 214.20	-5 0.13 61 -0	3.0 2.7 6 / 0.56 61.	12.0 / 5 0.00 0.19	2531. 2457. / 90		32.3 32.55 / -25.3 -17.2 10.6	-20.55 25.5	10.41		185-14 2
N 6821	1941.7 -657 -14.87 64.38 / 32.80 / 235.88	7B 0.14 66 +1	1.3 1.4 4 / 1.00	/ 0.00 0.57	1523. 1680. / 155 / 10 10 217	0.67 217 15	22.5 31.76 / -5.5 -8.0 20.3	9.2	9.37		-1-50-002 2
N 6822	1942.1 -1456 -18.39 57.11 / 25.33 / 229.11	10B 0.35 14 -12	15.0 17.3 1 / 1.00	8.66 / 1 0.00 0.69	-56. 66. / 81 / 6 12 105	2.78 105 13	0.7 24.23 / -0.3 -0.3 0.6	-15.57 3.5	8.42 8.47	1.12	-2-50-006 2
1945-18	1945.4 -1813 -20.44 53.77 / 22.48 / 228.23	9B 0.10 66 -0	2.4 2.3 6 / 0.60 58.	/ 0.19 0.37	1735. 1842. 1 / 190 / 30	0.48 29 / 1.05	24.4 31.94 / -9.6 -10.7 19.7	16.4	9.25		UGCA 416 / 594-17 -3-50-004 2
1950-58	1950.2 -5851 -31.00 14.61 / 338.60 / 214.38	5 0.13 61 -5	4.6 4.4 6 / 0.67 53.	/ 0.15 0.13	2144. 2050. 7 / 20		27.1 32.16 / -21.6 -14.8 6.8	34.8		0.37	IC 4901 / 142-50
N 6835	1951.8 -1242 -19.60 58.14 / 28.51 / 234.85	1BP 0.19 66 -1	2.6 2.3 4 / 0.35 76.	12.32 / 2 0.55 0.54	1583. 1718. / 9	0.83 11 32	22.8 31.79 / -6.9 -9.8 19.3	-19.47 15.3	9.98 9.55		-2-50-009 2
N 6836	1951.9 -1249 -19.68 58.02 / 28.41 / 234.79	9X 0.19 66 -1	1.4 1.6 4 / 1.00	/ 0.00 0.55	1628. 1762. / 151 / 30 30 217	0.53 217 15	23.3 31.84 / -7.1 -10.1 19.8	10.9	9.26		-2-50-010 2
2002-48	2002.9 -4832 -32.08 23.98 / 350.84 / 219.38	-5 0.45 61 -1	1.2 1.2 6 / 0.90 29.	/ 0.00 0.15	2930. 2890. / 90		35.5 32.75 / -25.2 -20.6 14.5	12.4			IC 4943 / 233-28 2
N 6861	2003.7 -4831 -32.21 23.95 / 350.88 / 219.52	-5 0.45 61 -1	2.2 2.0 6 / 0.56 61.	11.95 / 2 0.00 0.15	2859. 2819. / 90		35.5 32.75 / -25.1 -20.7 14.5	-20.80 20.7	10.51		IC 4949 / 233-32 2
2004-48	2004.7 -4821 -32.36 24.06 / 351.09 / 219.75	-2A 0.46 61 -1	2.1 1.7 6 / 0.35 76.	/ 0.00 0.14	2534. 2495. / 90		35.5 32.75 / -24.7 -20.6 14.3	17.6			NGC 6861D / 233-34 2
2005-62	2005.2 -6200 -32.85 11.07 / 334.88 / 215.32	3 0.15 19 -1	2.8 2.1 6 / 0.23 87.	/ 0.67 0.10	810. 703. 7 / 20		9.9 29.97 / -7.9 -5.6 1.9	6.1			IC 4951 / 143-10 2
N 6868	2006.3 -4831 -32.64 23.81 / 350.93 / 219.96	-5 0.47 61 -1	2.5 2.4 6 / 0.73 47.	11.69 / 6 0.00 0.14	2763. 2723. / 24		35.5 32.75 / -24.9 -20.9 14.3	-21.06 24.9	10.62		233-39 2
2006-06	2006.6 -627 -20.16 61.34 / 36.18 / 247.77	10 0.09 66 -0	2.9 2.8 2 / 0.57 60.	/ 0.22 0.35	1422. 1588. 1 / 98 / 8	0.84 11 / 1.08	21.0 31.61 / -3.8 -9.3 18.4	17.2	9.48		UGCA 417 / -1-51-000

(1) Name	(2) α / l / SGL	(3) δ / b / SGB	(4) Type / ρxyz / Group	(5) D25 / D25b,i / source	(6) d/D / i	(7) BT b,i / H-0.5 / B-H	(8) source / Ai-o / AB b	(9) Vh / Vo / Tel	(10) W20 / WR / WD	(11) ev / ew / source	(12) Fc / source-eF / fh	(13) R / μ	(14) SGX / SGY / SGZ	(15) MB b,i / Δ25	(16) log LB / log MH / log MT	(17) MH/LB / MH/MT / MT/LB	(18) UGC/ESO MCG	Alt. Names	AGN?	RCII
N 6870	2006.6 / 351.03 / 220.05	-4826 / -32.68 / 23.87	2A / 0.47 / 61 -1	2.4 / 2.1 / 6	0.52 / 64.	12.72	2 / 0.27 / 0.15	2757. / 2718.		50		35.5 / 32.75	-24.9 / -20.9 / 14.3	-20.03 / 21.8	10.20		233-41			2
N 6887	2013.5 / 345.72 / 219.49	-5257 / -34.06 / 19.26	4A / 0.28 / 61 -0 +1	2.9 / 2.5 / 6	0.41 / 72.	11.9	3 / 0.42 / 0.14	2692. / 2630. / 7		20		34.2 / 32.67	-24.9 / -20.5 / 11.3	-20.77 / 25.0	10.50		186-27			2
N 6890	2014.8 / 355.35 / 222.87	-4457 / -33.70 / 26.63	3A / 0.18 / 61 -0 +1	1.5 / 1.4 / 9	0.79 / 41.	12.95	2 / 0.08 / 0.03	2470. / 2450.		50		31.8 / 32.51	-20.8 / -19.3 / 14.3	-19.56 / -13.0	10.02		284-54 -7-41-023		S2	2
2015-39	2015.2 / 1.82 / 225.37	-3930 / -32.98 / 31.62	7A / 0.16 / 61 -0	1.1 / 1.1 / 6	0.92 / 27.	13.47	6 / 0.03 / 0.16	2719. / 2727.		25		35.3 / 32.74	-21.1 / -21.4 / 18.5	-19.27 / 11.3	9.90		340-12 -7-41-025	SCI 178		2
2019-44	2019.8 / 356.56 / 224.18	-4403 / -34.49 / 27.11	7X / 0.28 / 61 -2 +1	1.5 / 1.6 / 6	1.00	14.01	6 / 0.00 / 0.08	2970. / 2955.		14		38.2 / 32.91	-24.4 / -23.7 / 17.4	-18.90 / 17.8	9.75		285-05 -7-41-033			2
2020-44	2020.6 / 356.45 / 224.28	-4409 / -34.64 / 26.96	-2 / 0.31 / 61 -2 +1	2.4 / 2.1 / 6	0.43 / 70.	12.58	6 / 0.00 / 0.08	2942. / 2927.		90		37.9 / 32.89	-24.2 / -23.6 / 17.2	-20.31 / 23.2	10.32		285-07 -7-42-001	NGC 6902B, SCI 206		2
N 6902	2021.1 / 356.87 / 224.53	-4349 / -34.69 / 27.23	/ 0.35 / 61 -2 +1	6.8 / 6.4 / 6	0.67 / 53.	12.42	2 / 0.00 / 0.09	2767. / 2754.	342 / 383. / 385.	9 / 20 / 203	1.02 / 203 13	35.7 / 32.76	-22.6 / -22.2 / 16.3	-20.34 / 66.7	10.33 / 10.13 / 11.45	0.63 / 0.047 / 13.34	285-08 -7-42-002			2
U11557	2023.0 / 94.98 / 11.66	6002 / 12.82 / 42.64	8X / 0.08 / 40 -0	2.4 / 3.0 / 2	0.92 / 26.	12.3	3 / 0.03 / 1.20	1393. / 1684. / 1	115	10	0.71 / 12 / 1.11	23.7 / 31.87	17.0 / 3.5 / 16.0	-19.57 / 20.8	10.02 / 9.46	0.28	11557 10-29-005	A 2021		
N 6909	2024.2 / 352.81 / 223.55	-4712 / -35.53 / 23.92	-5 / 0.42 / 61 -0 +1	2.5 / 2.1 / 6	0.43 / 70.	12.71	6 / 0.00 / 0.07	2680. / 2649.		90		34.3 / 32.68	-22.7 / -21.6 / 13.9	-19.97 / 21.0	10.18		285-12			2
2028-48	2028.7 / 351.00 / 223.61	-4842 / -36.35 / 22.24	7B / 0.18 / 61 -0 +1	2.2 / 2.2 / 6	0.81 / 39.		/ 0.07 / 0.09	2430. / 2392. / 7		20		31.1 / 32.46	-20.8 / -19.8 / 11.8	/ 20.0			234-43			
2030-53	2030.3 / 345.46 / 221.86	-5309 / -36.58 / 18.10	5 P / 0.16 / 61 -0 +1	1.9 / 2.0 / 6	1.00		/ 0.00 / 0.12	2564. / 2503. / 7		20		32.4 / 32.56	-23.0 / -20.6 / 10.1	/ 18.9			186-62			
N 6925	2031.2 / 11.27 / 232.86	-3209 / -34.69 / 36.73	4A / 0.08 / 61 -0	4.6 / 3.6 / 6	0.25 / 84.	11.19 / 8.62 / 2.57	2 / 0.67 / 0.23	2799. / 2847. / 2	517 / 481. / 483.	15	1.07 / 18 / 1.03	36.6 / 32.82	-17.7 / -23.4 / 21.9	-21.63 / 38.5	10.84 / 10.20 / 11.41	0.23 / 0.061 / 3.70	463-04 -5-48-022			2
N 6946	2033.8 / 95.72 / 10.08	5959 / 11.68 / 41.97	6X / 0.07 / 14 -0	11.2 / 14.9 / 9	0.79 / 42.	7.92	1 / 0.08 / 1.63	46. / 338.	242 / 306. / 308.	6 / 10 / 72	2.32 / 25 / 5	5.5 / 28.70	4.0 / 0.7 / 3.7	-20.78 / 23.9	10.50 / 9.80 / 10.81	0.20 / 0.097 / 2.04	11597 10-29-006	ARP 29		2

Name	Coord A	Coord B	Type	Dim 1	Dim 2	Mag 1	Col a	Vel	Col b	Col c	Col d	Col e	Col f	Col g	Col h	Col i	Designation
N 6951	2036.6 / 100.90 / 13.88	6556 / 14.86 / 36.79	4X / 0.08 / 40 -0	3.7 / 4.5 / 2	0.91 / 28.	11.18	1 / 0.03 / 0.99	1422. / 1707.	335	10 / 10 / 217	0.92 / 217 10	24.1 / 31.91	18.7 / 4.6 / 14.4	-20.73 / 31.7	10.48 / 9.68	0.16	11604 11-25-002 2
2038-65	2038.9 / 329.67 / 217.36	-6550 / -35.98 / 6.15	10 / 0.08 / 61 -0	1.2 / 1.1 / 6	0.77 / 44.		/ 0.09 / 0.12	1622. / 1499. / 7		20		19.9 / 31.49	-15.7 / -12.0 / 2.1	6.4			IC 5028
N 6943	2039.8 / 325.95 / 216.09	-6856 / -35.30 / 3.32	6X / 0.13 / 62 +4	3.5 / 3.1 / 6	0.52 / 63.	11.6	5 / 0.26 / 0.15	3109. / 2972.		19		38.8 / 32.94	-31.3 / -22.8 / 2.2	-21.34 / 35.1	10.73		106-10
2041-46	2041.2 / 354.25 / 226.90	-4610 / -38.40 / 23.55	5 / 0.31 / 61 -3 +1	2.3 / 1.9 / 6	0.43 / 70.		/ 0.39 / 0.03	2691. / 2667. / 7		20		34.3 / 32.68	-21.5 / -23.0 / 13.7	19.0			074-06 2
2042-46	2042.6 / 354.43 / 227.20	-4602 / -38.64 / 23.55	5 P / 0.30 / 61 -3 +1	1.4 / 1.3 / 6	0.69 / 51.		/ 0.13 / 0.02	2725. / 2702. / 7		20		34.8 / 32.71	-21.7 / -23.4 / 13.9	13.2			285-48
2044-13	2044.2 / 34.06 / 252.16	-1302 / -31.36 / 50.25	10 / 0.44 / 14-12	2.3 / 2.0 / 4	0.50 / 65.		/ 0.29 / 0.14	-132. / 11. / 1	46 / 27. / 43.	7	0.48 / 15 / 1.04	0.7 / 24.23	-0.1 / -0.4 / 0.5	0.4	6.17 / 6.94	0.171	285-51 / DDO 210 / -2-53-003 2
N 6958	2045.4 / 4.50 / 232.15	-3811 / -38.60 / 30.07	-5 / -0.12 / 61 -0	2.2 / 2.2 / 6	0.80 / 41.	12.21	2 / 0.00 / 0.09	2742. / 2760.		150		35.4 / 32.74	-18.8 / -24.2 / 17.7	-20.53 / 22.7	10.40		341-15 -6-45-017 2
2047-69	2047.4 / 325.16 / 216.49	-6924 / -35.81 / 2.60	7 / 0.06 / 19 +1	5.0 / 3.6 / 6	0.18 / 90.	11.4 / 10.17 / 1.23	5 / 0.67 / 0.15	598. / 459. / 7	211 / 175. / 178.	10 / 10 / 301		6.7 / 29.11	-5.3 / -4.0 / 0.3	-17.71 / 7.0	9.28	3.31	IC 5052
2053-55	2053.2 / 341.56 / 223.62	-5555 / -39.63 / 14.12	7B / 0.07 / 61 -0	1.5 / 1.4 / 6	0.56 / 61.		/ 0.23 / 0.19	2092. / 2019. / 7		20		26.2 / 32.09	-18.4 / -17.5 / 6.4	10.7	9.80		074-15 2
U11651	2055.1 / 71.13 / 321.38	2546 / -12.59 / 59.73	8 / 0.06 / 64 -0	3.3 / 3.0 / 2	0.29 / 81.	12.6 / 10.88 / 1.72	3 / 0.67 / 0.84	1527. / 1804. / 1	290 / 256. / 258.	25	0.69 / 21 / 1.12	23.7 / 31.87	9.3 / -7.4 / 20.4	-19.27 / 20.8	9.90 / 9.44 / 10.60	0.35 / 0.070 / 4.97	187-35 / 11651 04-49-005
2056-72	2056.2 / 320.90 / 215.47	-7250 / -35.30 / -0.75	3 / 0.10 / 62 +4	3.4 / 2.6 / 6	0.28 / 82.		/ 0.67 / 0.15	3097. / 2942. / 7	284 / 251. / 253.	20		38.5 / 32.93	-31.3 / -22.3 / -0.5	29.2			IC 5071
2059-17	2059.7 / 31.44 / 251.91	-1700 / -36.36 / 44.80	4A / 0.06 / 60 -0	3.7 / 3.0 / 2	0.31 / 79.	11.2	/ 0.64 / 0.21	1479. / 1605. / 1		20	0.95 / 17 / 1.09	20.4 / 31.55	-4.5 / -13.8 / 14.4	17.9	9.57 / 10.51	0.114	047-19 / IC 5078, UGCA 419 / -3-53-021 2
N 7013	2101.4 / 75.13 / 327.89	2942 / -11.14 / 57.31	/ 0.08 / 65 +5	4.8 / 4.8 / 2	0.35 / 76.	12.95	3 / 0.00 / 1.21	767. / 1051.		15	0.78 / 11 75	14.2 / 30.76	6.5 / -4.1 / 11.9	-19.56 / 19.9	10.02 / 9.08	0.12	11670 -5-49-001 2
N 7007	2101.9 / 345.42 / 226.44	-5245 / -41.36 / 16.17	-2 / -0.14 / 61 -0	2.1 / 1.9 / 6	0.65 / 54.		6 / 0.00 / 0.04	2954. / 2897.		90		37.3 / 32.86	-24.7 / -26.0 / 10.4	-19.91 / 20.7	10.16		187-48 2

(1) Name	(2) α / ℓ / SGL	(3) δ / b / SGB	(4) Type / Q$_{xyz}$ / Group	(5) D$_{25}$ / D$_{25}^{b,i}$ / source	(6) d/D / i	(7) B$_T^{b,i}$ / H$_{-0.5}$ / B-H	(8) source / A$_B^{i-o}$ / A$_B^b$	(9) V$_h$ / V$_o$ / Tel	(10) W$_{20}$ / W$_R$ / W$_D$	(11) e$_v$ / e$_w$ / source	(12) F$_c$ source/e$_f$ / f$_h$	(13) R / μ	(14) SGX / SGY / SGZ	(15) M$_B^{b,i}$ / Δ$_{25}$	(16) log L$_B$ / log M$_H$ / log M$_T$	(17) M$_H$/L$_B$ / M$_H$/M$_T$ / M$_T$/L$_B$	(18) Alternate Names UGC/ESO MCG / AGN? RCII
2103-55	2103.9 / 342.21 / 225.36	-5509 / -41.25 / 13.99	10B / 0.06 / 61 -0	2.4 / 2.3 / 6	0.67 / 53.		/ 0.15 / 0.12	1360. / 1291. / 7		20		16.8 / 31.13	-11.5 / -11.6 / 4.1	/ 11.3			187-51
N 7020	2107.3 / 330.53 / 220.77	-6415 / -39.30 / 6.08	-2A / 0.23 / 62 -0 +1	3.4 / 2.8 / 6	0.43 / 70.	12.45	6 / 0.00 / 0.04	3029. / 2915.		90		37.8 / 32.89	-28.4 / -24.5 / 4.0	-20.44 / 30.9	10.37		NGC 7021 107-13 2
N 7029	2108.5 / 349.58 / 229.22	-4930 / -42.81 / 18.32	-5 / 0.18 / 61 -0	2.2 / 1.9 / 9	0.59 / 59.	12.69	2 / 0.00 / 0.00	2872. / 2832.		67		36.3 / 32.80	-22.5 / -26.1 / 11.4	-20.11 / 20.1	10.24		235-72 2
2112-65	2112.0 / 329.38 / 220.77	-6501 / -39.51 / 5.16	5 / 0.11 / 61 -0	2.6 / 1.8 / 6	0.13 / 90.		/ 0.67 / 0.03	1859. / 1741. / 7		20		22.7 / 31.78	-17.1 / -14.8 / 2.0	/ 11.9			107-16
U11707	2112.3 / 74.31 / 321.21	2632 / -15.03 / 55.80	8A / 0.09 / 65 +5	3.5 / 3.5 / 2	0.60 / 58.	13.4 / 12.09 / 1.31	3 / 0.20 / 0.58	908. / 1187. / 1	202 / 196. / 199.	10	1.14 / 8 / 1.13	15.7 / 30.98	6.9 / -5.5 / 13.0	-17.58 / 16.0	9.22 / 9.53 / 10.25	2.03 / 0.190 / 10.67	11707 04-50-001
N 7041	2113.2 / 350.73 / 230.45	-4834 / -43.69 / 18.64	-2X / 0.12 / 61-10	3.3 / 2.6 / 6	0.35 / 76.	11.99	2 / 0.00 / 0.06	1924. / 1889.		67		24.2 / 31.92	-14.6 / -17.7 / 7.7	-19.93 / 18.4	10.16		235-82 2
N 7049	2115.6 / 350.38 / 230.66	-4846 / -44.06 / 18.24	-2A / 0.12 / 61-10	3.9 / 3.4 / 6	0.50 / 65.	11.77	2 / 0.00 / 0.03	2190. / 2154.		67		27.6 / 32.20	-16.6 / -20.2 / 8.6	-20.43 / 27.4	10.36		236-01 2
2119-45	2119.9 / 354.14 / 233.05	-4559 / -45.11 / 20.02	4A / 0.19 / 61 +6	3.2 / 2.5 / 6	0.30 / 80.	11.8	6 / 0.67 / 0.06	2663. / 2641.		25		33.7 / 32.64	-19.0 / -25.3 / 11.5	-20.84 / 24.6	10.53		\001 A 2120 2
N 7059	2123.6 / 334.69 / 224.64	-6014 / -42.44 / 8.31	4X / 0.17 / 61+13	2.5 / 2.3 / 9	0.60 / 58.	12.92 / 9.89 / 3.03	2 / 0.19 / 0.06	1752. / 1657. / 7	352 / 370. / 372.	16 / 10 / 301		21.5 / 31.66	-15.1 / -14.9 / 3.1	-18.74 / 14.4	9.69 / / 10.76	11.76	287-13 2
2125-38	2125.2 / 5.41 / 239.32	-3805 / -46.39 / 25.49	5A / 0.17 / 63 -0	2.3 / 2.1 / 6	0.57 / 60.	12.58	6 / 0.22 / 0.13	2567. / 2587.		25		32.8 / 32.58	-15.1 / -25.5 / 14.1	-20.00 / 20.1	10.19		145-05 2 SCI 186
N 7064	2125.6 / 344.17 / 229.33	-5259 / -44.82 / 13.93	5B / 0.14 / 19 -4	4.1 / 2.9 / 9	0.17 / 90.	12.41	2 / 0.67 / 0.00	869. / 811.		12		10.6 / 30.14	-6.7 / -7.8 / 2.6	-17.73 / 9.0	9.28		342-50 -6-47-004 2
N 7070	2127.3 / 357.79 / 235.94	-4319 / -46.65 / 21.27	6A / 0.21 / 61 -6	2.2 / 2.2 / 9	0.91 / 28.	12.78	2 / 0.03 / 0.03	2393. / 2385.		15		30.3 / 32.41	-15.8 / -23.4 / 11.0	-19.63 / 19.5	10.04		188-09 2
N 7077	2127.4 / 56.02 / 281.27	212 / -33.22 / 49.62	-5 / 0.06 / 64 -0	0.8 / 0.8 / 2	1.00	13.9	3 / 0.00 / 0.18	842. / 1050.		220		13.3 / 30.62	1.7 / -8.4 / 10.1	-16.72 / 3.1	8.88		287-28 -7-44-016 2 ARAK 549 11755 00-54-028

Name									ID
N 7079	2129.3 356.31 235.55 / −4418 −46.93 20.28	−2B 0.19 61 −6	1.9 1.7 6 / 0.75 45.	12.48 / 0.00 0.02 2	2677. 2664. / 75		33.9 32.65 / −18.0 −26.2 11.8 / −20.17 16.8	10.26	287−36 −7−44−022 2
N 7083	2131.8 329.36 223.03 / −6407 −41.82 4.65	5A 0.31 62 −1	3.2 2.9 6 / 0.63 56.	11.6 8.80 2.80 / 0.18 0.06 5	3109. 2995. 7 / 408. 446. 448. / 16 10 301		38.7 32.94 / −28.2 −26.3 3.1 / −21.34 32.8	10.73 11.28	107−36 2 / 3.54
N 7090	2133.0 341.29 229.05 / −5447 −45.39 11.84	5B 0.13 19 −4	6.8 4.8 6 / 0.20 90.	10.43 9.45 0.98 / 0.67 0.01 2	866. 798. 7 / 229. 192. 195. / 10 10 301	0.53 18 1.05	10.4 30.09 / −6.7 −7.7 2.1 / −19.66 14.6	10.06 10.19	188−12 2 / 1.37
U11782	2135.7 63.65 291.88 / 845 −30.89 49.66	9B 0.09 64 −0	2.5 2.3 2 / 0.60 58.	13.7 / 0.20 0.13 3	1112. 1343. 1 / 146. 137. 141. / 15		17.0 31.15 / 4.1 −10.2 13.0 / −17.45 11.4	9.17 8.99 9.79	11782 01−55−006 / 0.66 0.158 4.18
N 7097	2137.1 358.37 237.77 / −4246 −48.49 20.49	−5 0.26 61 −6	1.9 1.7 6 / 0.75 45.	12.5 / 0.00 0.00 5	2404. 2398. / 90		30.4 32.41 / −15.2 −24.1 10.6 / −19.91 15.1	10.16	287−48 −7−44−029 2
N 7096	2137.4 328.98 223.50 / −6408 −42.36 4.25	1A 0.32 62 −1	1.5 1.5 6 / 1.00	12.5 / 0.00 0.05 5	2958. 2844. / 90		36.7 32.82 / −26.5 −25.2 2.7 / −20.32 16.1	10.32	IC 5121
N 7107	2139.2 354.90 236.44 / −4502 −48.58 18.57	8B 0.23 61 −0+11	1.8 1.8 6 / 0.91 28.	12.7 / 0.03 0.00 5	2198. 2180. 2 / 107 / 50	0.42 19 1.01	27.6 32.21 / −14.5 −21.8 8.8 / −19.51 14.5	10.00 9.30	107−46 2 / 0.20
N 7098	2139.3 316.44 216.48 / −7520 −36.68 −4.48	1X 0.07 55 −0	3.9 3.8 6 / 0.64 55.	/ 0.16 0.31	2376. 2210. 7 / 20		29.1 32.32 / −23.3 −17.2 −2.3 / 32.3		287−52 2
2144−35	2144.4 10.04 244.67 / −3507 −50.16 24.97	−2B 0.36 63 −2 +1	1.6 1.6 6 / 1.00	13.30 / 0.00 0.00 2	2610. 2644. / 90		33.3 32.61 / −12.9 −27.3 14.1 / −19.31 15.6	9.92	048−05 2
N 7125	2145.6 332.32 226.30 / −6057 −44.63 6.06	5X 0.31 62 −3 +1	3.0 2.8 9 / 0.68 52.	12.67 / 0.14 0.04 2	3085. 2986. 7 / 20		38.4 32.92 / −26.4 −27.6 4.1 / −20.25 31.4	10.29	IC 5131
N 7126	2145.7 332.43 226.37 / −6051 −44.69 6.12	2A 0.30 62 −3 +1	2.3 2.0 9 / 0.49 66.	13.02 / 0.30 0.04 2	3009. 2910. / 142 / 90		37.5 32.87 / −25.7 −27.0 4.0 / −19.85 21.9	10.13	403−27 −6−47−014 2
2145−81	2145.8 310.03 212.48 / −8146 −32.96 −9.55	5 0.10 55 −0	3.2 3.3 6 / 0.83 38.	/ 0.06 0.45	2561. 2368. 7 / 20		31.3 32.48 / −26.0 −16.6 −5.2 / 30.2		145−17 2 / 145−18 2 / 027−01
N 7137	2145.9 76.43 312.40 / 2156 −23.73 48.67	5XP 0.07 64 −0	1.7 1.7 2 / 0.83 38.	12.67 / 0.06 0.32 2	1685. 1952. 1 / 45 25 217	0.38 217 20 1.03	25.0 31.99 / 11.1 −12.2 18.8 / −19.32 12.4	9.92 9.18	11815 04−51−005 2 / 0.18
N 7135	2146.8 10.06 245.04 / −3507 −50.65 24.61	−2AP 0.32 63 −2 +1	2.7 2.5 9 / 0.74 46.	12.69 / 0.00 0.00 2	2718. 2751. / 30		34.7 32.70 / −13.3 −28.6 14.5 / −20.01 25.3	10.20	403−35 −6−48−001 2

(1) Name	(2) α / ℓ / SGL	(3) δ / b / SGB	(4) Type / Q_{xyz} / Group	(5) D_{25} / $D_{25}^{b,i}$ / source	(6) d/D / i	(7) $B_T^{b,i}$ / $H_{-0.5}$ / $B-H$	(8) source / $A_B^{l,o}$ / A_B^b	(9) V_h / V_o / Tel	(10) W_{20} / W_R / W_D	(11) e_v / e_w / source	(12) F_c source/e_f / f_h	(13) R / μ	(14) SGX / SGY / SGZ	(15) $M_B^{b,i}$ / Δ_{25}	(16) logL_B / logM_H / logM_T	(17) M_H/L_B / M_H/M_T / M_T/L_B	(18) Alternate Names UGC/ESO / MCG / AGN? RCII
U11820	2147.0 / 70.34 / 300.53	1401 / -29.49 / 47.93	9 / 0.15 / 64+12	1.6 / 1.6 / 2	0.70 / 50.		0.13 / 0.29	1108. / 1354. / 1	133 / 138. / 142.	10 / 8 / 226	0.71 / 8 / 1.02	17.1 / 31.17	5.8 / -9.9 / 12.7	8.0	9.18 / 9.64	0.340	11820 02-55-015
N 7141	2148.8 / 338.80 / 230.10	-5548 / -47.15 / 9.60	1 / 0.18 / 62 -0	4.6 / 4.2 / 6	0.67 / 53.		0.15 / 0.00	3002. / 2928. / 7		20		37.4 / 32.87	-23.7 / -28.3 / 6.2	45.9			189-07 / / 2
N 7144	2149.5 / 349.19 / 235.31	-4829 / -49.61 / 14.85	-5 / 0.29 / 61-12+11	3.5 / 3.5 / 9	1.00	11.62	2 / 0.00 / 0.00	1997. / 1961. / 1		102		24.8 / 31.98	-13.7 / -19.7 / 6.4	-20.36 / 25.3	10.34		237-11 / / 2
N 7145	2150.1 / 349.70 / 235.65	-4807 / -49.79 / 15.04	-5 / 0.33 / 61-12+11	2.5 / 2.5 / 9	0.91 / 28.	12.08	2 / 0.00 / 0.00	1918. / 1883. / 1		90		23.8 / 31.89	-13.0 / -19.0 / 6.2	-19.81 / 17.4	10.12		237-13 / / 2
2150-57	2150.6 / 335.91 / 228.87	-5751 / -46.55 / 7.93	10 / 0.16 / 61-13	2.1 / 2.0 / 6	0.80 / 41.		0.08 / 0.00	1835. / 1751. / 7		20		22.5 / 31.76	-14.7 / -16.8 / 3.1	13.1			145-25
N 7154	2152.4 / 10.21 / 245.94	-3503 / -51.79 / 23.81	10BP / 0.41 / 63 -2 +1	2.2 / 2.1 / 6	0.74 / 46.		0.10 / 0.00	2615. / 2648. / 1		28		33.3 / 32.61	-12.4 / -27.8 / 13.4	20.4			404-08 / -6-48-005 / 2
N 7155	2152.9 / 347.05 / 234.80	-4946 / -49.79 / 13.54	-2B / 0.31 / 61-12+11	2.2 / 2.2 / 6	0.92 / 27.	12.8	5 / 0.00 / 0.00	1893. / 1850. / 1		90		23.5 / 31.86	-13.2 / -18.7 / 5.5	-19.06 / 15.1	9.82		237-16 / / 2
2154-60	2154.4 / 332.11 / 227.36	-6033 / -45.77 / 5.63	10 / 0.18 / 61+13	1.0 / 0.9 / 6	0.80 / 41.		0.08 / 0.01	1685. / 1587. / 7		20		20.4 / 31.55	-13.8 / -15.0 / 2.0	5.4			146-2A
U11861	2155.7 / 111.05 / 13.19	7301 / 14.56 / 26.99	8X / 0.07 / 40 -0	3.6 / 5.1 / 2	0.76 / 45.	11.1	3 / 0.09 / 1.96	1489. / 1760. / 1	275 / 338. / 340.	10	1.15 / 8 / 1.17	24.7 / 31.97	21.5 / 5.0 / 11.2	-20.87 / 36.8	10.54 / 9.94 / 11.09	0.25 / 0.071 / 3.51	11861 12-20-000
N 7162	2156.6 / 356.37 / 239.87	-4333 / -51.90 / 17.44	5A / 0.27 / 61 -9	2.9 / 2.4 / 6	0.38 / 74.	12.89	2 / 0.48 / 0.00	2267. / 2255.		26		28.4 / 32.27	-13.6 / -23.4 / 8.5	-9.38 / -9.9	9.94		288-26 / -7-45-003 / 2
U11868	2156.7 / 75.39 / 306.60	1756 / -28.38 / 46.02	9B / 0.15 / 64-12	2.6 / 2.7 / 2	0.89 / 31.		0.04 / 0.27	1102. / 1357. / 1	155 / 244. / 246.	15	0.57 / 18 / 1.07	17.2 / 31.18	7.1 / -9.6 / 12.4	13.6	9.04 / / 10.37	0.047	II ZW 158 / 11868 03-56-001
2157-43	2157.5 / 356.62 / 240.12	-4322 / -52.09 / 17.45	9X / 0.27 / 61 -9	2.7 / 2.6 / 9	0.87 / 33.	13.03	2 / 0.05 / 0.00	2275. / 2264. / 2	124 / 180. / 183.	15	0.67 / 14 / 1.01	28.6 / 32.28	-13.6 / -23.6 / 8.6	-19.25 / 21.7	9.89 / 9.58 / 10.31	0.49 / 0.188 / 2.62	NGC 7162A / 288-28 -7-45-005
N 7166	2157.6 / 356.16 / 239.92	-4339 / -52.06 / 17.24	-2A / 0.26 / 61 -9	2.5 / 2.0 / 9	0.34 / 77.	12.82	2 / 0.00 / 0.00	2407. / 2395.		90		30.2 / 32.40	-14.5 / -25.0 / 9.0	-19.58 / 17.6	10.02		288-27 / -7-45-004 / 2

Name	Coord A	Coord B	Type	Dim	logD/PA	mag	—	V	—	N	—	Dist	—	M	mag2	extra	Other IDs
N 7177	2158.3 / 75.37 / 306.02	1730 / -28.96 / 45.60	3X / 0.14 / 64-12	3.2 / 3.0 / 2	0.64 / 55.	11.56	2 / 0.17 / 0.22	1188. / 1442.		17		18.2 / 31.30	7.5 / -10.3 / 13.0	-19.74 / 15.9	10.09		11872 03-56-003 2
N 7171	2158.3 / 43.44 / 266.81	-1331 / -47.92 / 35.66	3B / 0.13 / 63 -4	2.5 / 2.2 / 4	0.55 / 62.	12.67	2 / 0.24 / 0.09	2632. / 2770.		50		34.7 / 32.70	-1.6 / -28.1 / 20.2	-20.03 / 22.3	10.20		-2-56-005 2
N 7168	2158.9 / 343.32 / 233.88	-5159 / -49.95 / 11.30	-5 / 0.19 / 61 +7	2.1 / 2.0 / 6	0.67 / 53.	12.85	6 / 0.00 / 0.00	2783. / 2728.		90		34.7 / 32.70	-20.1 / -27.5 / 6.8	-19.85 / 20.3	10.13		237-26 2
N 7172	2159.1 / 15.12 / 249.38	-3207 / -53.06 / 24.68	2 P / 0.41 / 63 -1	2.1 / 1.8 / 6	0.52 / 64.	12.55	2 / 0.27 / 0.00	2651. / 2698.		40		33.9 / 32.65	-10.9 / -28.8 / 14.2	-20.10 / 17.8	10.23		466-38 -5-52-007 2
N 7173	2159.2 / 14.96 / 249.31	-3213 / -53.09 / 24.60	-5 / 0.35 / 63 -1	1.5 / 1.3 / 6	0.67 / 53.	13.05	2 / 0.00 / 0.00	2500. / 2547.		39		32.0 / 32.52	-10.3 / -27.2 / 13.3	-19.47 / 12.1	9.98		UGCA 422 / 466-39 -5-52-008 2
N 7174	2159.2 / 14.93 / 249.29	-3214 / -53.09 / 24.59	3 P / 0.30 / 63 -1	1.5 / 1.3 / 6	0.44 / 69.	13.15	2 / 0.36 / 0.00	2778. / 2824.		29		35.5 / 32.75	-11.4 / -30.2 / 14.8	-19.60 / 13.5	10.03		466-40 -5-52-010 2
N 7176	2159.2 / 14.93 / 249.29	-3214 / -53.09 / 24.59	-5 / 0.39 / 63 -1	1.4 / 1.3 / 6	0.85 / 36.	12.90	2 / 0.00 / 0.00	2525. / 2571.		29		32.3 / 32.54	-10.4 / -27.4 / 13.4	-19.64 / 12.3	10.05		UGCA 423 / 466-41 -5-52-011 2
2159-54	2159.3 / 339.96 / 232.24	-5419 / -49.12 / 9.62	10 / 0.16 / 61+13	2.0 / 1.7 / 6	0.46 / 68.		/ 0.34 / 0.00	1722. / 1654. / 7		20		21.0 / 31.61	-12.7 / -16.4 / 3.5	10.4			189-21
2159-12	2159.4 / 45.08 / 268.15	-1225 / -47.66 / 35.99	8B / 0.13 / 63 -4	2.0 / 1.9 / 2	0.82 / 39.		/ 0.07 / 0.09	2725. / 2868. / 1	182 / 237. / 239.	20	0.56 / 18 / 1.04	36.0 / 32.78	-0.9 / -29.1 / 21.1	20.0	9.67 / 10.51	0.145	UGCA 424 / -2-56-007
N 7180	2159.5 / 33.26 / 259.67	-2047 / -50.94 / 31.48	-2 / 0.07 / 61 -0	1.7 / 1.4 / 6	0.47 / 67.	13.34	2 / 0.00 / 0.06	1347. / 1451.		55		17.8 / 31.26	-2.7 / -15.0 / 9.3	-17.92 / 7.3	9.36		601-06 -4-52-008 2
2159-51	2159.6 / 343.91 / 234.29	-5132 / -50.22 / 11.54	10A / 0.37 / 14+12	4.6 / 4.2 / 6	0.67 / 53.	11.4	5 / 0.15 / 0.00	127. / 74.	535 / 497. / 498.	10	1.14 / 16 / 1.04	1.0 / 25.09	-0.6 / -0.8 / 0.2	-13.69 / 1.2	7.67		IC 5152 / 237-27 2
N 7184	2159.9 / 32.90 / 259.47	-2103 / -51.11 / 31.25	5B / 0.15 / 63 +3	6.1 / 4.3 / 6	0.20 / 90.	11.32 / 8.62 / 2.70	2 / 0.67 / 0.05	2625. / 2727. / 2		20		34.1 / 32.67	-5.3 / -28.7 / 17.7	-21.35 / 42.8	10.73 / 10.21 / 11.49	0.30 / 0.052 / 5.68	UGCA 425 / 601-09 -4-52-009 2
U11880	2200.2 / 77.29 / 308.90	1930 / -27.80 / 45.28	12BP / 0.07 / 64-12	1.6 / 1.7 / 2	1.00	13.6	3 / 0.00 / 0.21	1412. / 1671.		30		21.1 / 31.62	9.3 / -11.6 / 15.0	-18.02 / 10.5	9.40		IC 1420 / 11880 03-56-005
2200-34	2200.3 / 11.86 / 247.90	-3405 / -53.40 / 23.22	2B / 0.42 / 63 -1	2.0 / 1.6 / 6	0.33 / 77.	12.6	5 / 0.58 / 0.00	2584. / 2621.		90		32.8 / 32.58	-11.4 / -28.0 / 13.0	-19.98 / 15.3	10.18		IC 5156 / 404-25 -6-48-019 2

(1) Name	(2) α / ℓ / SGL	(3) δ / b / SGB	(4) Type / Q_{xyz} / Group	(5) D_{25} / D^i_{25} / source	(6) d/D / i	(7) $B_T^{b,i}$ / $H_{-0.5}$ / B-H	(8) source / A_B^{i-o} / A_B^b	(9) V_h / V_o / Tel	(10) W_{20} / W_R / W_D	(11) e_v / e_w / source	(12) F_c / source/e_F / f_h	(13) R / μ	(14) SGX / SGY / SGZ	(15) $M_B^{b,i}$ / Δ_{25}	(16) $\log L_B$ / $\log M_H$ / $\log M_T$	(17) M_H/L_B / M_H/M_T / M_T/L_B	(18) Alternate Names UGC/ESO MCG / AGN? RCII
N 7179	2201.1 / 327.01 / 225.31	-6417 / -44.51 / 2.41	4 / 0.28 / 62 -2 +1	1.8 / 1.5 / 6	0.48 / 67.		/ 0.32 / 0.01	2891. / 2774.		60		35.7 / 32.76	-25.1 / -25.3 / 1.5	-5.6			108-11 / 2
U11891	2201.5 / 93.61 / 341.66	4330 / -9.37 / 41.31	10 / 0.07 / 65 -0	3.7 / 5.0 / 2	0.77 / 44.	11.6	3 / 0.09 / 1.67	461. / 756. / 1	154 / 179. / 182.	7	1.36 / 226 2 / 1.42	10.5 / 30.10	7.5 / -2.5 / 6.9	-18.50 / 15.3	9.59 / 9.40 / 10.15	0.65 / 0.177 / 3.64	V ZW 380 / 11891 07-45-006
N 7196	2202.8 / 345.36 / 235.50	-5022 / -51.10 / 11.98	-5 / 0.19 / 61 -7	1.9 / 1.8 / 6	0.85 / 36.	12.55	2 / 0.00 / 0.00	3007. / 2959.		41		37.6 / 32.87	-20.8 / -30.3 / 7.8	-20.32 / 19.8	10.32		237-36 / 2
N 7192	2203.1 / 326.53 / 225.27	-6433 / -44.54 / 2.07	-5 / 0.28 / 62 -2 +1	1.7 / 1.7 / 6	1.00	12.21	6 / 0.00 / 0.00	2879. / 2761.		90		35.6 / 32.76	-25.0 / -25.3 / 1.3	-20.55 / 17.7	10.41		108-12 / 2
N 7200	2204.0 / 345.42 / 235.73	-5015 / -51.32 / 11.92	-5 / 0.24 / 61 -7	1.4 / 1.3 / 6	0.85 / 36.		/ 0.00 / 0.00	2937. / 2890.		90		36.7 / 32.82	-20.2 / -29.7 / 7.6	13.9			237-37 / 2
N 7204	2204.0 / 16.62 / 250.80	-3118 / -54.04 / 24.38	13 P / 0.35 / 63 -1	1.7 / 1.5 / 6	0.50 / 65.	10.9	/ 0.00 / 0.00	2630. / 2680.		90		33.5 / 32.63	-10.0 / -28.8 / 13.8	14.7			467-08 -5-52-029 / 2
U11909	2204.3 / 96.11 / 345.73	4700 / -6.85 / 39.59	12 P / 0.06 / 65 -0	3.1 / 3.2 / 2	0.26 / 84.		3 / 0.67 / 1.57	1108. / 1404. / 1	247 / 211. / 214.	10	0.91 / 9 / 1.22	18.8 / 31.38	14.1 / -3.6 / 12.0	-20.48 / 17.6	10.38 / 9.46 / 10.36	0.12 / 0.127 / 0.94	11909 08-40-001 / 2
N 7205	2205.1 / 334.86 / 230.39	-5740 / -48.36 / 6.70	4A / 0.17 / 61 -13	3.2 / 2.7 / 6	0.53 / 63.	11.3 / 8.86 / 2.44	5 / 0.26 / 0.00	1690. / 1605. / 7	334 / 332. / 334.	25 / 25 / 301		20.5 / 31.56	-13.0 / -15.7 / 2.4	-20.26 / 16.2	10.30 / 10.71	2.61	146-09 / 2
N 7217	2205.6 / 86.50 / 325.09	3107 / -19.70 / 43.52	2A / 0.15 / 65 +1	3.6 / 3.8 / 2	0.88 / 32.	10.64	1 / 0.04 / 0.42	946. / 1228.		16	0.40 / 36100	16.0 / 31.02	9.5 / -6.6 / 11.0	-20.38 / 17.8	10.34 / 8.81	0.03	11914 05-52-001 / 2
N 7213	2206.2 / 349.58 / 238.11	-4725 / -52.59 / 13.58	1A / 0.30 / 61 -11	2.1 / 2.1 / 6	1.00	11.35	2 / 0.00 / 0.00	1778. / 1745.		17		22.0 / 31.71	-11.3 / -18.1 / 5.2	-20.36 / 13.5	10.34		288-43 / L1.8 2
N 7218	2207.5 / 40.09 / 264.69	-1655 / -51.35 / 32.00	6B / 0.10 / 61+25	2.5 / 2.1 / 4	0.45 / 68.	12.11	2 / 0.35 / 0.08	1662. / 1783. / 1	273 / 253. / 255.	15 / 9 / 79	0.79 / 22 / 1.05	22.0 / 31.71	-1.7 / -18.6 / 11.7	-19.60 / 13.5	10.03 / 9.47 / 10.40	0.28 / 0.119 / 2.33	-3-56-008 / 2
2207-19	2207.5 / 36.81 / 262.52	-1906 / -52.16 / 30.85	10 / 0.11 / 61+25	2.2 / 1.8 / 6	0.37 / 74.	14.00	7 / 0.49 / 0.03	1738. / 1848. / 6	151	30	0.34 / 36 / 1.04	22.8 / 31.79	-2.6 / -19.4 / 11.7	-17.79 / 12.0	9.31 / 9.06	0.57	UGCA 426, DDO 211 / 601-25 -3-56-009 / 2
2207-46	2207.8 / 351.16 / 239.13	-4620 / -53.17 / 14.10	2 / 0.21 / 61 +8	3.0 / 2.1 / 6	0.16 / 90.		/ 0.67 / 0.00	2723. / 2695.		90		34.1 / 32.66	-17.0 / -28.4 / 8.3	20.9			IC 5171 / 288-46 / 2

2209-62	2209.5	-6219	5	2.6	0.06			1703.		20		20.5	-13.9					
	328.58	-46.39	0.15	1.8	90.		0.67	1595.				31.56	-15.1	10.8				
	227.38	3.10	61+13	6			0.05	7					1.1				146-14	
U11944	2209.6	1739	10	2.3	0.39			1733.	180	15	0.53	25.1	10.9					
	77.77	-30.66	0.11	1.9	73.		0.45	1985.	153.	11	226 13	32.00	-14.7	13.9	9.33	0.226		
	306.55	42.93	64 -0	2			0.16	1	156.	226	1.04		17.1		9.98		11944 03-56-018	
2210-46	2210.3	-4616	-2A	2.6	0.38	12.61	2	2000.		50		24.8	-12.2	-19.36	9.94		IC 5181	
	351.05	-53.60	0.32	2.1	74.		0.00	1972.				31.97	-20.7	15.2				
	239.48	13.82	61-11	9			0.00						5.9				289-01	2
2211-67	2211.2	-6706	5	3.8	0.15			1746.	390	20		21.0	-15.1				IC 5176	
	322.99	-43.69	0.12	2.6	90.	9.60	0.67	1615.	352.			31.61	-14.6	15.9				
	224.00	-0.29	61 -0	6			0.02	7	354.				-0.1		10.76		108-20	
N 7232	2212.6	-4606	-2B	2.5	0.33	12.9	5	1801.		50		22.3	-10.9	-18.84	9.73			
	351.12	-54.03	0.34	2.0	77.		0.00	1774.				31.74	-18.7	13.0				
	239.87	13.63	61-11	6			0.00						5.3				289-07	2
N 7233	2212.8	-4606		1.5	0.89			1832.		50		22.7	-11.1					
	351.10	-54.06	0.36	1.5	31.		0.00	1805.				31.78	-19.1	9.9				
	239.90	13.60	61-11	6			0.00						5.3				289-08	2
2212-45	2212.9	-4556	9B	1.7	0.90			2042.		20	0.69	25.3	-12.3				NGC 7232B	
	351.37	-54.13	0.25	1.7	29.		0.04	2016.			27	32.02	-21.3	12.6	9.50			
	240.04	13.70	61-11	6			0.00	2			1.01		6.0				289-09	2
2212-64	2212.9	-6438	5B	2.0	1.00			3062.		20		37.9	-26.3					
	325.57	-45.34	0.28	2.0			0.00	2943.				32.89	-27.2	22.1				
	225.94	1.25	62 -2 +1	6			0.00	7					0.8				108-23	
2213-47	2213.6	-4722	5	2.0	0.21			2748.		20		34.3	-17.2					
	348.99	-53.79	0.21	1.4	90.		0.67	2714.				32.67	-28.7	14.0				
	239.00	12.67	61 +8	6			0.00	7					7.5				289-10	
2213-21	2213.7	-2141	3B	2.3	0.93			2616.	170	30		33.8	-4.7				IC 1438	
	33.50	-54.36	0.17	2.3	25.		0.03	2712.				32.64	-29.4	22.7				
	260.93	28.23	63 -3	6			0.06	2			1.02		16.0				602-01 -4-52-029	
2214-21	2214.1	-2130	10 P	2.1	0.38	14.20	7	2571.	160	20	0.55	33.3	-4.5	-18.41	9.56	1.09	UGCA 427, DDO 212	
	33.85	-54.40	0.17	1.8	73.		0.46	2668.			39	32.61	-28.9	17.5	9.59			
	261.16	28.25	63 -3	6			0.07	2			1.01		15.7				602-03 -4-52-031	2
2214-45	2214.2	-4519	10	1.3	0.86			1822.		20		22.5	-10.7					
	352.26	-54.53	0.32	1.2	35.		0.05	1798.				31.76	-19.0	7.9				
	240.67	13.92	61-11	6			0.00	7					5.4				289-11	
2217-80	2217.8	-8015	5	2.8	0.34			1808.		20		21.7	-17.6					
	310.27	-34.92	0.11	2.4	77.		0.56	1620.				31.68	-12.1	15.2				
	214.46	-9.39	55 +5	6			0.44	7					-3.5				027-08	
2218-46	2218.3	-4619	6B	6.9	0.51	11.04	2	915.	206	10	1.51	11.1	-5.3	-19.18	9.86	0.55	IC 5201	
	350.23	-54.90	0.13	6.0	64.		0.27	886.	189.		6	30.22	-9.4	19.4	9.60	0.198		
	240.36	12.74	19 -5	9			0.00	2	192.		1.10		2.4		10.30	2.76	289-18	2

(1) Name	(2) α / ℓ / SGL	(3) δ / b / SGB	(4) Type / Q_{xyz} / Group	(5) D_{25} / $D_{25}^{b,i}$ / source	(6) d/D / i	(7) $B_T^{b,i}$ / $H_{-0.5}$ / B-H	(8) source / A_B^{i-o} / A_B^b	(9) V_h / V_o / Tel	(10) W_{20} / W_R / W_D	(11) e_v / e_w / source	(12) F_c / source/e_f / f_h	(13) R / μ	(14) SGX / SGY / SGZ	(15) $M_B^{b,i}$ / Δ_{25}	(16) log L_B / log M_H / log M_T	(17) M_H/L_B / M_H/M_T / M_T/L_B	(18) Alternate Names UGC/ESO / MCG / AGN? RCII
2219-48	2219.4 / 346.36 / 238.64	-4839 / -54.23 / 11.11	10 / 0.01 / 19 -5	1.3 / 1.3 / 6	1.00 /		0.00 / 0.00	706. / 664. / 7		20		8.4 / 29.62	-4.3 / -7.0 / 1.6	3.2			238-05
2220-42	2220.6 / 356.43 / 243.65	-4232 / -56.40 / 14.79	7B / 0.16 / 61 -0	2.4 / 2.2 / 6	0.70 / 50.		0.13 / 0.00	2427. / 2416. / 7		20		30.3 / 32.41	-13.0 / -26.2 / 7.7	19.5			289-26 -7-46-001
N 7280	2224.1 / 79.56 / 304.74	1554 / -34.24 / 39.29	-2 / 0.16 / 64 -3	2.0 / 2.0 / 2	0.69 / 50.	12.8	3 / 0.00 / 0.20	1846. / 2090.		31	-0.34 / 210 33	26.2 / 32.09	11.6 / -16.7 / 16.6	-19.29 / 15.3	9.91 / 8.50	0.04	12035 03-57-005 2
2224+15	2224.4 / 79.64 / 304.78	1555 / -34.28 / 39.22	10 / 0.16 / 64 -3	1.1 / 1.0 / 2	0.70 / 50.	14.9	3 / 0.13 / 0.20	1914. / 2158. / 1	125	20	-0.04 / 57 / 1.01	27.0 / 32.16	11.9 / -17.2 / 17.1	-17.26 / 7.9	9.10 / 8.82	0.53	UGCA 429 / 03-57-006
N 7292	2226.1 / 89.65 / 323.06	3003 / -23.17 / 39.24	10B / 0.18 / 65 -1	2.3 / 2.3 / 2	0.80 / 41.	12.6	3 / 0.08 / 0.31	993. / 1269. / 1	100 / 121. / 125.	10	0.72 / 11 / 1.06	16.2 / 31.05	10.0 / -7.6 / 10.3	-18.45 / 10.9	9.57 / 9.14 / 9.66	0.37 / 0.298 / 1.24	12048 05-53-003 2
2228-00	2228.0 / 65.65 / 285.18	-23 / -46.73 / 34.53	13 P / 0.03 / 64+11				0.08 / 0.16	1550. / 1737.		34		21.4 / 31.65	4.6 / -17.0 / 12.1				00-57-000 2
U12060	2228.2 / 92.15 / 327.50	3334 / -20.49 / 38.47	10B / 0.29 / 65 -1	2.0 / 2.2 / 2	1.00 /	13.8	3 / 0.00 / 0.38	887. / 1168. / 1	136	10 / 7 / 77	0.79 / 77 7 / 1.06	15.1 / 30.90	10.0 / -6.4 / 9.4	-17.10 / 9.7	9.03 / 9.15	1.31	12060 06-49-025 2
N 7302	2229.7 / 47.73 / 270.28	-1423 / -55.11 / 28.58	-2A / 0.10 / 63 -0	1.9 / 1.7 / 4	0.57 / 60.	13.05	2 / 0.00 / 0.15	2586. / 2713.		65		33.7 / 32.64	0.1 / -29.6 / 16.1	-19.59 / 16.7	10.03		IC 5228 / -2-57-013 2
U12069	2230.0 / 114.90 / 14.63	7615 / 15.88 / 23.27	8X / 0.07 / 42+20	2.0 / 2.4 / 2	0.68 / 52.	13.2	3 / 0.14 / 1.24	2369. / 2631. / 1	155 / 158. / 162.	15	0.46 / 25 / 1.05	35.9 / 32.78	32.0 / 8.3 / 14.2	-19.58 / 25.2	10.02 / 9.57 / 10.26	0.35 / 0.204 / 1.72	12069 13-16-000
U12074	2230.6 / 74.42 / 295.08	750 / -41.39 / 36.25	13 P / 0.11 / 64 -0	0.8 / 0.8 / 2	0.70 / 50.	13.8	3 / 0.00 / 0.28	1883. / 2100.		220		26.1 / 32.08	8.9 / -19.0 / 15.4	-18.28 / 6.1	9.50		ARAK 558 / 12074 01-57-008
N 7307	2231.0 / 357.94 / 245.97	-4111 / -58.61 / 14.07	5XP / 0.17 / 61+14	3.5 / 2.7 / 6	0.34 / 77.	12.3 / 11.15 / 1.15	5 / 0.56 / 0.00	2083. / 2077. / 2	274 / 243. / 245.	20	0.89 / 15 / 1.02	25.9 / 32.06	-10.2 / -22.9 / 6.3	-19.76 / 20.4	10.10 / 9.72 / 10.54	0.42 / 0.149 / 2.80	345-26 -7-46-003 2
U12082	2232.0 / 92.32 / 326.20	3237 / -21.72 / 37.78	9 / 0.31 / 65 -1	3.3 / 3.4 / 2	0.87 / 33.	13.81	7 / 0.05 / 0.32	803. / 1081. / 5	87 / 124. / 129.	5	0.85 / 5 / 1.20	13.9 / 30.72	9.1 / -6.1 / 8.5	-16.91 / 13.8	8.96 / 9.14 / 9.79	1.50 / 0.220 / 5.81	DDO 213 / 12082 05-53-006 2
N 7314	2233.0 / 27.15 / 259.21	-2618 / -59.74 / 22.04	4X / 0.08 / 61 -0	4.6 / 3.8 / 6	0.42 / 71.	11.14 / 9.11 / 2.03	2 / 0.40 / 0.01	1430. / 1499. / 2	317 / 295. / 297.	10	0.99 / 11 15 / 1.03	18.3 / 31.31	-3.2 / -16.7 / 6.9	-20.17 / 20.3	10.26 / 9.51 / 10.71	0.18 / 0.064 / 2.82	ARP 14 / 533-53 -4-53-018 2

Name	Coordinates	Type	Dimensions	Magnitudes	Velocities	Distance	M / abs	HI	Other names
N 7320	2233.7 3340 / 93.24 -21.02 / 327.47 37.32	7A / 0.31 / 65 -1	1.9 0.56 / 1.8 60. / 2	12.9 3 / 11.31 0.22 / 1.59 0.34	786. 192. 20 0.32 / 1066. 180. 12 217 15 / 1 184. 217 1.03	13.8 9.2 -17.80 / 30.70 -5.9 7.3 / 8.4	9.31 / 8.60 / 9.83	0.19 / 0.058 / 3.32	ARP 319, VV 288 / 12101 06-49-042 2
2233-03	2233.9 -309 / 63.97 -49.65 / 282.67 32.20	9 / 0.10 / 64+11	2.3 0.85 / 2.2 35. / 4	13.16 7 / 0.05 / 0.14	1692. 136. 10 0.58 / 1867. 187. 18 / 1 190. 1.06	22.9 4.3 -18.64 / 31.80 -18.9 14.7 / 12.2	9.65 / 9.30 / 10.17	0.45 / 0.134 / 3.36	ARP 3, DDO 214 / -1-57-016 2
N 7331	2234.8 3410 / 93.73 -20.72 / 328.07 37.04	4A / 0.33 / 65 -1	9.7 0.46 / 8.8 68. / 9	9.67 1 / 6.40 0.34 / 3.27 0.34	819. 530. 10 1.74 / 1099. 530. 6 217 6 / 4 532. 217 1.16	14.3 9.7 -21.10 / 30.77 -6.0 36.7 / 8.6	10.63 / 10.05 / 11.48	0.26 / 0.038 / 6.98	12113 06-49-045 2
N 7332	2235.0 2332 / 87.38 -29.67 / 314.71 37.36	-2 P / 0.12 / 65 -2 +1	3.9 0.27 / 3.0 82. / 2	11.72 2 / 0.00 / 0.13	1191. 11 / 1451.	18.2 10.2 -19.58 / 31.30 -10.3 15.9 / 11.1	10.02		12115 04-53-008 2
N 7339	2235.4 2332 / 87.47 -29.72 / 314.72 37.27	4X / 0.11 / 65 -2 +1	2.8 0.33 / 2.2 78. / 2	12.28 2 / 0.60 / 0.12	1276. 27 / 1536.	19.3 10.8 -19.15 / 31.43 -10.9 12.4 / 11.7	9.85		12122 04-53-009 2
N 7343	2236.3 3349 / 93.82 -21.18 / 327.59 36.77	3B / 0.07 / 65 +1	1.2 0.82 / 1.2 39. / 2	13.87 2 / 0.07 / 0.31	1216. 200 / 1495.	19.1 12.9 -17.54 / 31.41 -8.2 6.7 / 11.5	9.21		12129 06-49-059 2
2236-05	2236.5 -502 / 62.34 -51.35 / 280.89 30.91	10 / 0.07 / 65 -0	1.7 0.27 / 1.3 82. / 2	14.22 7 / 0.67 / 0.11	829. 82. 10 0.27 / 995. 64. 24 / 1 73. 1.02	12.0 1.9 -16.17 / 30.39 -10.1 4.6 / 6.1	8.66 / 8.43 / 8.73	0.59 / 0.496 / 1.18	UGCA 433, DDO 215 / -1-57-019 2
2238-45	2238.0 -4555 / 348.61 -58.22 / 242.80 10.27	3 / 0.13 / 61 -0	1.8 0.24 / 1.3 85. / 6	0.67 / 0.00	2828. 20 / 2797. / 7	35.2 -15.8 13.4 / 32.73 -30.8 / 6.3			289-48
U12151	2239.0 8 / 68.99 -48.39 / 286.83 32.10	10 / 0.10 / 64+11	2.9 0.70 / 2.8 50. / 2	0.13 / 0.20	1753. 193. 25 0.66 / 1939. 206. 226 10 / 1 209. 1.07	23.8 5.8 19.5 / 31.89 -19.3 / 12.7	9.41 / 10.38	0.108	12151 00-57-007
2239-45	2239.0 -4504 / 349.96 -58.72 / 243.61 10.62	1B / 0.29 / 61 -0+16	3.2 0.60 / 2.8 58. / 6	12.0 5 / 0.19 / 0.00	1743. 25 / 1716.	21.4 -9.3 -19.65 / 31.65 -18.8 17.5 / 3.9	10.05		IC 5240
N 7361	2239.6 -3020 / 19.27 -61.61 / 256.37 18.68	5 / 0.09 / 61 -0+16	3.9 0.24 / 2.8 85. / 6	12.27 2 / 10.90 0.67 / 1.37 0.00	1259. 233. 10 1.01 / 1306. 197. 36 11 / 2 199. 1.01	15.9 -3.5 -18.73 / 31.00 -14.6 13.0 / 5.1	9.68 / 9.41 / 10.17	0.54 / 0.177 / 3.03	290-02 2
2240-40	2240.5 -4007 / 359.03 -60.63 / 247.93 13.20	9 / 0.17 / 61+14	3.2 1.00 / 3.2 / 6	0.00 / 0.00	2153. 176. 15 0.76 / 2151. 15 / 2 1.01	26.7 -9.8 24.9 / 32.14 -24.1 / 6.1	9.61		IC 5237, UGCA 434 / 468-23 -5-53-027 2 / 345-46 -7-46-008
N 7363	2241.0 3344 / 94.67 -21.75 / 327.38 35.80	7X / 0.33 / 65 -1	1.6 0.78 / 1.6 43. / 2	14.0 3 / 0.08 / 0.34	830. 30 / 1108.	14.2 9.7 -16.77 / 30.77 -6.2 6.6 / 8.3	8.90		06-49-078 2
U12178	2242.6 610 / 75.96 -44.62 / 294.00 32.95	8X / 0.11 / 64 -0	3.2 0.55 / 2.9 61. / 2	12.9 3 / 0.23 / 0.17	1935. 242. 10 0.90 / 2142. 234. 10 / 1 236. 1.07	26.5 9.1 -19.22 / 32.12 -20.3 22.4 / 14.4	9.88 / 9.75 / 10.55	0.74 / 0.157 / 4.70	12178 01-58-004 2

(1) Name	(2) α ℓ SGL	(3) δ b SGB	(4) Type Q_xyz Group	(5) D_25 D^{b,i}_25 source	(6) d/D i	(7) B^{b,i}_T H_{-0.5} B-H	(8) source A^{i-o}_B A^b_B	(9) V_h V_o Tel	(10) W_20 W_R W_D	(11) e_v e_w source	(12) F_c source/e_f f_h	(13) R μ	(14) SGX SGY SGZ	(15) M^{b,i}_B Δ_25	(16) log L_B log M_H log M_T	(17) M_H/L_B M_T/L_B	(18) Alternate Names UGC/ESO MCG RCII / Alternate Names / AGN?
N 7371	2243.4 55.51 275.08	-1116 -56.42 26.91	 0.07 63 -0	1.5 1.5 4	1.00	12.64	2 0.00 0.16	2389. 2526.		45		31.2 32.47	2.5 -27.7 14.1	-19.83 13.7	10.12		-2-58-001 2
2243-65	2243.8 321.88 227.63	-6506 -47.46 -1.60	7 0.06 62 -0	3.5 2.4 6	0.11 90.		 0.67 0.00	2368. 2243. 7		20		28.8 32.30	-19.4 -21.3 -0.8	20.2			IC 5249 109-21
2247-89	2247.2 303.34 207.30	-8924 -27.92 -15.30	12 P 0.17 55 -0 +3	2.4 2.2 6	0.40 72.		 0.43 0.43	2527. 2305. 7		20		30.9 32.45	-26.5 -13.7 -8.2	19.9			NGC 2573B 001-09
U12212	2248.1 93.42 321.35	2852 -26.72 34.50	9 0.23 65 +1	2.5 2.5 2	0.75 45.		 0.10 0.24	896. 1163. 1	106. 117. 122.	8	0.58 226 10 1.06	14.7 30.84	9.5 -7.6 8.3	10.7	8.91 9.63	0.193	12212 05-53-019
U12221	2250.0 119.09 20.05	8238 20.95 19.27	7A 0.07 42 -0	2.5 2.2 2	0.36 75.	13.0	3 0.52 0.58	2057. 2302. 1	272	25	0.63 18 1.05	32.2 32.54	28.5 10.4 10.6	-19.54 20.7	10.01 9.65	0.43	12221 14-01-001
N 7410	2252.2 358.01 249.34	-3956 -62.83 11.42	-2B 0.40 61 -16	5.2 4.0 6	0.29 81.	11.25	2 0.00 0.05	1638. 1633.		18		20.1 31.52	-7.0 -18.5 4.0	-20.27 23.5	10.30		346-12 -7-47-002 2
N 7412	2252.9 351.91 246.83	-4255 -61.87 9.73	3B 0.42 61 -16	4.3 4.0 6	0.73 47.	11.80	2 0.11 0.00	1726. 1706. 2	184 203. 206.	20	0.86 22 1.02	21.1 31.62	-8.2 -19.1 3.6	-19.82 24.6	10.12 9.51 10.47	0.24 0.110 2.23	290-24 -7-47-004 2
N 7416	2253.1 65.86 281.81	-546 -54.98 26.80	3B 0.07 63 -0	3.3 2.5 4	0.23 86.	12.29	2 0.67 0.14	2770. 2927.		125		36.3 32.80	6.6 -31.8 16.4	-20.51 26.5	10.40		-1-58-004 2
N 7418	2253.8 3.49 251.81	-3717 -63.87 12.54	6X 0.24 61 -17+16	3.6 3.4 6	0.70 50.	11.82	2 0.13 0.05	1445. 1453.	248	20	0.88 7 24 1.04	17.8 31.25	-5.4 -16.5 3.9	-19.43 17.7	9.96 9.38	0.26	406-25 -6-50-013 2
2253-36	2253.8 5.23 252.50	-3630 -64.04 12.94	5 0.27 61 -17+16	4.3 3.0 6	0.20 90.	11.24	2 0.67 0.05	1659. 1671. 2	256	20	0.66 24 1.01	20.5 31.56	-6.0 -19.1 4.6	18.0	9.28		IC 5269B
2253-37	2253.9 4.04 252.04	-3702 -63.95 12.65	5B 0.13 61 -15	2.9 2.5 6	0.56 61.		 0.23 0.05	2050. 2059.		60		25.4 32.02	-7.6 -23.6 5.6	18.5			406-26 -6-50-012 2 NGC 7418A 406-27 -6-50-11B
2254-430	2254.3 351.37 246.83	-4305 -62.04 9.42	7B 0.17 19 -6	4.3 3.0 6	0.18 90.		 0.67 0.05	938. 917. 2	134 106. 112.	20	0.62 17 1.03	11.3 30.26	-4.4 -10.2 1.8	9.9	8.73 9.51	0.165	NGC 7412A 290-28 -7-47-000 2
2254-434	2254.4 350.23 246.34	-4340 -61.80 9.10	 0.37 61 -16	6.1 5.5 6	0.63 56.	11.40 8.08 3.32	2 0.00 0.00	1725. 1701.		17		21.0 31.61	-8.3 -19.0 3.3	-20.21 33.7	10.28		IC 5267 290-29 -7-47-007 2

2254-36	2254.4	-3644	-5	3.9	0.70	10.93	2	1624.		17		20.0	-5.9	-20.58	10.42		IC 1459, IC 5265
	4.65	-64.11	0.28	3.7	50.		0.00	1635.				31.51	-18.6	21.6			
	252.36	12.72	61-17+16	6			0.05						4.4				406-30 -6-50-016 2
N 7424	2254.5	-4121	6X	9.1	0.81	10.75	2	951.	177	8	1.79	11.5	-4.2	-19.55	10.01	0.79	
	354.77	-62.75	0.16	8.8	40.		0.07	938.	224.		5	30.30	-10.5	29.6	9.91	0.190	
	248.34	10.30	19 -6	9			0.05	2	226.		1.12		2.1		10.63	4.18	346-19 -7-47-008 2
2255-36	2255.0	-3618	-2A	1.7	0.50	13.52	6	2162.		90		26.9	-7.7	-18.63	9.64		IC 5269
	5.56	-64.32	0.14	1.4	65.		0.00	2175.				32.15	-25.0	11.0			
	252.81	12.83	61-15	6			0.05						6.0				406-32 -6-50-018 2
U12263	2255.0	7225	10	2.9	0.70			2671.	272	20	0.65	39.7	35.8				
	114.53	11.71	0.07	4.4	50.		0.13	2938.	306.		24	32.99	6.4	51.0	9.85	0.051	
	10.08	23.55	42+20	2			2.41	1	308.		1.14		15.8		11.14		12263 12-21-000
N 7437	2255.7	1402	7X	1.9	1.00	13.3	3	2117.	125	30	0.22	29.2	13.9	-19.03	9.80	0.22	
	85.88	-40.28	0.28	1.9			0.00	2345.			32	32.33	-20.7	16.2	9.15		
	303.89	31.47	64 -1	2			0.14	1			1.04		15.2				12270 02-58-041
U12281	2256.7	1319	8	3.3	0.07	14.0	3	2567.	305	20	0.85	34.9	16.3	-18.72	9.68	1.80	
	85.63	-41.02	0.14	2.4	90.		0.67	2793.	267.		16	32.72	-25.1	24.5	9.94	0.171	
	303.12	31.10	64 -2 +1	2			0.15	1	269.		1.06		18.0		10.70	10.56	12281 02-58-043
2256-37	2256.7	-3758	5X	2.4	0.67	11.70	2	1304.	200	20	0.61	16.0	-5.0	-19.32	9.92	0.13	IC 5273
	1.65	-64.25	0.21	2.2	53.	9.88	0.15	1308.			46	31.02	-14.9	10.3	9.02		
	251.52	11.69	61 -0+16	6		1.82	0.05	2			1.02		3.2				346-22 -6-50-020 2
N 7448	2257.6	1543	4A	2.7	0.51	11.70	2	2199.		20	1.01	30.3	15.2	-20.71	10.48	0.31	ARP 13
	87.57	-39.12	0.26	2.4	64.		0.28	2432.			11 20	32.41	-21.0	21.2	9.97		
	305.93	31.29	64 -1	9			0.17						15.8				12294 03-58-018 2
N 7457	2258.6	2953	-2A	4.5	0.62	11.28	2	705.		108		12.3	8.3	-19.18	9.86		
	96.22	-26.93	0.13	4.2	56.		0.00	971.				30.46	-6.3	15.1			
	322.57	32.21	65 +1	2			0.22						6.6				12306 05-54-026 2
N 7463	2259.4	1543	3XP	3.1	0.22	12.4	3	2445.	722	26	1.09	33.5	16.9	-20.22	10.28	0.72	
	88.05	-39.35	0.17	2.3	87.		0.67	2677.		50	217 15	32.62	-23.2	22.5	10.14		
	306.01	30.86	64 -1	2			0.20			217			17.2				12316 03-58-022 2
N 7464	2259.4	1542	-2	0.7	1.00	14.1	3	1870.		32		26.1	13.2	-17.98	9.38		ARAK 573
	88.03	-39.37	0.22	0.7			0.00	2102.				32.08	-18.1	5.3			
	305.99	30.86	64 -1	2			0.20						13.4				12315 03-58-023 2
N 7456	2259.4	-3951	6A	6.1	0.33	11.4	5	1211.	236	15	1.05	14.7	-4.9	-19.44	9.97	0.26	
	357.16	-64.16	0.13	4.8	78.	10.23	0.61	1205.	203.		18	30.84	-13.6	20.6	9.38	0.098	
	250.14	10.28	19 -6	6		1.17	0.05	2	206.		1.05		2.6		10.39	2.65	346-26 -7-47-011 2
N 7465	2259.5	1542	-2	1.3	0.60	13.1	3	1959.		23		27.2	13.7	-19.07	9.82		MARK 313
	88.06	-39.38	0.25	1.2	58.		0.00	2191.				32.17	-18.9	9.5			
	305.99	30.84	64 -1	2			0.20						14.0				12317 03-58-024 2
2259-37	2259.5	-3720	9	1.9	0.78			1378.	127	15	0.37	16.9	-5.0				
	2.72	-64.95	0.25	1.8	42.		0.08	1384.	149.		27	31.14	-15.8	8.9	8.83	0.117	
	252.36	11.53	61-17+16	6			0.05	2	153.				3.4		9.76		406-42 -6-50-000

(1) Name	(2) α ℓ SGL	(3) δ b SGB	(4) Type ρ_{xyz} Group	(5) D_{25} D_{25}^i source	(6) d/D i	(7) $B_T^{b,i}$ $H_{-0.5}$ B-H	(8) source A_B^{i-o} A_B^b	(9) V_h V_o Tel	(10) W_{20} W_R W_D	(11) e_v e_w source	(12) F_c source/e_f f_h	(13) R μ	(14) SGX SGY SGZ	(15) $M_B^{b,i}$ $\Delta_{25}^{b,i}$	(16) log L_B log M_H log M_T	(17) M_H/L_B M_H/M_T M_T/L_B	(18) Alternate Names UGC/ESO MCG AGN? RCII
N 7462	2300.0 354.38 249.09	-4107 -63.80 9.54	5B 0.13 19 -6	5.0 3.5 6	0.17 90.	12.1 10.24 1.86	5 0.67 0.05	1074. 1061. 2	199 163. 167.	10	0.91 13 1.02	13.0 30.57	-4.6 -12.0 2.2	-18.47 13.3	9.58 9.14 10.01	0.36 0.134 2.69	346-28 -7-47-013 2
N 7468	2300.5 88.75 306.76	1620 -38.97 30.70	13 P 0.29 64 -1	1.1 1.0 2	0.70 50.	13.7	3 0.00 0.22	2089. 2322.		12	0.61 12 32 1.04	28.9 32.31	14.9 -19.9 14.8	-18.61 8.4	9.64 9.53	0.79	MARK 314 12329 03-58-000 2
2300-46	2300.6 344.26 244.65	-4618 -61.49 6.79	9B 0.26 61-16	1.2 1.2 6	1.00		0.00 0.00	1533. 1494. 7		20		18.5 31.33	-7.8 -16.6 2.2	6.5			290-39
N 7479	2302.4 86.26 301.98	1203 -42.84 29.50	5B 0.17 64 -2 +1	3.9 3.8 9	0.76 45.	11.44	2 0.10 0.16	2382. 2602.	369 471. 472.	9 12 217	0.95 217 20	32.4 32.55	14.9 -23.9 15.9	-21.11 36.0	10.64 9.97 11.36	0.22 0.040 5.35	12343 02-58-060 2
U12344	2302.5 90.65 309.28	1827 -37.39 30.53	7B 0.18 64 -4	2.3 2.0 2	0.44 70.		0.37 0.17	1633. 1871. 1	164 140. 145.	10	0.63 17 1.04	23.2 31.83	12.6 -15.5 11.8	13.5	9.36 9.89	0.298	12344 03-58-029
U12350	2302.8 89.53 307.15	1635 -39.04 30.19	9 0.27 64 -1	2.9 2.4 2	0.37 74.	13.5	3 0.49 0.23	2138. 2371. 1	223 193. 196.	15	0.55 16 1.05	29.5 32.35	15.4 -20.3 14.8	-18.85 20.7	9.73 9.49 10.35	0.57 0.138 4.14	12350 03-58-032 2
N 7497	2306.7 91.40 308.81	1754 -38.36 29.46	5B 0.23 64 -4	4.1 3.4 2	0.38 74.	12.2 9.87 2.33	3 0.47 0.16	1710. 1945.	304 277. 279.	15 8 217	1.14 217 7 1.10	24.1 31.91	13.2 -16.4 11.9	-19.71 23.9	10.08 9.90 10.73	0.67 0.151 4.47	12392 03-59-002 2
N 7496	2307.0 347.83 247.49	-4342 -63.80 7.13	3B 0.51 61-16	3.5 3.6 6	1.00	11.55	0.00 0.05	1657. 1629. 2	169	15	0.68 17 1.02	20.1 31.51	-7.6 -18.4 2.5	-19.96 21.1	10.18 9.29	0.13	291-01 -7-47-020 2 L/S
2307+18	2307.6 91.81 309.14	1810 -38.23 29.29	10 0.22 64 -4	1.1 1.1 2	1.00		0.00 0.15	1785. 2021. 1	88	10	0.09 20 1.01	25.2 32.00	13.8 -17.0 12.3	8.1	8.89		UGCA 436 03-59-005
N 7507	2309.4 23.43 261.05	-2849 -68.03 13.77	-5 -0.09 61 -0+16	2.1 2.2 6	1.00	11.22	0.00 0.08	1548. 1593.		12		19.3 31.43	-2.9 -18.5 4.6	-20.21 12.4	10.28		469-19 -5-54-022 2
N 7518	2310.6 83.91 295.88	603 -48.96 26.19	1X 0.00 65 -7	1.6 1.6 2	0.94 23.	14.1	3 0.02 0.21	643. 839.		190		10.0 30.00	3.9 -8.1 4.4	-15.90 4.7	8.55		MARK 527, TURN 137A
U12423	2310.6 83.99 295.97	608 -48.89 26.21	5 0.06 65 -7	3.4 2.5 2	0.14 90.	13.2	3 0.67 0.21	886. 1082.		78		13.0 30.57	5.1 -10.5 5.7	-17.37 9.5	9.14		TURN 137B 12423 01-59-013
N 7537	2312.0 82.76 294.02	414 -50.65 25.38	4A 0.20 64 -9	2.1 1.7 2	0.28 81.	12.93 10.68 2.25	2 0.67 0.19	2661. 2850.		25		35.3 32.74	13.0 -29.1 15.1	-19.81 17.5	10.12		TURN 138A 12442 01-59-016 2

```
Name     Coord1        Coord2/Coord3         Type  Dim          B/idx          Vel          col3          single  col5           DistMod/XYZ        AbsMag    Mags            Indices        Designations

N 7531   2312.0  -4353  346.40 -64.48 6.25    4A    5.4 4.4 6    11.46 9.02 2.44  1598. 1568. 2  341. 316. 318.  15   1.30 9 1.03    19.3 31.43          -19.97    10.18 9.87 10.86  0.49 0.104 4.74  291-10 -7-47-025 2
                        247.77                0.46  0.39 73.     2 0.46 0.05                                                          -7.3 -17.7 2.1      24.8
                        61-16

N 7541   2312.2   416   82.85 -50.65 25.34    4B    3.2 2.7 2    11.74 9.01 2.73  2506. 2695.                    64   1.13 217 10    33.4 32.62          -20.88    10.54 10.18       0.43            TURN 138B
                        294.07                0.15  0.36 75.     2 0.52 0.20                                                          12.3 -27.6 14.3     26.3                                             12447 01-59-017 2
                        64 -9

N 7552   2313.5  -4253  348.07 -65.24 6.49     2B   3.5 3.5 6    11.31            1609. 1583. 2  280             50   0.99 18 1.03    19.5 31.45          -20.14    10.25 9.57       0.21            IC 5294
                        248.78                0.52  0.89 31.     2 0.04 0.05                                                          -7.0 -18.1 2.2      19.9
                        61-16

N 7582   2315.6  -4239  348.08 -65.69 6.26     2B   4.5 3.9 9    11.06            1459. 1433.                    15                  17.6 31.23          -20.17    10.26                            291-12 -7-47-028 2
                        249.17                0.39  0.50 65.     2 0.29 0.05                                                          -6.2 -16.3 1.9      20.0
                        61-16

N 7590   2316.2  -4231  348.21 -65.85 6.22     4A   2.7 2.2 9    11.60            1437. 1412.                    34                  17.3 31.19          -19.59    10.03                            S2
                        249.34                0.33  0.37 74.     2 0.49 0.05                                                          -6.1 -16.1 1.9      11.1                                             291-16 -7-47-029 2
                        61-16

N 7606   2316.5   -846  69.09 -61.29 20.31      3A  5.2 4.2 4    11.03            2227. 2362.                    54                  28.9 32.31          -21.28    10.70                            347-33 -7-47-030 2
                        280.97                0.07  0.38 74.     2 0.48 0.05                                                          5.2 -26.6 10.0      35.4
                        63 +5

N 7599   2316.6  -4232  348.09 -65.90 6.15      5B  4.9 3.8 9    11.30            1663. 1638.                    73                  20.2 31.52          -20.22    10.28                            -2-59-012 2
                        249.36                0.43  0.31 79.     2 0.66 0.05                                                          -7.1 -18.8 2.2      22.4
                        61-16

N 7625   2318.0   1657  93.90 -40.47 26.65     1AP  1.6 1.6 2    12.73            1641. 1869.    197             8    0.64 56 16     23.0 31.81          -19.08    9.82 9.36        0.35            ARP 212, III ZW 103
                        308.21                0.17  1.00          2 0.00 0.07                   20                                    12.7 -16.2 10.3     10.7                                             VV 280
                        64 +5 +4                                                                210                                                                                                     12529 03-59-038 2

2318-42  2318.1  -4200  348.86 -66.42 6.15      10  2.4 2.4 6                     16. -7. 7                      20                  1.0 25.00             0.7                                     347-08 -7-47-034
                        249.97                0.37  0.87 34.     0.05 0.05                                                           -0.3 -1.0 0.1
                        14-13

N 7632   2319.4  -4246  346.92 -66.21 5.59     -2B  2.2 1.9 6                     1600. 1573.                    60                  19.3 31.43           10.7                                     IC 5313
                        249.39                0.52  0.50 65.     0.00 0.05                                                           -6.8 -18.0 1.9                                                 291-21 -7-47-035 2
                        61-16

N 7640   2319.7   4034  105.24 -18.94 27.65     5B  9.6 7.8 2    10.21 8.89 1.32  369. 643.      263. 226. 229.  6    2.00 25 5      8.6 29.68           -19.47    9.98 9.87 10.46  0.77 0.255 3.04  12554 07-48-002 2
                        334.78                0.18  0.25 85.     1 0.67 0.46                                    15                   6.9 -3.3 4.0        19.6
                        65 -4                                                                                  25

U12578   2321.8    -23  81.61 -55.83 21.74    12 P  1.7 1.6 2    14.5             2696. 2863. 1  127 135. 139.   15   0.41 17 1.02   35.4 32.75          -18.25    9.49 9.51 9.94   1.04 0.368 2.82  12578 00-59-038
                        289.93                0.27  0.72 48.     3 0.12 0.13                                                          11.2 -30.9 13.1     16.5
                        64+10

U12588   2322.3   4104  105.91 -18.65 27.13      8  1.9 2.1 2    12.8             425. 698. 1    116             15   0.52 16 1.05   9.3 29.84           -17.04    9.01 8.46        0.28            12588 07-48-005
                        335.32                0.18  1.00          3 0.00 0.46                                                         7.5 -3.4 4.2        5.7
                        65 -4

2323-32  2323.7  -3240  11.87 -70.84 9.32       10  1.5 1.4 6                     62. 82.                        10   0.54 222 20    1.0 24.97             0.4      6.54                            UGCA 438, UKS 2323-326
                        258.90                0.45  0.88 32.     0.04 0.05                                                           -0.2 -1.0 0.2                                                 407-18 -5-55-012
                        14-13
```

(1) Name	(2) α / ℓ / SGL	(3) δ / b / SGB	(4) Type / Q_{xyz} / Group	(5) D_{25} / $D_{25}^{b,i}$ / source	(6) d/D / i	(7) $B_T^{b,i}$ / $H_{-0.5}$ / B-H	(8) source / A_B^{i-o} / A_B^b	(9) V_h / V_o / Tel	(10) W_{20} / W_R / W_D	(11) e_v / e_w / source	(12) F_c source/e_F / f_h	(13) R / μ	(14) SGX / SGY / SGZ	(15) $M_B^{b,i}$ / $\Delta_{25}^{b,i}$	(16) log L_B / log M_H / log M_T	(17) M_H/L_B / M_H/M_T / M_T/L_B	(18) Alternate Names UGC/ESO MCG AGN? RCII
2324+18	2324.0 / 96.19 / 309.62	1800 / -40.15 / 25.41	13 P / 0.09 / 64 +5 +4	0.5 / 0.5 / 2	0.89 / 31.		0.00 / 0.07	1538. / 1766.		43		21.8 / 31.69	12.6 / -15.2 / 9.4	3.2			UGCA 439, MARK 324 / 03-59-000 2
2324-37	2324.2 / 357.81 / 254.44	-3737 / -69.48 / 7.08	10 / 0.18 / 19 -7	1.5 / 1.2 / 6	0.33 / 77.		0.58 / 0.05	690. / 686. / 2	103	25	0.49 / 24 / 1.01	8.2 / 29.57	-2.2 / -7.8 / 1.0	2.9	8.32		347-17 -6-51-005
N 7661	2324.3 / 316.34 / 229.42	-6533 / -49.60 / -5.45	5 / 0.07 / 55 -0	1.9 / 1.7 / 6	0.65 / 54.		0.16 / 0.00	1992. / 1858. / 7		20		23.8 / 31.88	-15.4 / -18.0 / -2.3	11.8			110-11
U12613	2326.0 / 94.75 / 305.87	1428 / -43.55 / 24.32	10A / 0.46 / 14+12	4.6 / 4.2 / 2	0.63 / 55.	12.37	1 / 0.17 / 0.09	-178. / 39. / 1	48 / 33. / 48.	7 / 5 / 217	0.74 / 217 7 / 1.20	1.0 / 25.00	0.6 / -0.7 / 0.4	-12.63 / 1.2	7.24 / 6.74 / 7.59	0.31 / 0.142 / 2.21	PEG IRR, DDO 216 / 12613 02-59-046 2
2326-41	2326.0 / 347.71 / 250.99	-4136 / -67.88 / 5.02	4A / 0.40 / 61-16	2.7 / 2.7 / 6	0.97 / 17.	12.2	5 / 0.01 / 0.05	1512. / 1488. / 2	142	15	0.38 / 28 / 1.02	18.1 / 31.29	-5.9 / -17.1 / 1.6	-19.09 / 14.3	9.83 / 8.90	0.12	IC 5325
U12632	2327.6 / 106.78 / 334.88	4043 / -19.31 / .26.14	9 / 0.18 / 65 -4	4.6 / 4.9 / 2	0.82 / 39.	12.13	7 / 0.07 / 0.50	424. / 695. / 1	131 / 166. / 170.	7	1.26 / 6 / 1.23	9.2 / 29.82	7.5 / -3.5 / 4.1	-17.69 / 13.2	9.27 / 9.19 / 10.02	0.83 / 0.146 / 5.68	347-18 -7-48-004 2 / DDO 217 / 12632 07-48-007 2
N 7689	2329.9 / 325.78 / 239.79	-5422 / -59.40 / -1.16	6X / 0.10 / 55 -0	2.7 / 2.5 / 9	0.71 / 49.	12.03	2 / 0.12 / 0.00	1981. / 1895.		14		23.8 / 31.88	-12.0 / -20.5 / -0.5	-19.85 / 17.4	10.13		192-07 2
N 7690	2330.2 / 328.66 / 241.97	-5158 / -61.34 / -0.16	3A / 0.09 / 61 -0+16	2.1 / 1.7 / 6	0.38 / 73.	12.2	5 / 0.46 / 0.00	1392. / 1317.		50		16.4 / 31.07	-7.7 / -14.5 / 0.0	-18.87 / 8.1	9.74		240-06 2
2331-36	2331.8 / 359.42 / 256.23	-3622 / -71.37 / 6.24	7A / 0.18 / 19 -7	10.5 / 10.4 / 6	0.93 / 25.	11.17	2 / 0.03 / 0.05	707. / 706. / 2	119	8	1.62 / 5 / 1.08	8.4 / 29.62	-2.0 / -8.1 / 0.9	-18.45 / 25.5	9.57 / 9.47	0.79	IC 5332
U12682	2332.4 / 98.57 / 309.92	1757 / -41.00 / 23.43	10 / 0.07 / 64 -0	1.7 / 1.6 / 2	0.83 / 38.	13.88	7 / 0.06 / 0.10	1397. / 1621. / 1	125 / 162. / 166.	25	0.49 / 17 / 1.03	20.0 / 31.50	11.8 / -14.0 / 7.9	-17.62 / 9.3	9.24 / 9.09 / 9.85	0.71 / 0.174 / 4.09	408-09 -6-51-012 2 / DDO 218 / 12682 03-60-007 2
U12690	2333.2 / 87.16 / 292.15	56 / -56.31 / 19.39	10 / 0.33 / 64-10	2.0 / 2.1 / 2	1.00		0.00 / 0.14	2613. / 2779. / 1	105	15 / 11 / 226	0.32 / 226 15 / 1.05	34.2 / 32.67	12.2 / -29.9 / 11.4	21.0	9.39		12690 00-60-015
N 7713	2333.6 / 353.84 / 254.70	-3812 / -70.88 / 5.15	6B / 0.18 / 19 -7	4.6 / 3.9 / 6	0.42 / 71.	11.1	5 / 0.40 / 0.05	698. / 687. / 2	236	40	1.09 / 14 / 1.02	8.2 / 29.56	-2.1 / -7.8 / 0.7	-18.46 / 9.3	9.58 / 8.92	0.22	347-28 -6-51-013 2
N 7714	2333.7 / 88.23 / 293.15	153 / -55.56 / 19.53	3BP / 0.33 / 64-10	1.9 / 1.8 / 2	0.75 / 45.	12.8	3 / 0.10 / 0.16	2811. / 2981.		20		36.9 / 32.83	13.7 / -32.0 / 12.3	-20.03 / 19.4	10.20		ARP 284, MARK 538 / VV 51 / 12699 00-60-017 2 L/S

Name	Coordinates	Type	Dim	b/a, PA	mag	col	V₁	N	col	col	col	col	V₂	col	col	Notes
N 7715	2333.8 153 / 88.26 −55.57 / 293.16 19.51	10 P / 0.35 / 64−10	3.1 / 2.2 / 2	0.15 / 90.	13.8	3 / 0.67 / 0.16	2764. / 2934.	—	64	—	36.3 / 32.80	13.5 / −31.5 / 12.1	−19.00 / 23.3	9.79	—	ARP 284, VV 51 / 12700 00−60−018 2
2333−39	2333.8 −3904 / 351.51 −70.49 / 253.93 4.76	7 / 0.26 / 61 −0+16	4.6 / 4.0 / 6	0.50 / 65.	—	0.29 / 0.05	1553. / 1538. / 7	—	20	—	18.7 / 31.36	−5.2 / −17.9 / 1.6	21.8	—	—	347−29 −7−48−009
N 7716	2334.0 2 / 86.62 −57.18 / 291.29 18.94	3B / 0.26 / 64−10	2.0 / 2.0 / 9	0.89 / 30.	12.80	2 / 0.04 / 0.11	2572. / 2734. / 1	239 / 400. / 401.	20	0.65 / 23 / 1.05	33.7 / 32.64	11.6 / −29.7 / 10.9	−19.84 / 19.7	10.13 / 9.71 / 10.96	0.38 / 0.056 / 6.80	12702 00−60−019 2
U12709	2334.9 8 / 87.07 −57.22 / 291.46 18.76	9 / 0.35 / 64−10	2.9 / 2.7 / 2	0.64 / 55.	14.01	7 / 0.17 / 0.12	2677. / 2839. / 1	155 / 152. / 155.	10	0.30 / 32 / 1.07	35.1 / 32.73	12.2 / −30.9 / 11.3	−18.72 / 27.7	9.68 / 9.39 / 10.27	0.51 / 0.133 / 3.87	DDO 219 / 12709 00−60−022 2
U12710	2335.0 1743 / 99.22 −41.45 / 309.78 22.78	10 P / 0.09 / 64 −0	1.6 / 1.5 / 2	0.64 / 55.	14.2	3 / 0.17 / 0.08	2521. / 2743. / 1	155 / 152. / 155.	15	0.47 / 21 / 1.02	34.2 / 32.67	20.2 / −24.2 / 13.2	−18.47 / 15.0	9.58 / 9.54 / 10.00	0.91 / 0.344 / 2.64	12710 03−60−015
2335−48	2335.1 −4800 / 332.97 −64.88 / 245.90 0.81	5 / 0.08 / 60 −0	5.4 / 3.8 / 6	0.10 / 90.	9.25	0.67 / 0.05	2844. / 2786. / 7	554 / 516. / 517.	20	—	34.9 / 32.72	−14.3 / −31.9 / 0.5	38.7	11.48	—	240−11
2335+29	2335.2 2951 / 104.30 −30.09 / 322.92 24.29	13 P / 0.06 / 65 −0	0.5 / 0.5 / 2	0.82 / 39.	14.9	3 / 0.00 / 0.21	1308. / 1559.	—	70	—	19.6 / 31.46	14.2 / −10.8 / 8.1	−16.56 / 2.9	8.82	—	UGCA 441, MARK 328 / A2335, ZW COMPAC / 05−55−000 2
U12713	2335.7 3026 / 104.84 −29.57 / 323.57 24.23	0.01 / 65 +4	1.3 / 1.2 / 2	0.54 / 62.	14.51	2 / 0.00 / 0.24	289. / 541. / 1	115	30	0.37 / 22 / 1.02	6.9 / 29.19	5.0 / −3.7 / 2.8	−14.68 / 2.4	8.06 / 8.05	0.96	B2 2335+30 / 12713 05−55−044 2
N 7721	2336.2 −648 / 79.69 −63.12 / 284.63 16.36	5A / 0.08 / 63 +5	2.9 / 2.4 / 4	0.38 / 73.	11.76 / 9.69 / 2.07	2 / 0.46 / 0.07	2015. / 2148. / 1	325 / 299. / 301.	20	1.13 / 13 / 1.06	26.1 / 32.08	6.3 / −24.2 / 7.4	−20.32 / 18.3	10.32 / 9.96 / 10.68	0.44 / 0.194 / 2.27	−1−60−017 2
N 7723	2336.4 −1314 / 69.26 −67.91 / 278.35 14.18	3B / 0.13 / 63 −6	3.9 / 3.6 / 4	0.61 / 57.	11.59	2 / 0.19 / 0.07	1860. / 1966.	—	38	—	23.7 / 31.88	3.3 / −22.8 / 5.8	−20.29 / 24.9	10.31	—	−2−60−005 2
N 7727	2337.3 −1234 / 70.94 −67.61 / 279.08 14.20	1XP / 0.13 / 63 −6	3.3 / 3.4 / 4	1.00	11.46	2 / 0.00 / 0.09	1821. / 1929.	—	27	—	23.3 / 31.84	3.6 / −22.3 / 5.7	−20.38 / 23.1	10.34	—	ARP 222, VV 67 / −2−60−008 2
U12732	2338.1 2557 / 103.72 −33.99 / 318.75 23.26	9 / 0.15 / 65 −3	2.9 / 3.0 / 2	1.00	12.7	3 / 0.00 / 0.14	756. / 998. / 1	140	8	1.19 / 226 3 / 1.10	12.4 / 30.46	8.5 / −7.5 / 4.9	−17.76 / 10.9	9.30 / 9.38	1.20	12732 04−55−042
N 7731	2339.0 328 / 91.60 −54.82 / 295.14 18.68	1B / 0.26 / 64+10	1.6 / 1.5 / 2	0.70 / 50.	13.4	3 / 0.13 / 0.17	2799. / 2972.	—	88	—	36.8 / 32.83	14.8 / −31.6 / 11.8	−19.43 / 16.1	9.96	—	TURN 143A, ZW COMPACT / 12737 00−60−034 2
N 7741	2341.4 2548 / 104.51 −34.37 / 318.69 22.51	6B / 0.14 / 65 −3	4.1 / 3.8 / 2	0.69 / 51.	11.69	1 / 0.13 / 0.12	750. / 990. / 4	206 / 220. / 222.	10 / 7 / 79	1.10 / 217 7 / 1.04	12.3 / 30.45	8.6 / −7.5 / 4.7	−18.76 / 13.6	9.70 / 9.28 / 10.28	0.38 / 0.099 / 3.86	12754 04−55−050 2

(1) Name	(2) α / ℓ / SGL	(3) δ / b / SGB	(4) Type / ρ_{xyz} / Group	(5) D_{25} / $D_{25}^{b,i}$ / source	(6) d/D / i	(7) $B_T^{b,i}$ / $H_{-0.5}$ / B–H	(8) source / A_B^{i-o} / A_B^b	(9) V_h / V_o / Tel	(10) W_{20} / W_R / W_D	(11) e_v / e_w / source	(12) F_c / source,e_f / f_h	(13) R / μ	(14) SGX / SGY / SGZ	(15) $M_B^{b,i}$ / $\Delta_{25}^{b,i}$	(16) $\log L_B$ / $\log M_H$ / $\log M_T$	(17) M_H/L_B / M_H/M_T / M_T/L_B	(18) UGC/ESO	MCG	AGN? RCII
2341-32	2341.4 / 10.60 / 260.82	-3215 / -74.58 / 6.08	9 / 0.17 / 14-13	3.9 / 2.8 / 6	0.20 / 90.		/ 0.67 / 0.05	267. / 282. / 2	112 / 88. / 95.	8	1.06 / 8 / 1.02	3.3 / 27.60	-0.5 / -3.2 / -0.4	/ 2.7	8.10 / 8.78	0.206	UGCA 442 / 471-06	-5-56-001	2
N 7742	2341.7 / 97.42 / 302.52	1029 / -48.75 / 19.75	3A / 0.10 / 64 -6	2.0 / 2.0 / 2	0.91 / 28.	12.08	2 / 0.03 / 0.14	1622. / 1818.		30		22.2 / 31.73	11.2 / -17.6 / 7.5	-19.65 / 13.0	10.05		12760	02-60-010	2
N 7743	2341.8 / 96.96 / 301.67	939 / -49.53 / 19.54	-2B / 0.13 / 64 -6	2.9 / 2.9 / 9	0.83 / 38.	12.02	2 / 0.00 / 0.18	1802. / 1995.		65		24.4 / 31.93	12.0 / -19.5 / 8.1	-19.91 / 20.7	10.16		12759	02-60-011	2
N 7744	2342.4 / 339.11 / 250.80	-4312 / -69.24 / 1.60	-2X / 0.08 / 60 -0	2.5 / 2.4 / 6	0.80 / 41.	12.30	6 / 0.00 / 0.05	2990. / 2952.		90		36.9 / 32.83	-12.1 / -34.8 / 1.0	-20.53 / 25.9	10.40		IC 5348 / 292-17	-7-48-017	2
2345-57	2345.1 / 319.27 / 237.94	-5721 / -57.99 / -4.33	7 / 0.10 / 55 -0	3.9 / 2.7 / 6	0.20 / 90.		/ 0.67 / 0.00	1905. / 1801. / 7		20		22.7 / 31.78	-12.0 / -19.2 / -1.7	17.9			149-01		
N 7755	2345.4 / 15.38 / 262.48	-3049 / -75.71 / 5.84	5B / 0.06 / 60 -0	3.5 / 3.4 / 6	0.78 / 43.	11.6	5 / 0.09 / 0.05	2968. / 2988. / 2	299 / 384. / 385.	15	0.87 / 31 / 1.03	37.0 / 32.84	-4.8 / -36.5 / 3.8	-21.24 / 36.7	10.69 / 10.01 / 11.20	0.21 / 0.065 / 3.22	UGCA 443 / 471-20	-5-56-014	2
U12791	2346.3 / 105.84 / 319.00	2557 / -34.56 / 21.43	10 / 0.13 / 65 -3	1.8 / 1.5 / 2	0.39 / 73.	15.06	7 / 0.45 / 0.14	799. / 1037. / 1	118 / 98. / 103.	20	0.36 / 22 / 1.02	12.9 / 30.55	9.0 / -7.9 / 4.7	-15.49 / 5.7	8.39 / 8.58 / 9.20	1.56 / 0.242 / 6.44	DDO 220 / 12791	04-56-011	2
2346-38	2346.8 / 349.83 / 255.88	-3803 / -73.17 / 2.83	10 / 0.19 / 19 -7	18.5 / 15.0 / 6	0.36 / 75.		/ 0.52 / 0.05	657. / 642.		10		7.7 / 29.43	-1.9 / -7.4 / 0.4	33.7			348-09		
N 7764	2348.4 / 341.55 / 253.25	-4101 / -71.55 / 1.42	10 / 0.18 / 61 -0+16	2.1 / 2.0 / 6	0.80 / 41.	12.57	2 / 0.08 / 0.05	1706. / 1676. / 2	190	60	0.42 / 41 / 1.01	20.5 / 31.56	-5.9 / -19.7 / 0.5	-18.99 / 12.0	9.79 / 9.04	0.18	293-04	-7-48-027	2
2349-52	2349.4 / 322.48 / 242.35	-5251 / -62.24 / -3.20	10 / 0.01 / 19 -0 +7	2.4 / 1.7 / 6	0.17 / 90.		/ 0.67 / 0.05	577. / 492. / 7		20		6.1 / 28.93	-2.8 / -5.4 / -0.3	3.0			149-03		
U12843	2353.0 / 104.64 / 310.55	1739 / -42.95 / 18.55	8X / 0.25 / 64 -7	2.9 / 2.5 / 2	0.47 / 67.	13.1	3 / 0.33 / 0.09	1777. / 1989. / 1	208 / 186. / 189.	20	0.71 / 11 / 1.06	24.5 / 31.95	15.1 / -17.7 / 7.8	-18.85 / 17.9	9.73 / 9.49 / 10.25	0.57 / 0.172 / 3.33	12843	03-30-039	
U12846	2353.3 / 104.92 / 311.06	1808 / -42.51 / 18.58	9 / 0.26 / 64 -7	2.0 / 2.0 / 2	0.91 / 28.		/ 0.03 / 0.07	1744. / 1958. / 1	110	10	0.22 / 226 15 / 1.04	24.1 / 31.91	15.0 / -17.3 / 7.7	14.1	8.98		12846	03-01-002	
2353-81	2353.7 / 305.38 / 214.87	-8151 / -35.30 / -13.40	5 / 0.16 / 55 -5	3.5 / 3.2 / 6	0.50 / 65.		/ 0.29 / 0.38	1925. / 1725. / 7		20		23.0 / 31.80	-18.3 / -12.8 / -5.3	21.5			012-10		

Name													Identification
U12856	2354.2 1633 104.57 −44.09 309.47 18.06	10B 0.27 64 −7	2.5 0.38 2.1 74. 2	13.3 3 0.47 0.07	1781. 1990. 1	182 154. 158.	25	0.61 17 1.04	24.5 31.95	14.8 −18.0 7.6	−18.65 15.0	9.65 9.39 10.01	0.54 0.237 2.30	ARP 262, VV 255 12856 03-01-004
N 7793	2355.3 −3251 4.59 −77.19 261.36 3.14	8A 0.25 14−13	9.2 0.70 8.6 50. 9	9.52 2 7.87 0.13 1.65 0.05	231. 236. 4	194 208. 211.	7	1.81 36 5 1.13	2.8 27.21	−0.4 −2.7 0.2	−17.69 7.0	9.27 8.70 9.95	0.27 0.057 4.76	349-12 −6-01-009 2
N 7798	2356.9 2029 106.88 −40.48 313.66 18.17	12 0.13 64 −8	1.5 0.94 1.5 23. 2	12.6 3 0.02 0.12	2403. 2621. 1	92	15	0.03 37 1.02	32.6 32.57	21.4 −22.4 10.2	−19.97 14.3	10.18 9.06	0.08	MARK 332 12884 03-01-010 2
N 7800	2357.0 1432 104.66 −46.21 307.54 16.99	13 0.27 64 +7	2.5 0.72 2.4 48. 2	12.8 3 0.00 0.07	1748. 1949. 1	232 262. 264.	15	0.93 12 1.06	23.9 31.89	13.9 −18.1 7.0	−19.09 16.7	9.83 9.69 10.52	0.72 0.146 4.95	12885 02-01-007 2
2359-15	2359.2 −1545 75.70 −73.60 277.82 8.14	10 0.45 14+12	10.2 0.36 8.3 75. 9	10.82 1 0.51 0.07	−124. −42. 1		20	1.88 5 1.56	1.0 25.00	0.1 −1.0 0.1	−14.18 2.4	7.86 7.88	1.04	UGCA 444, DDO 221, WLM −3-01-015 2

Table II
Reordering of the Catalog by Group Affiliations

The 2367 nearby galaxies are now listed by group affiliation. The listing specifies the membership of the 36 clouds, the 254 associations, and the 336 groups of two or more members. There are 31 galaxies that are not identified with any of these units. For details regarding the way group memberships were determined, see ref. (26).

Column 1: Group affiliation   (column 4c in Table I)
Column 2: Galaxy name   (column 1a in Table I)
Column 3: Morphological type   (column 4a in Table I)
Column 4: Absolute blue magnitude   (column 15a in Table I)
Column 5: Systemic velocity   (column 9b in Table I)

Note: The absolute magnitude of our Galaxy, in Group 14-12, is inferred from our maximum rotation velocity.

Group ID	Name	Type	$M_B^{b,i}$	V_o	Group ID	Name	Type	$M_B^{b,i}$	V_o
Virgo Cluster and Southern Extension					11 -1	N 4472	-5	-21.82	847.
					11 -1	N 4473	-5	-20.19	2205.
11 -1 +1	1216+14	13 P		-314.	11 -1	N 4474	-2	-18.61	1455.
11 -1	1217+12	13 P		187.	11 -1	N 4476	-2A	-18.06	1899.
11 -1	1220+12	13 P		-40.	11 -1	N 4477	-2B	-19.87	1190.
11 -1	1223+15	13 P		392.	11 -1	N 4478	-5	-19.08	1403.
11 -1	1228+12	-5 P	-16.88	1408.	11 -1	N 4479	-2B	-17.77	749.
11 -1	1228+12	13 P		1171.	11 -1	N 4483	-2B	-17.93	903.
11 -1	1230+09	-5	-16.13	1200.	11 -1	N 4486	-5 P	-21.64	1181.
11 -1	1230+11	3XP	-15.48	27.	11 -1	N 4492	1A	-17.93	1639.
11 -1	1230+13	13 P	-16.33	300.	11 -1	N 4498	7B	-18.93	1448.
11 -1	N 4124	-2A	-19.13	1551.	11 -1	N 4501	3A	-21.23	2217.
11 -1	N 4168	-5	-18.93	2342.	11 -1	N 4503	-2B	-19.12	1335.
11 -1	N 4178	8B	-19.77	285.	11 -1	N 4519	7B	-18.91	1136.
11 -1	N 4189	6X	-18.72	2039.	11 -1	N 4522	4	-18.83	2241.
11 -1	N 4192	2X	-21.08	-220.	11 -1	N 4523	9X	-16.80	198.
11 -1	N 4206	4A	-19.05	616.	11 -1	N 4526	-2X	-20.55	354.
11 -1	N 4212	4A	-19.56	-163.	11 -1	N 4535	5X	-20.60	1873.
11 -1	N 4216	3X	-20.96	55.	11 -1	N 4540	6X	-18.93	1224.
11 -1	N 4237	5X	-18.87	871.	11 -1	N 4548	3B	-20.28	406.
11 -1	N 4254	5A	-20.84	2323.	11 -1	N 4550	-2	-18.82	274.
11 -1	N 4262	-2B	-18.81	1290.	11 -1	N 4551	-5	-18.37	903.
11 -1	N 4267	-2B	-19.41	1177.	11 -1	N 4552	-5	-20.40	165.
11 -1	N 4294	6B	-19.00	327.	11 -1	N 4564	-5	-19.30	942.
11 -1	N 4298	5A	-19.34	1044.	11 -1	N 4567	4A	-19.20	2139.
11/-1	N 4299	7B	-18.30	150.	11 -1	N 4568	4A	-19.81	2168.
11 -1	N 4302	5	-19.34	1044.	11 -1	N 4569	2X	-21.27	-383.
11 -1	N 4307	1	-18.63	1207.	11 -1	N 4570	-2	-19.43	1634.
11 -1	N 4318	-5	-17.13	-300.	11 -1	N 4571	6A	-19.38	282.
11 -1	N 4321	4X	-21.13	1522.	11 -1	N 4578	-2A	-18.86	2196.
11 -1	N 4340	-2B	-19.17	790.	11 -1	N 4579	3X	-20.67	1729.
11 -1	N 4342	-2	-17.59	609.	11 -1	N 4595	5	-18.43	601.
11 -1	N 4350	-2A	-19.28	1120.	11 -1	N 4596	-2B	-19.67	1939.
11 -1	N 4351	2A	-18.46	2214.	11 -1	N 4608	-2B	-19.04	1789.
11 -1	N 4365	-5	-20.52	1073.	11 -1	N 4612	-2X	-18.23	1740.
11 -1	N 4371	-2B	-19.30	897.	11 -1	N 4621	-5	-20.45	340.
11 -1	N 4374	-5	-20.95	854.	11 -1	N 4623	-2	-17.83	1873.
11 -1	N 4377	-2A	-18.55	1265.	11 -1	N 4638	-2	-19.19	1006.
11 -1	N 4379	-5	-18.67	971.	11 -1	N 4639	3X	-19.10	917.
11 -1	N 4380	2A	-18.93	872.	11 -1	N 4647	5X	-19.34	1285.
11 -1	N 4382	-2AP	-21.11	717.	11 -1	N 4649	-5	-21.36	1127.
11 -1	N 4383	-2 P	-18.63	1545.	11 -1	N 4651	5A	-20.07	748.
11 -1	N 4387	-5	-18.31	432.	11 -1	N 4654	6X	-20.31	978.
11 -1	N 4388	3 P	-19.93	2527.	11 -1	N 4660	-5	-19.30	943.
11 -1	N 4394	3B	-19.49	889.	11 -1	N 4689	5A	-19.62	1715.
11 -1	N 4406	-5	-21.15	-419.	11 -1	N 4694	-2BP	-18.93	1121.
11 -1	N 4411	7X	-17.59	1186.	11 -1	N 4698	2A	-20.08	905.
11 -1	N 4413	3B	-18.33	14.	11 -1	N 4710	-2A	-19.36	1076.
11 -1	N 4417	-2B	-19.13	733.	11 -1	N 4754	-2B	-19.74	1393.
11 -1	N 4419	1B	-19.72	-342.	11 -1	N 4762	-2B	-20.04	874.
11 -1	N 4421	1B	-18.82	1625.	11 -1	U 7209	4A	-18.03	2144.
11 -1	N 4424	2BP	-19.11	366.	11 -1	U 7249	10	-16.63	541.
11 -1	N 4425	-2	-18.45	1804.	11 -1	U 7279	9	-17.03	1883.
11 -1	N 4429	-2A	-20.08	1028.	11 -1	U 7307	10		1091.
11 -1	N 4435	-2B	-19.51	792.	11 -1	U 7352	8	-17.23	2363.
11 -1	N 4438	0AP	-20.36	182.	11 -1	U 7470	9BP	-17.53	-517.
11 -1	N 4442	-2B	-19.84	489.	11 -1	U 7513	5	-17.63	898.
11 -1	N 4450	2A	-20.36	1899.	11 -1	U 7546	7A	-18.11	1170.
11 -1	N 4451	-2A	-17.83	600.	11 -1	U 7547	10	-16.23	1016.
11 -1	N 4452	-2	-18.03	130.	11 -1	U 7557	9	-17.93	833.
11 -1	N 4458	-5	-18.34	308.	11 -1	U 7563	10	-17.23	2271.
11 -1	N 4459	-2A	-19.86	1039.	11 -1	U 7666	10	-17.03	2489.
11 -1	N 4461	-2B	-19.21	1811.	11 -1	U 7737	10		598.
11 -1	N 4464	-5	-17.53	1102.	11 -1	U 7781	9	-17.93	978.
11 -1	N 4467	-5	-15.73	1377.	11 -1	U 7784	10B	-17.53	1055.
11 -1	N 4469	0B	-18.73	404.	11 -1	U 7822	10	-16.99	1995.

Group ID	Name	Type	$M_B^{b,i}$	V_0
11 +2 +1	1221+04	-2	-16.87	829.
11 +2 +1	N 4457	0X	-19.55	752.
11 +2 +1	N 4586	1A	-19.17	722.
11 +2 +1	N 4630	9B	-18.16	587.
11 +2 +1	N 4636	-5	-20.68	869.
11 +2 +1	N 4665	0B	-19.36	678.
11 +2 +1	N 4688	6B	-18.26	891.
11 +2 +1	N 4701	6A	-18.83	624.
11 +2 +1	N 4713	7X	-19.21	560.
11 +2 +1	N 4765	1	-18.83	687.
11 +2 +1	N 4772	1A	-18.77	982.
11 +2 +1	N 4808	6A	-19.40	668.
11 +2 +1	N 4809	10 P	-17.63	843.
11 +2 +1	N 4810	10 P	-17.37	779.
11 +2 +1	N 4900	5B	-19.09	870.
11 +2 +1	U 7943	5	-17.59	743.
11 +2 +1	U 8053	8X	-17.43	621.
11 +2 +1	U 8074	9	-17.29	811.
11 -3 +1	N 4532	10B	-19.09	1909.
11 -3 +1	U 7739	10	-16.20	1931.
11 -4 +1	1229+04	10		1659.
11 -4 +1	N 4420	5	-18.22	1557.
11 -4 +1	N 4496	9B	-19.03	1625.
11 -4 +1	N 4527	4X	-19.72	1614.
11 -4 +1	N 4536	4X	-20.02	1748.
11 -4 +1	U 7512	10	-16.09	1379.
11 -4 +1	U 7612	9B	-16.82	1452.
11 +4 +1	N 4303	4X	-20.71	1464.
11 +4 +1	U 7354	-5	-16.52	1461.
11 +4 +1	U 7685	8B	-18.34	1405.
11 -5 +1	N 4339	-5	-19.68	1169.
11 -5 +1	N 4423	9	-18.21	984.
11 +5 +1	N 4061	-5	-16.62	1545.
11 +6 +1	N 4643	0B	-20.50	1233.
11 +6 +1	U 7911	9B	-18.16	1066.
11 +7 +1	N 4064	1BP	-19.32	959.
11 +7 +1	N 4065	-5	-17.52	1122.
11 -8 +1	N 4632	5A	-18.76	1572.
11 -8 +1	N 4666	5X	-19.77	1395.
11 +9 +1	N 4753	13	-20.05	1102.
11 +9 +1	U 8041	7B	-18.46	1218.
11 -0 +1	N 4037	5BP	-18.78	844.
11 -0 +1	N 4293	0B	-20.04	824.
11 -0 +1	N 4561	8B	-17.76	1430.
11 -0 +1	N 4580	4X	-19.44	1187.
11 -0 +1	N 4635	7A	-18.66	944.
11 -0 +1	N 4758	3BP	-18.15	1196.
11 -0 +1	N 4845	1A	-19.56	1124.
11 -0 +1	N 4866	-2A	-19.06	1912.
11 -0 +1	N 4880	-2A	-18.68	1499.
11 -0 +1	U 7239	10	-18.18	1126.
11 -0 +1	U 7332	10	-17.33	805.
11 -0 +1	U 7477	13 P	-17.59	657.
11 -0 +1	U 8011	10		747.
11 -0 +1	U 8036	5B	-19.21	893.
11 -0 +1	U 8285	9	-17.01	825.
11 -0 +1	U 8385	9X	-18.00	1079.

Group ID	Name	Type	$M_B^{b,i}$	V_0
11-10+10	1242-08	10		1225.
11-10	1246-09	9B	-19.93	1164.
11-10	1252-10	10		1171.
11-10	1255-09	5		1341.
11-10	N 4699	3X	-21.74	1358.
11-10	N 4700	7B	-20.30	1248.
11-10	N 4742	-5	-19.81	1168.
11-10	N 4781	7B	-20.56	1113.
11-10	N 4790	5B	-19.57	1204.
11-11+10	1241-05	9	-19.33	1287.
11-11+10	1243-05	10		1337.
11-11+10	1246-04	10		1206.
11-11+10	1250-04	10		1278.
11-11+10	1250-06	10		1398.
11-11+10	N 4697	-5	-21.67	1169.
11-11+10	N 4731	6B	-21.17	1358.
11-11+10	N 4775	7A	-20.32	1426.
11-12+10	1302-073	9B		988.
11-12+10	N 4951	6X	-20.13	1048.
11-12+10	N 4958	-2B	-20.30	984.
11-13+10	1256-11	10		1153.
11-13+10	1257-12	10B		1428.
11-14+10	1233-07	9		838.
11-14+10	1251-11	10		667.
11-14+10	N 4487	6X	-20.19	878.
11-14+10	N 4504	6A	-19.94	847.
11-14+10	N 4594	1AP	-22.98	962.
11-15+10	1258-15	10		1220.
11-15+10	N 4856	0B	-20.27	1087.
11+16+10	1230-04	8		1151.
11+16+10	N 4546	-2B	-20.32	882.
11-17+10	1302-075	7B	-18.26	1417.
11-17+10	1309-06	8B		1359.
11-17+10	N 4942	6A		1617.
11-17+10	N 4981	4X	-20.42	1549.
11-17+10	N 4995	3X	-20.51	1577.
11 -0+10	1254-03	10		1411.
11 -0+10	1301-03	9		1249.
11 -0+10	1303-17	10		1296.
11 -0+10	1307-10	5		1073.
11 -0+10	1316-08	10	-18.01	1183.
11 -0+10	1335-09	9	-19.13	1184.
11 -0+10	N 4597	9B	-19.15	904.
11 -0+10	N 4684	-2B	-19.59	1461.
11 -0+10	N 4691	3BP	-20.12	989.
11 -0+10	N 4818	2X	-20.66	1012.
11 -0+10	N 4984	1X	-20.04	1101.
11 -0+10	N 5088	5A	-19.74	1324.
11-18+18	N 5170	3A	-21.00	1347.
11-18	N 5247	5A	-20.95	1216.
11-19+19	1312-22	10		1330.
11-19	N 5042	5X		1210.

Group ID	Name	Type	$M_B^{b,i}$	V_0
11-20+20	N 5037	1A	-20.21	1730.
11-20	N 5054	3A	-21.18	1587.
11-21+21	1322-19	6X		1815.
11-21	1323-21	10		1273.
11-21	N 5084	-2	-20.41	1558.
11-21	N 5087	-2A	-20.51	1666.
11-21	N 5134	1A	-20.09	1532.
11-22+22	1321-24	10		1880.
11-22	1325-27	10		1656.
11-22	N 5061	-5	-21.19	1775.
11-22	N 5085	5A	-20.71	1780.
11-22	N 5101	0B	-20.95	1679.
11-23+22	1304-28	7B		2019.
11-23+22	N 4965	9A		2068.
11 0+22	1324-29	10 P		1705.
11-24+24	1212+06	-5 P	-17.73	1844.
11-24	N 4215	-2A	-19.63	1980.
11-24	N 4224	1A	-20.08	2543.
11-24	N 4233	-2	-19.65	2117.
11-24	N 4235	1A	-20.75	2487.
11-24	N 4260	1B	-20.29	1734.
11-24	N 4261	-5	-21.41	2089.
11-24	N 4270	-2	-19.63	2233.
11-24	N 4273	5B	-20.58	2234.
11-24	N 4281	-2	-20.52	2488.
11-24	N 4324	-2A	-20.13	1582.
11-24	N 4378	1A	-20.53	2447.
11+24	N 3976	5X	-20.98	2381.
11+24	N 4234	10	-19.12	1952.
11+24	N 4412	4BP	-19.75	2185.
11-25+24	1221+00	10		1928.
11-25+24	N 4385	1B	-19.78	2011.
11+26+24	1229-02	13 P		2100.
11+26+24	N 4348	3	-20.74	2055.
11+26+24	N 4454	1B	-19.87	2220.
11-27+27	1247-10	8		2251.
11-27	1249-09	6		2108.
11-27	N 4658	5B	-20.86	2250.
11-28+28	1334-11	7X		2349.
11-28	N 5254	5A		2194.
11-29+29	N 4593	3B	-21.58	2560.
11-29	N 4602	4X	-21.39	2417.
11-30+30	N 4891	3 P		2389.
11-30	N 4899	5X	-20.82	2497.
11-30	N 4902	3B	-21.14	2563.

Group ID	Name	Type	$M_B^{b,i}$	V_0
11-31+31	1310-15	8A		2348.
11-31	N 5017	-5	-19.83	2384.
11-31	N 5044	-5	-21.01	2548.
11-31	N 5049	-2	-19.49	2588.
11-32+32	N 5161	6X	-21.12	2196.
11-32	N 5188	1BP	-20.49	2169.
11+32	1310-32	8B		2178.
11-33+33	1305-22	9		2434.
11-33	1306-24	10		2687.
11-33	1307-23	10 P		2783.
11-33	1308-23	5B		2672.
11-34+34	N 4428	5X	-20.78	2874.
11-34	N 4433	4X	-20.81	2755.
11-35+35	1355-29	3A	-21.46	2496.
11-35	1358-30	7X		2428.
11-35	N 5464	10B	-20.07	2528.
11+36+36	U 8084	8B	-18.70	2670.
11+36	U 8153	6	-19.23	2779.
11+36	U 8263	5B	-18.94	2949.
11 -0	1258-04	10		2855.
11 -0	1309-11	6B		1963.
11 -0	1402-24	6B		2191.
11 -0	1427-34	7B	-20.15	2858.
11 -0	N 4653	6X	-20.19	2506.
11 -0	N 4939	4A	-22.43	2973.
11 -0	N 4947	3B	-20.71	2194.
11 -0	N 5018	-5	-21.46	2729.
11 -0	N 5077	-5	-20.52	2682.
11 -0	N 5147	8B	-19.45	1020.

Ursa Major Cloud

Group ID	Name	Type	$M_B^{b,i}$	V_0
12 -1 +1	1135+48	9BP	-17.15	744.
12 -1	1156+46	10		1210.
12 -1	N 3718	1BP	-20.19	1068.
12 -1	N 3726	5X	-20.35	914.
12 -1	N 3729	1BP	-19.25	1117.
12 -1	N 3769	3B	-19.95	773.
12 -1	N 3782	6XP	-18.25	793.
12 -1	N 3870	-2		822.
12 -1	N 3877	5A	-20.25	954.
12 -1	N 3893	5X	-20.27	1043.
12 -1	N 3913	7A	-18.45	1053.
12 -1	N 3917	6A	-19.55	1056.
12 -1	N 3924	9	-16.55	1077.
12 -1	N 3928	13 P	-18.15	1057.
12 -1	N 3938	5A	-20.26	852.
12 -1	N 3949	4A	-19.92	868.
12 -1	N 3953	5B	-20.63	1139.
12 -1	N 3972	4A	-19.25	947.

Group ID	Name	Type	$M_B^{b,i}$	V_o
12 -1	N 3982	4X	-19.45	1208.
12 -1	N 3990	-2	-17.75	820.
12 -1	N 3992	4B	-20.75	1142.
12 -1	N 4010	9	-19.15	968.
12 -1	N 4013	4	-19.55	883.
12 -1	N 4026	-2	-19.47	958.
12 -1	N 4051	4X	-20.25	763.
12 -1	N 4085	5X	-18.84	794.
12 -1	N 4088	5XP	-20.47	844.
12 -1	N 4100	5A	-20.15	1153.
12 -1	N 4102	3X	-19.02	953.
12 -1	N 4111	-2A	-19.65	842.
12 -1	N 4138	-2A	-18.85	1091.
12 -1	N 4142	6B	-18.23	1257.
12 -1	N 4143	-2X	-19.05	831.
12 -1	N 4157	3X	-20.15	854.
12 -1	N 4183	6A	-18.85	987.
12 -1	N 4217	3	-19.85	1098.
12 -1	N 4218	-2	-18.82	1462.
12 -1	N 4220	-2A	-18.95	1052.
12 -1	N 4346	-2B	-18.95	922.
12 -1	N 4389	6BP	-18.75	786.
12 -1	U 6399	9	-17.75	873.
12 -1	U 6446	7A	-18.45	725.
12 -1	U 6628	9A	-18.25	901.
12 -1	U 6667	5	-17.75	1051.
12 -1	U 6713	9	-17.05	966.
12 -1	U 6816	10B	-17.03	998.
12 -1	U 6818	13 P	-17.25	857.
12 -1	U 6840	9B	-16.90	1102.
12 -1	U 6917	9B	-18.65	996.
12 -1	U 6923	10	-17.55	1172.
12 -1	U 6930	7X	-18.45	851.
12 -1	U 6956	9B		995.
12 -1	U 6962	6X	-18.39	827.
12 -1	U 6973	2 P	-18.62	756.
12 -1	U 6983	6B	-18.35	1166.
12 -1	U 7089	9	-18.05	826.
12 -1	U 7218	10 P	-16.75	882.
12 -2 +1	N 3631	5A	-20.69	1239.
12 -2 +1	N 3657	5XP	-19.55	1293.
12 -2 +1	U 6251	9X	-16.78	1006.
12 -3 +1	N 3733	5	-19.72	1278.
12 -3 +1	N 3756	5X	-19.96	1382.
12 -3 +1	N 3804	5A	-18.96	1479.
12 -3 +1	N 3850	5B	-18.19	1264.
12 -3 +1	N 3898	2A	-20.10	1272.
12 -3 +1	N 3998	-2A	-20.12	1238.
12 -3 +1	U 6616	7A	-18.37	1261.
12 -3 +1	U 6706	9 P	-18.71	1525.
12 -3 +1	U 6912	13 P	-17.73	1468.
12 -4 +1	U 6448	13 P	-17.30	1121.
12 -4 +1	U 6534	6	-19.41	1400.
12 -5 +1	N 3945	-2B	-20.30	1341.
12 -5 +1	N 4036	-2	-20.56	1510.
12 -5 +1	N 4041	4A	-20.08	1360.
12 -5 +1	U 6682	9		1439.
12 -5 +1	U 7009	10	-17.93	1250.
12 -5 +1	U 7056	6B	-18.33	1406.
12 +5 +1	N 4121	-5	-17.74	1600.
12 +5 +1	N 4125	-5	-21.35	1495.
12 +5 +1	U 7020	-2	-17.92	1586.

Group ID	Name	Type	$M_B^{b,i}$	V_o
12 -6 +1	1204+40	10		914.
12 -6 +1	N 4145	7X	-20.21	1048.
12 -6 +1	N 4151	2X	-20.46	1022.
12 -6 +1	N 4156	3B	-17.02	759.
12 -6 +1	N 4369	1A	-19.87	1091.
12 -6 +1	U 7125	9	-18.71	1098.
12 -6 +1	U 7175	9	-17.50	1202.
12 -6 +1	U 7207	10		1077.
12 -6 +1	U 7257	9	-17.62	967.
12 +6 +1	N 3930	5X	-19.03	936.
12 +6 +1	N 3941	-2B	-19.98	940.
12 +6 +1	U 6955	10	-18.46	938.
12 -0 +1	1222+67	13 P	-15.65	926.
12 -0 +1	1244+48	12		1214.
12 -0 +1	N 3359	5B	-20.59	1131.
12 -0 +1	N 3556	6B	-20.77	785.
12 -0 +1	N 4800	3A	-18.72	831.
12 -0 +1	U 5846	10	-16.10	1122.
12 -0 +1	U 5979	10A	-17.17	1261.
12 -0 +1	U 6447	13 P	-17.77	1425.
12 -0 +1	U 7534	10B	-17.08	845.
12 -0 +1	U 8146	6	-17.56	805.
12 -7 +7	N 2985	2A	-20.65	1431.
12 -7	N 3027	7B	-19.30	1211.
12 -7	U 5455	10		1438.
12 -7	U 5612	8B	-18.93	1161.
12 -8 +7	N 3252	6B	-18.09	1299.
12 -8 +7	N 3403	4A	-19.37	1359.
12 -9 +9	N 3259	4X	-19.59	1867.
12 -9	U 5776	13 P	-17.37	1746.
12-10+10	N 2591	5	-19.70	1509.
12-10	N 2655	0X	-21.02	1624.
12-10	N 2715	5X	-20.08	1304.
12-10	N 2748	4A	-19.88	1562.
12+10	N 3057	8B	-18.71	1717.
12+10	U 4238	7B	-19.43	1714.
12+10	U 4841	7X	-18.60	1289.
12-11+11	0635+75	13 P		972.
12-11	N 2146	2BP	-20.24	1108.
12-11	U 3371	10		1003.
12+11	U 3137	3	-18.27	1193.
12-12+12	N 3610	-5	-20.73	1869.
12-12	N 3613	-5	-20.99	2154.
12-12	N 3619	0A	-19.63	1748.
12-12	N 3642	4A	-20.66	1729.
12-12	N 3669	9	-20.19	2041.
12-12	N 3683	5B	-19.87	1783.
12+12	1130+55	13 P		1815.

Group ID	Name	Type	$M_B^{b,i}$	V_o		Group ID	Name	Type	$M_B^{b,i}$	V_o
12-13+12	N 3445	9X	-19.79	2116.		Ursa Major Southern Spur				
12-13+12	N 3458	-2X	-19.15	1891.						
						13 -1 +1	N 3073	-5 P	-18.03	1132.
12 -0+12	N 3471	1	-19.89	2188.		13 -1	N 3079	9B	-21.01	1200.
12 -0+12	U 6027	10		1824.		13 -1	N 3206	6B	-19.31	1245.
						13 -1	U 5459	5	-18.74	1184.
12-14+14	N 3065	-2A	-19.61	2117.		13 +1	1023+56	13 P		880.
12-14	N 3066	4XP	-19.08	2204.		13 +1	N 3264	9B	-18.16	1023.
						13 +1	N 3310	4XP	-20.09	1064.
12+15+15	U 2953	2AP	-20.91	1092.		13 +1	U 5460	7B	-18.39	1151.
12+15	U 3203	12 P	-18.79	937.		13 +1	U 5720	13 P	-18.48	1519.
12-16+16	N 3780	5A	-21.16	2491.		13 -2 +2	N 3353	13 P	-17.93	949.
12-16	N 3888	4X	-20.17	2507.		13 -2	U 5848	9	-16.86	914.
12-17+17	N 2551	0A	-19.68	2375.		13 -3 +3	U 5998	13 P	-17.49	1320.
12-17	U 4390	7B	-18.52	2326.		13 -3	U 6029	13 P	-18.03	1442.
12+17	N 2633	3B	-19.99	2302.		13 -4 +4	N 2685	-2BP	-19.30	965.
						13 -4	U 4683	10		999.
12-18+18	1124+54	13 P		2979.		13 -5 +5	0917+64	13 P	-17.36	1582.
12-18	N 3656	-5 P	-20.24	2910.		13 -5	N 2805	7X	-20.72	1851.
12+18	N 3549	5A	-21.27	2944.		13 -5	N 2814	13	-18.02	1778.
						13 -5	N 2820	5BP	-19.62	1691.
12-19+19	U 2800	10		1385.		13 -5	N 2880	-2B	-19.60	1621.
12-19	U 2813	10		1611.						
12-19	U 2855	5X	-20.63	1410.		13 -6 +5	N 2654	1	-19.89	1478.
12-20+19	U 2729	0		2140.		13 -6 +5	N 2726	1	-19.50	1526.
12-20+19	U 2765	0		1881.		13 -6 +5	N 2742	5A	-19.83	1393.
						13 -6 +5	N 2768	-5	-21.13	1503.
12+20+19	0335+66	0		1697.		13 -0 +5	0946+55	13 P	-16.69	1559.
12+20+19	U 3144	10B	-18.28	1830.		13 -0 +5	N 2950	-2B	-20.02	1452.
12 -0+19	U 3190	5A	-19.44	1766.		13 -7 +7	1138+35	10 P	-17.75	1510.
12 -0+19	U 3317	10		1424.		13 -7	1143+35	13 P	-18.00	1346.
12 -0+19	U 3439	5	-19.37	1685.		13 -7	N 3755	5XP	-19.40	1575.
						13 -7	N 3813	5A	-19.99	1474.
12-21+21	0848+73	13 P		2556.		13 +7	U 6531	8B	-17.88	1580.
12-21	U 4576	3B	-19.65	2744.						
12+21	N 2544	-2	-20.78	2949.		13 -8 +8	1127+37	12 P	-18.07	2014.
						13 -8	N 3652	5BP	-20.52	2100.
12 -0	0939+76	13 P	-17.99	2526.		13 -8	N 3665	-2A	-20.86	2012.
12 -0	1243+71	13 P	-16.62	1228.						
12 -0	N 2787	-2B	-18.93	759.		13 -9 +9	N 3900	-2A	-20.20	1666.
12 -0	U 3580	1A	-19.97	1356.		13 -9	N 3912	1	-19.59	1724.
12 -0	U 3730	13 P	-18.98	2868.						
12 -0	U 3975	8B	-19.05	2639.		13-10+10	N 3694	13 P	-19.25	2250.
12 -0	U 4173	10	-17.31	1054.		13-10	U 6491	8	-18.64	2523.
12 -0	U 4888	5A	-20.38	2417.		13-11+11	N 4158	0	-20.35	2480.
12 -0	U 5688	9B	-18.88	2063.		13-11	U 7170	5	-18.69	2382.
						13+12+11	1209+16	13 P	-19.10	2400.
						13+12+11	N 4152	5X	-20.35	2086.

Group ID	Name	Type	$M_B^{b,i}$	V_0
13 -0	1008+58	13 P	.	2270.
13 -0	1013+45	13 P	-16.77	1668.
13 -0	1039+48	13 P	-17.30	1585.
13 -0	1046+52	13	-18.40	2410.
13 -0	1149+35	13 P		2168.
13 -0	N 2549	-2A	-19.46	1168.
13 -0	N 3225	5	-19.73	2227.
13 -0	N 3320	6	-20.38	2367.
13 -0	N 3448	13	-19.79	1468.
13 -0	N 3583	3B	-21.05	2188.
13 -0	N 3614	5X	-20.91	2381.
13 -0	N 3687	1X	-20.25	2373.
13 -0	N 3689	5X	20.24	2698.
13 -0	N 3786	1XP	-20.40	2744.
13 -0	N 3788	2XP	-20.11	2322.
13 -0	N 4162	4A	-20.94	2454.
13 -0	N 4275	12	-19.42	2307.
13 -0	U 4121	9B	-17.11	1180.
13 -0	U 5518	10		2109.
13 -0	U 5953	13 P	-18.73	1804.
13 -0	U 6162	6	-19.30	2268.
13 -0	U 6637	13 P	-18.92	2937.

Coma - Sculptor Cloud

Group ID	Name	Type	$M_B^{b,i}$	V_0
14 -1 +1	1242+28	12	-15.38	958.
14 -1	N 4062	5A	-18.27	764.
14 -1	N 4136	5X	-18.23	606.
14 -1	N 4150	-2A	-17.48	235.
14 -1	N 4173	7B	-17.53	1113.
14 -1	N 4203	-2X	-18.38	1094.
14 -1	N 4245	0B	-17.74	881.
14 -1	N 4251	-2B	-18.40	999.
14 -1	N 4274	2B	-19.22	914.
14 -1	N 4278	-5	-18.82	642.
14 -1	N 4283	-5	-17.02	1124.
14 -1	N 4308	-5	-15.83	602.
14 -1	N 4310	-2XP	-17.13	893.
14 -1	N 4314	1B	-18.65	878.
14 -1	N 4359	5B	-17.43	1256.
14 -1	N 4393	7X	-17.53	741.
14 -1	N 4414	5A	-19.12	723.
14 -1	N 4448	2B	-18.43	686.
14 -1	N 4494	-5	-19.23	1289.
14 -1	N 4525	5	-17.53	1134.
14 -1	N 4559	6X	-20.07	810.
14 -1	N 4562	6	-16.53	1377.
14 -1	N 4565	3A	-20.34	1213.
14 -1	U 7300	10		1202.
14 -1	U 7673	10		639.
14 +1	1244+26	-5 P	-15.14	870.
14 +1	N 4020	7	-16.73	742.
14 +1	N 4455	7	-16.45	611.
14 +1	U 6900	10	-14.26	541.
14 +1	U 7007	9		786.
14 -2 +1	N 4725	2XP	-20.65	1199.
14 -2 +1	N 4747	5BP	-17.80	1183.
14 +2 +1	N 4670	0BP	-17.24	1068.
14 +3 +3	N 4826	2A	-19.15	394.
14 +3	U 8024	10	-14.11	380.

Group ID	Name	Type	$M_B^{b,i}$	V_0
14 -4 +4	N 3985	9B	-16.68	600.
14 -4	N 4096	5X	-19.39	644.
14 -4	N 4242	8X	-17.91	580.
14 -4	N 4248	13	-16.13	556.
14 -4	N 4258	4X	-20.59	521.
14 -4	N 4288	7B	-16.43	602.
14 -4	N 4460	-2B	-17.23	624.
14 -4	N 4485	10B	-17.66	724.
14 -4	N 4490	7BP	-19.55	629.
14 -4	N 4618	9B	-18.12	602.
14 -4	N 4625	9XP	-16.69	667.
14 -4	N 4707	9	-16.12	569.
14 -4	U 7267	9	-15.47	561.
14 -4	U 7271	7B	-15.47	599.
14 -4	U 7408	10A	-15.86	530.
14 -4	U 7608	10	-15.72	601.
14 -4	U 7639	10	-15.32	462.
14 -4	U 7690	10 P	-16.37	598.
14 -4	U 7699	5B	-16.76	539.
14 -4	U 7719	8	-15.25	728.
14 -4	U 7774	7	-15.57	574.
14 -4	U 7950	10	-14.76	612.
14 -5 +5	N 5023	5	-17.29	484.
14 -5	N 5055	4A	-20.14	575.
14 -5	N 5194	4AP	-20.75	573.
14 -5	N 5195	13 P	-19.32	658.
14 -5	N 5229	5B	-15.82	476.
14 -5	U 8313	5B	-15.83	699.
14 -5	U 8331	10A	-14.29	358.
14 -5	U 8683	10	-15.74	748.
14 -6 +6	N 4534	8	-17.75	830.
14 -6	N 4631	7B	-20.12	631.
14 -6	N 4656	9BP	-19.21	667.
14 -6	N 4687	13 P	-15.51	724.
14 -6	U 7916	10		640.
14 -7 +7	1223+48	13 P	-13.86	367.
14 -7	1236+33	10		329.
14 -7	N 4144	6X	-16.87	331.
14 -7	N 4163	10A	-13.62	188.
14 -7	N 4190	10	-14.04	255.
14 -7	N 4214	10X	-17.57	309.
14 -7	N 4244	6A	-17.60	276.
14 -7	N 4395	9A	-17.14	332.
14 -7	N 4449	10B	-17.66	262.
14 -7	N 4736	2A	-19.37	368.
14 -7	U 6817	10	-14.34	269.
14 -7	U 7559	10B	-13.15	252.
14 -7	U 7577	10	-14.63	254.
14 -7	U 7599	9	-13.01	312.
14 -7	U 7605	10	-13.03	329.
14 -7	U 7698	10	-14.85	343.
14 -7	U 7866	10	-15.08	396.
14 -7	U 7949	10		371.
14 -7	U 8188	9	-15.62	380.
14 -7	U 8215	10		314.
14 -7	U 8308	10		259.
14 -7	U 8320	10B	-15.45	291.
14 +8 +7	U 8651	10	-13.86	287.
14 +8 +7	U 8760	10	-13.87	273.
14 +8 +7	U 8833	10 P		303.

Group ID	Name	Type	$M_B^{b,i}$	V_0
14 -9 +9	N 5204	9A	-16.79	346.
14 -9	N 5238	8X	-15.16	368.
14 -9	N 5457	6X	-20.45	379.
14 -9	N 5474	6AP	-17.58	423.
14 -9	N 5477	9	-14.76	443.
14 -9	N 5585	7X	-17.96	466.
14 -9	U 8508	10A	-12.73	202.
14 -9	U 8837	10B	-14.82	283.
14 -9	U 9405	10		393.
14-10+10	0818+71	10		262.
14-10	N 2366	10B	-16.69	248.
14-10	N 2403	6X	-19.68	261.
14-10	N 2976	5AP	-16.13	142.
14-10	N 3031	2A	-18.29	96.
14-10	N 3034	13 P	-19.42	352.
14-10	N 3077	13 P	-16.13	148.
14-10	U 4305	10	-17.34	305.
14-10	U 4459	10	-12.33	144.
14-10	U 5139	10	-15.24	289.
14-10	U 5423	10	-15.06	479.
14-10	U 5666	9X	-16.37	177.
14+10	N 4236	8B	-17.27	163.
14+10	U 6456	13 P	-11.45	96.
14+10	U 8201	10	-14.58	201.
14-11+11	0232+59	-5		224.
14-11	0238+59	4		209.
14-11	0355+66	0		262.
14-11	0427+63	0		66.
14-11	0509+62	10	-16.56	264.
14-11	N 1560	7A	-16.38	164.
14-11	N 1569	10B	-16.37	87.
14-11	U 2847	6X	-21.56	229.
14-12+12	GALAXY	4	-20.5 *	0.
14-12	0051-73	10B	-16.10	-30.
14-12	0057-33	-5		162.
14-12	0237-34	-5	-11.01	-51.
14-12	0524-69	9B	-18.19	12.
14-12	1927-17	10		30.
14-12	2044-13	10		11.
14-12	N 147	-5	-14.49	89.
14-12	N 185	-5 P	-14.87	39.
14-12	N 205	-5 P	-15.60	-1.
14-12	N 221	-5	-15.34	35.
14-12	N 224	3A	-20.67	-59.
14-12	N 598	6A	-18.31	3.
14-12	N 6822	10B	-15.57	66.
14-12	U 192	10B	-16.26	-83.
14-12	U 668	10B	-14.30	-125.
14-12	U 9749	-5	-7.60	17.
14-12	U10822	-5	-8.20	-31.
14+12	1008-04	10B	-13.85	114.
14+12	2159-51	10A	-13.69	74.
14+12	2359-15	10	-14.18	-42.
14+12	N 404	-2A	-16.11	178.
14+12	N 3109	10	-16.62	130.
14+12	U 5364	10	-12.51	-25.
14+12	U 5373	10B	-14.27	129.
14+12	U 8091	10	-11.61	165.
14+12	U 9128	10	-12.38	196.
14+12	U12613	10A	-12.63	39.

Group ID	Name	Type	$M_B^{b,i}$	V_0
14-13+13	0005-34	10		198.
14-13	0040-22	10	-14.28	388.
14-13	0047-21	10		331.
14-13	2318-42	10	-25.00 X	-7.
14-13	2323-32	10	-24.97 X	82.
14-13	2341-32	9		282.
14-13	N 55	9B	-18.13	106.
14-13	N 247	7X	-17.98	190.
14-13	N 253	5X	-20.02	260.
14-13	N 300	7A	-16.88	97.
14-13	N 7793	8A	-17.69	236.
14+13	0142-43	10		293.
14+13	N 625	9B	-16.35	317.
14-14+14	0236-61	10		329.
14-14	N 1313	7B	-18.60	254.
14+14	0255-54	7B	-17.64	403.
14+14	0331-50	5		455.
14+14	N 1311	9		389.
14-15+15	1243-33	10		365.
14-15	1324-41	10		311.
14-15	1334-27	10		409.
14-15	1342-41	10		338.
14-15	1346-35	10		141.
14-15	N 4945	6B	-20.65	325.
14-15	N 5102	-2A	-17.59	251.
14-15	N 5128	-2 P	-20.97	323.
14-15	N 5236	5X	-20.31	337.
14-15	N 5253	-3 P	-16.66	233.
14-15	N 5264	10	-15.68	300.
14-15	N 5408	10		318.
14+15	1331-45	10		612.
14-16+15	1424-46	10		204.
14-16+15	1457-48	10		412.
14-16+15	1510-46	5		356.
14+16+15	1616-60	10		440.
14+16+15	1742-64	10B	-16.06	162.
14-17+17	1259-16	10		569.
14-17	1300-17	10 P	-16.52	578.
14+17	N 5068	6X	-18.93	510.
14-18+18	N 4517	6A	-19.58	1004.
14-18	N 4592	8A	-18.50	954.
14+19+19	U 3974	10	-13.21	153.
14+19	U 4115	10A	-13.51	214.
14+20+20	0731-68	9		253.
14+20	1215-79	6		185.
14+20	1409-65	3A	-21.23	214.
14+20	N 2915	10B	-15.25	191.

(handwritten annotation at right of 14-13 rows: "from μ column (13)")

Group ID	Name	Type	$M_B^{b,i}$	V_0	Group ID	Name	Type	$M_B^{b,i}$	V_0
14 -0	N 3738	10 P	-16.11	314.	15 -0 +7	U 5740	9X	-15.87	707.
14 -0	N 4068	10A	-15.33	302.					
14 -0	N 4204	8B	-16.78	810.	15 -9 +9	N 3600	1	-18.41	740.
14 -0	N 4605	5BP	-17.41	279.	15 -9	N 3675	3A	-19.84	805.
14 -0	N 4941	2X	-17.31	593.	15 -9	U 6161	8B	-17.14	793.
14 -0	N 6503	6A	-18.64	305.					
14 -0	N 6946	6X	-20.78	338.	15 -10+10	N 2681	0X	-19.59	761.
14 -0	U 5918	10	-14.06	465.	15 -10	U 4499	8X	-17.54	749.
14 -0	U 6541	10	-14.41	314.	15 -10	U 4514	5B	-17.36	759.
14 -0	U 6782	10		477.					
14 -0	U 7321	6	-14.79	362.	15+10	0930+55	13 P		826.
14 -0	U 9240	10A	-14.52	281.	15+10	N 2841	3A	-20.58	686.
					15+10	U 5151	13 P	-15.78	450.
Leo Spur									
					15-11+10	N 2541	6A	-18.37	596.
15 -1 +1	N 3299	8X	-15.46	465.	15-11+10	N 2552	9A	-17.47	574.
15 -1	N 3351	3B	-19.26	649.					
15 -1	N 3368	2X	-19.62	766.	15+11+10	N 2500	7B	-17.96	572.
15 -1	N 3377	-5	-18.55	595.	15+11+10	N 2537	10BP	-17.63	479.
15 -1	N 3379	-5	-19.39	752.	15+11+10	U 4278	7	-17.45	591.
15 -1	N 3384	-2B	-18.73	641.					
15 -1	N 3412	-2B	-18.14	737.	15 -0+10	U 4148	7	-16.24	747.
15 -1	N 3489	-2X	-17.92	576.	15 -0+10	U 4426	10		396.
15 -1	U 5889	9	-14.40	449.					
					15+12+12	N 2337	10B	-17.38	467.
15 -2 +1	N 3593	0A	-17.01	510.	15+12	U 3817	10		471.
15 -2 +1	N 3623	1X	-19.81	692.					
15 -2 +1	N 3627	3X	-19.66	623.	15 -0	N 2344	5B	-18.57	1002.
15 -2 +1	N 3628	3 P	-19.96	734.	15 -0	N 3621	7A	-19.87	470.
					15 -0	U 3273	9	-18.69	734.
15 -3 +1	1053+06	13 P	-16.18	1050.	15 -0	U 3475	9	-16.87	512.
15 -3 +1	N 3423	6A	-18.68	851.	15 -0	U 3860	10	-14.63	365.
					15 -0	U 3966	10	-15.47	368.
15 -4 +1	1102+29	13 P		592.					
15 -4 +1	N 3486	5X	-18.61	632.	Centaurus Spur				
15 -4 +1	N 3510	9B	-16.83	664.					
					16 -1 +1	1357-45	5B	-19.11	1300.
15 -5 +1	1000-05	7		449.	16 -1	1415-47	7		1087.
15 -5 +1	N 3115	-2	-19.18	475.	16 -1	N 5530	4A	-20.04	1013.
					16 -1	N 5643	5X	-20.97	1019.
15 -0 +1	0948+08	10 P		397.					
15 -0 +1	N 2903	4X	-19.85	451.	16 +1	1420-49	10		1347.
15 -0 +1	N 3239	9BP	-17.33	639.					
15 -0 +1	N 3274	6XP	-15.97	481.	16 -2 +2	1412-43	5B		1691.
15 -0 +1	N 3344	4X	-18.47	513.	16 -2	N 5483	5A	-20.26	1578.
15 -0 +1	N 3365	5	-17.95	810.					
15 -0 +1	N 3495	7	-18.74	980.	16 -3 +2	1418-46	4		1460.
15 -0 +1	N 3521	4X	-19.88	628.	16 -3 +2	1423-45	10		1758.
15 -0 +1	U 5272	10	-15.12	469.					
15 -0 +1	U 5340	10 P	-14.43	441.	16 -0 +2	1428-43	5A	-20.35	1770.
15 -0 +1	U 5764	10	-14.89	543.					
					16 -4 +4	1317-35	8		1248.
15 -6 +6	N 2683	3A	-18.96	373.	16 -4	1324-37	7		1235.
15 -6	U 4787	8	-16.07	509.	16 -4	N 5121	-2A	-19.72	1324.
15 +6	U 4704	8	-16.01	586.	16 -5 +5	1348-33	7B		1206.
					16 -5	N 5398	8BP	-19.01	1053.
15 +7 +7	N 3104	10A	-17.36	617.					
15 +7	N 3184	6X	-19.34	604.					
15 +7	N 3198	5B	-19.62	686.					
15 +7	N 3319	6B	-18.71	754.					
15 +8 +7	N 3432	9B	-18.38	600.					
15 +8 +7	U 5829	10B	-15.94	608.					

Group ID	Name	Type	$M_B^{b,i}$	V_0
16 -0	1455-47	10		878.
16 -0	1505-52	7		1230.
16 -0	1516-41	3B		1191.
16 -0	1525-42	10		1195.
16 -0	N 5556	7X	-19.81	1236.

Triangulum Spur

Group ID	Name	Type	$M_B^{b,i}$	V_0
17 -1 +1	N 891	3A	-19.96	712.
17 -1	N 925	7X	-19.66	710.
17 -1	N 949	7A	-17.85	774.
17 -1	N 959	9	-17.49	767.
17 -1	N 1003	6A	-18.61	794.
17 -1	N 1023	-2B	-20.16	787.
17 -1	N 1058	5A	-17.87	676.
17 -1	U 1807	10		815.
17 -1	U 1865	9	-16.23	742.
17 -1	U 2014	10		737.
17 -1	U 2023	10	-17.22	763.
17 -1	U 2034	10	-17.14	752.
17 -1	U 2259	8B	-16.89	742.
17 +1	N 746	10B	-17.94	915.
17 +1	U 2080	6X	-19.05	1074.
17 -2 +2	N 278	3XP	-19.62	883.
17 -2	U 731	10		875.
17 -3 +3	N 1012	13 P	-18.59	1123.
17 -3	U 2017	10		1157.
17 -3	U 2053	10	-16.25	1173.
17 -4 +4	0140+19	10		648.
17 -4	N 628	5A	-20.32	798.
17 -4	N 660	1BP	-19.58	982.
17 -4	U 891	9X	-15.55	785.
17 -4	U 1176	10		770.
17 -4	U 1195	10 P	-17.67	895.
17 -5 +5	N 672	6B	-18.65	592.
17 -5	U 1249	9B	-17.61	506.
17 +6 +6	N 784	8	-16.96	362.
17 +6	U 1281	8	-16.49	343.
17 -7 +7	0059-07	8X		1161.
17 -7	0102-06	7X	-19.00	1172.
17 -0	0153+05	13 P	-15.56	900.
17 -0	N 428	9X	-19.14	1260.
17 -0	N 1036	13 P	-17.05	889.
17 -0	N 1156	10B	-17.46	477.
17 -0	U 1072	-2	-16.23	900.
17 -0	U 1561	10 P	-15.83	740.
17 -0	U 2082	5	-18.55	834.
17 -0	U 2684	10		417.

Perseus Cloud

Group ID	Name	Type	$M_B^{b,i}$	V_0
18 -1 +1	N 1169	3X	-21.33	2564.
18 -1	U 2459	8		2640.
18 -2 +1	N 1171	5	-20.99	2906.
18 -2 +1	U 2531	8A	-21.17	2872.
18 -0 +1	N 1282	-5	-19.44	2343.
18 -0	U 2463	9X	-18.84	2051.

Pavo - Ara Cloud

Group ID	Name	Type	$M_B^{b,i}$	V_0
19 -1 +1	1823-67	9B	-17.85	593.
19 -1	1847-64	10		870.
19 -1	1850-64	10		657.
19 -1	1908-62	10		823.
19 -1	1911-62	10		772.
19 -1	2005-62	3		703.
19 -1	N 6684	-2B	-19.06	741.
19 -1	N 6744	4X	-21.39	717.
19 +1	2047-69	7	-17.71	459.
19 -2 +2	1643-57	13 P		694.
19 -2	1654-60	5		980.
19 -2	1713-57	9		952.
19 -2	N 6300	5BP	-20.33	958.
19 -3 +2	1702-62	5		1353.
19 -3 +2	N 6215	5A	-20.85	1406.
19 -3 +2	N 6221	5B	-21.04	1332.
19 -4 +4	N 7064	5B	-17.73	811.
19 -4	N 7090	5B	-19.66	798.
19 -5 +5	2218-46	6B	-19.18	886.
19 -5	2219-48	10		664.
19 -6 +6	2254-430	7B		917.
19 -6	N 7424	6X	-19.55	938.
19 -6	N 7456	6A	-19.44	1205.
19 -6	N 7462	5B	-18.47	1061.
19 -7 +7	2324-37	10		686.
19 -7	2331-36	7A	-18.45	706.
19 -7	2346-38	10		642.
19 -7	N 7713	6B	-18.46	687.
19 -8 +7	N 24	5A	-17.77	595.
19 -8 +7	N 45	8A	-17.82	511.
19 -0 +7	-2349-52	10		492.
19 -0	N 244	13 P	-16.65	995.

Group ID	Name	Type	$M_B^{b,i}$	V_0		Group ID	Name	Type	$M_B^{b,i}$	V_0
Leo Cloud						21 -8 +8	N 3245	-2A	-20.07	1202.
						21 -8	N 3254	4A	-20.31	1309.
21 -1 +1	N 3596	5X	-20.31	1084.		21 -8	N 3277	2A	-19.51	1402.
21 -1	N 3605	-5	-18.03	600.		21 -8	U 5662	3B	-17.52	1269.
21 -1	N 3607	-2A	-20.54	840.						
21 -1	N 3608	-5	-19.94	1118.						
21 -1	N 3626	-2A	-20.40	1383.		21 -9 +9	N 3395	6XP	-20.00	1595.
21 -1	N 3655	5A	-19.96	1384.		21 -9	N 3396	10BP	-20.04	1649.
21 -1	N 3659	9B	-19.45	1200.		21 -9	N 3430	5X	-20.21	1533.
21 -1	N 3681	4X	-19.67	1152.		21 -9	N 3442	13 P	-19.10	1713.
21 -1	N 3684	4A	-19.69	1075.						
21 -1	N 3686	4B	-19.94	1076.		21 +9	U 5706	10		1466.
21 -1	U 6320	13 P	-18.17	1050.						
						21 -10+10	N 3156	-5	-18.54	994.
21 +1	U 6258	10	-17.58	1382.		21 -10	N 3166	0X	-20.22	1203.
						21 -10	N 3169	1AP	-20.44	1051.
21 -2 +1	N 3501	5	-18.89	1040.		21 -10	U 5522	6	-18.20	1065.
21 -2 +1	U 6151	9	-17.18	1241.		21 -10	U 5539	10B	-17.56	1097.
21 -2 +1	U 6171	10 P	-17.71	1118.						
						21+10	0952+08	13 P	-16.65	1122.
21 -3 +1	N 3370	5A	-19.85	1212.		21+10	1014-03	8B		1109.
21 -3 +1	N 3443	7A	-18.02	1028.		21+10	N 3044	5B	-19.74	1145.
21 -3 +1	N 3447	7X	-19.01	965.		21+10	U 5708	7X	-18.24	1007.
21 -3 +1	N 3454	5B	-18.34	1033.						
21 -3 +1	N 3455	3X	-19.06	987.		21-11+11	N 3611	1AP	-19.51	1490.
21 -3 +1	N 3507	3B	-20.08	879.		21-11	N 3630	-2	-19.37	1363.
21 -3 +1	U 5947	10 P	-17.16	1158.		21-11	N 3640	-5	-20.78	1198.
21 -3 +1	U 6007	10BP		923.		21-11	N 3664	9BP	-18.94	1216.
21 -3 +1	U 6035	10B	-16.74	968.		21-11	U 6345	10B	-18.95	1450.
21 -3 +1	U 6112	7	-18.16	932.						
						21-12+12	0945+33	13 P	-18.04	1443.
21 -4 +1	N 3773	-2AP	-18.25	904.		21-12	0954+33	13 P	-17.48	1470.
21 -4 +1	N 3810	5A	-20.11	882.		21-12	N 2964	4X	-19.90	1260.
						21-12	N 2968	3 P	-19.39	1559.
21 +4 +1	N 3666	5A	-19.55	947.		21-12	N 2970	-2	-17.73	1629.
21 +4 +1	N 3705	2X	-20.15	891.		21-12	N 3003	4B	-20.45	1439.
21 +4 +1	U 6655	13 P	-16.96	750.		21-12	N 3011	-2	-17.91	1417.
						21-12	N 3021	5B	-19.40	1490.
21 -5 +1	N 3338	5A	-20.80	1174.		21-12	N 3067	2X	-19.68	1414.
21 -5 +1	N 3346	6B	-19.35	1137.		21-12	U 5393	9B	-18.02	1407.
21 -5 +1	N 3389	5A	-19.82	1138.						
						21-13+12	N 3026	10	-19.14	1431.
21 -0 +1	N 3437	5X	-19.65	1212.		21-13+12	N 3032	-2X	-19.36	1439.
21 -0 +1	N 3547	5 P	-19.25	1414.						
21 -0 +1	U 6633	-2	-18.85	1650.		21-14+12	U 5349	9	-18.44	1361.
						21-14+12	U 5391	9	-18.36	1550.
21 -6 +6	N 3162	4X	-19.71	1215.						
21 -6	N 3177	3A	-18.75	1122.		21-15+12	N 2793	9BP	-18.30	1636.
21 -6	N 3185	1B	-18.93	1146.		21-15+12	N 2859	-2B	-20.38	1557.
21 -6	N 3187	5BP	-18.88	1500.						
21 -6	N 3190	1AP	-20.29	1216.		21 -0+12	0944+39	13 P	-17.81	1617.
21 -6	N 3193	-5	-20.05	1277.		21 -0+12	1003+29	13 P	-17.37	1341.
21 -6	N 3226	-5 P	-19.57	1270.		21 -0+12	N 2893	0B	-18.84	1650.
21 -6	N 3227	1XP	-20.17	1081.		21 -0+12	N 2955	3AP	-19.04	1721.
21 -6	N 3287	7B	-18.97	1058.		21 -0+12	U 5478	10	-17.21	1326.
21 -6	N 3301	0B	-19.61	1244.						
21 -6	U 5588	13 P	-18.12	1350.		21-16+16	N 2798	1BP	-19.67	1710.
21 -6	U 5675	9		1001.		21-16	N 2799	9BP	-18.69	1739.
21 +6	N 3098	-2	-18.99	1257.		21-17+16	N 2844	1A	-18.54	1477.
21 +6	U 5833	-2	-17.02	1214.		21-17+16	N 2852	-2	-18.73	1880.
						21-17+16	N 2853	1	-18.65	1792.
21 -7 +7	N 3504	2X	-20.38	1478.						
21 -7	N 3512	5X	-19.01	1338.						
21 +7	N 3414	-2BP	-20.23	1357.						

Group ID	Name	Type	$M_B^{b,i}$	V_0
21-18+18	N 2770	5A	-20.96	1908.
21-18	N 2778	-5	-19.46	2017.
21-18	N 2780	12BP	-18.69	2173.
21-18	U 4777	10	-18.25	2019.
21-18	U 4837	9	-17.71	1848.
21+19+19	U 4260	10	-18.55	2284.
21+19	U 4393	12B	-19.82	2154.
21+20+20	0913+53	13 P	-17.86	2286.
21+20	N 2701	4XP	-20.04	2391.
21 -0	0825+52	13 P		1773.
21 -0	N 2712	3B	-19.85	1857.
21 -0	N 2776	5X	-20.77	2643.
21 -0	N 2782	1XP	-20.79	2529.
21 -0	N 3020	6B	-19.30	1305.
21 -0	N 3041	5X	-19.77	1295.
21 -0	N 3294	5A	-20.19	1570.
21 -0	N 3629	6A	-19.64	1471.
21 -0	U 4543	8A	-18.78	1983.
21 -0	U 4922	9A	-19.55	2025.
21 -0	U 5172	9A	-18.74	2635.
21 -0	U 5633	8B	-17.75	1264.
21 -0	U 6670	10B	-18.08	838.

Crater Cloud

Group ID	Name	Type	$M_B^{b,i}$	V_0
22 -1 +1	1152-16	8		1601.
22 -1	1155-22	10		1561.
22 -1	1203-22	10B		1498.
22 -1	1210-20	10		1340.
22 -1	N 3956	5A	-20.20	1439.
22 -1	N 3957	-2A	-19.29	1620.
22 -1	N 3981	4 P	-19.87	1499.
22 -1	N 4024	-5	-19.46	1482.
22 -1	N 4027	8BP	-20.54	1461.
22 -1	N 4033	-5	-19.49	1312.
22 -1	N 4038	10BP	-21.06	1446.
22 -1	N 4039	10 P		1429.
22 -1	N 4050	3B	-20.49	1700.
22 +1	N 3955	13	-19.18	1117.
22 +1	N 3962	-5	-20.73	1621.
22 +1	N 4094	6	-19.84	1232.
22 -2 +1	1138-09	6X		1540.
22 -2 +1	1142-09	4B		1536.
22 -2 +1	N 3892	-2B	-19.77	1532.
22 -0 +1	N 4462	2B	-20.51	1658.
22 -3 +3	1213-11	10		985.
22 -3	1214-11	10		1098.

Group ID	Name	Type	$M_B^{b,i}$	V_0
22 -4 +4	1146-28	9B		1692.
22 -4	1150-28	8		1461.
22 -4	1158-24	8		1581.
22 -4	1202-27	9		1553.
22 -4	1203-27	5B	-20.43	1618.
22 -4	N 3885	1A	-20.24	1706.
22 -4	N 3904	-5	-20.11	1369.
22 -4	N 3923	-5	-21.23	1546.
22 -4	N 3936	5	-20.51	1789.
22 +4	N 3673	3B	-20.20	1696.
22 +4	N 3717	2A	-20.75	1477.
22 -5 +4	1207-29	5A	-20.90	1907.
22 -5 +4	N 4105	-5	-20.36	1654.
22 -5 +4	N 4106	-2B	-20.30	1944.
22 -6 +6	1106-28	10		1239.
22 -6	N 3585	-5	-20.93	1236.
22 -7 +7	N 3637	-2B	-19.65	1649.
22 -7	N 3672	5A	-21.17	1654.
22 -7	N 3732	1X	-19.05	1514.
22 -8 +8	1139-06	10 P		1515.
22 -8	N 3818	-5	-19.40	1317.
22 -9 +9	1150-03	10		1499.
22 -9	N 3952	9 P	-19.96	1458.
22 +9	U 6780	7AP	-19.26	1574.
22-10+10	N 4030	4A	-20.27	1311.
22-10	U 7053	10		1311.
22+10	U 6877	13 P	-17.34	1050.
22-11+11	N 4116	8B	-19.73	1177.
22-11	N 4123	4B	-20.25	1207.
22+11	N 4179	-2	-20.17	1144.
22+11	U 7178	10		1205.
22+12+12	N 4045	4	-19.97	1827.
22+12	U 6903	6X	-19.22	1748.
22+12	U 6998	9		1789.
22 -0	1041-09	7		1858.
22 -0	1048-19	6B		1808.
22 -0	1107-23	4A	-20.05	1841.
22 -0	1220-13	10		974.
22 -0	N 3321	5		2262.
22 -0	N 3887	4B	-19.91	998.
22 -0	N 4129	5	-19.25	1034.

Group ID	Name	Type	$M_B^{b,i}$	V_0	Group ID	Name	Type	$M_B^{b,i}$	V_0
Centaurus Cloud					23 -0	1229-51	5		2376.
					23 -0	1428-55	9		2903.
23 -1 +1	1239-40	5		2397.	23 -0	1441-49	9		2100.
23 -1	1242-40	13 P		2238.	23 -0	N 3882	5		1544.
23 -1	1242-40	2A		2261.	23 -0	N 4219	4A	-21.07	1739.
23 -1	1247-41	-2		2449.	23 -0	N 4444	3		2666.
23 -1	N 4603	5A	-21.93	2323.	23 -0	N 4930	5B		2364.
23 -1	N 4645	-5	-21.03	2412.	23 -0	N 5365	-2B	-20.39	2232.
23 -1	N 4650	0B		2397.					
23 -1	N 4677	-2	-20.10	2978.					
23 -1	N 4696	-5	-22.05	2575.	Lynx Cloud				
23 -1	N 4730	-5		2034.					
23 -1	N 4743	-2		2820.	24 -1 +1	N 2273	1B	-20.76	1961.
23 -1	N 4767	-5	-21.13	2901.	24 -1	U 3504	6X	-20.60	2222.
					24 -1	U 3530	7X	-19.70	2224.
23 -2 +2	1317-47	5		2622.	24 -1	U 3598	10B	-18.51	2110.
23 -2	N 5064	3A	-21.18	2752.					
					24 +1	U 3685	3B	-20.33	1913.
23 +2	1307-46	1		2843.	24 +1	U 3826	7X	-19.27	1847.
23 +2	1323-47	5		2402.					
23 +2	N 5156	1B	-21.08	2757.					
					24 -2 +2	U 3574	6A	-19.42	1545.
23 -0 +2	1245-45	10		2935.	24 -2	U 3647	10B	-17.35	1475.
23 -0 +2	N 5011	-5	-21.10	2838.					
23 -3 +3	1337-48	6A		2628.	24 -3 +3	N 2460	1A	-19.52	1553.
23 -3	1346-48	5		2715.	24 -3	U 4093	3B	-17.93	1646.
23 -3	N 5266	-2A	-21.29	2865.					
					Antlia - Hydra Cloud				
23 -4 +3	1357-48	5		2661.					
23 -4 +3	N 5333	-2		2538.	31 -1 +1	N 3312	2AP	-21.33	2442.
					31 -1	N 3314	0BP		2762.
23 -0 +3	1406-45	9B		2427.					
23 -0 +3	1408-49	5		2697.	31 -2 +2	1027-350	-3		1609.
23 -0 +3	N 5489	1		2773.	31 -2	1027-351	-5		1498.
					31 -2	N 3257	-2X		2779.
23 -5 +5	1212-35	5B		2445.	31 -2	N 3258	-5	-20.32	2564.
23 -5	1221-34	5A	-21.14	2487.	31 -2	N 3260	-5		2169.
23 -5	N 4304	3B	-20.32	2395.	31 -2	N 3268	-5	-20.28	2518.
					31 -2	N 3273	-2A		2176.
23 -0 +5	1225-39	-5	-21.23	2730.					
23 -0 +5	1235-35	5X	-20.22	2700.	31 +2	1026-31	4B	-19.94	2859.
					31 +2	1036-37	5		2766.
					31 +2	N 3223	3A	-21.71	2617.
23 -6 +6	N 5670	-2		2737.	31 +2	N 3241	2A	-19.64	2594.
23 -6	N 5688	5		2628.	31 +2	N 3275	3B	-21.04	2957.
23 -7 +7	1252-44	10		1985.	31 -3 +2	N 3347	3B	-20.81	2642.
23 -7	N 4835	5X	-21.05	1947.	31 -3 +2	N 3358	0X	-20.37	2629.
23 -8 +7	1252-49	7B		1888.	31 -4 +2	0953-32	6X	-20.98	2731.
23 -8 +7	1253-50	-5	-20.70	2031.	31 -4 +2	0957-34	7		2613.
					31 -4 +2	N 3038	1A	-20.95	2426.
23 -0 +7	1244-53	10		1622.	31 -4 +2	N 3087	-5	-20.69	2375.
23 -0 +7	1256-49	10		1653.	31 -4 +2	N 3095	5X	-20.97	2438.
					31 -5 +2	0957-29	5		2197.
23 -9 +9	1259-50	5		1131.	31 -5 +2	1001-27	5X	-20.27	2533.
23 -9	N 4976	-5	-21.05	1132.	31 -5 +2	N 3078	-5	-20.63	2230.
					31 -5 +2	N 3089	2X	-20.03	2375.
					31 +5 +2	1008-25	9		2243.
					31 +5 +2	N 3001	4X	-20.36	2190.
					31 +5 +2	N 3054	3X	-20.55	1925.
					31 +5 +2	N 3081	1X	-20.05	2146.
					31 +5 +2	N 3203	-2A	-20.01	2153.

Group ID	Name	Type	$M_B^{b.i}$	V_0		Group ID	Name	Type	$M_B^{b.i}$	V_0
31 −0 +2	0936−38	7		2175.		31−18+18	0940−05	8BP		1734.
31 −0 +2	N 3208	4A	−19.80	2738.		31−18	0941−05	7A		1808.
31 −0 +2	N 3390	3	−20.45	2578.		31−18	N 2974	−5	−20.63	1786.
31 −0 +2	N 3449	0X	−20.58	2994.						
						31−19+19	0949+01	10		1662.
31 −6 +6	N 3256	13 P	−21.42	2595.		31−19	N 3023	6XP	−19.70	1683.
31 −6	N 3263	6B		2724.		31−19	U 5238	7X	−18.40	1592.
31 −6	N 3318	3X	−20.99	2622.						
						31−20+19	U 5224	8B	−17.67	1756.
31 −7 +6	1026−39	7B		2530.		31−20+19	U 5249	6B	−19.01	1695.
31 −7 +6	N 3250	−5	−21.26	2582.						
						31 −0+19	N 2967	5A	−20.31	1963.
31 −8 +6	1010−47	5		2202.						
31 −8 +6	1014−48	5		2447.		31−21+21	0815−29	7		1370.
31 −8 +6	1021−47	5		2380.		31−21	N 2559	5BP		1296.
						31−21	N 2566	2B		1377.
31 +9 +6	1056−49	7B		2455.						
31 +9 +6	1108−48	2B		2434.		31+21	0833−31	8		1266.
31 −0 +6	N 3261	5B	−21.12	2280.		31−22+21	0831−21	10		1511.
						31−22+21	N 2613	3A	−21.24	1411.
31−10+10	1107−37	−5	−20.13	2611.						
31−10	N 3557	−5	−22.18	2877.		31 −0+21	0859−26	9		1682.
31−10	N 3564	−2	−20.12	2561.		31 −0+21	0908−32	10		1251.
31−10	N 3568	3B		2172.						
						31−23+23	N 2775	2A	−20.13	965.
31+10	1113−33	3B	−19.65	2738.		31−23	U 4797	9	−16.92	1140.
31−11+10	1125−36	4B	−19.60	2710.						
31−11+10	N 3706	−5	−21.14	2780.		31−24+24	U 3912	10B	−17.96	1073.
						31−24	U 3946	10	−17.39	1028.
31+11+10	N 3783	1B	−20.53	2627.						
						31 −0	1000−21	7X		2806.
31−12+12	0935−22	7X		2143.		31 −0	N 2525	5B	−19.89	1354.
31−12	N 2935	4X	−21.74	2009.		31 −0	N 2663	−5		2020.
31−12	N 2983	−2B	−19.72	1752.		31 −0	N 2708	3A		1795.
31−12	N 2986	−5	−20.70	2132.		31 −0	N 2888	−5	−19.35	1952.
						31 −0	N 2962	−2X	−19.77	1937.
31−13+12	0930−16	2X		1868.		31 −0	N 3055	5X	−19.67	1615.
31−13+12	N 2907	1A	−19.55	1810.		31 −0	U 3964	8X	−17.66	1271.
						31 −0	U 4508	13 P	−17.93	1736.
31−14+12	0909−14	10	−17.86	1803.		31 −0	U 5245	8	−17.24	1220.
31−14+12	N 2763	6	−19.53	1640.						
31−14+12	N 2781	1X	−20.23	1778.		Cancer − Leo Cloud				
31+14+12	N 2848	5X	−19.78	1789.		32 −1 +1	0859+26	10 P		2694.
						32 −1	N 2735	13 P	−18.97	2534.
31−15+12	N 2992	1 P	−20.51	1965.		32 −1	N 2750	5X	−20.62	2600.
31−15+12	N 2993	1 P	−19.60	1970.						
						32 +1	N 2764	−2	−19.59	2523.
31−16+12	0917−12	10		1703.						
31−16+12	0918−12	10		1663.		32 +2 +2	N 2608	3B	−19.95	2063.
31−16+12	N 2855	0A	−19.80	1660.		32 +2	U 4559	4	−18.94	2027.
31+17+12	0922−24	4XP	−18.88	2138.		32 +3 +3	0955+13	13 P	−19.03	2700.
31+17+12	N 2865	−5	−20.70	2443.		32 +3	N 3153	5	−20.12	2672.
						32 +3	U 5425	5	−19.45	2577.
31 −0+12	0909−19	7		1914.		32 +3	U 5458	12	−19.34	2700.
31 −0+12	N 2811	0B	−20.53	2260.						
31 −0+12	N 2815	*3B	−20.68	2016.						
31 −0+12	N 2902	−2A	−19.46	1816.						

Group ID	Name	Type	$M_B^{b,i}$	V_o		Group ID	Name	Type	$M_B^{b,i}$	V_o
32 -4 +4	N 3367	5B	-21.21	2913.		34 -3 +3	0505-16	9B		1889.
32 -4	N 3419	-2X	-19.79	2894.		34 -3	0509-14	6A	-19.91	1823.
						34 -3	N 1832	4B	-20.17	1772.
32 +4	N 3300	-2X	-20.06	2867.						
32 +4	N 3433	5A	-20.58	2582.		34 -4 +4	N 1888	5BP	-20.38	2160.
						34 -4	N 1889	-5		2322.
32 -0	0923+19	13 P	-18.68	2381.						
32 -0	0931+11	13 P	-18.57	2358.		34 +5 +4	N 1784	5B	-20.37	2182.
32 -0	N 2577	-3	-19.14	2054.		34 +5 +4	N 1843	7		2470.
32 -0	N 2872	-5	-20.59	2825.						
32 -0	N 2940	-3	-18.59	2824.		34 -6 +6	0618-20	12 P		1767.
32 -0	N 2990	5	-20.29	2994.		34 -6	N 2227	6B	-19.92	1999.
32 -0	U 6157	8A	-19.07	2859.						
						34 +6	0601-20	7B		1776.

Carina Cloud

Group ID	Name	Type	$M_B^{b,i}$	V_o		Group ID	Name	Type	$M_B^{b,i}$	V_o
33 -1 +1	N 2842	-2B		2490.		34 +7 +7	N 2280	6A	-21.03	1651.
33 -1	N 2887	-5		2561.		34 +7	N 2325	-5	-20.97	1854.
33 -2 +1	0913-601	5		2620.		34 -8 +8	N 2179	0A	-19.54	2549.
33 -2 +1	0913-603	5		2701.		34 -8	N 2207	4XP	-21.94	2465.
						34 -8	N 2223	4X	-20.85	2492.
33 -0 +1	0944-63	5		2618.						
33 -0 +1	N 2714	-2		2421.		34 +8	0618-16	6A	-20.21	2642.
						34 +8	N 2196	2A	-20.68	2119.
33 -3 +3	0916-62	5		1831.						
33 -3	0926-60	12 P		1918.		34 -0	0440-08	9		2414.
						34 -0	0449-02	12 P		2100.
33 +3	1001-64	1		1867.		34 -0	0502+01	13 P	-18.57	3000.
						34 -0	0506-09	10		2561.
33 -4 +4	0729-61	4		2920.		34 -0	0508-02	13 P		2729.
33 -4	N 2417	4A	-21.27	2891.		34 -0	0525-16	6B	-19.30	2002.
						34 -0	0541-19	5B		2559.
33 +4	0743-58	5		2615.		34 -0	0550-17	6A		2934.
						34 -0	N 1625	3B	-20.52	2953.
33 +5 +5	0621-59	3		1994.		34 -0	N 1690	-5	-18.31	2100.
33 +5	0639-58	3 P		2325.		34 -0	N 1954	5B		2977.

Virgo - Libra Cloud

Group ID	Name	Type	$M_B^{b,i}$	V_o
33 -0	0733-46	5		2593.
33 -0	0737-55	3		2530.
33 -0	0822-60	5		2291.
33 -0	0919-68	5		2072.
33 -0	N 2601	-2A		2953.

Group ID	Name	Type	$M_B^{b,i}$	V_o
41 -1 +1	1503+01	-5	-18.21	2193.
41 -1	N 5806	3X	-20.36	1298.
41 -1	N 5813	-5	-20.77	1879.
41 -1	N 5831	-5	-19.95	1682.
41 -1	N 5838	-2A	-20.60	1430.

Lepus Cloud

Group ID	Name	Type	$M_B^{b,i}$	V_o
34 -1 +1	0613-26	10		1571.
34 -1	0615-27	3		1465.
34 -1	N 2217	-2B	-20.17	1382.

41 -1	N 5845	-5	-19.30	1786.
41 -1	N 5846	-5	-21.16	1785.
41 -1	N 5850	3B	-20.69	2532.
41 -1	N 5854	-2B	-19.73	1632.
41 -1	N 5864	-2B	-19.47	1633.
41 -1	U 9760	7	-18.47	2023.

| 34 +1 | 0602-26 | 7X | | 1593. |
| 34 +1 | N 2139 | 6X | -19.91 | 1632. |

| 34 -2 +2 | 0539-22 | 10 | | 1532. |
| 34 -2 | N 1964 | 3X | -20.44 | 1480. |

| 34 +2 | 0516-21 | 10 | | 1661. |

Group ID	Name	Type	$M_B^{b,i}$	V_0		Group ID	Name	Type	$M_B^{b,i}$	V_0
41 -2 +1	N 5690	5	-20.45	1729.		41-10+10	1511-15	7A	-20.94	2229.
41 -2 +1	N 5691	1XP	-20.00	1850.		41-10	N 5878	3A	-20.96	2057.
41 -2 +1	N 5692	13 P	-19.16	1800.		41-10	N 5915	2BP	-20.84	2229.
41 -2 +1	N 5701	0B	-20.35	1498.						
41 -2 +1	N 5705	7B	-18.99	1737.		41+10	1510-17	8B		2210.
41 -2 +1	N 5713	4XP	-20.60	1871.						
41 -2 +1	N 5740	3X	-19.85	1558.						
41 -2 +1	N 5746	3X	-21.73	1785.		41-11+11	1515-23	-5	-18.67	2294.
41 -2 +1	N 5750	0B	-20.13	2000.		41-11	N 5898	-5	-20.49	2173.
41 -2 +1	N 5774	7X	-19.51	1587.		41-11	N 5903	-5	-20.88	2438.
41 -2 +1	N 5775	5B	-20.69	1580.						
41 -2 +1	U 9385	10		1633.		41+11	1510-20	9		2199.
41 -2 +1	U 9432	10	-17.83	1556.		41+11	1523-22	4X		2251.
41 -2 +1	U 9483	12	-19.20	1627.						
						41-12+12	N 5426	5AP	-20.33	2296.
41 +2 +1	1450-03	8X		1924.		41-12	N 5427	5AP	-20.94	2483.
41 +2 +1	N 5584	6X	-20.29	1597.						
41 +2 +1	N 5792	3B	-20.83	1914.		41-13+12	N 5468	6X	-21.09	2763.
						41-13+12	N 5493	-2 P	-20.67	2555.
41 -3 +1	N 5566	2B	-21.23	1489.						
41 -3 +1	N 5574	-2B	-19.13	1685.						
41 -3 +1	N 5576	-5	-20.40	1497.		41-14+14	N 5595	5X	-21.03	2568.
41 -3 +1	N 5577	4A	-19.37	1455.		41-14	N 5597	6X	-20.66	2573.
41 -3 +1	N 5638	-5	-20.12	1676.						
41 -3 +1	N 5668	7A	-20.10	1563.						
41 -3 +1	U 9310	8B	-18.10	1829.		41+15+15	1457-13	10		2726.
41 -3 +1	U 9380	10	-18.17	1692.		41+15	N 5728	1X	-21.67	2879.
						41+15	N 5757	3B	-20.68	2678.
41 +3 +1	N 5645	7B	-19.42	1371.		41+15	N 5796	-5	-20.91	2871.
41 -4 +1	N 5496	7	-20.11	1488.		41+16+16	N 5984	7B	-19.08	1199.
41 -4 +1	N 5506	9		1690.		41+16	U 9912	8B	-17.47	1117.
41 -4 +1	U 9057	8B	-19.03	1510.		41+16	U10014	10	-16.48	1203.
41 -5 +1	1430+08	-5 P		2186.		41 -0	1442-20	10		2259.
41 -5 +1	N 5665	9BP	-20.54	2264.		41 -0	N 5324	4A	-21.02	2954.
						41 -0	N 5334	5B	-19.66	1311.
41 -6 +1	N 5861	5X	-20.80	1821.		41 -0	N 5600	4 P	-19.67	2425.
41 -6 +1	N 5885	5X	-20.75	1968.		41 -0	U 8507	13 P	-17.89	1014.
						41 -0	U 8839	10	-18.43	971.
41 -0 +1	1436-08	10		1761.		41 -0	U 9126	10	-17.70	2294.
41 -0 +1	1446-09	10BP	-19.52	1790.		41 -0	U 9164	13 P	-17.96	2551.
41 -0 +1	N 5507	-5		2169.						
41 -0 +1	N 5669	6X	-19.88	1376.						
41 -0 +1	N 5812	-5	-20.58	2027.		Canes Venatici - Camelopardalis Cloud				
41 -0 +1	N 5921	4B	-20.67	1508.						
41 -0 +1	N 5964	7B	-19.46	1492.		42 -1 +1	1349+40	13 P	-17.89	2462.
41 -0 +1	U 9169	10	-18.90	1273.		42 -1	N 5290	4BP	-20.99	2673.
41 -0 +1	U 9249	7	-17.87	1361.		42 -1	N 5296	-2	-18.19	2384.
41 -0 +1	U 9275	8X	-18.72	1311.		42 -1	N 5297	3X	-21.39	2507.
41 -0 +1	U 9500	9B		1698.		42 -1	N 5313	3	-20.49	2696.
						42 -1	N 5320	5	-20.49	2709.
						42 -1	N 5326	1	-20.29	2653.
41 -7 +7	N 5300	5A	-19.52	1123.		42 -1	N 5350	4B	-20.76	2410.
41 -7	N 5348	3	-18.89	1411.		42 -1	N 5353	-2	-20.84	2116.
41 -7	N 5360	13BP	-17.70	1131.		42 -1	N 5362	3 P	-20.19	2361.
41 -7	N 5363	13	-20.57	1081.		42 -1	N 5371	4X	-21.57	2654.
41 -7	N 5364	4AP	-21.17	1348.		42 -1	N 5383	3BP	-20.96	2355.
						42 -1	U 8733	5B	-19.49	2447.
41 +7	U 8995	8X	-18.07	1210.		42 -1	U 8742	10		2352.
						42 -1	U 8798	8		2388.
41 -8 +7	N 5248	4X	-21.07	1112.						
41 -8 +7	U 8575	10	-17.30	1122.		42 +1	1359+37	13	-18.20	2789.
41 -8 +7	U 8614	10	-18.73	1004		42 +1	U 8611	7X	-18.66	2741.
41 +9 +9	1444-17	5		2122						
41 +9	N 5756	5 P	-20.76	2106						

Group ID	Name	Type	$M_B^{b,i}$	V_0	Group ID	Name	Type	$M_B^{b,i}$	V_0
42 −2 +1	N 5169	3B	−18.95	2585.	42 −12+12	U 7164	9X		2288.
42 −2 +1	N 5173	−5	−19.40	2507.	42 −12	U 7168	13 P	−17.69	2344.
42 −2 +1	N 5198	−5	−20.26	2592.					
42 −2 +1	U 8597	7B	−18.84	2540.	42+12	N 4128	−2A	−20.10	2481.
42 −0 +1	N 5347	2B	−19.48	2364.	42 −0+12	1205+67	13 P	−18.34	2391.
42 −0 +1	N 5529	5	−21.31	2971.	42 −0+12	N 4256	3A	−21.26	2680.
42 −0 +1	N 5630	9	−20.18	2777.	42 −0+12	N 4545	6B	−20.59	2860.
42 −0 +1	U 9324	9B	−19.03	2877.	42 −0+12	U 7941	7	−19.59	2445.
42 −3 +3	N 5660	5X	−20.60	2483.	42 −13+13	N 4127	4	−19.87	2006.
42 −3	N 5676	4A	−21.45	2273.	42 −13	N 4291	−5	−20.22	1968.
42 −3	N 5682	3B	−19.12	2414.	42 −13	N 4319	2B	−20.25	1868.
42 −3	N 5689	0B	−20.05	2353.	42 −13	N 4331	10 P	−18.33	1756.
42 −3	N 5714	5	−19.38	2382.	42 −13	N 4386	−5	−19.94	1866.
42 −3	U 9426	10		2463.	42 −13	N 4589	−5	−20.64	2007.
					42 −13	N 4750	2A	−20.29	1698.
42 +3	N 5633	3A	−20.34	2458.	42 −13	U 7872	10		2073.
42 −4 +3	N 5448	1X	−20.69	2101.	42+13	N 3329	1A	−20.11	2056.
42 −4 +3	N 5480	5AP	−19.92	2045.	42+13	N 5452	7X	−19.39	2272.
42 −4 +3	N 5481	−2	19.86	2350.					
42 −4 +3	U 9083	8	−17.82	2044.	42 −14+14	N 3735	5A	−21.46	2853.
					42 −14	U 6711	13 P	−19.26	2700.
42 −5 +3	N 5783	5X	−19.85	2497.					
42 −5 +3	U 9663	10	−18.71	2594.	42+14	N 3516	−2B	−20.59	2709.
42 −6 +3	N 5422	0	−19.45	1986.					
42 −6 +3	N 5473	−2X	−20.29	2156.	42+15+15	U 3496	10		1795.
42 −6 +3	N 5485	−2AP	−20.18	2137.	42+15	U 4646	6X	−20.24	1822.
42 −6 +3	U 8892	10	−18.27	1901.					
42 −7 +3	N 5631	−2A	−20.07	2144.	42 −16+16	N 2268	4X	−20.81	2432.
42 −7 +3	N 5678	3X	−20.96	2391.	42 −16	N 2276	5X	−21.19	2619.
42 −7 +3	U 9391	8B	−17.85	2097.	42 −16	N 2300	−5	−20.79	2168.
					42 −16	U 3522	13 P	−18.31	2346.
42 +7 +3	N 5687	−2	−20.08	2283.					
42 +7 +3	U 9477	10		2510.	42 −17+16	N 2336	4X	−22.00	2386.
					42 −17+16	U 3834	5X	−20.61	2225.
42 −8 +3	1331+60	10		2226.					
42 −8 +3	1348+62	10		2299.	42 −0+16	N 2732	−2	−19.68	2206.
42 −8 +3	N 5308	−2	−20.36	2133.					
42 −8 +3	N 5322	−5	−21.66	2061.	42 −18+18	N 1530	3B	−21.32	2661.
42 −8 +3	N 5376	3X	−19.93	2224.	42 −18	U 3130	3X	−20.04	2683.
42 −8 +3	U 8716	13 P		2222.					
					42 −19+18	U 2411	9		2767.
42 +8 +3	1350+64	13 P	−17.06	1828.	42 −19+18	U 2824	13 P	−21.46	2722.
42 +8 +3	N 5205	12	−19.68	1940.					
42 +8 +3	U 8894	9		1943.	42 −0+18	0321+72	10		2284.
					42 −0+18	U 2620	9X	−19.37	2464.
42 −0 +3	N 5377	1B	−20.69	1951.					
42 −0 +3	N 5486	5 P	−18.34	1559.					
42 −0 +3	U 8441	10	−18.84	1661.	42+20+20	U12069	8X	−19.58	2631.
42 −0 +3	U 8639	10	−17.94	1834.	42+20	U12263	10		2938.
42 −0 +3	U 8658	4X	−19.44	2172.					
42 +9 +9	N 5894	9B	−20.55	2679.	42 −0	1258+64	0		2125.
42 +9	U 9638	10	−18.50	2461.	42 −0	1331+69	13 P		1529.
					42 −0	N 3147	4A	−21.73	2881.
					42 −0	N 4290	3B	−20.72	2885.
42+10+10	N 5985	3X	−21.42	2718.	42 −0	N 4814	3A	−20.27	2663.
42+10	U 9837	5X	−19.47	2857.	42 −0	N 5283	−2	−18.98	2875.
					42 −0	N 5375	2B	−20.69	2444.
42+11+11	N 4194	−2BP	−20.01	262?	42 −0	U 8355	−2	−19.59	2834.
42+11	N 4384	1 P	−19.62	241?	2 −0	U 9764	8B	−18.67	2448.
					2 −0	U10054	8B	−18.76	1725.
					2 −0	U12221	7A	−19.54	2302.

Group ID	Name	Type	$M_B^{b.i}$	V_o
Canes Venatici Spur				
43 -1 +1	1253+34	10		766.
43 -1	1324+38	10		1237.
43 -1	N 4861	9B	-18.87	885.
43 -1	N 5005	4X	-21.27	1070.
43 -1	N 5014	1	-18.79	1157.
43 -1	N 5033	5A	-21.03	931.
43 -1	N 5107	6B	-18.49	1010.
43 -1	N 5112	6B	-19.76	1046.
43 -1	U 8246	6B	-17.69	855.
43 -1	U 8261	10		903.
43 -1	U 8280	5	-17.90	865.
43 -1	U 8303	10	-18.02	1002.
43 -1	U 8323	10 P	-16.76	888.
43 +1	1309+26	12	-17.09	898.
43 +1	N 5273	-2A	-19.22	1090.
43 +1	U 8578	13 P	-16.05	878.
43 -2 +1	N 5145	12	-18.98	1313.
43 -2 +1	U 8365	7B	-17.65	1294.
43 -2 +1	U 8449	9		1324.
43 -2 +1	U 8489	8X	-18.05	1401.
43 -0 +1	1339+30	13 P		1113.
43 -0 +1	U 8333	10		950.
43 -3 +3	U 9560	13 P	-17.07	1317.
43 -3	U 9562	13 P	-17.58	1366.
43 -0	N 5301	3A	-20.31	1673.
43 -0	N 5523	6A	-19.36	1099.
43 -0	N 5727	8X	-18.45	1594.
43 -0	U 8588	9	-17.80	1551.
43 -0	U 9242	6	-18.70	1551.
Draco Cloud				
44 -1 +1	1535+55	13 P	-16.08	860.
44 -1	N 5866	-2A	-20.08	875.
44 -1	N 5879	4A	-19.48	962.
44 -1	N 5907	5A	-20.55	854.
44 -1	N 5963	12 P	-18.75	854.
44 -1	U 9769	8A	-17.04	1029.
44 -1	U 9776	10		1018.
44 -2 +1	N 6015	6A	-20.05	1048.
44 -2 +1	U10031	9B		1099.
44 -0 +1	N 6140	5A	-19.54	1147.
44 -0 +1	U 9893	13 P	-16.07	840.
44 -3 +3	U10669	10		682.
44 -3	U10736	8X	-17.24	735.
44 -4 +4	N 6207	5A	-19.54	1048.
44 -4	N 6255	5B	-18.62	1124.
44 -5 +5	N 6340	0A	-19.98	1441.
44 -5	N 6395	5 P	-19.58	1415.
44 -6 +5	N 6236	6X	-19.73	1525.
44 -6 +5	N 6248	6B	-18.93	1364.
44 -7 +5	N 6412	5A	-19.69	1573.
44 -7 +5	U10792	10		1481.
44 -0 +5	N 6217	4B	-20.19	1592.
44 -8 +8	N 6643	5A	-20.79	1744.
44 -8	N 6654	0B	-20.12	2063.
44 -8	U11193	10	-17.69	1786.
44 -8	U11332	7BP		1824.
44 -0	1437+37	13 P		661.
44 -0	N 5608	10 P	-17.32	780.
44 -0	N 5832	3BP	-18.20	660.
44 -0	N 5949	5A	-17.84	645.
44 -0	N 6239	3BP	-18.91	1160.
44 -0	N 6689	7A	-18.73	750.
44 -0	U 7490	9A	-17.13	636.
44 -0	U 9211	10	-16.64	819.
44 -0	U 9992	10	-15.84	646.
44 -0	U10310	10B	-17.51	917.
44 -0	U10608	8B	-16.55	1325.
44 -0	U10806	8B	-18.27	1175.
44 -0	U10991	10		1712.
44 -0	U11000	13 P	-17.00	900.
Coma Cloud				
45 -1 +1	N 5012	5X	-20.23	2620.
45 -1	N 5016	5X	-19.43	2614.
45 -1	U 8290	9 P	-18.62	2599.
45 +1	U 8409	8A	-19.05	2817.
45 +2 +2	N 4793	5X	-20.90	2498.
45 +2	N 4961	6B	-19.25	2543.
45 -0	N 5089	5 P	-19.22	2186.
45 -0	N 5116	5B	-20.17	2854.
45 -0	N 5117	5B	-19.41	2461.

Group ID	Name	Type	$M_B^{b,i}$	V_0
Fornax Cluster and Eridanus Cloud				
51 -1 +1	0314-35	7		1438.
51 -1	0323-362	9B		1694.
51 -1	0323-363	7B		870.
51 -1	0323-37	0		1922.
51 -1	0334-34	2		1471.
51 -1	0335-35	-2	-17.84	1778.
51 -1	0338-35	10B		1879.
51 -1	0341-36	9		741.
51 -1	0344-35	9		1777.
51 -1	0352-36	-5	-18.71	1223.
51 -1	N 1310	5A		1573.
51 -1	N 1316	-2XP	-21.47	1631.
51 -1	N 1317	0X	-19.20	1800.
51 -1	N 1326	-2B	-19.84	1221.
51 -1	N 1341	5B	-18.14	1720.
51 -1	N 1350	2B	-19.98	1648.
51 -1	N 1351	-5	-18.34	1350.
51 -1	N 1365	3B	-21.26	1492.
51 -1	N 1374	-5	-18.84	1105.
51 -1	N 1375	-2	-17.94	523.
51 -1	N 1379	-5	-19.07	1239.
51 -1	N 1380	-2A	-20.04	1663.
51 -1	N 1381	-2A	-18.80	1629.
51 -1	N 1386	-2A	-19.04	696.
51 -1	N 1387	-2	-19.14	1091.
51 -1	N 1389	-5	-18.75	927.
51 -1	N 1399	-5	-20.29	1294.
51 -1	N 1404	-5	-19.94	1759.
51 -1	N 1427	-5	-19.20	1416.
51 -1	N 1428	-2		77.
51 -1	N 1437	1B	-18.95	1022.
51 +2 +1	0317-32	3B		1161.
51 +2 +1	N 1339	-5	-18.22	1161.
51 +2 +1	N 1344	-5	-19.46	1107.
51 +2 +1	N 1366	-2	-17.45	914.
51 +2 +1	N 1406	5B	-18.46	931.
51 -3 +1	0236-27	-2B	-17.88	1345.
51 -3 +1	N 1079	0XP	-18.83	1377.
51 -3 +1	N 1097	3B	-20.79	1181.
51 -0 +1	0308-33	10		1009.
51 -0 +1	N 1292	5A	-18.79	1261.
51 -0 +1	N 1425	3A	-19.86	1376.
51 -4 +4	0322-21	3B	-17.85	1241.
51 -4	0347-27	5		1404.
51 -4	N 1315	-2B		1639.
51 -4	N 1325	4A	-19.46	1503.
51 -4	N 1331	-5	-17.07	1233.
51 -4	N 1332	-2	-20.07	1471.
51 -4	N 1345	12 P	-17.30	1444.
51 -4	N 1353	3X	-19.42	1421.
51 -4	N 1371	1X	-19.73	1360.
51 -4	N 1385	6B	-19.78	1390.
51 -4	N 1395	-5	-20.34	1586.
51 -4	N 1398	2B	-20.57	1281.
51 -4	N 1401	-2		1461.
51 -4	N 1415	0X	-18.84	1399.
51 -4	N 1426	-5	-18.85	1332.
51 -4	N 1439	-5	-18.95	1560.
51 -4	N 1482	1 P		1542.
51 -5 +4	0311-25	10		1638.
51 -5 +4	0316-23	10		1440.
51 -5 +4	0316-26	4		1699.
51 -5 +4	N 1201	-2A	-19.95	1630.
51 -5 +4	N 1255	4X	-20.09	1600.
51 -5 +4	N 1302	0B	-20.11	1599.
51 -6 +4	0308-22	5		1244.
51 -6 +4	N 1187	5B	-19.76	1314.
51 -7 +4	0307-20	9B		1702.
51 -7 +4	0321-19	6 P		1753.
51 -7 +4	N 1179	6X	-19.64	1716.
51 -7 +4	N 1232	5X	-21.11	1607.
51 -7 +4	N 1297	-5	-18.71	1475.
51 -7 +4	N 1300	4B	-20.42	1502.
51 -8 +4	N 1407	-5	-21.02	1716.
51 -8 +4	N 1440	-2X	-18.77	1437.
51 -8 +4	N 1452	0B	-19.06	1805.
51 +8 +4	N 1400	-2A	-16.55	389.
51 -9 +4	0403-17	10		1777.
51 -9 +4	0404-17	9A		1754.
51 -0 +4	0304-14	10 P		1480.
51 -0 +4	0314-24	10		1982.
51 -0 +4	0325-17	-5 P	-16.94	1753.
51 -0 +4	0330-17	10		1876.
51 -0 +4	0331-21	4B	-19.53	1760.
51 -0 +4	N 1145	5		1911.
51 -0 +4	N 1172	-5	-18.50	1485.
51 -0 +4	N 1359	9BP	-19.49	1883.
51 -0 +4	N 1461	-2A	-18.51	1356.
51 -10+10	0447-29	9	-18.13	1290.
51 -10	0450-28	10		1279.
51 -11+11	N 1357	1A	-19.62	1964.
51 -11	N 1421	4X	-21.12	2021.
51 -12+11	0323-16	9		1804.
51 -12+11	0327-15	9		1816.
51 -12+11	N 1309	4A	-20.17	2070.
51 -13+13	0310-05	13 P		2231.
51 -13	0312-04	10	-18.33	2197.
51 -14+14	N 1199	-5	-20.09	2528.
51 -14	N 1209	-5	-20.36	2624.
51+15+15	N 1015	1B	-20.30	2655.
51+15	N 1090	5B	-20.51	2786.

Group ID	Name	Type	$M_B^{b,i}$	V_0		Group ID	Name	Type	$M_B^{b,i}$	V_0
51 -0	0223-10	9X		2091.		52 -8 +7	0146-10	6X		2021.
51 -0	0306-23	12		2803.		52 -8 +7	N 681	2X	-19.17	1779.
51 -0	0320-11	7B		2755.		52 -8 +7	N 701	5B	-19.25	1869.
51 -0	0450-25	9X	-17.90	1202.						
51 -0	0513-30	6B		1279.		52 -9 +7	0146-12	9	-17.85	1644.
51 -0	0558-28	9B	-16.99	1166.		52 -9 +7	0147-13	10	-16.34	1739.
51 -0	N 895	6A	-20.11	2309.		52 -9 +7	0154-12	8B		1877.
51 -0	N 986	2B	-20.17	1868.		52 -9 +7	N 720	-5	-20.38	1676.
51 -0	N 1222	13 P		2473.						
51 -0	N 1241	3B	-19.83	2134.		52 -0 +7	N 755	3B		1669.
51 -0	N 1320	2		2977.						
51 -0	N 1640	3B	-18.97	1452.		52-10+10	N 864	5X	-20.20	1633.
51 -0	U 2460	4B	-19.57	2721.		52-10	U 1670	9	-16.77	1690.
Cetus - Aries Cloud						52-11+11	N 676	0	-20.65	1601.
						52-11	N 718	1X	-19.28	1755.
52 -1 +1	0238-06	10		1336.						
52 -1	N 988	5B		1504.						
52 -1	N 991	5X	-19.17	1541.		52-12+12	N 470	4A	-20.05	2481.
52 -1	N 1022	1B	-19.20	1506.		52-12	N 474	-2A	-20.62	2636.
52 -1	N 1035	5A	-18.68	1232.		52-12	N 488	3A	-21.36	2380.
52 -1	N 1042	6X	-19.91	1371.		52-12	N 493	5	-19.91	2436.
52 -1	N 1051	9B		1300.		52-12	N 520	13 P	-20.23	2271.
52 -1	N 1052	-5	-19.81	1456.		52-12	U 871	10		2294.
52 -1	N 1084	5A	-20.26	1401.						
52 -1	N 1110	10		1325.		52-13+12	U 1102	13 P	-18.13	2086.
52 -1	N 1140	10 P	-18.77	1484.		52-13+12	U 1133	10		2067.
52 -2 +1	0246-02	10		1103.		52 -0+12	N 514	5X	-20.24	2615.
52 -2 +1	0248+04	12 P	-16.01	1026.		52 -0+12	N 524	-2A	-21.16	2596.
52 -2 +1	N 1055	3	-19.48	1030.		52 -0+12	U 634	9X	-18.09	2347.
52 -2 +1	N 1068	3A	-21.39	1176.		52 -0+12	U 711	5	-18.72	2088.
52 -2 +1	N 1073	5B	-19.44	1242.		52 -0+12	U 750	13 P	-18.66	2647.
52 -2 +1	U 2162	10		1226.						
52 -2 +1	U 2275	9	-17.28	1061.		52-14+14	N 770	1B	-18.75	2583.
52 -2 +1	U 2302	9	-16.83	1133.		52-14	N 772	3A	-21.80	2605.
						52-14	U 1546	5	-18.17	2499.
52 -3 +1	0228-04	9		1648.						
52 -3 +1	N 936	-2B	-20.08	1392.		52-15+14	N 691	5A	-20.74	2807.
52 -3 +1	N 941	6X	-18.67	1660.		52-15+14	U 1276	8B	-19.60	2895.
52 -3 +1	N 955	3	-19.07	1566.						
52 -3 +1	U 1839	8	-16.91	1567.		52 -0+14	U 1197	10	-18.63	2942.
						52 -0+14	U 1547	10B	-18.64	2788.
52 -4 +1	0255-02	9B		1542.		52 -0+14	U 1551	8B	-20.05	2817.
52 -4 +1	N 1087	5X	-20.21	1545.						
52 -4 +1	U 2345	9B		1521.						
						52 -0	0314+03	-3	-15.66	900.
52 -5 +5	0202-06	8B		1397.		52 -0	N 357	0B	-19.84	2622.
52 -5	N 779	3X	-20.01	1435.		52 -0	N 450	6X	-19.46	1853.
						52 -0	N 473	0X	-19.77	2406.
52 -6 +6	0219-20	9 P		1546.		52 -0	N 578	5X	-20.09	1622.
52 -6	0223-21	9	-17.39	1510.		52 -0	N 613	4B	-20.53	1449.
52 -6	0224-24	8B		1459.		52 -0	N 803	5	-19.61	2214.
52 -6	N 899	10		1524.		52 -0	N 821	-5	-20.21	1874.
52 -6	N 907	7BP		1683.		52 -0	N 918	5X	-19.44	1625.
52 -6	N 908	5A	-20.80	1466.		52 -0	N 972	2 P	-20.19	1677.
						52 -0	N 1253	6B	-19.71	1696.
52 -7 +7	0123-06	5		2033.		52 -0	N 1337	6A	-19.36	1199.
52 -7	N 584	-5	-20.68	1931.						
52 -7	N 596	-5	-20.22	1962.						
52 -7	N 600	7B	-18.90	1895.						
52 -7	N 615	3A	-20.19	1910.						
52 -7	N 636	-5	-19.71	1990.						

Dorado Cloud

Group ID	Name	Type	$M_B^{b,i}$	V_0
53 -1 +1	0405-55	10		851.
53 -1	0416-56	7		1152.
53 -1	N 1515	1X	-19.47	961.
53 -1	N 1533	-2B	-18.76	576.
53 -1	N 1536	5B	-17.58	1377.
53 -1	N 1543	-2B	-19.07	1183.
53 -1	N 1546	-2	-18.14	974.
53 -1	N 1549	-5	-19.77	937.
53 -1	N 1553	-2A	-20.17	1064.
53 -1	N 1566	4X	-20.45	1262.
53 -1	N 1574	-2A	-19.34	669.
53 -1	N 1596	-2A	-18.74	1252.
53 -1	N 1617	1	-19.72	818.
53 +1	N 1705	-2AP	-16.23	409.
53 -2 +1	0415-60	-2XP	-18.04	907.
53 -2 +1	0421-63	9		1072.
53 -2 +1	N 1559	6B	-20.13	1068.
53 -3 +1	0454-62	10		1170.
53 -3 +1	N 1672	3B	-19.84	1077.
53 -3 +1	N 1688	5B	-18.35	995.
53 -3 +1	N 1703	3B		1290.
53 -3 +1	N 1824	9B		1012.
53 +3 +1	0445-57	10		978.
53 -4 +1	N 1892	3		1115.
53 -4 +1	N 1947	-2 P	-18.72	866.
53 -4 +1	N 2082	4B	-17.61	849.
53 -5 +1	N 1473	7B		1118.
53 -5 +1	N 1511	5AP	-19.40	1127.
53 -6 +1	0542-52	10		832.
53 -6 +1	N 2101	10 P		939.
53 -6 +1	N 2104	4		898.
53 +6 +1	0535-52	10		1030.
53 -0 +1	0311-57	12B		951.
53 -0 +1	0450-61	7		743.
53 -0 +1	0507-63	10		1225.
53 -0 +1	0539-58	5B		1088.
53 -0 +1	0607-61	5		942.
53 -0 +1	N 1796	5BP	-17.52	759.
53 -0 +1	N 1809	12		1060.
53 -0 +1	N 1853	7		1145.
53 -7 +7	0317-49	5		856.
53 -7	0334-44	3		903.
53 -7	0347-49	7		787.
53 -7	0349-38	10		715.
53 -7	0357-46	10		712.
53 -7	N 1411	-2A	-18.55	867.
53 -7	N 1433	1B	-19.67	889.
53 -7	N 1448	6A	-19.92	1000.
53 -7	N 1483	4		954.
53 -7	N 1487	9 P	-17.62	677.
53 -7	N 1493	6B	-18.47	870.
53 -7	N 1494	4A	-18.53	928.
53 -7	N 1510	-5 P	-16.58	782.
53 -7	N 1512	1B	-19.02	725.
53 -7	N 1527	-2	-18.12	832.

Group ID	Name	Type	$M_B^{b,i}$	V_0
53 -8 +7	0324-52	6	-17.80	882.
53 -8 +7	0330-52	5B	-18.63	892.
53 -9 +7	0307-41	7		804.
53 -9 +7	N 1291	0B	-20.26	678.
53 -0 +7	N 1249	6B	-18.55	828.
53-10+10	0506-38	10		807.
53-10	0510-33	10	-16.87	729.
53-10	N 1792	4A	-20.13	1003.
53-10	N 1808	0X	-19.52	782.
53-10	N 1827	6		836.
53-11+10	0515-37	5		1130.
53-11+10	0519-37	9		1079.
53-12+10	0508-31	9	-17.21	778.
53-12+10	N 1800	-2XP	-16.26	522.
53 -0+10	0539-35	9		1044.
53 -0+10	N 1679	10B		870.
53 -0+10	N 1879	10X	-17.77	1038.
53 -0+10	N 1963	5		1098.
53-13+13	N 1531	-5 P	-17.77	998.
53-13	N 1532	2B	-20.02	1047.
53-13	N 1537	-3	-19.29	1181.
53-14+14	0603-33	5		553.
53-14	N 2188	9B	-17.87	515.
53+14	N 2090	5A	-18.54	707.
53-15+15	0737-50	9B		789.
53-15	0756-49	5		826.
53-16+15	0726-45	10		722.
53-16+15	N 2427	8X	-19.01	689.
53 -0+15	0748-54	10 P		756.
53 -0+15	0824-53	12		758.
53-17+17	0720-72	10		1235.
53-17	N 2397	5B	-19.17	1027.
53-17	N 2434	-5	-19.48	1137.
53-17	N 2442	4B	-20.66	1195.
53+17	0616-70	7		1035.
53+17	0730-74	10		881.
53+17	0810-74	10		980.
53+17	0839-74	7		1088.
53-18+18	0946-74	7		896.
53-18	N 3059	5B	-19.75	990.
53-19+19	0756-76	5		1498.
53-19	0804-76	10		1486.
53-20+20	N 1637	5X	-18.33	639.
53-20	U 3174	10	-15.63	588.

Group ID	Name	Type	$M_B^{b,i}$	V_o		Group ID	Name	Type	$M_B^{b,i}$	V_o
53-21+21	1007-66	5BP		1093.		54 -5 +5	N 3511	5A	-20.15	854.
53-21	N 3136	-5	-20.40	1391.		54 -5	N 3513	5B	-19.46	945.
53-22+21	0857-68	5B		1159.		54 -0	1038-23	10	-17.46	941.
53-22+21	N 2788	12		1255.						
53-22+21	N 2836	12		1422.		**Apus Cloud**				
53 -0+21	1111-69	5		1063.		55 -1 +1	1512-72	10		2997.
						55 -1	1522-73	6		2759.
53 -0	0338-45	10		1389.		55 -1	N 5833	5		2782.
53 -0	0346-80	7B		1406.						
53 -0	0350-71	7		1206.		55 -2 +1	1439-77	5		2513.
53 -0	0352-49	10		1391.		55 -2 +1	N 5612	-2A	-20.20	2538.
53 -0	0407-45	-5	-18.90	1264.						
53 -0	0419-21	7A		768.		55 -0 +1	1418-75	5		2324.
53 -0	0430-49	5		1641.		55 -0 +1	1505-75	2		2632.
53 -0	0435-52	9B		1421.		55 -0 +1	N 5967	6X	-20.81	2695.
53 -0	0454-56	10		1540.						
53 -0	0548-14	8BP		728.		55 -3 +3	1809-85	10 P		2303.
53 -0	0553+03	13 P	-17.17	688.		55 -3	N 6438	-5 P	-20.23	2220.
53 -0	0609-21	10		641.						
53 -0	0618-08	5		561.		55 -4 +3	1409-87	3		2027.
53 -0	0630-52	10		921.		55 -4 +3	1510-87	10		2046.
53 -0	0644-47	4		782.						
53 -0	0649-52	6		808.		55 -0 +3	2247-89	12 P		2305.
53 -0	0715-57	6		812.						
53 -0	1003-75	5		1517.		55 -5 +5	0000-80	10		1757.
53 -0	1148-75	10		1567.		55 -5	2353-81	5		1725.
53 -0	N 685	5X	-19.02	1223.						
53 -0	N 1507	9B	-18.33	810.		55 +5	0017-77	9B		1622.
53 -0	N 1518	8B	-18.15	810.		55 +5	2217-80	5		1620.
53 -0	N 1744	7B	-18.06	562.						
53 -0	N 2310	-2	-18.63	917.		55 -0	0018-64	10		1622.
53 -0	U 3303	9		437.		55 -0	0038-63	7BP		1575.
						55 -0	0048-66	7B		1513.
Antlia Cloud						55 -0	0621-86	1		1638.
						55 -0	1317-77	7		2416.
54 -1 +1	0925-31	10		794.		55 -0	1350-82	6		2221.
54 -1	0931-32	9		640.		55 -0	1454-82	6		2294.
54 -1	0941-31	10		922.		55 -0	1706-77	5		2730.
54 -1	0942-31	8X	-18.31	970.		55 -0	2145-81	5		2368.
54 -1	0943-30	10		720.		55 -0	2345-57	7		1801.
54 -1	N 2997	5X	-20.74	798.		55 -0	N 406	3A	-19.30	1334.
54 -1	N 3056	-2A	-18.02	768.		55 -0	N 7098	1X		2210.
						55 -0	N 7661	5		1858.
54 -2 +1	1001-26	7A		686.		55 -0	N 7689	6X	-19.85	1895.
54 -2 +1	N 3113	8		810.						
54 -2 +1	N 3137	7		831.		**Telescopium - Grus Cloud**				
54 -2 +1	N 3175	3XP	-19.55	849.						
						61 -1 +1	2002-48	-5		2890.
54 -3 +1	0907-22	10		453.		61 -1	2004-48	-2A		2495.
54 -3 +1	0911-19	10		509.		61 -1	N 6861	-5	-20.80	2819.
54 -3 +1	0919-22	10		579.		61 -1	N 6868	-5	-21.06	2723.
54 -3 +1	N 2784	-2A	-18.72	434.		61 -1	N 6870	2A	-20.03	2718.
54 -3 +1	N 2835	5B	-19.66	620.						
						61 -2 +1	2019-44	7X	-18.90	2955.
54 -0 +1	0834-26	10 P		604.		61 -2 +1	2020-44	-2	-20.31	2927.
54 -0 +1	0907-33	10		846.		61 -2 +1	N 6902	0B	-20.34	2754.
54 -0 +1	1033-24	8B	-17.78	785.						
54 -0 +1	1033-36	10		675.		61 -3 +1	2041-46	5		2667.
						61 -3 +1	2042-46	5 P		2702.
54 +4 +4	1038-48	5B		762.						
54 +4	1105-46	7		772.						

Group ID	Name	Type	$M_B^{b,i}$	V_0		Group ID	Name	Type	$M_B^{b,i}$	V_0
61 -0 +1	2028-48	7B		2392.		61-13+13	2150-57	10		1751.
61 -0 +1	2030-53	5 P		2503.		61-13	N 7205	4A	-20.26	1605.
61 -0 +1	N 6887	4A	-20.77	2630.						
61 -0 +1	N 6890	3A	-19.56	2450.		61+13	2154-60	10		1587.
61 -0 +1	N 6909	-5	-19.97	2649.		61+13	2159-54	10		1654.
						61+13	2209-62	5		1595.
61 -4 +4	1852-541	-2A		2991.		61+13	N 7059	4X	-18.74	1657.
61 -4	1852-542	-5	-20.75	2535.						
61 -4	1852-543	-2	-20.25	2675.						
61 -4	N 6707	3		2676.		61+14+14	2240-40	9		2151.
61 -4	N 6708	-2		2539.		61+14	N 7307	5XP	-19.76	2077.
61 +4	1828-57	7B		2622.		61-15+15	2253-37	5B		2059.
61 +4	1911-54	5B	-20.80	2615.		61-15	2255-36	-2A	-18.63	2175.
61 -5 +5	1931-57	5		1819.		61-16+16	2254-434	0A	-20.21	1701.
61 -5	1938-58	10		1831.		61-16	2300-46	9B		1494.
61 -5	1950-58	5		2050.		61-16	2326-41	4A	-19.09	1488.
61 -5	N 6810	1A	-20.61	1900.		61-16	N 7410	-2B	-20.27	1633.
						61-16	N 7412	3B	-19.82	1706.
61 +5	1902-59	5		1740.		61-16	N 7496	3B	-19.96	1629.
61 +5	1931-61	10		1687.		61-16	N 7531	4A	-19.97	1568.
						61-16	N 7552	2B	-20.14	1583.
						61-16	N 7582	2B	-20.17	1433.
61 -6 +6	N 7070	6A	-19.63	2385.		61-16	N 7590	4A	-19.59	1412.
61 -6	N 7079	-2B	-20.17	2664.		61-16	N 7599	5B	-20.22	1638.
61 -6	N 7097	-5	-19.91	2398.		61-16	N 7632	-2B		1573.
61 +6	2119-45	4A	-20.84	2641.		61-17+16	2253-36	5		1671.
						61-17+16	2254-36	-5	-20.58	1635.
						61-17+16	2259-37	9		1384.
61 -7 +7	N 7196	-5	-20.32	2959.		61-17+16	N 7418	6X	-19.43	1453.
61 -7	N 7200	-5		2890.						
						61 -0+16	0003-41	7B		1502.
61 +7	N 7168	-5	-19.85	2728.		61 -0+16	2239-45	1B	-19.65	1716.
						61 -0+16	2256-37	5X	-19.32	1308.
						61 -0+16	2333-39	7		1538.
61 +8 +8	2207-46	2		2695.		61 -0+16	N 7361	5	-18.73	1306.
61 +8	2213-47	5		2714.		61 -0+16	N 7507	-5	-20.21	1593.
						61 -0+16	N 7690	3A	-18.87	1317.
61 -9 +9	2157-43	9X	-19.25	2264.		61 -0+16	N 7764	10	-18.99	1676.
61 -9	N 7162	5A	-19.38	2255.						
61 -9	N 7166	-2A	-19.58	2395.		61-18+18	0031-31	9	-17.74	1579.
						61-18	N 134	4X	-21.09	1572.
61-10+10	N 7041	-2X	-19.93	1889.		61-18	N 148	-2	-18.29	1534.
61-10	N 7049	-2A	-20.43	2154.						
						61-19+18	N 254	-2B	-18.51	1425.
61-11+11	2210-46	-2A	-19.36	1972.		61-19+18	N 289	4B	-19.94	1614.
61-11	2212-45	9B		2016.						
61-11	2214-45	10		1798.		61-20+18	0033-25	8X	-18.87	1574.
61-11	N 7213	1A	-20.36	1745.		61-20+18	N 150	4BP	-20.12	1601.
61-11	N 7232	-2B	-18.84	1774.						
61-11	N 7233	0B		1805.		61-21+21	0043-11	9X	-17.31	1688.
						61-21	0045-10	5B		1423.
61-12+11	N 7144	-5	-20.36	1961.		61-21	N 255	4X	-19.26	1674.
61-12+11	N 7145	-5	-19.81	1883.						
61-12+11	N 7155	-2B	-19.06	1850.		61-22+21	N 178	9B	-18.56	1544.
						61-22+21	N 210	3X	-20.06	1700.
61 -0+11	N 7107	8B	-19.51	2180.						
						61-23+21	N 274	-2XP	-19.00	1820.
						61-23+21	N 275	6BP	-18.97	1822.
						61-23+21	N 337	7B	-19.93	1729.

Group ID	Name	Type	$M_B^{b,i}$	V_0		Group ID	Name	Type	$M_B^{b,i}$	V_0
61-24+21	0049-00	13 P	-16.98	1800.		63 -2 +1	2144-35	-2B	-19.31	2644.
61-24+21	0049-01	13 P	-16.46	1500.		63 -2 +1	N 7135	-2AP	-20.01	2751.
						63 -2 +1	N 7154	10BP		2648.
61 -0+21	N 157	4X	-20.85	1751.						
						63 -3 +3	2213-21	3B		2712.
61+25+25	2207-19	10	-17.79	1848.		63 -3	2214-21	10 P	-18.41	2668.
61+25	N 7218	6B	-19.60	1783.						
						63 +3	N 7184	5B	-21.35	2727.
61 -0	0007-18	10		1611.						
61 -0	0020-53	5		1329.		63 -4 +4	2159-12	8B		2868.
61 -0	1811-58	5		2664.		63 -4	N 7171	3B	-20.03	2770.
61 -0	1830-58	6B	-20.80	2134.						
61 -0	1939-60	4		2409.		63 +5 +5	N 7606	3A	-21.28	2362.
61 -0	1941-54	-5	-20.55	2457.		63 +5	N 7721	5A	-20.32	2148.
61 -0	2015-39	7A	-19.27	2727.						
61 -0	2038-65	10		1499.		63 -6 +6	N 7723	3B	-20.29	1966.
61 -0	2053-55	7B		2019.		63 -6	N 7727	1XP	-20.38	1929.
61 -0	2103-55	10B		1291.						
61 -0	2112-65	5		1741.		63 -0	2125-38	5A	-20.00	2587.
61 -0	2211-67	5		1615.		63 -0	N 7302	-2A	-19.59	2713.
61 -0	2220-42	7B		2416.		63 -0	N 7371	0A	-19.83	2526.
61 -0	2238-45	3		2797.		63 -0	N 7416	3B	-20.51	2927.
61 -0	N 216	13	-17.67	1597.						
61 -0	N 6305	-2		2631.						
61 -0	N 6925	4A	-21.63	2847.		Pegasus Cloud				
61 -0	N 6958	-5	-20.53	2760.						
61 -0	N 7007	-2	-19.91	2897.		64 -1 +1	N 7437	7X	-19.03	2345.
61 -0	N 7029	-5	-20.11	2832.		64 -1	N 7448	4A	-20.71	2432.
61 -0	N 7180	-2	-17.92	1451.		64 -1	N 7463	3XP	-20.22	2677.
61 -0	N 7314	4X	-20.17	1499.		64 -1	N 7464	-2	-17.98	2102.
61 -0	U 328	9B		2107.		64 -1	N 7465	-2	-19.07	2191.
						64 -1	N 7468	13 P	-18.61	2322.
						64 -1	U12350	9	-18.85	2371.
Pavo - Indus Spur										
						64 -2 +1	N 7479	5B	-21.11	2602.
62 -1 +1	N 7083	5A	-21.34	2995.		64 -2 +1	U12281	8	-18.72	2793.
62 -1	N 7096	1A	-20.32	2844.						
						64 -3 +3	2224+15	10	-17.26	2158.
62 -2 +1	2212-64	5B		2943.		64 -3	N 7280	-2	-19.29	2090.
62 -2 +1	N 7179	4		2774.						
62 -2 +1	N 7192	-5	-20.55	2761.		64 -4 +4	2307+18	10		2021.
						64 -4	N 7497	5B	-19.71	1945.
62 -3 +1	N 7125	5X	-20.25	2986.		64 -4	U12344	7B		1871.
62 -3 +1	N 7126	2A	-19.85	2910.						
						64 +5 +4	2324+18	13 P		1766.
62 -0 +1	N 7020	-2A	-20.44	2915.		64 +5 +4	N 7625	1AP	-19.08	1869.
62 +4 +4	2056-72	3		2942.		64 -6 +6	N 7742	3A	-19.65	1818.
62 +4	N 6943	6X	-21.34	2972.		64 -6	N 7743	-2B	-19.91	1995.
62 -0	2243-65	7		2243.		64 -7 +7	U12843	8X	-18.85	1989.
62 -0	N 7141	1		2928.		64 -7	U12846	9		1958.
						64 -7	U12856	10B	-18.65	1990.
Pisces - Austrinus Spur										
						64 +7	N 7800	13	-19.09	1949.
63 -1 +1	2200-34	2B	-19.98	2621.		64 +7	U 99	9		1935.
63 -1	N 7172	2 P	-20.10	2698.						
63 -1	N 7173	-5	-19.47	2547.						
63 -1	N 7174	3 P	-19.60	2824.						
63 -1	N 7176	-5	-19.64	2571.						
63 -1	N 7204	13 P		2680.						

Group ID	Name	Type	$M_B^{b,i}$	V_0
64 -8 +8	N 7798	12	-19.97	2621.
64 -8	N 7817	4X	-20.79	2532.
64 -9 +9	N 7537	4A	-19.81	2850.
64 -9	N 7541	4B	-20.88	2695.
64-10+10	N 7714	3BP	-20.03	2981.
64-10	N 7715	10 P	-19.00	2934.
64-10	N 7716	3B	-19.84	2734.
64-10	U12690	10		2779.
64-10	U12709	9	-18.72	2839.
64+10	N 7731	1B	-19.43	2972.
64+10	U12578	12 P	-18.25	2863.
64+11+11	2228-00	13 P		1737.
64+11	2233-03	9	-18.64	1867.
64+11	U12151	10		1939.
64-12+12	N 7177	3X	-19.74	1442.
64-12	U11868	9B		1357.
64-12	U11880	12BP	-18.02	1671.
64+12	U11820	9		1354.
64 -0	N 7077	-5	-16.72	1050.
64 -0	N 7137	5XP	-19.32	1952.
64 -0	U 260	5	-19.56	2305.
64 -0	U11651	8	-19.27	1804.
64 -0	U11782	9B	-17.45	1343.
64 -0	U11944	10		1985.
64 -0	U12074	13 P	-18.28	2100.
64 -0	U12178	8X	-19.22	2142.
64 -0	U12682	10	-17.62	1621.
64 -0	U12710	10 P	-18.47	2743.

Pegasus Spur

Group ID	Name	Type	$M_B^{b,i}$	V_0
65 -1 +1	N 7292	10B	-18.45	1269.
65 -1	N 7320	7A	-17.80	1066.
65 -1	N 7331	4A	-21.10	1099.
65 -1	N 7363	7X	-16.77	1108.
65 -1	U12060	10B	-17.10	1168.
65 -1	U12082	9	-16.91	1081.
65 +1	N 7217	2A	-20.38	1228.
65 +1	N 7343	3B	-17.54	1495.
65 +1	N 7457	-2A	-19.18	971.
65 +1	U12212	9		1163.
65 -2 +1	N 7332	-2 P	-19.58	1451.
65 -2 +1	N 7339	4X	-19.15	1536.
65 -3 +3	N 7741	6B	-18.76	990.
65 -3	U12732	9	-17.76	998.
65 -3	U12791	10	-15.49	1037.

Group ID	Name	Type	$M_B^{b,i}$	V_0
65 -4 +4	N 7640	5B	-19.47	643.
65 -4	U12588	8	-17.04	698.
65 -4	U12632	9	-17.69	695.
65 +4	U12713	0	-14.68	541.
65 +5 +5	N 7013	0A	-19.56	1051.
65 +5	U11707	8A	-17.58	1187.
65 -6 +6	N 14	10	-18.44	1059.
65 -6	N 7814	2A	-19.98	1250.
65 -6	U 17	9	-15.53	1081.
65 -6	U 122	10	-16.41	1048.
65 +6	N 100	5	-17.69	1035.
65 +6	U 156	10	-17.30	1318.
65 +6	U 191	9	-17.46	1320.
65 -7 +7	N 7518	1X	-15.90	839.
65 -7	U12423	5	-17.37	1082.
65 -0	2236-05	10	-16.17	995.
65 -0	2335+29	13 P	-16.56	1559.
65 -0	U 300	10		1488.
65 -0	U11891	10	-18.50	756.
65 -0	U11909	12 P	-20.48	1404.

Sagittarius Cloud

Group ID	Name	Type	$M_B^{b,i}$	V_0
66 -1 +1	N 6835	1BP	-19.47	1718.
66 -1	N 6836	9X		1762.
66 +1	N 6814	4X	-20.41	1698.
66 +1	N 6821	7B		1680.
66 -0	1945-18	9B		1842.
66 -0	2006-06	10		1588.

Serpens Cloud

Group ID	Name	Type	$M_B^{b,i}$	V_0
71 -1 +1	N 5951	5	-19.65	1860.
71 -1	N 5953	-5	-19.59	2147.
71 -1	N 5954	5	-19.33	2066.
71 -1	N 5962	5A	-20.66	2048.
71 -2 +1	N 5970	5B	-20.55	2038.
71 -2 +1	U 9941	10	-18.01	1929.
71 -3 +3	U10061	10		2182.
71 -3	U10086	13 P	-18.57	2400.
71 -4 +4	N 6070	6A	-20.72	2058.
71 -4	U10288	5	-19.89	2099.
71 -4	U10290	9	-18.94	2039.
71 +4	U10133	5X	-19.41	1978.

Group ID	Name	Type	$M_B^{b,i}$	V_o
71 -5 +5	U 9977	5	-19.40	1944.
71 -5	U 9979	10	-18.58	2006.
71 -6 +6	1534+30	13 P	-17.57	1788.
71 -6	N 5961	13 P	-18.48	1800.
71 -6	N 5974	13 P	-18.24	1875.
71 -0	1559+18	13 P	-18.14	2638.
71 -0	N 6035	5X	-18.43	2357.
71 -0	N 6106	5A	-19.58	1542.
71 -0	N 6118	6A	-20.62	1629.
71 -0	N 6181	5X	-20.92	2515.
71 -0	U10020	7A	-19.22	2195.
71 -0	U10041	8X	-19.13	2224.
71 -0	U10058	9B		2282.
71 -0	U10419	10		2786.

Bootes Cloud

Group ID	Name	Type	$M_B^{b,i}$	V_o
72 -1 +1	N 5929	-5		2606.
72 -1	N 5930	3XP	-20.41	2873.
72 -1	U 9857	10B		2491.
72 -1	U 9858	4X	-20.54	2774.
72 +1	N 5899	5X	-20.79	2701.
72 +1	U 9936	9	-18.15	2800.
72 -0	1452+42	13 P	-18.69	2669.
72 -0	1544+46	13 P	-18.45	2970.
72 -0	N 5981	5	-19.12	1921.

Ophiuchus Cloud

Group ID	Name	Type	$M_B^{b,i}$	V_o
73 -1 +1	N 6570	9B	-20.29	2478.
73 -1	N 6574	4X	-20.76	2509.
73 +1	N 6555	5X	-20.85	2424.
73 +1	U11141	8	-19.14	2426.
73 -2 +2	N 6509	6	-19.85	1974.
73 -2	U11074	7	-19.13	2059.
73 -2	U11093	5	-19.99	2122.
73 -3 +3	N 6384	4X	-21.31	1805.
73 -3	U10862	5B	-19.35	1834.
73 -0	U10805	9B	-17.28	1715.
73 -0	U11105	8	-18.87	2435.
73 -0	U11113	6X	-18.66	2547.
73 -0	U11152	8B	-19.73	2935.

Isolated

Group ID	Name	Type	$M_B^{b,i}$	V_o
20 +1 +1	U 3587	12	-19.07	1173.
20 +1	U 3658	10		1089.
20 -0	U 3390	8X		1534.
20 -0	U 3463	4X	-20.55	2811.
20 -0	U 5707	6X	-19.49	2816.
30 +1 +1	N 2339	4X	-20.77	2166.
30 +1	U 3691	6A	-20.09	2102.
30 +2 +1	N 2357	5	-19.59	2201.
30 +2 +1	U 3808	7B	-18.80	2317.
30 -0	N 2377	3X		2242.
30 -0	N 3246	8X	-19.62	1977.
30 -0	U 3234	10		1370.
40 -0	1517-36	4B	-20.70	2940.
40 -0	N 6000	6		2029.
40 -0	N 6951	4X	-20.73	1707.
40 -0	U11557	8X	-19.57	1684.
40 -0	U11861	8X	-20.87	1760.
50 -0	0144-58	10		2055.
60 -1 +1	U 608	8X	-18.85	3000.
60 -1	U 625	4X	-21.14	2852.
60 -0	0016-19	9		2116.
60 -0	2059-17	4A		1605.
60 -0	2335-48	5		2786.
60 -0	N 48	5XP	-18.11	2036.
60 -0	N 7744	-2X	-20.53	2952.
60 -0	N 7755	5B	-21.24	2988.
60 -0	U 477	9	-19.15	2843.
70 -1 +1	1604+41	13 P	-17.90	2221.
70 -1	U10200	13 P	-18.93	2137.
70 +1	U10095	10		2055.
70 -0	1634+52	13 P	-17.87	2883.
70 -0	1638+37	13 P	-17.68	2100.
70 -0	1742+40	13 P	-17.82	1800.
70 -0	N 6339	6B	-19.84	2336.
70 -0	N 6368	3	-21.26	2920.
70 -0	N 6667	2XP	-20.88	2847.
70 -0	N 6703	-2	-20.56	2591.
70 -0	N 6764	4B	-20.54	2691.
70 -0	U 9762	9		2396.
70 -0	U11124	6B	-19.15	1852.
70 -0	U11220	10		1706.
70 -0	U11283	8BP	-19.19	2230.

Table III
The Affiliation of Rich Clusters in Supercluster Complexes

Members of a sample of 382 rich clusters with redshifts less than 30,000 kilometers/second (10% of the velocity of light) are assigned to components of structure on very large scales. Sixty-four percent are assigned to one of five *supercluster complexes* (see ref. 27). It is possible to identify substructure within these complexes. Supercluster complex and substructure affiliations are recorded in Table III.

Column 1: Abell or alternative cluster name. An asterisk indicates the cluster is in a sufficiently dense region in the core of the specific complex that a pathway can be found to every other cluster with an asterisk in the same complex with cluster-to-cluster steps of less than 40 megaparsecs. The plus signs designate association with the incompletely surveyed Indus region in the southern celestial hemisphere.
Column 2: Redshift or systemic velocity, as a fraction of the velocity of light.
Columns 3, 4, 5: Supergalactic coordinates, in megaparsecs. Relativistic corrections are required, so proper distances are given, assuming an Einstein-de Sitter Universe and a Hubble Constant of 75 kilometers/second/megaparsec (ref. 27).

PISCES - CETUS SUPERCLUSTER COMPLEX

Pisces - Cetus Supercluster

Name	Redshift	SGX	SGY	SGZ
Abell				
* 14	0.0640	3.	-229.	10.
* 44	0.0559	119.	-161.	33.
* 74	0.0672	16.	-238.	-10.
* 75	0.0617	158.	-152.	35.
* 76	0.0416	82.	-131.	15.
77	0.0719	203.	-145.	47.
79	0.0927	214.	-228.	45.
* 85	0.0518	54.	-182.	2.
* 86	0.0610	18.	-218.	-11.
104	0.0822	216.	-182.	38.
* 114	0.0566	19.	-203.	-19.
* 117	0.0536	57.	-187.	-10.
* 119	0.0440	70.	-147.	-2.
* 133	0.0604	22.	-215.	-28.
* 147	0.0438	80.	-141.	-7.
* 150	0.0596	139.	-164.	1.
* 151	0.0526	41.	-186.	-25.
* 154	0.0658	165.	-167.	4.
* 158	0.0645	160.	-166.	3.
* 160	0.0447	112.	-122.	0.
* 168	0.0452	78.	-147.	-14.
* 171	0.0706	172.	-181.	-3.
* 179	0.0547	146.	-136.	-3.
* 189	0.0334	63.	-108.	-14.
* 193	0.0482	106.	-141.	-15.
* 195	0.0422	115.	-107.	-6.
* 225	0.0692	181.	-165.	-21.
* 240	0.0618	132.	-175.	-35.
* 245	0.0790	160.	-219.	-48.
* 246	0.0700	142.	-198.	-44.
* 257	0.0706	171.	-178.	-38.
* 261	0.0467	79.	-147.	-42.
277	0.0947	122.	-283.	-92.
* 279	0.0815	144.	-233.	-71.
281	0.0880	122.	-263.	-86.
* 2665	0.0559	92.	-171.	60.

Perseus - Pegasus Chain

Name	Redshift	SGX	SGY	SGZ
Abell				
* 71	0.0220	67.	-48.	17.
* 262	0.0161	57.	-26.	-2.
* 347	0.0187	69.	-21.	-7.
* 426	0.0183	67.	-15.	-17.
* 2506	0.0331	57.	-90.	66.
* 2572	0.0395	84.	-101.	68.
* 2589	0.0421	87.	-111.	68.
* 2593	0.0433	85.	-118.	69.
* 2634	0.0312	82.	-70.	48.
* 2657	0.0414	75.	-125.	51.
* 2666	0.0265	72.	-61.	36.

Pegasus - Pisces Chain

Name	Redshift	SGX	SGY	SGZ
Abell				
16	0.0838	144.	-245.	55.
2618	0.0705	161.	-161.	103.
2622	0.0621	155.	-131.	93.
2625	0.0609	136.	-148.	87.
2626	0.0573	130.	-139.	83.
2630	0.0675	135.	-176.	92.
2675	0.0726	136.	-203.	76.
2694	0.0958	163.	-268.	83.
2700	0.0978	136.	-292.	74.
non-Abell				
0008.0+1056	0.0890	168.	-244.	72.

Sculptor Region

Name	Redshift	SGX	SGY	SGZ
Abell				
194	0.0178	31.	-61.	-9.
260	0.0348	116.	-61.	-7.
397	0.0325	89.	-69.	-49.
400	0.0232	53.	-59.	-41.
non-Abell				
0023.0-3318	0.0497	-23.	-181.	-8.
0105.1-4708	0.0230	-28.	-81.	-21.
0107.6-4610	0.0240	-28.	-85.	-22.
0237.5-3124	0.0210	-1.	-70.	-41.
0410.6-6252	0.0170	-36.	-40.	-38.
0621.6-6426	0.0265	-65.	-40.	-66.
2345.0-2824	0.0279	-10.	-105.	13.
2349.1-2839	0.0370	-13.	-138.	14.
2349.3-3442	0.0310	-22.	-115.	7.

Virgo - Hydra - Centaurus Supercluster

Name	Redshift	SGX	SGY	SGZ
Abell				
1060	0.0114	-27.	23.	-27.
non-Abell				
1228.3+1240	0.0039	-4.	15.	-1.
1246.0-4103	0.0107	-38.	16.	-8.
1346.3-3012	0.0145	-49.	29.	3.
1400.7-3344	0.0138	-48.	24.	5.
1611.0-6045	0.0176	-67.	-10.	8.
1957.2-3840	0.0186	-45.	-40.	40.

AQUARIUS SUPERCLUSTER COMPLEX

Aquarius - Capricornus Region

Name	Redshift	SGX	SGY	SGZ
Abell				
2366	0.0542	3.	-145.	134.
2377	0.0808	-8.	-214.	182.
2382	0.0648	-23.	-185.	137.
2384	0.0943	-51.	-261.	179.
2388	0.0615	58.	-143.	158.
2399	0.0587	7.	-165.	134.
2400	0.0881	-7.	-241.	183.
2410	0.0806	2.	-224.	169.
2412	0.0735	-42.	-219.	132.
2415	0.0597	18.	-168.	133.
2420	0.0838	-3.	-240.	163.
2440	0.0904	56.	-245.	180.
2448	0.0810	26.	-240.	146.
2457	0.0597	54.	-171.	120.
2459	0.0736	-3.	-231.	118.
2462	0.0698	-8.	-223.	107.
2469	0.0656	97.	-164.	136.

Aquarius Region

Name	Redshift	SGX	SGY	SGZ
Abell				
* 2528	0.0955	-16.	-307.	101.
* 2538	0.0817	-6.	-269.	89.
* 2556	0.0865	-12.	-285.	85.
* 2559	0.0796	24.	-260.	94.
* 2566	0.0821	-5.	-273.	81.
* 2597	0.0852	39.	-278.	89.
* 2638	0.0825	47.	-273.	71.
* 2670	0.0745	54.	-251.	52.

Name	Redshift	SGX	SGY	SGZ

HERCULES - CORONA BOREALIS SUPERCLUSTER COMPLEX

Hercules Supercluster

Name	Redshift	SGX	SGY	SGZ
Abell				
* 1890	0.0570	-77.	175.	77.
* 1899	0.0536	-43.	172.	82.
* 1913	0.0533	-46.	169.	85.
* 1983	0.0441	-39.	133.	87.
* 1991	0.0586	-43.	173.	115.
* 2004	0.0640	-22.	188.	129.
* 2020	0.0578	-80.	156.	114.
* 2022	0.0564	-7.	166.	119.
* 2040	0.0456	-66.	121.	97.
* 2052	0.0348	-52.	93.	77.
* 2055	0.0530	-79.	135.	114.
* 2063	0.0337	-47.	89.	78.
* 2107	0.0421	-22.	111.	108.
* 2147	0.0356	-30.	83.	100.
* 2148	0.0442	-10.	107.	123.
* 2151	0.0371	-27.	86.	106.
* 2152	0.0374	-30.	86.	106.
* 2162	0.0320	2.	77.	93.
* 2197	0.0303	25.	70.	87.
* 2199	0.0303	22.	70.	88.

Bootes Supercluster

Name	Redshift	SGX	SGY	SGZ
Abell				
1691	0.0722	48.	244.	57.
1749	0.0590	32.	202.	60.
* 1775	0.0696	-13.	235.	73.
* 1781	0.0762	2.	254.	84.
* 1793	0.0849	14.	276.	97.
* 1795	0.0616	-12.	209.	71.
* 1800	0.0724	-7.	241.	84.
* 1813	0.0962	33.	303.	116.
* 1825	0.0618	-35.	206.	75.
* 1827	0.0668	-34.	221.	82.
* 1831	0.0733	-9.	241.	93.
* 1873	0.0776	-10.	248.	110.
* 1927	0.0740	-21.	229.	122.
1939	0.0883	-30.	262.	149.
2021	0.0994	-43.	270.	193.

Corona Borealis Supercluster

Name	Redshift	SGX	SGY	SGZ
Abell				
* 2056	0.0763	-9.	208.	168.
* 2061	0.0768	2.	209.	170.
* 2065	0.0721	-11.	196.	162.
* 2067	0.0732	3.	199.	164.
* 2079	0.0662	-5.	179.	154.
* 2089	0.0743	-9.	195.	174.
* 2092	0.0669	5.	178.	158.
* 2124	0.0654	26.	168.	160.
2177	0.0769	-12.	163.	214.
non-Abell				
* 1521.0+2835	0.0829	-9.	221.	183.

Corona Borealis - Hercules Supercluster

Name	Redshift	SGX	SGY	SGZ
Abell				
2108	0.0919	-65.	218.	215.
2110	0.0978	5.	241.	226.
2142	0.0899	-11.	208.	227.
2159	0.0979	-67.	198.	256.
2175	0.0978	8.	204.	260.
2178	0.0928	-20.	189.	252.
2244	0.0970	43.	164.	281.
2249	0.0809	41.	135.	243.
2253	0.0884	67.	145.	258.
non-Abell				
1550.0+2019	0.0899	-50.	206.	223.

LEO SUPERCLUSTER COMPLEX

Leo Supercluster

Name	Redshift	SGX	SGY	SGZ
Abell				
* 754	0.0528	-36.	99.	-161.
* 780	0.0545	-49.	103.	-163.
* 838	0.0502	-33.	117.	-139.
* 957	0.0440	-32.	123.	-102.
* 978	0.0527	-57.	139.	-121.
* 979	0.0550	-63.	142.	-127.
992	0.0533	22.	171.	-90.
* 993	0.0533	-53.	143.	-120.
* 999	0.0318	-1.	103.	-62.
* 1016	0.0321	-5.	104.	-63.
* 1069	0.0630	-82.	167.	-129.
1100	0.0464	16.	159.	-62.
* 1139	0.0386	-36.	123.	-67.
* 1142	0.0353	-15.	121.	-53.
* 1177	0.0316	6.	114.	-36.
* 1185	0.0304	18.	111.	-28.
1213	0.0468	27.	166.	-38.
1216	0.0524	-71.	159.	-80.
* 1228	0.0350	30.	127.	-22.
* 1257	0.0339	30.	123.	-18.
* 1267	0.0321	12.	118.	-24.
1308	0.0481	-69.	150.	-63.
* 1314	0.0341	55.	116.	-2.
1334	0.0559	-83.	173.	-68.
* 1367	0.0215	-3.	81.	-15.
* 1656	0.0232	1.	88.	13.

Leo - Coma Supercluster

Name	Redshift	SGX	SGY	SGZ
Abell				
1020	0.0650	-12.	198.	-120.
1032	0.0794	-43.	227.	-152.
1145	0.0677	-4.	225.	-87.
1149	0.0710	-41.	225.	-104.
1171	0.0704	-61.	218.	-105.
1238	0.0716	-76.	222.	-94.
1337	0.0826	-53.	271.	-74.
1346	0.0970	-85.	304.	-91.
1356	0.0698	-46.	235.	-61.
1362	0.0966	-77.	306.	-84.
1371	0.0673	-26.	233.	-49.
1385	0.0831	-51.	276.	-63.
1390	0.0796	-47.	267.	-58.
1474	0.0791	-42.	271.	-33.
1526	0.0797	-53.	272.	-19.
1541	0.0892	-85.	293.	-22.
1552	0.0845	-68.	284.	-14.
1569	0.0784	-44.	270.	1.
1589	0.0718	-34.	251.	8.
1638	0.0621	-31.	220.	19.

Sextans Supercluster

Name	Redshift	SGX	SGY	SGZ
Abell				
744	0.0729	54.	177.	-179.
763	0.0840	55.	203.	-200.
819	0.0759	14.	192.	-183.
858	0.0881	-9.	221.	-207.
912	0.0888	-48.	223.	-202.
933	0.0958	-53.	244.	-208.
954	0.0922	-58.	239.	-196.

Name	Redshift	SGX	SGY	SGZ		Name	Redshift	SGX	SGY	SGZ
URSA MAJOR SUPERCLUSTER COMPLEX						480	0.0473	85.	-83.	-128.
						484	0.0386	51.	-77.	-111.
Ursa Major Supercluster						496	0.0320	30.	-61.	-100.
						500	0.0666	22.	-127.	-200.
Abell						505	0.0543	182.	73.	24.
757	0.0517	114.	137.	-63.		509	0.0836	137.	-102.	-233.
1003	0.0520	95.	160.	-37.		514	0.0731	29.	-128.	-222.
1169	0.0582	82.	192.	-26.		526	0.0541	100.	-56.	-160.
* 1218	0.0792	134.	241.	-9.		527	0.0794	261.	92.	5.
* 1270	0.0689	124.	211.	3.		533	0.0472	11.	-80.	-154.
1275	0.0603	55.	209.	-26.		539	0.0205	40.	-17.	-66.
* 1291	0.0530	102.	164.	7.		548	0.0410	-7.	-49.	-145.
* 1318	0.0566	105.	176.	7.		553	0.0670	205.	70.	-101.
* 1377	0.0514	96.	161.	12.		559	0.0757	240.	113.	-21.
* 1383	0.0603	108.	188.	13.		564	0.0779	242.	123.	-21.
* 1400	0.0778	136.	235.	19.		566	0.0984	289.	151.	-63.
* 1436	0.0646	117.	198.	23.		568	0.0751	177.	102.	-167.
* 1452	0.0631	100.	202.	18.		569	0.0196	60.	33.	-33.
1461	0.0537	59.	187.	6.		576	0.0381	117.	69.	-45.
* 1468	0.0844	127.	262.	24.		582	0.0581	148.	99.	-111.
* 1507	0.0592	117.	176.	33.		592	0.0624	63.	83.	-198.
* 1534	0.0698	139.	200.	44.		595	0.0666	181.	128.	-86.
* 1616	0.0833	131.	250.	59.		623	0.0871	24.	114.	-276.
* 1767	0.0701	123.	201.	79.		628	0.0835	164.	167.	-170.
* 1783	0.0766	117.	224.	90.		634	0.0267	79.	60.	-26.
1904	0.0708	78.	211.	110.		644	0.0704	-11.	92.	-232.
						671	0.0494	86.	112.	-113.
Draco Supercluster						680	0.0790	149.	179.	-148.
						690	0.0788	119.	178.	-172.
Abell						695	0.0687	116.	161.	-142.
834	0.0707	187.	166.	-9.		779	0.0226	37.	66.	-43.
* 945	0.0917	231.	211.	10.		924	0.0989	122.	285.	-122.
* 1186	0.0791	209.	175.	44.		991	0.0880	27.	265.	-144.
* 1254	0.0628	160.	155.	31.		1035	0.0792	107.	245.	-70.
* 1279	0.0551	134.	147.	23.		1097	0.0793	65.	257.	-80.
* 1412	0.0839	209.	193.	55.		1126	0.0852	-1.	271.	-113.
1496	0.0941	172.	265.	46.		1187	0.0791	89.	258.	-42.
* 1500	0.0720	184.	167.	54.		1190	0.0794	94.	257.	-39.
* 1681	0.0908	210.	213.	84.		1365	0.0763	38.	263.	-30.
* 1741	0.0745	175.	179.	77.		1399	0.0913	-127.	271.	-87.
* 1851	0.0864	199.	197.	101.		1609	0.0891	-2.	304.	27.
1859	0.0988	166.	260.	126.		1630	0.0649	-86.	215.	2.
non-Abell						1631	0.0508	-122.	139.	-15.
* 1101.0+7042	0.0843	209.	201.	31.		1644	0.0449	-114.	120.	-12.
						1650	0.0845	-139.	257.	2.
						1651	0.0825	-147.	246.	-0.
						1736	0.0350	-108.	76.	-1.
OUTSIDE FIVE MAJOR SUPERCLUSTER COMPLEXES						1750	0.0860	-150.	252.	42.
						1773	0.0776	-123.	235.	56.
Abell						1809	0.0788	-114.	240.	73.
21	0.0948	247.	-189.	82.		1836	0.0363	-90.	98.	29.
102	0.0657	108.	-208.	8.		1837	0.0376	-92.	102.	31.
116	0.0669	109.	-212.	-0.		2028	0.0772	-105.	197.	152.
134	0.0694	102.	-224.	-11.		2029	0.0767	-112.	192.	151.
272	0.0877	268.	-136.	-17.		2033	0.0821	-116.	204.	161.
278	0.0896	269.	-145.	-23.		2048	0.0945	-140.	223.	184.
358	0.0576	62.	-177.	-92.		2168	0.0619	94.	141.	144.
376	0.0489	166.	-60.	-35.		2184	0.0550	73.	125.	138.
399	0.0715	174.	-150.	-106.		2198	0.0798	74.	171.	207.
401	0.0748	183.	-154.	-110.		2241	0.0635	24.	115.	195.
404	0.0632	215.	-55.	-45.		2247	0.0392	115.	70.	57.
407	0.0470	159.	-55.	-44.		2248	0.0663	176.	117.	107.
410	0.0897	173.	-204.	-152.		2250	0.0654	54.	115.	196.
415	0.0788	87.	-209.	-156.		2255	0.0800	166.	141.	173.
419	0.0406	18.	-121.	-90.		2256	0.0601	165.	104.	94.
423	0.0797	87.	-208.	-162.		2271	0.0568	156.	98.	92.
449	0.0803	268.	76.	26.		2295	0.0508	127.	81.	109.
450	0.0607	174.	-81.	-105.		2296	0.0617	170.	99.	102.
465	0.0855	171.	-150.	-187.		2301	0.0874	210.	125.	175.
478	0.0900	192.	-124.	-207.		2319	0.0564	94.	29.	179.

Name	Redshift	SGX	SGY	SGZ	Name	Redshift	SGX	SGY	SGZ
OUTSIDE MAJOR COMPLEXES (continued)					2104.3-2541	0.0385	-48.	-103.	88.
					+ 2113.1-5937	0.0588	-151.	-145.	35.
non-Abell					+ 2126.1-5102	0.0797	-170.	-207.	74.
0022.2-5712	0.0450	-83.	-142.	-26.	+ 2130.7-6215	0.0555	-144.	-139.	22.
0035.0-0126	0.0735	106.	-236.	18.	+ 2131.0-5351	0.0772	-171.	-200.	60.
0258.3-3714	0.0360	-16.	-111.	-76.	2137.0-2245	0.0331	-28.	-99.	71.
0317.9-5403	0.0550	-80.	-146.	-111.	+ 2142.1-5150	0.0535	-117.	-150.	45.
0329.2-5247	0.0569	-79.	-148.	-120.	+ 2143.1-5732	0.0743	-172.	-193.	40.
0341.9-5349	0.0575	-84.	-144.	-125.	+ 2147.0-5533	0.0675	-152.	-181.	42.
0427.8-5350	0.0384	-61.	-87.	-97.	+ 2150.6-5805	0.0760	-174.	-198.	36.
0430.5-6135	0.0545	-107.	-115.	-121.	+ 2152.3-5549	0.0376	-89.	-107.	23.
0600.2-4002	0.0490	-53.	-62.	-160.	+ 2156.5-5624	0.0754	-167.	-202.	39.
0608.5-3335	0.0350	-27.	-37.	-123.	2158.2-6011	0.0996	-224.	-248.	33.
0626.5-5422	0.0502	-98.	-63.	-143.	+ 2201.1-5822	0.0405	-97.	-114.	17.
0820.0+0647	0.0809	44.	139.	-241.	+ 2201.2-5019	0.0365	-76.	-110.	29.
0935.6-2007	0.0335	-51.	61.	-98.	+ 2212.7-5148	0.0680	-135.	-196.	41.
1251.6-2845	0.0553	-162.	116.	-27.	+ 2213.5-5250	0.0532	-110.	-157.	31.
1326.0-3100	0.0479	-151.	92.	-3.	+ 2218.2-5523	0.0396	-88.	-118.	18.
1843.0+4530	0.0918	135.	79.	271.	+ 2221.0-5644	0.0590	-129.	-168.	21.
+ 2006.0-5639	0.0585	-162.	-123.	59.	+ 2222.4-5605	0.0789	-164.	-219.	29.
+ 2029.1-6312	0.0763	-210.	-161.	42.	+ 2228.5-5500	0.0752	-152.	-214.	28.
+ 2035.9-6132	0.0709	-192.	-155.	44.	+ 2231.8-5243	0.0554	-110.	-167.	25.
+ 2047.8-5255	0.0446	-113.	-110.	49.	2244.0+3925	0.0820	211.	-102.	161.
+ 2048.7-5208	0.0467	-116.	-116.	53.	2339.0+2632	0.0952	226.	-193.	127.
2059.0-2813	0.0379	-54.	-101.	85.	2356.0-6112	0.0959	-186.	-264.	-40.
+ 2104.0-3955	0.0520	-99.	-138.	85.					

REFERENCES

1. Aaronson, M., J. Huchra, J. R. Mould, R. B. Tully, J. R. Fisher, H. van Woerden, W. M. Goss, P. Chamaraux, U. Mebold, B. Siegman, G. Berriman, and S. E. Persson. *Astrophysical Journal Supplement*, **50**, 241, 1982.

2. Aaronson, M., J. R. Mould, and J. Huchra. *Astrophysical Journal*, **237**, 655, 1980.

3. Arakelian, M. A. *Soobshcheniya Byurakanskoj Observatorii*, **47**, 3, 1975.

4. Burstein, D., and C. Heiles. *Astrophysical Journal*, **225**, 40, 1978.

5. Cesarsky, D. A., S. Lausten, J. Lequeux, H. E. Schuster, and R. M. West. *Astronomy and Astrophysics*, **61**, L31, 1977.

6. Chamaraux, P., W. M. Goss, U. Mebold, R. B. Tully, and H. van Woerden, unpublished.

7. de Vaucouleurs, G., and A. de Vaucouleurs. *Reference Catalogue of Bright Galaxies*. Austin: University of Texas Press, 1964 (BGC).

8. de Vaucouleurs, G., A. de Vaucouleurs, and R. Buta. *Astronomical Journal*, **86**, 1429, 1981.

9. de Vaucouleurs, G., A. de Vaucouleurs, and H. G. Corwin, Jr. *Second Reference Catalogue of Bright Galaxies*. Austin: University of Texas Press, 1976 (RCII).

10. Dreyer, J. L. E. *New General Catalogue of Nebulae and Clusters of Stars*, 1888. Reprinted in *Memoirs of the Royal Astronomical Society*, 1962 (NGC).

11. Fisher, J. R., and R. B. Tully. *Astrophysical Journal Supplement*, **47**, 139, 1981.

12. Fouqué, P., and G. Paturel. *Astronomy and Astrophysics Supplement*, **53**, 351, 1983.

13. Fouqué, P., and G. Paturel. *Astronomy and Astrophysics*, **150**, 192, 1985.

14. Holmberg, E. B. *Meddelande Lunds Observatorium*, Series II, No. 136, 1958.

15. Karachentseva, V. *Soobshcheniya Byurakanskoj Observatorii*, **39**, 61, 1968.

16. Lauberts, A. *ESO/Uppsala Survey of the ESO (B) Atlas*, Garching bei München, Federal Republic of Germany: European Southern Observatory, 1982 (ESO).

17. Laustsen, S., W. Richter, J. van der Lans, R. M. West, and R. N. Wilson. *Astronomy and Astrophysics*, **54**, 639, 1977.

18. Lo, K. Y., and W. L. W. Sargent. *Astrophysical Journal*, **227**, 756, 1979.

19. Longmore, A. J., T. G. Hawarden, B. L. Webster, W. M. Goss, and U. Mebold. *Monthly Notices of the Royal Astronomical Society*, **184**, 97P, 1978.

20. Nilson, P. Uppsala General Catalogue of Galaxies. Uppsala, Sweden: *Societatis Scientiarum Upsaliensis*, 1973 (UGC).

21. Nilson, P. *Catalogue of Selected Non-UGC Galaxies*, Uppsala Astronomical Observatory Report No. 5, 1974.

22. Reif, K., U. Mebold, W. M. Goss, H. van Woerden, and B. Siegman. *Astronomy and Astrophysics*, **50**, 451, 1982.

23. Rubin, V. C., W. K. Ford, Jr., N. Thonnard, M. S. Roberts, and J. A. Graham. *Astronomical Journal*, **81**, 687, 1976.

24. Rubin, V. C., S. Moore, and F. C. Bertiau. *Astronomical Journal*, **72**, 59, 1967.

25. Sandage, A. *Astronomical Journal*, **183**, 711, 1978.

26. Tully, R. B. *Astrophysical Journal*, **321**, 280, 1987.

27. Tully, R. B. *Astrophysical Journal*, **323** (December 1), 1987.

28. Tully, R. B. *Astronomical Journal*, 1988.

29. Tully, R. B., and J. R. Fisher. *Nearby Galaxies Atlas*. Cambridge: Cambridge University Press, 1987 (NBG Atlas).

30. Tully, R. B., and P. Fouqué. *Astrophysical Journal Supplement*, **58**, 67, 1985.

31. Tully, R. B., and E. J. Shaya. *Astrophysical Journal*, **281**, 31, 1984.

32. Turner, E. L. *Astrophysical Journal*, **208**, 20, 1976.

33. van den Bergh, S. *Astrophysical Journal*, **131**, 215, 1960.

34. van den Bergh, S. *Astronomical Journal*, **71**, 922, 1966.

35. Vorontsov-Velyaminov, B. A., A. A. Krasnogorskaya, and V. P. Arkipova. *Morphological Catalogue of Galaxies*, Vols. 1–5. Moscow: Sternberg Institution, 1962–74 (MCG).

36. Zwicky, F., E. Herzog, P. Wild, M. Karpowicz, and C. T. Kowal. *Catalogue of Galaxies and Clusters of Galaxies*, Vols. 1–6. Pasadena: California Institute of Technology, 1961–68 (CGCG).